Slope Safety Preparedness for Impact of Climate Change

Slope Safety Preparedness for Impact of Climate Change

Editors

Ken Ho

Geotechnical Engineering Office, Civil Engineering and Development Department, Hong Kong Government, Hong Kong

Suzanne Lacasse

Norwegian Geotechnical Institute, Oslo, Norway

Luciano Picarelli

Department of Civil Engineering, Design, Building and Environment, DIcDEA, Seconda Università di Napoli, Aversa, Italy

CRC Press
Taylor & Francis Group
Boca Raton London New York

CRC Press is an imprint of the
Taylor & Francis Group, an **informa** business
A BALKEMA BOOK

Picture (top): Hurricane Katrina making landfall near New Orleans, USA on 29 August 2005 (Courtesy of NASA).
Picture (bottom): 2011 flows in Nova Friburgo, Rio de Janeiro State, Brazil (courtesy of Professor Willy A. Lacerda, Federal University of Rio de Janeiro).

CRC Press
Taylor & Francis Group
6000 Broken Sound Parkway NW, Suite 300
Boca Raton, FL 33487-2742

First issued in paperback 2019

ISBN-13: 978-1-138-03230-9 (hbk)
ISBN-13: 978-0-367-88553-3 (pbk)

Library of Congress Cataloging-in-Publication Data

Names: Ho, Ken, editor.
Title: Slope safety preparedness for impact of climate change / editors, Ken Ho, Geotechnical Engineering Office, Department of Civil Engineering and Development, Hong Kong SAR Government, Suzanne Lacasse, Norwegian Geotechnical Institute, Oslo, Norway, Luciano Picarelli, Department of Civil Engineering, Design, Home Building and Environment, DIcDEA, Italy.
Description: Leiden, The Netherlands : CRC Press/Balkema, [2017] | Includes bibliographical references and index.
Identifiers: LCCN 2016037641 (print) | LCCN 2016040117 (ebook) | ISBN 9781138032309 (hbk : alk. paper) | ISBN 9781315387789 (eBook) | ISBN 9781315387789 (ebook)
Subjects: LCSH: Slopes (Soil mechanics)–Stability. | Slope stabilization. | Soil conservation. | Climatic changes–Risk management.
Classification: LCC TA749 .S545 2016 (print) | LCC TA749 (ebook) | DDC 624.1/51363–dc23
LC record available at https://lccn.loc.gov/2016037641

Visit the Taylor & Francis Web site at
http://www.taylorandfrancis.com

and the CRC Press Web site at
http://www.crcpress.com

Table of contents

Foreword

It gives me great pleasure to write this note in support of this excellent volume on the topic of the impact of climate change on landslides. There is compelling scientific evidence that our climate is changing. The question is should we worry about it? Yes, the ice is melting and the water level is going up; this impacts a number of civil engineering issues including coastal infrastructure. The question raised in this book is related to the preparedness of various countries to address the impact on landslides.

I appreciate the fact that this group of engineers did not follow this quote from Eric Holdeman regarding the way some people approach such problems:

- It won't happen
- If it happens, it won't happen to me
- If it happens, and it happens to me, it won't be so bad
- If it happens, and it happens to me, and it's bad, there is nothing I can do to stop it anyway

I wish to thank the FedIGS Joint Technical Committee No. 1 on Landslides for this initiative. I am very impressed by the strong international response with over 20 countries contributing to the book including Australia, Brazil, Canada, China, Colombia, Hong Kong, India, Indonesia, Italy, Japan, Malaysia, Norway, Russia, Singapore, Spain, Switzerland, Taiwan, Thailand, UK, US, Vietnam. This extensive world representation makes the book even more valuable to practitioners and professors.

A special thanks goes to my distinguished colleague Luciano Picarelli for his work on the FedIGS Joint Technical Committee on Landslide not only for this document but more generally for his stellar leadership of the committee. Another special thanks goes to the editors of this document Ken Ho, Suzanne Lacasse, and Luciano Picarelli who no doubt spent countless hours preparing this manuscript for publication. And then a special thanks to all the authors of the papers summarizing their countries' landslide problems associated with climate change and their approach to solve it.

Our geotechnical engineering profession is very fortunate to have so many volunteers who care enough about their profession to dedicate a lot of their time to its progress while trying to balance this voluntary work with their regular work, their family time, not to forget their fitness time. Volunteers such as you should be commended for your ability to multitask in a world where time is a commodity becoming increasingly difficult to manage.

My very best wishes to all of you. I can assure you that this book will be on my shelf as soon as it comes out.

Jean-Louis Briaud
President of FedIGS, Federation of International Geo-engineering Societies
Distinguished Professor, Texas A&M University

Preface

The JTC1 *Natural Slopes and Landslides* is a joint technical committee of the Federation of International Geo-engineering Societies (FedIGS). FedIGS is the alliance of the learned societies ISSMGE, ISRM, IAEG and IGS. In a meeting during the 11th International Symposium on Landslides in Banff, Canada in 2012, the committee agreed to a proposal by the Hong Kong member to prepare a technical report on "Slope Safety Preparedness of the Effects of Climate Change". The report would focus on the impact of climate change on geo-hydrological hazards with special emphasis on landslides. The basic idea was to invite national groups to report on the status of preparedness for climate change impact on landslides in their countries. The fundamental questions were: "Are we prepared for the possible impact of climate change?" and "What actions are being or should be undertaken to face this challenge?" The committee the Hong Kong team to coordinate the work and report.

The idea was successful. Two years later, a number of well-documented papers were produced. It was then decided to organize an international Forum on the topic. The forum was held at the Seconda Università di Napoli, jointly organized by the University and the Euromediterranean Center for Climate Changes (CMCC) with the scientific coordination of the Hong Kong team led by Ken Ho, Suzanne Lacasse and myself as JTC1 chair, took place on November 17 and 18, 2015. Well attended and animated by original and stimulating data and in-depth discussions, the Forum had a significant impact. To promulgate the collective insight on this important and challenging topic, all of the high quality reports presented at the Forum are collated in this book, together with an overview paper on the state-of-the-art (2016).

In my opinion, the volume is a unique and stimulating collection of contributions on a timely topic that the media are widely and perhaps quite elusively debating. In fact, given the lack of quantitative data and the difficulties in developing rational and reliable climate change and related landslide scenarios because of the considerable uncertainties in the analysis parameters, the key elements to formulate the appropriate measures for landslide risk mitigation are not obvious. The book provides a thorough framework for the current state of the knowledge and studies as well as the initiatives that are being undertaken in 21 regions in the world.

We hope that this book will prove to be a useful support to all those who need information on a fundamental subject that is strongly related to the safety and prosperity of mankind in the coming centuries.

Luciano Picarelli
Chairman of the JTC1
Natural Slopes and Landslides, FedIGS

Chapter 1

Preparedness for climate change impact on slope safety

K.K.S. Ho
Geotechnical Engineering Office, Hong Kong SAR Government, Hong Kong

S. Lacasse
Norwegian Geotechnical Institute, Oslo, Norway

L. Picarelli
DIcDEA, Seconda Università di Napoli, Aversa, Italy

ABSTRACT

This book presents the approaches adopted by different countries and regions over the world, describing the best practice in four areas: (1) Assessment of the impact of climate change on rainfall patterns and other meteorological events that can cause slope instability; (2) Evaluation of the severity of the impact of climate change on slope instability and landslide risk; (3) Implications of climate change impact on slope design, landslide risk management, risk mitigation strategy and risk adaptation measures; and (4) Need for further work on slope safety preparedness. Thirteen (13) country/region reports covering a total of 17 areas in the world summarize their state-of-practice: Brazil, Canada, Hong Kong, Italy, Japan, Norway, the Russian Federation, the Southeast Asia region (including Indonesia, Malaysia, Singapore, Thailand and Vietnam), Spain, Switzerland, Taiwan, the UK and the USA. In addition, four countries, Australia, China, Columbia and India, filled out a questionnaire on their country's practice on the same topic. This overview paper consolidates the contributions and key observations made in the country/region reports and summarizes some of the most urgent research needs and emerging trends.

Landslide risk management in a changing or extreme climate context represents a formidable challenge. The insight from the country/region reports reinforces the need for a comprehensive slope safety system that manages landslide risk in a holistic manner using both "hard" engineering measures and "soft" non-engineering approaches. The challenges posed by extreme weather events, which may be exacerbated by climate change, require urgent attention and careful assessment from both a technical and a strategy and policy perspective. Scenario studies indicate that the number, scale and mobility of the landslides associated with extreme weather and climate change could be unprecedented. Slopes in permafrost, mountains with glaciers, coastal slopes, natural slopes underlain by weak granular materials susceptible to debris flows/slides, and slopes prone to progressive deterioration with time (degradation in strength or development of distress) are identified as vulnerable settings and potential critical hazards that will likely respond strongly to climate change with an increase in frequency, magnitude and perhaps runout of landslides. Hence landslide risk will increase. Conventional risk assessment methodologies are not well suited for dealing with the risk posed by low probability – high impact (extreme) events. Stress testing provides a complementary approach that may be used for managing the risk to constructed facilities and critical infrastructure. In stress tests, the focus is on the performance of the system under extreme event scenarios. This is a rapidly evolving field of new research initiatives.

The review finds that slope safety preparedness for the impact of climate change varies considerably from one country/region to another. There is much room for cross-fertilization of ideas and insights, as well as joint development of strategies and best practice. Disasters often manifest themselves as fast events, but the vulnerability for disasters is built up slowly, and can be the result of neglecting to be adequately prepared. Focus needs to shift from prevention-mitigation to building resilience and reducing risks. There is an increasing awareness amongst the engineering profession that reliance on engineering works alone in combating the climate change impact is neither practicable nor cost-effective. Instead, due emphasis should be placed on improving warning systems, enhancing emergency preparedness and response, community resilience and recovery. For enhanced preparedness and resilience to take root, the risk management strategy has to be embraced by the community. Hence, effective public education and strong government support are essential.

1.1 JTC1 INITIATIVE ON CLIMATE CHANGE IMPACT ON SLOPE SAFETY

The Joint Technical Committee No. 1 (JTC1) on Natural Slopes and Landslides was established by the Federation of International Geo-engineering Societies (FedIGS), which today comprises the International Society of Soil Mechanics and Geotechnical Engineering (ISSMGE), the International Society for Rock Mechanics (ISRM), the International Association of Engineering Geology (IAEG) and the International Geosynthetics Society (IGS). In 2012, JTC1 launched one of its flagship initiatives on the subject of slope safety preparedness for the impact of climate change to take stock of the status and approaches adopted by different countries and regions and to establish the best practice from an international perspective in four areas:

(a) Assessment of the impact of climate change on rainfall patterns and other meteorological factors that can cause slope instability;
(b) Evaluation of the severity of the impact of climate change on slope instability and landslide risk;
(c) Implications of climate change impact on slope design, landslide risk management, risk mitigation strategy and risk adaptation measures; and
(d) Need for further work on slope safety preparedness.

The JTC1 project culminated in 13 country/region reports covering a total of 17 areas summarizing their state-of-practice: Brazil, Canada, Hong Kong, Italy, Japan, Norway, the Russian Federation, the Southeast Asia region (including Indonesia, Malaysia, Singapore, Thailand and Vietnam), Spain, Switzerland, Taiwan, the UK and the USA. In addition, four countries, namely Australia, China, Columbia and India, completed detailed questionnaire on slope safety preparedness for the impact of climate change in their countries. A majority of the key players working on the subject of climate change impact on slope safety has therefore contributed to the project.

The country/region reports had the following objectives:

(a) Provide an overview of the landslide-related activities in their countries/regions;
(b) Present specific climate change projections and highlight examples of research into the possible impact in terms of increased/decreased landslide activities; and
(c) Discuss, for the respective country/region, the regulatory framework, multi-disciplinary hazard assessment, landslide risk management, adaptation measures and priority areas for further work.

Key contributions to the topic of climate change impact and preparedness strategies associated with landslide risk have also been made by two recently completed European collaborative research studies, namely the CHANGES (*www.itc.nl/changes*) and Safe-Land projects (*esdac.jrc.ec.europa.eu/projects/safeland*). The JTC1 project also sought the expertise of meteorological specialists to consolidate the state-of-the-art and state-of-practice within climate science and climate modelling, and to summarize the latest findings of the national climate change models.

This overview paper summarises the contributions, insights and key observations made in the country/region reports. Readers should refer to the individual reports for a detailed description of the work being pursued in different parts of the world.

1.2 KEY FINDINGS OF DIFFERENT COUNTRIES/REGIONS

Climate change is likely to modify the frequency, size and intensity of climatic events that can cause landslides. Such climatic events include, among others, intense and prolonged rainfall, coastal storm waves and tides, melting of snow and ice and rain-on-snow (Coe, 2016). Climate change may also lead to an increase in forest fires which, in turn, can increase the susceptibility to landslide from subsequent rainfall, or in flooding, which can cause subsequent landslides due to undercutting of unstable banks or instability as the flood waters recede.

Table A1 in Annex A summarises in some detail the status of development work in the different countries/regions with respect to: (1) the engineering and geological profession's understanding of climate change impact on slope safety, (2) the approaches adopted for climate projections, and the evaluation of critical hazard scenarios, (3) key vulnerabilities due to the climate impact, and (4) the extent of consideration of landslide risk management and adaptation measures in response to climate change. More details on the above issues are given in the respective country/region reports. The information in Table A1 suggests the following:

- A large number of countries and regions have completed downscaling studies to assess climate change.
- Most of the countries and regions predict increasing landslide hazards and therefore increasing risk to population associated with the effect of climate change. A majority of the countries/regions are more concerned with extreme events rather than the climate change effects per se (which may or may not be responsible for

an increase in the occurrence of extreme events). The expected landslide concerns vary however greatly from one country or region to the other, depending on the weather patterns, soil types and serious landslides experienced in the past, and the nature of the potential triggers (earthquake, floods, rainfall, etc.)
- The more sizeable countries or regions with widely varying climates have a variety of key vulnerabilities. Especially in these countries, the impact of climate change may be reflected by increased number of landslides and increased intensity in some parts, whilst other part of the country/region will be little affected.
- In all cases where permafrost and ice are present, the global warming will result in increased landslide hazards and risk to the population.
- In terms of specific landslide risk management or adaptation measures for climate change, the different countries and regions are at different stages of preparedness for climate change. Hong Kong is the region that has come furthest in terms of holistic and systematic assessment of impact of extreme weather events on slope safety through stress testing, based on which further steps are being taken to enhance emergency preparedness and response. All countries and regions acknowledge that they need to develop an improved understanding of the interrelationship among climate effects (e.g. rainfall, evapotranspiration, forest fires, permafrost thaw, etc.), hydro-mechanical properties, failure modes, and slope instability. All also acknowledge the difficulty in preparing realistic scenarios for improved preparedness. Many countries and regions are interested in early warning systems as an effective mitigation measure, so further development and sharing of experience in this area are warranted. Many countries and regions are presently focusing on improved communication and are looking for international cooperation for increased competence and improved preparedness.

The salient points and pertinent observations are summarized in the following sections.

1.3 PREDICTION OF CHANGING WEATHER PATTERN

According to the United Nation's Intergovernmental Panel on Climate Change (IPCC), the earth's atmospheric carbon dioxide concentration, the main driver of global climate change in the last century, has increased by over 40% since pre-industrial times, which is the highest in the last 800,000 years. The IPCC also concluded that human influence has been detected as a causal effect for the warming of the atmosphere and ocean, changes in the global water cycle, reduction in snow and ice, rise of global mean sea level and more frequent extreme climate. The effects of climate change have already emerged, and many locations around the globe are experiencing raised temperatures, more frequent extreme rainfall and rising sea level.

A top list of the causes and effects of global warming include:

Causes

- Carbon dioxide emissions from fossil fuel burning power plants and burning gasoline for transportation.

- Methane emissions from animals, agriculture such as rice paddies, and from Arctic seabeds
- Deforestation, especially tropical forests for wood, pulp, and farmland.
- Increase in usage of chemical fertilizers on crop lands.

Effects

- Rise in sea levels worldwide.
- More severe storms.
- Crop failures, extinction of species and disappearance of coral reefs
- Increased landslide activity in vulnerable regions

1.3.1 Methodology

Climate change predictions usually employ a two-stage modelling approach, namely:

1. an Atmosphere-Ocean General Circulation Model (AOGCM), which incorporates the key global physical and chemical processes; and
2. a Regional Climate Model (RCM), which translates the AOGCM coarse general solutions into more refined higher scale solutions, accounting for the influence of orography on local atmospheric phenomena.

Such process driven models, which are calibrated against past climate behaviour, represent the state-of-the-art in climate science and are now commonly adopted in most locations instead of extrapolations using trend analysis. Various techniques are available for local climate projections: (a) dynamic downscaling, (b) statistical downscaling and (c) pattern scaling.

a) Dynamic downscaling involves a regional climate model (RCM) of a finer resolution (typically 50 km or less) than the coarser resolution AOGCM (e.g. 200 km × 200 km), but it uses the AOGCM output as boundary condition to produce predictions at a finer scale. The approach is physics-based. Being computationally very demanding, the RCM model often uses only one AOGCM as input, rendering it difficult to assess the uncertainty in the projection. Generating a large ensemble of projections using a single AOGCM and a single RCM (to assess uncertainty) is possible, but this would be expensive and time consuming.
b) Statistical downscaling is a method of projecting climate change from relatively coarser-resolution AOGCMs by developing quantitative relationship between large-scale atmospheric variables (predictors) and regional/local-scale variables (predicands), assuming that the statistical relationships also hold for future climate conditions. Climate projection for a specific area can then be obtained using AOGCM output as predictors. This technique is far less computationally demanding than dynamic downscaling.
c) Pattern scaling has been used in climate change projections for different time periods and emission scenarios of local scale that are not simulated by AOGCMs.

1.3.2 Predictions and uncertainties

Country-specific or region-specific climate predictions, including sensitivity analyses, have been done by many countries. Some countries have already updated their assessments using the latest IPCC Fifth Assessment Report (AR5) results (IPCC 2014), whilst others presented findings based on the AR4 results (IPCC 2007), as the updating work using AR5 results was still in progress. Climate change predictions are usually made for the near future and the end of century scenarios. In most cases, average temperature, evapotranspiration and long duration rainfall (albeit with comparatively more uncertainty as duration increases, and typically with increasing precipitation during the rainy season and with decreasing precipitation during the dry season) are projected to increase as years pass by. In some locations, the projections suggest an increase in the intensity of typhoons and an increase in associated extreme rainfall.

The frequency of severe rainfalls and hot days is projected to increase in many locations, which will impact rain-induced landslides and the stability of otherwise stable permafrost slopes. The sea water level is projected to rise, with varying degrees in different locations, which can also have implications for coastal slope stability.

In the different locations, the climate models adopted mainly differ in their resolution, depending on whether dynamical downscaling or statistical downscaling of the global climate models (AOGCM) is adopted to obtain rainfall projections for small areas. The resolution of an RCM may still not be sufficient to capture the local weather forcing (e.g. cloud microphysics and small-scale turbulence) on the scale of a few km. Furthermore, there are likely to be larger uncertainties associated with climate change models for tropical monsoon or cyclone rainfall (especially in Asia) as compared with non-tropical areas, as meso-scale phenomena, such as squall lines and thunderstorm systems which can result in very intense rainfall, cannot be realistically simulated by the current climate models.

Climate science is by nature intrinsically uncertain and the degree of uncertainty cannot be easily quantified. Uncertainties include future greenhouse gas emissions, status of glacier and ice sheet, choice of models, ability of climate models to simulate physical processes, downscaling methodology and stability of the statistical downscaling in the future etc.

There appears to be a consensus amongst the meteorological profession that rainfall predictions are generally subject to much higher variability than temperature predictions. Current projections of changes in rainfall have mainly tended to focus on the predicted changes in long-term mean conditions, instead of the frequency and magnitude of extreme events that can generate, locally and intermittently, very intense rainfall. This has major practical significance, as experience shows that changes in mean annual precipitation generally correlate poorly with the temporal and spatial scale of shallow (<3 m) landslides and debris flows which tend to be triggered by short- to medium-duration intense rainstorms.

1.3.3 Potential landslide risk and current climate models

Based on the observations from larger rain-induced landslide events in recent years, the landslide risk associated with extreme, record-breaking, rainfall events is likely to be significant and may dominate the risk profile in some areas. It is therefore essential to keep in mind the potential occurrence of extreme rainfall events which can cause catastrophic landslides and pose significant risk to human life and health in cities and other developed areas located at the foot of mountains and hills.

Although the science of climate modelling has made significant advances in the last two decades, notable limitations in climate models still exist. As one example, small-scale physical, biological and chemical processes cannot be fully described by mathematical equations.

In light of the uncertainties and limitations associated with the current climate models, quantitative predictions of climate change scenarios are not sufficiently reliable at the present time for direct use in landslide predictions and impact assessments, despite the availability of more advanced and higher resolution climate models. The current projected climate change scenarios are also constrained by the limited projection period of around 50 to 100 years. As the past is no longer an adequate predictor for the future, the direct use of the projections from the climate models is subject to considerable uncertainties. Nevertheless, the current climate change predictions at large, taken more as a qualitative assessment of the likely trends, do point to differing degrees of potential adverse effects of climate change on slope safety, as explained in the country/region reports. In many cases, the impact of climate change is increased landslide risk.

As such, the engineering geology and geotechnical engineering professions need to pay due attention to extreme rainfall event scenarios and the need for as reliable as possible meteorological predictions of climate change and occurrence of extreme weather, especially extreme rainfall events. The profession needs to be aware that the current capability of the available climate models is now subject to an even greater degree of uncertainty with respect to the prediction of extreme rainfall events under a changing climate. Further development is warranted in this area.

There is also a need for enhanced interaction and collaboration amongst the engineering geology/geotechnical engineering and meteorological professions. The need of the landslide practitioner for its landslide assessment studies could then be conveyed more clearly. Relevant issues, such as spatial and temporal resolution and spatial coherence of projected rainfall scenarios, could also be addressed. The climate projections could be made more informative and adapted for the subsequent geographical implications for slope safety.

1.4 CHANGING LANDSLIDE PATTERN UNDER CHANGING CLIMATE

For countries of significant spatial extent, the climate predictions can vary markedly from site to site, which can lead to a range of projected climate change impact. The climate change patterns predicted by the climate models will vary significantly from one region to another, such that some locations may experience an adverse increase in both landslide frequency and magnitude (size and travel distance) of landslides, other places may experience beneficial decrease, whilst certain areas may experience a mixture of changes (e.g. decrease in frequency and increase in magnitude). Also, depending on the geographical and geomorphological settings, the impact of climate change on slope safety can vary substantially even over small distances.

1.4.1 Types of landslides

A landslide is any slope movement induced by a change in boundary conditions or by natural soil/rock deterioration. Different and variously named landslide types can be found, displaying different mechanisms and movement rates which depend on slope morphology, material involved and trigger. The most popular classification is the one proposed by Varnes (1978), recently updated by Hungr *et al.* (2014), who defined six fundamental landslide classes: i) slope deformations, i.e. typically slow slope movements not necessarily associated with a general slope failure; ii) falls, consisting of rock blocks falling from extremely steep slopes; iii) topples, characterised by a typical forward overturning of sub-vertical rock columns or plates; iv) slides, consisting of soil/rock masses brought to failure by a mechanical process of shear deformation, which slide downslope along one or more slip surfaces; v) flows, which main feature is a post-failure flow-like movement, often channelized, as the landslide consists of an essentially viscous fluid mass; vi) spreads, characterized by nearly horizontal stretching of a soil/rock mass subjected to intense basal shear strains or by multiple retrogressive sliding (Hungr *et al.*, 2014). Each of the fundamental landslide types can in turn be classified into sub-types according to the nature of the material involved or special features of the failure or the post-failure mechanism. As an example, rock avalanches, debris flows, flow slides and earthflows are all landslide in the fundamental flow class.

Weather is the most common trigger for landslides, depending though on the ground material that is susceptible to slide. Weather as a trigger implies that climate changes in many countries can affect the impact of landslides, possibly leading to a change in recurrent types of movements. In particular, as suggested by many country and region reports in this book, major changes might affect the more superficial soil/rock layers, leading to shallow landslides such as rock falls, topples, shallow slides, rock avalanches, debris flows, flow slides and earthflows. This can be explained by the weather related phenomena, mostly precipitation or snowmelt, essentially affecting the water content and the pore water pressure regime within the uppermost few metres of the soil mantle. The role of permeability of the soil mass therefore becomes prominent. In addition, in some cases the uppermost layers can be subjected to an increase in the rate of natural process of soil/rock deterioration induced by cyclic changes in the water content and temperature.

Many of the country and region reports suggest that the distribution of event typologies, potentially triggered by weather forcing, may change. In particular, as discussed by Picarelli *et al.* (2016), climate change may lead either to a decrease in the return time of fast-moving flow-like landslides in unsaturated granular soils, or to a slow-down, albeit moderate, of active slides in clayey soils.

1.4.2 Key climate change factors

As discussed above, changes in the rainfall regime, and in the sea level, may exacerbate the occurrence of shallow landslides (generally in the top 3 m or so). However, in some cases deep-seated landslides may also occur as delayed response failures after prolonged periods of intense rainfall, especially in deeply fractured materials characterised by rather high overall permeability. The frequency of rock fall and rock avalanche may also increase due to climate change.

Changes in landslide susceptibility will result from the combined changes in the preparatory factors (e.g. antecedent groundwater conditions, vegetation, etc.) and triggers (e.g. rainstorm events, freeze-thaw cycles, flooding). Climate change is likely to modify the above factors as well as other possible controlling variables, such as land-use or vegetation cover, but it may not necessarily result in a notable increase in landslide frequency.

In overall terms, current findings (e.g. Cloutier *et al.*, 2016; Coe, 2016; Dijkstra *et al.*, 2016) suggest that permafrost slopes, mountains with glaciers, coastal slopes, slopes mantled with unsaturated granular materials, and slopes that are prone to progressive deterioration with time due to repeated wetting-drying cycles and/or successive heavy rainstorms (referred to as slope conditioning by Dijkstra *et al.*, 2016), will probably be strongly and adversely affected by climate change. Other slide types that may also be adversely affected include quick clay slides, flows and spreads (e.g. in Norway, Sweden and Canada) usually induced by river erosion.

1.4.3 Issues hampering understanding of climate change impact on landslide occurrence

Three issues hamper rapid advances in the understanding of the impact of climate change on landslide occurrence (Coe, 2016):

- Landslide activity is inherently difficult to forecast even with a stationary climate;
- Future changes in human activities and land use (e.g. deforestation), which may or may not be related to climate change, could have a comparable or even greater impact on landslide activity than climate change effect alone; and
- Current projections of changes in precipitation appear somewhat inconsistent and haphazard. In many locations, focus is set on the changes in longer-term mean conditions rather than on the frequency and magnitude of extreme weather events with severe rainfall that can be due to climate change.

1.4.4 What is an extreme event?

The U.S. National Science Foundation defines an extreme event as "a physical occurrence that with respect to some class of related occurrences, is either notable, rare, unique, profound, or otherwise significant in terms of its impacts, effects, or outcomes." The Intergovernmental Panel on Climate Change (IPCC) has the following, more quantitative definition for an extreme events "... An event that is rare at a particular place and time of year. Definitions of "rare" vary, but an extreme weather event would normally be as rare as or rarer than the 10th or 90th percentile of the observed probability density function ..." (IPCC, 2012).

An example of an extreme event is the Great East Japan earthquake (Tōhoku earthquake) and tsunami of 11th March 2011. This magnitude 9.0 (Mw) earthquake was the most powerful earthquake ever recorded to have hit Japan, and the fourth most powerful earthquake in the world since modern record-keeping began in 1900. One of the catastrophic consequences of this event was the Fukushima Dai-ichi nuclear power plant accident.

Another example of a "usual" natural hazard event leading to extreme consequences was the 2010 eruptions of Eyjafjallajökull volcano in Iceland (Gudmundsson *et al.*, 2012). These relatively small volcanic eruptions caused enormous disruption to air travel across western and northern Europe over an initial period of six days in April 2010. During the period 14–20 April, ash covered large areas of northern Europe when the volcano erupted. About 20 countries closed their airspace for commercial jet traffic and it affected about 10 million travellers (WENRA 2011).

1.4.5 Extreme weather events

Extreme weather events (such as very intense rainfall) are expected to have a definite negative impact on slope stability. In the present context, extreme weather events refer to events that are very severe in terms of their intensity (either as the extreme of historical distribution or more severe than that observed in the (incomplete) historical records), and at the same time very rare in occurrence.

The impact of extreme weather events (near-'black swans' events), which may be exacerbated by climate change, is considered as a major risk concern in some regions (e.g. Hong Kong, Taiwan, etc.) especially if they have experienced some extreme events earlier, while work on extreme events is still in its infancy in many locations. This highlights the need for increased efforts on the effect of climate change on extreme weather events (in particular, rainfall) by the meteorological community. There is, however, considerable uncertainty in predicting the future changes in extreme rainstorms (e.g. hurricanes) and the regional climatic phenomena (e.g. the El Nino Southern Oscillation). The challenges in making sufficiently reliable projections of extreme weather events are daunting.

An extreme weather event can also be a natural aleatory phenomenon within the natural and intrinsic variability of the weather system. Extreme events therefore need to be considered irrespective of whether climate change prevails or not. This view is arguable. Opinions are split as to whether it is significant to seek to distinguish the two. From the perspective of engineering, mitigation and adaptation, this distinction (although rigorously elegant) is probably inconsequential, as the need to address extreme weather conditions is reinforced according to current wisdom and based on empirical observations.

Although difficult to quantify, it is a widely held view that extreme weather events can become more frequent and more severe due to climate change. This appears to be corroborated, at least in part, by observations of increasing 'extreme' rainfalls in many locations in recent years (with local historical rainfall records broken at a faster rate as well as by a larger margin as compared to the past). A number of locations with hilly terrain (e.g. Brazil, Taiwan, Hong Kong, etc.) have experienced extreme rainfall associated with typhoons or other climate anomalies, leading to significant numbers landslides and casualties in developed areas.

1.4.6 Extreme landslide scenarios

From a landslide risk management perspective, extreme rainfall scenarios might result in a notable increase in the number, volume, runout and area affected by landslides, which can be of unprecedented dimensions. This needs to be duly factored into the assessment, particularly for natural terrain landslides. The increased scale has major implications for landslide risk management: usually appropriate landslide mitigation measures (e.g. check dams or debris-resisting barriers) may be overwhelmed by landslides that are larger than considered in the design.

Extreme landslide scenarios associated with extreme rainfall events tend to share five distinctive features that are of primary concern to risk and emergency management:

a) Widespread landslides, including possibly massive landslides, with increased propensity, scale and mobility of the moving masses and resulting in serious damage and potentially significant casualties in developed areas.
b) A large number of landslides and widely affected population, together with breakdown of infrastructure and the ongoing nature of landslide hazards, pose an extreme challenge to the emergency response and recovery work.
c) There is a lack of understanding of extreme landslide scenarios, partly due to the scarcity of their earlier occurrence and thereby the limited attention given to studying extreme landslide events.
d) Preparedness for the emergency management of extreme landslides is generally inadequate, which would aggravate the consequence of the disaster.
e) Extreme rainfall events tend to affect a comparatively much smaller region than a severe earthquake. Severe earthquakes may also cause extreme landslides, especially when preceded by severe rainy periods. An example was the M7.4 Khait earthquake in Tajikistan (Strom, 2016). There were thousands of fatalities, many of which were persons buried by landslides. Hence, the total landslide damage and death toll in extreme rainfall events would not be as drastic as those in extreme earthquake events. However, it is noteworthy that the landslide impact of an extreme rainfall event in a local area, e.g. a densely populated urban area, could be just as serious as the landslide impact from an extreme earthquake event. On the other hand, natural hillsides are generally subject to an ongoing degradation process. Increased nature terrain landslide activity may occur after an extreme rainfall event, although this will perhaps not be as severe as after an extreme earthquake event.

1.4.7 Approaches to quantify impact of climate change on slope safety

Because the needs of the stakeholders differ, different approaches to quantify the impact of climate change on slope safety are required. A private house owner is concerned about local safety whereas a railway asset manager, a government department or a policy minister will need to know the vulnerable connections, other vulnerabilities and the overall scale and nature of the landslide problem. A different framework is needed in terms of impact assessments and communication of the findings to the different stakeholders. To quantify the impact of climate change on slope safety, three broad approaches are emerging, namely:

1. Conventional tools of advanced mechanics-based numerical models (coupling atmospheric environment and rainfall with seepage and slope material response);

2. Scenario projections using systems-based correlations calibrated against slope response in past severe rainfall events; and
3. Stress testing

1.4.7.1 Mechanics-based models

To assess the impact of climate change (e.g. changes in rainfall intensity and pattern, temperature and groundwater level), some countries (e.g. Italy, Singapore, Thailand and United Kingdom) have resorted to the use of numerical models that account for the appropriate flux boundary (with due consideration of runoff, evaporation, desiccation cracks, etc.) combined with unsaturated soil mechanics. Field measurements of instrumented slopes, some with provisions to apply different simulated rainfall, are used to verify and calibrate the numerical models of transient seepage and slope responses.

Such approach is mechanics-based and enables sensitivity analyses and study of potential changes in failure mode and scale for different rainfall patterns. However, it is difficult to incorporate in a realistic manner in the model the variability of the ground condition and, in particular, possible local adverse geological features. The approach tends to be better suited to the study of individual slopes than for the assessment of the regional response of a large number of slopes in a given rainstorm, especially in the presence of highly variable natural terrain. In particular, the above approach cannot be used effectively to help evaluate the scale of the problem in a system, e.g. for a risk manager or owner of a large number of slopes.

1.4.7.2 Systems-based scenario assessment

The alternative approach is a scenario-based assessment. This novel approach was developed by the Geotechnical Engineering Office (GEO) in Hong Kong (Ho *et al.*, 2016). It uses the Probable Maximum Precipitation (PMP) concept, which is generally accepted and well understood in engineering circles, in establishing plausible extreme rainstorm scenarios.

Describing rational scenarios in the most vulnerable contexts for impact assessments calls for a multidisciplinary approach and it can only give, at best, approximate answers. Hence, it may not be appropriate to seek for an overly precise answer to a fuzzy question, given many uncertainties. A pragmatic approach was adopted in Hong Kong by considering suitable bounds for extreme events to obtain an order of magnitude estimate of the size of the problem.

In this approach, plausible extreme rainfall scenarios are defined as a Level 1 event and a Level 2 event respectively, both benchmarked against the PMP in order to put the severity of the extreme rainfalls into perspective. Level 1 event corresponds to the most severe rainstorm observed in more than 100 years of rainfall record transposed to the most vulnerable region of a densely urbanized area. Level 2 event corresponds to a simulated rainstorm that is more severe (hitherto not yet observed, but plausible) than the historical extreme, projected to the end of the 21st century with allowance made for climate change effect (e.g. allowing for the projected increase in tropical cyclone rainfall towards the end of 21st century).

The PMP for Hong Kong was established with input from hydrological experts using state-of-the-art techniques (e.g. storm transposition with suitable orographic corrections) from which appropriate rainstorms from meteorological similar regions can be defined as credible extreme rainfall scenarios (details in Ho *et al.*, 2016). The likely slope response (i.e. extreme landslide scenarios) is based on making projections from the available rainfall-landslide density correlation that was derived from past data on rainfall (e.g. automatic rain gauges) and landslides (e.g., aerial photographs and field mapping). Good quality field data on past landslides can also be obtained from systematic landslide investigation programme. Thus, the projections of slope responses make the best use of the available local empirical data and hence can minimise the uncertainty where possible. Ho *et al.* (2016) emphasized, however, the considerable residual uncertainty associated with predicting the actual slope response under extreme rainfall, because the number, volume and mobility of landslides will be unprecedented.

With the empirical relationship between landslide density and normalized rainfall from past severe rainstorms, and available information on the scale and mobility of landslides under different rainfall conditions, one can make a reasonable assessment of the likely extreme landslide scenarios corresponding to different level rainstorms. The use of normalized rainfall (e.g. rainfall at a given locality divided by the mean annual rainfall at that locality based on past rainfall records over a 30-year reference period) is to allow for the spatial variability of rainfall. Reference to the correlation of rainfall with landslide density caters to the need for a systems risk approach in terms of the total number of landslides and the likely distribution of the scales and mobility of the corresponding landslide debris. This scenario approach is more helpful than the first approach for risk managers and policy administrators.

As mentioned, the residual uncertainty remains significant. This must be borne in mind in the interpretation of the results. In particular, the quantification of risk to life associated with extreme rainfall events should not be done without qualification (Ho *et al.*, 2016), as the estimate may be misleading and can be under- or over-estimated, given the lack of experience with the actual slope performance under extreme rainfall.

1.4.7.3 Key findings from scenario-based assessments

Scenario-based assessments indicate that natural terrain landslides (mainly comprising shallow debris flows or debris slides in granular colluviums or saprolites[1]) will be more frequent and dominate the risk profile, as the natural terrain responds much more strongly in terms of number, scale and mobility of landslides at more intense rainfall levels as compared with man-made slopes. The risk profile changes with increasing rainfall.

Apart from the escalated scale of the landslides due to extreme rainfall, other forms of less commonly observed landslide modes may be triggered by extreme rainfall and have significant impact on the community. Low-frequency, high-consequence landslides may include very sizeable, deep-seated failures that are very rare, and may catch the stakeholders by surprise. The Hong Kong landslide impact approach with Level 1 and Level 2 extreme rainfalls can also serve as a stress test (see next section) on

[1] Saprolites: chemically weathered rock. Saprolites form from the deep weathering of the bedrock surface. Conditions for the formation of the deep weathering include a relief flat enough to prevent erosion and to allow leaching of the products of chemical weathering, long periods of tectonic stability and humid tropical to temperate climate.

the prevailing landslide emergency system in terms of its ability or otherwise to cope with the projected serious landslide incidents.

Based on the stress test, it was concluded that under Level-1 rainfall, the existing emergency system would be stretched to its limit. This highlights the need for further streamlining, building in more redundancies and addressing the potential bottlenecks of the current system. Under Level-2 rainfall, the very large number of natural terrain landslides predicted (a total of 50,000 landslides, up to 9,000 of which could impact on existing buildings and roads) would completely overwhelm the government's landslide emergency system (which could at most cater for about 300 serious reported landslides). This emphasizes the need for the formulation of a new strategy for crisis management under extreme rainfall conditions. The extreme landslide scenarios will result in major impact to a densely developed city like Hong Kong. As an example, the following are envisaged:

- With high density of landslide on natural terrain, many landslides will 'merge' to form larger landslides, particularly along drainage lines, resulting in very large and long-runout debris flows.
- Due to the possible occurrence of very large and mobile landslides, the knocking-down of multi-storey buildings cannot be ruled out, resulting in multiple causalities and significant building damages.
- Evacuation of a large number of buildings, including residential, commercial, schools or critical facilities such as hospitals and fire stations, due to actual landslide damage or threat of immediate landslide danger, may be required.
- Most of the roads along the toe of sizeable natural hillsides and the access to subway may be affected and blocked by landslide debris. Other public transportation would be inoperable.
- Apart from landslides, uncontrolled surface water flow and overland flow could happen at multiple locations, causing more landslides and turning roads into rivers.
- Landslide debris could also damage telecommunication, water, electricity and gas supplies and severe the lifelines.
- Parts of the local communities with sole access are particularly vulnerable to isolation from land transport due to blockage or damage of the road networks.
- It is likely that landslide impact under such major rainstorm would be compounded by other hazards occurring at the same time, e.g. major flooding.

1.4.7.4 Stress testing

Conventional strategies for managing the risk posed by natural and/or man-made hazards rely increasingly on quantitative risk assessment (QRA). The QRA methods estimate the various components of risk, namely hazard (temporal probability of occurrence of a threat), vulnerability and exposure of elements of risk, and utility or value of the elements at risk. The methods do this quantitatively and provide a value estimate of the expected loss. As mentioned, one of the challenges in the management of risk associated with extreme events is that the mechanism triggering an extreme event may be different from those triggering the more frequent events. Climate change has introduced substantial non-stationarity into risk management decisions. Non-stationarity is the realization that past experiences may no longer be a reliable predictor of the future character and frequency of events; it applies both to hazards and to the corresponding response of the systems. As climate change is expected to change the frequency, magnitude, and other characteristics of extreme events, some of which will be associated with extreme impacts, risk management strategies must accommodate a shifting distribution of the latter.

The conventional design approach implicitly accepts that there is a "residual" risk, which could be "neglected" because the probability of that risk being realized is extremely small. This residual or neglected risk can be due to "extreme events", which have a longer return period than the return period for the design load (denoted with blue stars in Fig. 1.1), or they could be due to the uncertainty in the prediction models and lack of knowledge of the mechanisms at work (denoted with red stars in Fig. 1.1). Both types of events pose a risk to the integrity and performance of the system, a risk which is implicitly accepted and knowingly neglected in conventional engineering design. Nevertheless, these events can occur, and when they do, they are referred to

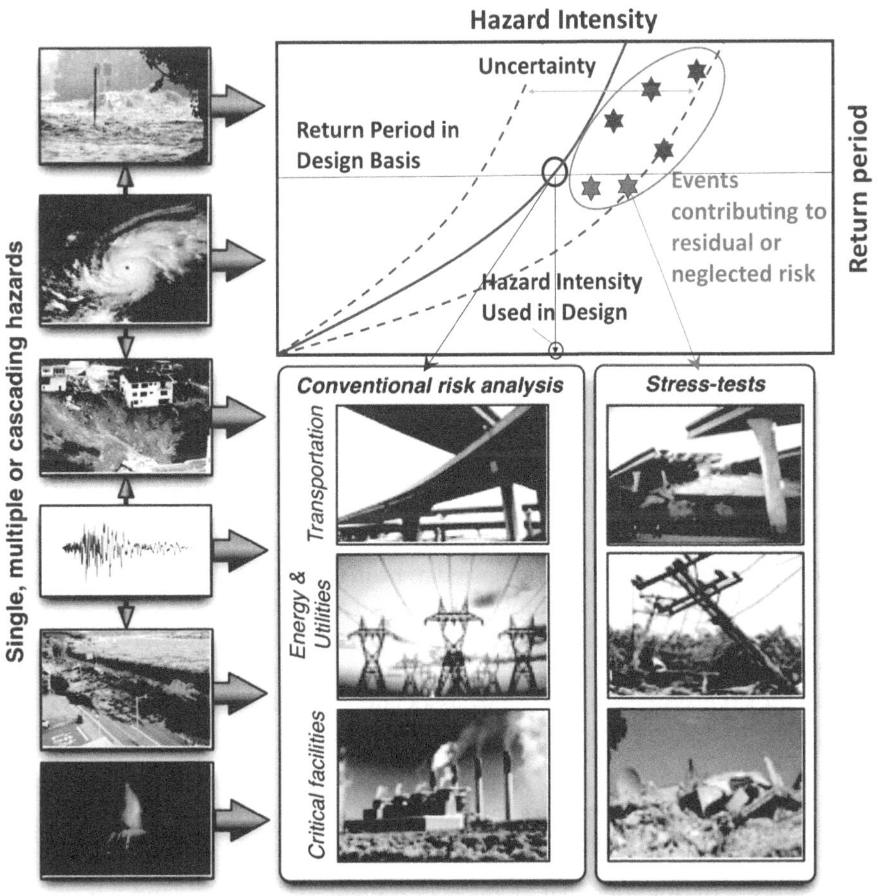

Figure 1.1 Residual (neglected) risks in conventional reliability-based design (Nadim, 2016).

as extreme events. Therefore, the conventional engineering design is not suitable for dealing with the risks posed by extreme events.

Stress testing is a procedure used to determine the stability and robustness of a system or entity. It involves testing the specific system or entity to beyond its normal operational capacity, often to a breaking point, in order to observe its performance/reaction to a pre-defined internal or external pressure or force. Stress tests have been used for many years in air traffic safety, in particular for airplanes and helicopters. In recent years, stress testing has often been associated with methodologies to assess the vulnerability of a financial system or specific components of it, such as banks. A number of analytical tools have been developed in this area and have been frequently used since the late 1990s (e.g. Borio *et al.*, 2012). However, stress testing has also been criticized in the field of forecasting due to the use of historical data taken from periods of time that are considered to be insufficient for reliable predictions.

Stress testing has been applied to the comprehensive safety and risk assessment of Nuclear Power Plants, in particular in the aftermath of the 2011 Fukushima Daiichi accident. In particular, the accident highlighted three areas of potential weakness in present safety approaches: (1) inadequacy of safety margins in the case of extreme external events (especially natural hazards); (2) lack of robustness with respect to events that exceed the design basis; and (3) ineffectiveness of current emergency management under highly unfavourable conditions. These issues were the focus of the stress tests imposed on all nuclear power plants in Europe in 2011 and 2012 (WENRA, 2011).

Nadim (2016) proposed stress testing as a complement to traditional risk assessment for managing the risk posed by extreme events, focusing on the challenges faced by civil engineers in general, and geotechnical engineers in particular, in managing the risk posed to critical infrastructure by extreme natural hazard events. Most risk evaluations are based on probability estimates using historical data and consequence models that try to estimate the impact of unwanted future hazard situations. For natural hazards, historical data may in some cases be sparse or highly uncertain. Similarly, simplified models of highly complex situations may yield forecasts containing significant uncertainty. Both of these situations may therefore neglect risks that should be introduced into the evaluations.

A stress test is an examination of the safety of a system under those particularly unfavourable scenarios that fall outside the design basis specified by the regulatory regime, by the operational institution or by the stakeholders. The stress test can assess the system's response to scenarios expected to be in the residual and neglected risk areas (Fig. 1.1). Stress testing is not a substitute for conventional risk or safety assessments, but it provides additional valuable insight for extreme situations. What stress tests and conventional risk or safety assessments have in common is that they both rely on a description of the system, which helps associate the state of the system and a set of consequences under any given or potential scenario.

1.4.8 Does climate change need to be a priority in landslide risk assessment?

There is a body of opinions that changes in human activity and land use will have a much greater impact on landslide risk than climate change, especially given the likely time frame involved in climate change, and hence the effect of climate change on slope safety is not considered to be of a high priority. The European projects SafeLand and CHANGES looked into these aspects.

The above view is understandable especially in those locations that do not have a comprehensive slope safety system to manage landslide risk. The lack of an effective regulatory regime to manage landslide risk associated with new developments, and insufficient funding by the respective government to deal with existing developments that are subject to high landslide risk, may cloud the awareness and assessment of the importance of climate change on slope safety.

The current wisdom is that a comprehensive slope safety system would entail a holistic and multi-pronged approach to combat landslide risk, involving the use of both hard engineering measures and soft non-engineering approaches. In circumstances with large gaps in basic landslide risk management, the need to address the fundamental gaps takes priority. However, it should be cautioned that the potential significant impact of climate change, in particular extreme weather events, should not be underestimated, as the consequences may be unprecedented.

The increasing occurrence of freak weather and very severe rainstorms and unusual weather in unseasoned times in many locations highlights the fact that climate change impact is not that remote in time. It is critical not to allow complacency or ignorance to lead one into developing misconceptions or preconceived ideas about the potential impact of climate change on slope safety. It is important to pursue the assessment of climate change impact on slope safety in parallel with the influence of human action on land use and vulnerability to slope failure together with the effect of demographical changes on landslide risk. The focus should be on identifying the critical risk components associated with key landslide scenarios, together with other initiatives to enhance landslide risk management in general. The fundamental problems hampering real progress lie in an insufficient understanding of the slope response, especially under extreme weather conditions, as well as lack of data and experience.

Most of the countries responding to the JTC1 project initiative have undertaken a qualitative assessment of the critical hazard scenarios and key vulnerabilities in terms of the nature and intensity of the landslides associated with climate change. Hong Kong, as a leading region in this area, has completed a rigorous and systematic quantitative assessment of the nature and scale of the landslide hazard and risk associated with extreme weather events (Ho *et al.*, 2016). Using a systems approach, Hong Kong was able to identify the critical hazard scenario, key vulnerabilities, the changing risk profile as the rainstorm developed and the implications for landslide risk management.

1.4.8.1 Safeland project

The need to protect people and property in view of the changing pattern of landslide hazard and risk caused by climate change, human activity and changes in demography, and the need for societies in Europe to live with the risk associated with natural hazards, formed the basis for the 2009–2012 European SafeLand project "Living with landslide risk in Europe: Assessment, effects of global change, and risk management strategies". The project involved 27 partners from 12 European countries, and had international collaborators and advisers from mainland China, Hong Kong, India, Japan and USA. SafeLand also involved 25 End-Users from 11 countries. SafeLand was coordinated by the Norwegian Geotechnical Institute's (NGI) Centre of Excellence "International

Centre for Geohazards (ICG)" (http://safeland-fp7.eu/). Nadim and Kalsnes (2014) present the results of the project in more detail. The SafeLand conclusion was that climate change, human activity and change in land use and demography all need to be considered in the assessment of landslide risk, and that climate impact on slope safety need to be given high priority.

The SafeLand project provided:

- Guidelines on landslide triggering processes and run-out modelling.
- Empirical methods for predicting the characteristics of threshold rainfall events for triggering of precipitation-induced landslides, and for assessing the changes in landslide frequency as a function of changes in the demography and population density.
- Guidelines for landslide susceptibility, hazard and risk assessment and zoning.
- Methodologies for the assessment of physical and societal vulnerability.
- Identification of landslide hazard and risk hotspots in Europe. The maps show the location of the areas with highest landslide risk and allow a ranking of the countries by exposed area and population.
- Simulation of regional and local climate change over regions of Europe at spatial resolutions of 10 × 10 km and 2.8 × 2.8 km. The simulations were used for an extreme value analysis of trends in heavy precipitation events, and subsequent effects on landslide hazard and risk.
- Guidelines for the use of remote sensing, monitoring and early warning.
- Prototype web-based "toolbox" of innovative mitigation measures, including over 60 structural and non-structural landslide risk mitigation options.
- Case histories and "hotspots" of European landslides covering almost all types of landslide in Europe from seven countries (Italy, France, Norway, Switzerland, Austria, Andorra, and Romania).
- Stakeholder workshops and participatory processes to involve the population exposed to landslide risk in the selection of the most appropriate risk mitigation measure(s).

1.4.8.2 CHANGES Project

The CHANGES Marie Curie Training Network education network (Changing Hydro-meteorological Risks – as Analyzed by a New Generation of European Scientists) aimed at developing an advanced understanding of how global changes (environmental and climate changes and socio-economical change) affect the temporal and spatial patterns of hydro-meteorological hazards and associated risks in Europe. The project focused on the assessment and modelling of the changes, and incorporating them in sustainable risk management strategies, including spatial planning, emergency preparedness and risk communication. The work was inter-disciplinary and inter-sectoral, with stakeholder participation. The main objectives of the project were to:

1) Provide high-level training, teaching and research in the field of hazard and risk management in a changing environmental context to European young scientists;
2) Reduce the fragmentation of research on natural processes; and

3) Develop a methodological framework combined with modelling tools for probabilistic multi-hazard risk assessment taking into account changes in hazard scenarios (related to climate change) and exposed elements at risk.

The network consisted of 11 full partners from seven European countries. The network was run by ITC, Faculty of Geo-Information, Science and Earth Observation of the University of Twente in The Netherlands, and employed 17 Early Stage Researchers from all over the world.

Scientific advances were made on four main themes: (1) Analysis of the changes in hydro-meteorological hazards as a result of environmental changes (climate change and changes in land use/land cover and ecosystems), with regional and local scale probabilistic hazard assessments for floods, landslides and debris flows at four sites; (2) Evaluation of the environmental changes and vulnerability of elements at risk, triggered by global change, including climate change and interacting with economic development, leading to changes in exposed elements at risk; (3) Integration of the techniques for hazard assessment with the results of the exposure and vulnerability analysis, into a platform for Quantitative Risk Assessment (QRA) using multi-hazard risk assessment techniques; (4) Obtaining an advanced understanding on the temporal and spatial patterns of hydro-meteorological hazards and associated risks in sustainable risk management strategies, focusing on spatial planning and emergency preparedness, response and rescue activities. Instruments and tools for risk management were developed to cover the entire disaster cycle, also considering the complexity of changing environments. The outcomes were used for risk communication purposes, achieved by designing specific risk visualization and communication tools, aimed at local authorities, planning agencies, civil protection organizations, and the general public.

A "Risk Change Spatial Decision Support system for the Analysis of Changing Hydro-meteorological Risk" was developed. The Spatial Decision Support System analyses the effect of risk reduction planning alternatives on reducing the risk now and in the future, and support decision makers in selecting the best alternatives. The decision support system is composed of a number of integrated modules. It is available online, and can be accessed through the URL: http://changes.itc.utwente.nl/RiskChanges.

1.5 LANDSLIDE RISK MANAGEMENT

1.5.1 Quantifying the uncertainties

Nearly all the national/regional reports mentioned the importance and difficulties in quantifying the uncertainties. The current limited understanding and the lack of data are the major constraints today. The approaches adopted so far are mostly not amenable to verification by other means. The physical processes involved (e.g. cloud microphysics, small-scale turbulence, etc.) are generally not well understood and often not modelled in detail. Thus, both model uncertainty and parameter uncertainty can be significant despite the use of state-of-the-art climate models, not to mention the fact that there may also be other "unknown unknowns".

There is a perception that the level of uncertainty in the prediction of the impact of climate change on deep-seated (thicker than ~5 m) landslides is less as compared to shallow landslides, as the former are generally more sensitive to longer term precipitation patterns, which climate projections can more readily provide. On the other hand, shallow landslides and debris flows in unsaturated granular materials are highly dependent on individual short-duration rainstorms, which are more difficult to forecast. On a similar vein, the temperature predictions are subject to less uncertainty than rainfall, and hence it follows that the predicted behaviour of permafrost slopes may be subjected to less uncertainty.

1.5.2 Insights from scenario-based assessments

To review slope safety preparedness for climate change impact, it is necessary to have a sufficient understanding of the scale of the problem as well as the critical hazard scenarios. In the case of rain-induced landslides, the spatial distribution, areal extent and the characteristics (e.g. duration and intensity) of severe rainstorms, together with the corresponding slope response and the nature of facilities affected by the landslides, will affect the scale of the problem (i.e. the risk level). In practice, a systems approach is called for in conducting a proper assessment of the nature and overall scale of the problem to the society at large. Most countries, however, have not progressed far with such systems risk assessment. The majority of the preliminary assessments made are qualitative in nature.

The findings in Hong Kong have put the potential impact of extreme rainfall events in perspective, for the conditions and settings in Hong Kong. Over the years, Hong Kong has implemented a comprehensive slope safety system, which is a framework for systematic and multi-pronged management of landslide risk using both engineering (hard) and non-engineering (soft) approaches. This has largely evolved into an effective model for landslide risk management. In general, prevailing slope safety systems are often geared towards managing the risk of heavy rainstorms that occur relatively frequently (say, from several times a year to once every several years), and they are not specifically designed to deal with, or tested against, extreme rainfall events.

Given the nature of extreme events and the limitations of the available technology and resources, the key findings in Hong Kong have shown that it would not be pragmatic to aim to prevent the extensive slope failures under extreme landslide scenarios. Taking cognizance of the infrequent occurrence of extreme landslide scenarios, emergency management would constitute the principal strategy for tackling extreme landslide scenarios in Hong Kong.

Such approach can also be taken for dealing with non-landslide-related damages (e.g. structural failures) under extreme earthquake events. The approach is now beginning to be promoted to deal with flooding risk associated with climate change, i.e. involving a shift in the paradigm in enhancing the resilience of the community facing extreme events.

There remains fuzziness on the number, volume and mobility of landslides under an extreme weather event. Hence, risk quantification will be fraught with major difficulties. Notwithstanding this, the use of scenario-based assessments and a form of stress testing of a slope safety system can provide considerable insight and a pragmatic basis for reviewing appropriate climate change adaptation strategy.

For countries that already have a slope safety system for landslide risk management, the focus should be on further enhancing community resilience and system resilience, together with improving crisis preparedness against the impact of climate change, especially that associated with extreme weather events.

Sound land-use planning is a fundamental component of landslide risk management to prevent that residential buildings, schools, hospitals, etc. be erected landslide-susceptible areas, consistent with the usual climate change approaches. Problems arise with densely populated cities where space is at a premium and where many existing developments are already located in areas that are exposed to landslides (typical situations are found in Hong Kong and the densely populated areas of Japan, Italy, Southern America). In such circumstances, other risk management approaches are called for in managing the landslide risk in such circumstances. In managing extreme weather events, it is essential to have a well-considered emergency and crisis management plan, and to raise public awareness through dialog about the limitations of engineering works, which can at best mitigate the impact but not prevent or eradicate the adverse consequences.

Improving the robustness of the slope safety and/or protection system and enhancing the resilience of the community are of the essence. Improving system robustness entails improving the structural mitigation measures and the landslide warning or alert systems, enhancing communication amongst various emergency organizations and providing up to date information and pragmatic advice to the public in an effective manner, improving sharing of emergency information, building in more redundancies to address potential bottlenecks, sourcing additional contingency resources in emergency management, streamlining emergency management procedures, ensuring more effective emergency response and strengthening crisis preparedness, etc. Enhancing community resilience entails engaging the community, public education, facilitating a proper understanding of the reality and nature of landslide risk particularly under extreme weather conditions, promoting preparedness, self-protection and neighbourhood support, the community maintaining vigilance of signs of landslide hazards, conducting drills of emergency evacuation where appropriate, etc.

In terms of managing the risk of extreme events, given the nature and scale of the problem as well as the many inherent uncertainties, an adaptive risk management approach is probably more relevant, as opposed to further enhancing the design standards of engineering measures.

The country/region reports also provide insight in several topics from strategies for managing extreme events, new technologies, institutional issues, international cooperation, coordination and enhanced communication. Some of the more salient aspects are summarized below.

1.5.3 Managing extreme weather events

In the context of protecting the community from the adverse consequences of landslide disasters, the following strategies are pertinent in the management of extreme weather events:

a) *Avoidance*, with land-use planning, warning or alert systems, and public education.

b) *Prevention*, such as preventive activities to protect from disasters, e.g. enforcing slope investigation, design, construction, supervision and maintenance standards.
c) *Mitigation*, with the implementation of engineering measures to reduce the impact of landslides, e.g. retrofitting substandard slopes, or adding mitigation measures.
d) *Preparedness*, focusing on procedures, human resource management, emergency systems, training of the vulnerable community for a prompt response etc.
e) *Response*, involving search and rescue, evacuation and provision of basic humanitarian needs, relief measures, emergency inspections for identification of any imminent danger, safe settlement of evacuated people etc.
f) *Recovery*, starting after the immediate threat to life has been dealt with, to bring the affected area back to the normal and carry out repair or mitigation works.

Items (a) to (c) may be construed as broad risk reduction or control strategies whereas items (d) to (f) relate mainly to emergency management. Resilience embraces items (b) to (e) and reflects the ability of a system to withstand shocks and stresses whilst maintaining its essential functions. Resilient systems are also more amenable to repair and recovery afterwards. Picarelli *et al.* (2008) presented an illustration of the Italian approach to such problems.

1.5.4 Institutional issues and addressing the policy gap

A number of country/region reports note that the state institutions should have a stronger focus on developing procedures and strategies for dealing with increasing landslide hazard. However, there is often a lack of competence and/or capacity at the local government level. Yet, the local authorities are those responsible for implementing mitigation or preventive measures and for disaster management. Extreme weather events often have a local character, but there is a need for a national course of action and support to the local authorities for a strategy to operate in an optimal manner. Taiwan (Lin *et al.*, 2016), on the other hand, being neither a member of the UN Framework Convention on Climate Change nor a signatory of the Kyoto Protocol, has adopted the three main global warning issues form the IPCC Climate Change report, for their government policy, as illustrated in Figure 1.2.

Public policy decision-making and risk management must be risk-based and rational. Suitable methodologies are needed to address both the portfolio risk as well as the risk posed by a specific, individual slope affecting a high-consequence facility. Careful consideration should be given to examining whether there is a need to pursue formal probabilistic approaches, or whether simplified approaches such as risk-based scenarios assessments would suffice in providing a reasonable assessment of the likely scale of the problem and the critical hazard scenarios. It should be borne in mind that a successful outcome can be achieved probably only through a multi- and cross-disciplinary approach, with continued interaction between scientific and engineering approaches. Decisions are difficult to achieve solely on the local level, as they need support and competence form the highest level.

At present, there is a major policy gap since the impact of climate change on landslide risk is not well understood. Slope safety implications are not factored into the existing policy addressing climate change. Landslide risks are typically not considered in the national climate change adaptation plans. In this regard, Hong Kong has set

Figure 1.2 Three main global warming issues from the IPCC Climate Change Report adopted by Taiwan government (Lin *et al.*, 2016).

an example: the findings of the scenario-based assessment of the impact of extreme rainfall events on slope safety were promulgated to the general public and widely reported in the local press, thus recognizing the need to take appropriate adaptive risk management actions to cater for extreme rainfall events allowing for the potential impact of climate change. This is a good illustration of the results of the assessment of climate change impact on slope safety leading to the acceptance by government and becoming an integral part of the government strategy.

1.5.5 Application of improved technology and methods

Recent advances in know-how, information technology and technological solutions are gradually providing new tools for establishing scenarios capable of accounting for the major variables involved, e.g. analysis of historical landslide data geographically and their relationship with rainfall, new and more powerful sensors to measure atmospheric variables, improved monitoring of the ground for characterization and landslide warning purposes, and powerful numerical codes to simulate the complex ground-atmosphere interaction, the saturated/unsaturated soil behaviour and the post-failure slope movements. To assist with projections of slope response and landslide characteristics under increasing rainfall intensity, it is advisable to compile a catalogue of significant landslides as well as proceed with careful analysis of storm-based events with extensive landslides with a view to deriving a relationship between rainfall and landslide occurrence.

It is of critical importance to consider possible concurrent occurrence of multiple hazards and the development of integrated strategies including the entire system(s), the interdependencies amongst different system and sub-system components, and local distributed capacity, for more effective emergency response (i.e. through adopting a systems approach).

1.5.6 National and international cooperation

The JTC1 project provided a platform for consolidation of the state-of-the-art and cross-fertilisation of ideas and insights. It would be useful to keep up with the momentum and consider organising in due course special multi-disciplinary conferences to discuss all aspects of landslide risk management in coping with climate change impact, including progress, preferably involving government officials that are responsible for formulating climate change related policies in roundtable discussion of strategic issues. To further strengthen landslide risk management and adaptation to climate change impact, consideration should be given to the following:

a) Promote technical cooperation on, e.g., the modelling of the slope-atmosphere interaction, the high velocity and long runout landslides and the development of inexpensive and effective slope monitoring and early warning systems.
b) Exchange relevant information among regions. The exchange should include data, models and experience of governance and mitigation measures, as well as understanding of capability and limitations of landslide prevention and mitigation works. Various forms of communication such as workshops, forums, networks and media channels and websites should be considered.

Landslides and their impact are difficult to predict, and they have both direct and indirect consequences. Small changes in trigger magnitude and frequency can have large impacts, e.g. in quick clays where the safety factor is often low. To improve predictions, the scientific community should be encouraged to accumulate as much information on antecedent landslide conditions as possible and compile landslide inventories over the past 100–200 years. Knowledge and best practice sharing is necessary, given the many-faceted challenges. The profession should collect databases of reliable information to establish the relationship(s) between rainfall and landslide, both at the local and regional levels.

1.5.7 Coordination with flood risk management experts

As for landslides, risk associated with floods can greatly increase with climate change. It is increasingly accepted in the flood management community that a paradigm shift from reliance on defence to resilience is necessary, as one cannot hope to protect all communities from flooding due to every conceivable event. Resilience in that context is about recognising that flooding will occur, understanding what the consequences might be, mitigating the impacts, warning and preparing for the flooding, responding to events, and finally recovering from the events. All these stages are important and they involve structural measures and non-structural approaches and actions taken before, during and after a flood event.

There are many parallels between landslides and flooding, also in terms of the new challenges posed by climate change. Hence, there is much room for cross-fertilisation of ideas between the two professions. A second reason for enhanced collaboration between the two professions is related to the likelihood of cascading effects of landslides on flooding and vice versa, together with the impact of the concurrent occurrence of several hazards (landslide and flooding, and possibly other hazards such as earthquake, storm surge, tree fall from slopes, etc.), which need to be duly considered in emergency preparedness and response plans. Such cross-disciplinary coordination and cooperation is highly recommended and effectively required if a holistic national policy is to be achieved.

1.5.8 Enhanced communication

Many country/region reports highlighted that communicating the message effectively is of paramount importance for appropriate adaptation in a changing or extreme climate. Improved communication is needed at three levels: (i) on the cross-disciplinary scientific level, (ii) with the stakeholders and (iii) not the least, with the general public. Some of the country/region reports noted that communication is poor among the groups actively involved in landslide risk assessment. Continuing the dialogue is imperative to provide the insights required to shape an environment that is more resilient in a climate change/extreme event context. The awareness of politicians and the general public should be raised in connection with the impact of climate change, including the possible occurrence of extreme weather events as these could trigger widespread and massive landslides leading to major calamity in the developed areas.

Several groups around the world have formed cross-disciplinary teams to work on climate impact on landslides. In some countries, disasters were the catalyst for such cooperation and broader multi-disciplinarity. Focus should be set on enhanced inter-country collaboration. A task-force could be established to help country representatives share information and progress. A website could also be the repository for accumulated data.

1.5.9 Formulating strategies for a way forward

The various country/region reports show that slope safety preparedness for the impact of climate change varies considerably from one to the other. Around the globe, catastrophic landslides induced by extreme rainfall have been reported in recent years: the landslide in Xiaolin village during the passage of Typhoon Morakot in Taiwan in 2009 caused over 400 fatalities; the debris flow in Gansu in 2010 led to more than 1,400 fatalities; the rain-induced landslides in Brazil in 2011 caused more than 900 fatalities; the torrential rain in South Korea in 2011 brought about more than 30 landslide-related fatalities; Tropical Storm Washi brought flash floods and severe landslides, resulting in 1,010 fatalities in the Philippines in 2011; Typhoon Wipha struck eastern Japan in 2013, causing over 40 landslide-related fatalities. The increasing number of casualties associated with freak weather events has further emphasized the need to address the impact of extreme rainfall events on landslide risk.

In general, the lack of a good understanding of the key vulnerabilities and of the scale of the slope safety problem due to climate change prevents the formulation of a specific landslide risk management strategy tailored specifically for climate change impact. In the short term, this has, in some locations, resulted in a reactive approach based essentially on the use of hard engineering measures, which may be appropriate for the local high-risk settings. However, such measures are not tenable as a solution to combat extreme weather event impact, as they are neither affordable nor appropriate from either a technical or environmental point of view.

Many countries are not yet in a position to assess rigorously the potential effect of climate change on slope safety in terms of identifying the critical hazard scenarios and ensuing consequences. Hence, most of the suggestions made for enhancing landslide risk management by the participating countries and regions are mainly generic and not targeted specifically at enhancing slope safety preparedness for climate change and extreme weather events. Nevertheless, it is of interest to examine how one might approach slope safety preparedness for climate change in a more rigorous and scientific manner, to help formulate strategies for a way forward.

Each country or region has its unique ground conditions, slope and infrastructures assets and climate conditions. Hence the observations or assessments of one are not be directly transferable to the other. However, the key is to derive insights from the approaches and findings in other places to assist with developing a suitable way forward in identifying the critical hazard scenarios and assessing the scale of the landslide hazards and consequences.

From a landslide risk management perspective, it is important to consider all the pertinent landslide risk scenarios associated with climate change. Some countries may be concerned about the progressive deterioration of old, engineered (i.e. man-made) slopes along railway lines given the repeated wetting and drying cycles coupled with more severe rainfall that may be caused by climate change. However, other critical scenarios (e.g. extreme rainfall events) arising from climate change also need to be duly considered. In so far as landslide risk management is concerned, extreme events that need to be considered are those which would cause many sizeable and/or highly mobile landslides, with the consequences mentioned earlier.

1.6 EMERGING NEEDS

There is a need for improved prediction (and reduced uncertainty in prediction) of the impact of climate change on extreme weather events. This may take a long time to achieve, and does not depend on the geotechnical community for progress. In the meantime, there is also need for geo-research on material properties, behaviour and models.

1.6.1 Need for more geotechnical research

There is a consensus in the country/region reports on the need for more geo-research on:

- Spatial and temporal characterization of soil parameters, using typical sites.
- Development of mechanically-based tools to predict the consequences of short-term forecasted rainstorms.
- Conversion of site-specific data into regional-scale information, accounting for the variability where feasible.
- Integration of critical climate scenarios into a multi-hazard systems model for holistic risk assessment.

- Extreme weather events should be comprehensively documented in detail with respect to the overall slope response, landslide debris runout and landslide emergency management, together with the lessons learnt.

Following the Great East Japan earthquake and tsunami of March 2011, leading to the Fukushima Dai-ichi nuclear accident, the European Commission initiated collaborative research projects to develop methods for stress testing of critical infrastructure and management of the risk posed by rare, extreme events and by cascading hazards or events. The key European research projects and their objectives are briefly described in Section A1 in Annex A.

1.6.2 Enhanced emergency preparedness for extreme weather events

It is imperative that the community remains highly vigilant of the possibility of serious landslides under a changing climate and be prepared to deal with extreme rainfall events. It is neither practical nor cost-effective to rely on engineering solutions to manage the risk posed by extreme rainfall. The engineering approach needs to be supplemented by other measures involving enhanced emergency preparedness, response and recovery. To enhance slope safety preparedness for extreme rainfall, one could consider improving early warning systems, improving emergency response and crisis management systems, developing temporary landslide evacuation plans for vulnerable communities, scheduling large-scale drills of emergency evacuations, and using landslide detection and alert systems to supplement landslide risk mitigation works in high-risk areas, etc. A major challenge in preparedness is effective and timely issue of warnings, with a minimum of false or unnecessary alarms.

1.7 CONCLUSIONS

This overview paper consolidates the key observations and insights from the contributions made by 21 countries/regions in addressing the vexed question of potential impact of climate change on landslides. This subject has major public safety implications and generally constitutes a significant gap in government policy on climate change at the present time. There is no doubt that the climate-slope response system involves major uncertainties, and hence landslide risk management in a changing/extreme climate context represents a formidable challenge.

Changes in the climate over the next decades will alter the hydro-mechanical behaviour of many sub-surface systems, and will lead to increased landslide risk in many regions. The several ways climate change may impact engineered slopes and natural hillsides and the temporal, spatial and institutional dimensions of the problems, make adaptation and preparedness more difficult (El-Zein, 2016).

This review reinforces the need for a comprehensive slope safety system that manages landslide risk in a holistic manner using both hard engineering measures and soft non-engineering approaches. The acute challenges posed by the potential impact of extreme weather events, which may be exacerbated by climate change, require special attention and careful assessment. This needs to be addressed not just from a technical but also strategy and policy perspective, given that the number, scale and mobility of the corresponding landslides could be unprecedented, and that their assessments are fraught with considerable uncertainty.

The review has taken stock of the latest assessment tools, best practice in landslide risk management, new methods under research and national and regional strategies or policies that are under development. In terms of potential critical hazard scenarios, slopes in permafrost, mountains with glaciers, coastal slopes, natural slopes underlain by weak granular materials susceptible to debris flows/slides, and slopes prone to progressive deterioration with time (degradation in strength or development of distress) are identified as vulnerable settings that will likely respond strongly to climate change with, in many cases, projected increase in frequency, magnitude and runout of landslide events. Hence the landslide risk will increase due to climate change.

For geomechanics research to succeed in developing an improved understanding of landslide occurrence associated with climate impact, it is important to enlarge the geotechnical engineer's and geoscientist's sphere of interest and influence across many disciplines.

Conventional risk assessment methodologies are not well suited for dealing with the risk posed by low probability – high impact (extreme) events. Stress testing provides a complementary approach that may be used for managing the risk posed by extreme events to constructed facilities and critical infrastructure. In stress tests, the focus is on the performance of the system under consideration subject to extreme event scenarios. The aim is to identify the weaknesses in the system and evaluate how the system can be made more robust under extreme events without focusing much on the probability of the extreme event scenarios happening. A key challenge is the development of scenarios that form the basis for the stress tests. The appropriate treatment of expert judgment and uncertainty, both in terms of the specification of scenarios and interpretation of the test results, is another key challenge. This is a rapidly evolving field. The new research initiatives in Europe and elsewhere are likely to provide the tools and methodologies needed for stress testing of critical infrastructure in the near future.

Stress testing is not a substitute for conventional risk or safety assessments, but it provides additional valuable insight for extreme situations. What stress tests and conventional risk or safety assessments have in common is that they both rely on a description of the system, which helps associate the state of the system and a set of consequences under any given or potential scenario.

The present review finds that slope safety preparedness for the impact of climate change varies considerably from one country/region to another. There is much room for cross-fertilization of ideas and insights, as well as joint development of strategies and best practice. Disasters can manifest themselves as fast events, but the vulnerability for disasters is built up slowly, and can be the result of neglecting to be adequately prepared. Focus has shifted from prevention-mitigation to building resilience and reducing risks.

There is an increasing awareness amongst the engineering profession that reliance on engineering works alone in combating the climate change impact is neither practicable nor cost-effective. Instead, due emphasis should be placed on improving warning systems, enhancing emergency preparedness and response, community resilience and recovery. For enhanced preparedness and resilience to take root, the risk management strategy has to be embraced by the community. Hence, effective public education and strong government support are essential.

Promoting a continued high-level discussion and technical cooperation on an international scale is of paramount importance, especially when accounting for the increasing interest of mass media and the population that could be at risk due to landslides caused related to climate changes.

ACKNOWLEDGEMENTS

This paper is published with the permission of the Head of the Geotechnical Engineering Office and the Director of Civil Engineering and Development, Hong Kong SAR Government. The authors wish to thank J. Coe and A. Strom for their careful review of the manuscript and their useful suggestions. The authors are also grateful to all the authors of the country and region reports in this book.

REFERENCES

Borio, C., Drehmann, M. & Tsatsaronis, K. (2012) *Stress-Testing Macro Stress Testing: Does It Live Up to Expectations?* BIS Working Papers, No. 369, January.

Cloutier, C., Locat, J., Geertsema, M., Jakob, M. & Schnorbus, M. (2016) Slope safety preparedness for effects of climate change. In: *Potential Impacts of Climate Change on Landslides Occurrence in Canada*. Country Report for Canada. In this publication.

Coe, J.A. (2016) *Landslide Hazards and Climate Change: A Perspective From the United States*. Country Report for the US. In this publication.

Dijkstra, T.A., Jenkins, G.O., Gunn, D., Dashwood, C., Dankers, R., Nixon, N., Petley, D., Gibson, A. & Winter, M.G. (2016) *Landslides and Climate Change in the United Kingdom*. Country Report for the UK. In this publication.

El-Zein, A. (2016) Resilience and vulnerability to climate change challenges of temporal and geographical scales for geotechnical engineering. *Australian Geomechanics*, 51(2), pp. 65–76.

Gudmundsson, M.T., Thordarson, T., Höskuldsson, Á., Larsen, G., Björnsson, H., Prata, F.J., Magnússon, E., Högnadóttir, T., Petersen, G.N., Hayward, C.L., Stevenson, J.A. & Jónsdóttir, I. (2012) Ash generation and distribution from the April-May 2010 eruption of Eyjafalljökull, Iceland. *Scientific Reports*, 2, 572.

Ho, K.K.S., Sun, H.W., Wong, A.C.W., Yam, C.F. & Lee, S.M. (2016) *Enhancing Slope Safety Preparedness for Extreme Rainfall and Potential Climate Change Impacts in Hong Kong*. Region Report for Hong Kong. In this publication.

Hungr, O., Leroueil, S. & Picarelli, L. (2014) Varnes classification of landslides types, an update. *Landslides*, 11 (2), 167–194.

IPCC (2007) Fourth Assessment Report AR4. Climate Change 2007: Synthesis Report. In: Core Writing Team, Pachauri, R.K. & Reisinger, A. (eds.) *Contribution of Working Groups I, II and III to the Fourth Assessment Report of the Intergovernmental Panel on Climate Change*. Geneva, IPCC. 104 pp.

IPCC (2012) *IPCC Special Report on Managing the Risks of Extreme Events and Disasters to Advance Climate Change Adaptation (SREX)*. Available from: http://ipccwg2.gov/SREX/report/

IPCC (2014) Fifth Assessment Report AR5. Climate Change 2014: Synthesis Report. In: Core Writing Team, Pachauri, R.K. & Meyer, L.A. (eds.) *Contribution of Working Groups I, II and III to the Fifth Assessment Report of the Intergovernmental Panel on Climate Change*. Geneva, IPCC. 151 pp.

Lin, M.L., Lin, S.C. & Lin, Y.C. (2016) *Review of Landslide Occurrence and Climate Change in Taiwan.* Region Report for Taiwan. In this publication.

Liu, Z., Nadim, F., Garcia-Aristizabal, A., Mignan, A., Fleming, K. & Quan Luna, B. (2015) A three-level framework for multi-risk assessment. *Georisk Assessment and Management of Risk for Engineered Systems and Geohazards.* Available from: 10.1080/17499518.2015.1041989.

Nadim, F. (2016) Challenges in managing the risk posed by extreme events. In: Huang, H.W., Li, J., Zhang, J. & Chen, J.B. (eds.) *Keynote Paper. 6th Asian-Pacific Symposium on Structural Reliability and Its Applications (APSSRA6), 28-30 May 2016, Shanghai, China.*

Nadim, F. & Kalsnes, B.G. (2014) Progress of living with landslide risk in Europe. Plenary Lecture. World Landslide Forum 3 *Landslide Risk Mitigation: Towards a Safer Geo-Environment.* Beijing. Chp in "*Landslide Science for a Safer Geoenvironment*". 1: The International Programme on Landslides (IPL). pp. 3–20. ISBN: 978-3-319-04999-1.

Picarelli, L., Versace, P., de Riso, R. & Palmieri, M. (2008) Landslide disaster management in Italy. In: *Proceedings of 2007 Int. Forum on Landslide Disaster Management, Hong Kong, 10–12 December 2007, The Hong Kong Institution of Engineers*, Vol. 1. pp. 281–318. ISBN: 978-962-7619-29-1.

Picarelli, L., Comegna, L., Gariano, S., Guzzetti, F., Mercogliano, P., Rianna, G. Santini, M. & Tommasi, P. (2016) *Potential Climate Changes in Italy and Consequences on Land Stability.* Country report for Italy. In this publication.

Strom, A.L. (2016) Personal communication.

Varnes, D.J. (1978) Slope movement types and processes. In: Schuster, R.L. & Krizek, R.J. (eds.) *Landslides, Analysis and Control, Special Report 176: Transportation Research Board.* Washington, DC, National Academy of Sciences. pp. 11–33.

WENRA (West European Nuclear Regulators Association) (2011) "Stress tests" specifications—Proposal by the WENRA Task Force, 21 April 2011. Available from: http://www.oecd-nea.org/nsd/fukushima/documents/WENRA20110421StressTestsSpecifications2011-04-21.pdf.

ANNEX A

The annex contains two parts:

1) Table A1 summarizing the key findings from the country/region reports on climate change assessment, impact of climate change on slope stability and the specific landslide risk management and adaptation measures for climate change, and
2) Section A1 on key European research projects on stress testing of critical infrastructure for rare, extreme events and on methods for multi-hazard and multi-risk assessment.

Table A1 Studies on impact of climate change on slope safety in different countries and regions.

Country/ region report	*Climate change assessment*	*Findings from climate models*	*Impact of climate change on slopes/critical landslide scenarios*	*Specific landslide risk management/ adaptation measures for climate change*
Brazil	Details not provided	–	– Reduced frequency of light rain days (disturbing the soil-water-biota interaction), and increasing the frequency of heavy rain days (inducing more landslides). – Shallow translational landslide is the dominant failure mechanism.	– Past landslide disasters triggered by extreme rainfall showed the spatial variability of landslide responses. Detailed thematic susceptibility mapping to account for rainfall characteristics and variable conditions (hillslope hydrology, soil properties, geology, geomorphology, geotechnics, vegetation and land use).
Canada	Downscaling studies completed	– Despite the uncertainties associated with rainfall predictions, projections tend towards an increase in rainfall intensity.	– Potential impacts of climate change on landslide occurrence are very uncertain due to not fully understood weather impact and very complex, not uniform, and uncertain climate change. – Climate change impact on landslides will depend on the slope failure mechanisms – Rise in mean air temperature will cause retreat and degradation of mountain permafrost, making the slopes more vulnerable to landslides – Permafrost thaw will lead to mass movements. – Increased rainfall will amplify the retrogressive process of thaw slumps. – Increase in floods will undercut banks and quick clays and lead to rapid drawdown, both causing landslides. – Increase in extreme precipitation will increase the frequency and possibly magnitude of debris flows, shallow debris slides, rock avalanches and rockfalls. – Natural hazards associated with glaciers will reduce.	– Improving the understanding of current situation. – Integrating impact of human activities and climate change and adapting practices to reduce overall landslide risk.

Hong Kong	Statistical downscaling completed	– Number of extremely wet year will increase. – Annual rainfall and temperature will increase.	– Both natural terrain (debris flows and debris slides/avalanches on slopes with unsaturated granular materials) and man-made slopes will be strongly affected by severe rainfall of short-duration. – Scenario of main concern is extreme rainfall causing multiple natural terrain landslides of unprecedented number, magnitude and runout.	– Scenario-based assessment and stress testing of entire slope safety system; consider concurrent multiple hazard scenarios; consider low-frequency, large-magnitude landslide events. – Strengthening and streamlining existing landslide emergency system and enhancing landslide warning system and preparedness. – Develop new strategy for landslide crisis management under extreme rainfall. – Continue to implement systematically robust risk mitigation measures for existing slopes posing a high risk.
Indonesia	Downscaling studies completed	– Increase in rainfall intensity.	– Climate change will have an impact on the frequency, distribution and failure mode of landslides, and may reactivate old landslides. – Increased frequency of failures expected in clay shales and expansive soils.	– Instrumentation of vulnerable sites. – Numerical analysis to model and understand mechanism of rain-induced slope failure. – Warning and evacuation orders to be issued as necessary.
Italy	Dynamic downscaling studies and examination of paleo evidence with dating techniques.	– Increase in frequency and intensity of heavy rainstorms although the mean annual precipitation may decrease. – Increase in frequency of rapid snowmelt and rain-on-snow events.	– More severe extreme rainfall will result in an increase in landslide occurrence, particularly rapid, shallow landslides on slopes in relatively permeable soils (e.g. unsaturated granular material). – Increase in frequency and magnitude of debris flows. – Decrease in the susceptibility of slow moving landslides in stiff overconsolidated clays.	– Improve slope monitoring strategy, tools for advance warning, and landslide early warning systems. – Consider implications of climate impact on landslide mitigation and stabilisation works and on landslide risk assessment and management. – Risk reduction through both structural (hard) measures in high risk areas and non-structural (soft) measures. – Targeted public education and information dissemination.

(*continued*)

Table A1 Continued.

Country/ region report	*Climate change assessment*	*Findings from climate models*	*Impact of climate change on slopes/critical landslide scenarios*	*Specific landslide risk management/ adaptation measures for climate change*
Japan	Downscaling studies completed.	– Increasing trend of extreme rainfall events. – Extreme rainfall can be associated with typhoons and other climatic anomalies.	– Increase in number of landslides due to increased frequency of extreme rainfall. – Most of the landslides on hillsides will occur as either channelized debris flows with small and shallow landslides, or deep-seated and sizeable landslides.	– Designation of warning zones for urbanisation near mountains. – Prompt and effective implementation of landslide countermeasures.
Malaysia	Downscaling studies completed.	– Increase in rainfall intensity, duration and frequency.	– Heavy rainfall will trigger shallow landslides.	– Use bioengineering techniques such as live poles to stabilize shallow failure through root binding and evapotranspiration,
Norway	Dynamic downscaling studies completed (bias correction down to $1 \times 1\ km^2$).	– Noticeable increase in average temperature, increased precipitation and increased periods of intense precipitation.	– Intense rainfall will cause flash floods and debris flows. – More prolonged rainfall will increase the occurrence of larger soil slides and quick clay slides due to river erosion.	– Strengthen support to municipalities in the application of latest know-how and innovative solutions for local civil protection and for the proper evaluation of natural hazards.
Russian Federation	Details not provided.	– Increase or decrease of precipitation in different areas. – Increased snow melt rate. – More frequent abnormal intense rainfall and extreme weather events.	– Rise of the zero isotherm could result in permafrost degradation leading to ice-rock avalanches and active rock glaciers or permafrost related landslides. – More frequent and larger regional landslide.	– Review of existing regional landslide hazard assessments. – Accounting for new geological risks in technical planning. Risk assessment to consider climate phenomenon of low probability of occurrence (0.5% to 1%); number of such evens will tend to increase due to climate change.

Singapore	Statistical downscaling studies completed.	– Increase in extreme weather events including intense rainfall, high temperature, strong wind and floods. – Increase in temperature will affect the evaporation rate which may lead to formation of desiccation cracks and increased surface infiltration.	– Heavy rainfall will trigger shallow landslides.	– Review building control regulations. – Establish a Slope Hazard Analysis and Repository System to identify locations susceptible to slope failure and prioritise slope works to be carried out. – Conduct periodic inspection on slopes. – Develop slope risk management framework, conduct public communication and introduce a systematic incident management work flow. – Develop a common portal and platform for government agencies to access and share data with a view to enhancing collaboration. – Prepare a slope susceptibility map. – Explore citizen engagement and crowd-sourcing data in GIS for enhancing operational decisions.
Spain	Both statistical and dynamic downscaling studies completed.	– Warming in the 21st century across the country, with permafrost melting, snow coverage reduction, altitudinal vegetation displacement and marked contrast between winter and summer precipitation. – Intensity of summer heat waves will increase.	– Increase in intense rainfall will increase shallow landslides, debris flows and rockfalls. – More frequent and larger landslide events expected. – Landslide susceptibility could increase due to vegetation cover deterioration and more frequent forest fires. – Torrential flooding can induce new failures by undercutting slopes. – Increased winter rainfall could favour reactivation of some big rotational slides and earthflows; the reduced seasonal and inter-annual rainfall may cause large movements in dormant landslides.	– Urban and land-use planning to avoid developments in landslide susceptible areas. – Implementation of best practices, including reforestation of bare slopes. – Spanish Land Bill enacted in 2008 envisages the preparation of natural hazard maps for landslides and development of emergency plans

(continued)

Table A1 Continued.

Country/region report	*Climate change assessment*	*Findings from climate models*	*Impact of climate change on slopes/critical landslide scenarios*	*Specific landslide risk management/adaptation measures for climate change*
		– Increased likelihood of drought events. – Reduction of precipitation but increased irregularity and more frequent extreme events.		
Switzerland	Downscaling studies completed.	– Noticeable shift of seasonality of mean precipitation – Frequency of extreme rainfall events may increase – Storm events may be more violent in the mountainous regions. – Freezing level in the summer will be on average 500 m lower than the present level. – Wintertime temperature will rise by 3°C to 4°C in the 21st century and even more in the summer.	– Glacier retreat will be accelerated and contribute to the melting of the upper layers of the permafrost zones, which will increase landslide hazards, in particular debris flows. – More intense floods will cause debris flows and mud flows on slopes with increased frequency and intensity. – Progressive melting of morainic deposits at high altitudes will increase the number and scale of debris flows. – Snow avalanches can be adversely affected by climate change. – Due to increase in evapotranspiration, the frequency of large slides may decrease slightly.	– Update or develop new hazard maps. – Consider the effects of climate change on hazard maps. – Integrate hazard maps in local town and country planning documents and procedures. – Natural hazard management strategy, including studies and implementation of protective works as assessed using a protection objective matrix. – Improve knowledge and coordination of cantons and counties and include elaboration of more reliable climatic and hydrological scenarios and assess the impact of climate change on the management of natural hazards. – Transition from a pure defence against natural hazards to integrated risk management. – Early warning system for shallow slides and debris flows. – Review if past protective measures are adequate to cope with the effects of climate change in the future. – Review design standards for protective measures.

Taiwan	Statistical downscaling projections done based on AR4 scenario.	– Increase in temperature at the end of the 21st century between 2°C and 3°C. – Greater disparity of wet and dry seasons. – Most of the models predict an increasing mean precipitation in the wet season in the summer and decreasing precipitation in the dry season, and an increase in the intensity of rainstorms. – Number of typhoons will decrease, but intensity and associated extreme rainfall will increase.	– Majority of past landslide disasters with large casualties were caused by torrential rain associated with typhoons. – Hillsides are more susceptible after severe earthquakes, with more landslides and more significant failures during post-earthquake heavy rainstorms – Climate change will lead to increased magnitude of landslide disasters, increased susceptibility of cascading hazards, increased recovery difficulties and the occurrence of compound disasters. – Design codes can become insufficient in terms of the disaster prevention.	– Adaptation strategy based on risk avoidance (i.e. avoiding high-risk locations) and risk reduction (i.e. increasing regional resilience to withstand the impacts of climate). – Interdisciplinary strategies, including: • Implementing regional planning and management, especially in urban regions, • Enhancing disaster prevention and avoidance capacity, • Executing comprehensive river basin management, • Prioritising regions with high-risk due to climate change, • Enhancing integration of environmental monitoring and disaster warning systems, • Creating a platform for information exchange, • Evaluating vulnerability of current critical infrastructure, • Strengthening capacity for responding to extreme climate, publicising risk information and carrying out disaster-reduction education and drills in high risk areas.
Thailand	Downscaling studies completed.	– More intense rainfall.	– More localised landslides and more destructive, widespread debris flows.	– Instrumentation of critical slopes to understand the relationship among climate, hydromechanical properties of unsaturated soils and slope instability. – Examination of the development of dynamic hazard maps,and early landslide warning.

(continued)

Table A1 Continued.

Country/ region report	*Climate change assessment*	*Findings from climate models*	*Impact of climate change on slopes/critical landslide scenarios*	*Specific landslide risk management/ adaptation measures for climate change*
				– Use of bioengineering techniques for slope stabilisation and erosion control. – Evaluation protocol and guide to good land management practice for hillslope development. – Public education and community capacity building to enhance adaptation and preparedness to landslides.
United Kingdom	Dynamic downscaling with probabilistic projections.	– Increase in summer and winter temperatures. – Increase in winter rainfall and decrease in summer rainfall. – Increase in number of heavy rainfall (short duration, high intensity). – Rising sea level.	– Increase in the failure of engineered slopes (cuts and embankments) involving small-scale slumps or flows (possible progressive deterioration of old slopes), – Increase in frequency and magnitude of debris flows in coarser-grained material slopes in the winter. – Extended periods of exceptionally dry weather could lead to more wildfires and associated post-wildfire debris flow. – Increase in the vulnerability of coastal slopes.	– Maps with locations of vulnerable slopes (no climate change impact yet). – Engineering new slope assets requires assessment of long-term performance under changes in rainfall and more frequent extreme events. – Engineering natural and engineered slopes requires improved asset management and resilience quantification. – Develop tools for reduction of uncertainty and risk to population.
United States	Downscaling studies completed.	– Frequency, intensity and size of events causing landslides are affected. – Precipitation during extreme storms likely to become more intense and less frequent. – Evapotranspiration and drought to likely increase in some areas.	– Adverse changes in landslide activity will be greatest in areas with permafrost, glaciers, and sea ice. – Effects of climate change will be a combination of adverse and beneficial because of different settings. – Increase in sea level will impact all coastal areas and steep and prone to landslide shorelines. – Large uncertainty in future changes in extreme storms (e.g. hurricanes) and regional climatic phenomena. – Increased evapotranspiration, drought and wildfires may lead to a decrease in deep-seated landslides,	– Need for advances in predicting changes in the frequency, size and intensity of future extreme storms and regional climatic phenomena. – Some states have developed adaptation strategies or climate change response plans, with (1) coordinated data collection, analysis and monitoring, (2) improvements to existing infrastructure to protect and extend its life, and (3) building resilience

		– Number and extent of wildfires and length of wildfire seasons to increase. – Further contraction of snow, ice cover and permafrost. – Sea level rise.	whilst increase in wildfires may lead to an increase in post-wildfire debris flows, especially under heavy rainfall.	into new infrastructure for a changing climate.
Vietnam	Downscaling studies completed.	– Increase in temperature, rainfall and rise in sea level. – Increase occurrence of extreme weather events.	– Increase in rainfall and rise in sea water level has caused an increase in landslide frequency and flash floods.	– Develop landslide hazard maps.
Questionnaires				
Australia	National downscaling studies completed[2].	– Increase in short duration rainstorms. – More intense peak rainfall and less rainfall during less intense times.	– Increase in landslides expected, including shallow failures on steep hill slopes. – The science of predicting the effects of climate change and the frequency of landslides from rainfall is considered not sufficiently advanced at this time to warrant consideration of climate change when carrying out landslide zoning studies.	– For zones identified as vulnerable to landslides, it is required to consider the impact of climate change on the proposed development. – Started expressing changes to Intensity-Frequency-Duration plots on rainfall recurrence curves.
China	Downscaling studies completed.	– Concern about very sizeable debris flows during increased rainfall, especially in areas recently struck by strong earthquakes.	– Debris flows impacting on development represents the critical landslide scenario.	– Urban and rural landslide risk management, in particular in new areas, and reduction in rainfall threshold for rain-induced landslides after strong earthquakes are essential. – Landslide prevention and mitigation system, landslide emergency management and landslide monitoring network are needed. – forecasting system for landslides and early warning system (with community participation). – Relocation of communities with high landslide risk.

(continued)

Table A1 Continued.

Country/ region report	*Climate change assessment*	*Findings from climate models*	*Impact of climate change on slopes/critical landslide scenarios*	*Specific landslide risk management/ adaptation measures for climate change*
				– Engineering works to reduce landslide risk. – Enhanced information exchange, technical cooperation h and mutual support to international counterparts.
Columbia	Statistical downscaling completed.	– Average temperature will continue to rise. – Annual rainfall will increase in some regions and decrease in others.	– Concern with the impact of extreme weather events on slope stability.	– Landslide hazard inventories. – Zoning studies completed (methodology and scale should be improved). – Improve early warning system (especially national rain gauge network).
India	Details not provided.	- - -	– Concern with climate change adversely affecting extreme weather events, the latter being not well understood and appearing to be random events. – Numerous recent catastrophic landslides appear to be linked to extreme weather events, with rainfall data demonstrating apparently changing rainfall patterns and new rainfall intensity peaks; however, in-depth and definitive scientific studies are lacking.	– Established National Disaster Management Authority, State Disaster Management Authorities, District level Disaster Management Authorities, and National Disaster Response Force. – Funding projects on community-centric early warning. – Public education and training to enhance preparedness. – Work in progress on (a) integration of slope safety management with land planning, (b) slope safety management under multi-hazards, (c) slope safety management in high altitude areas including snow avalanches, (d) slope safety management in urban areas, (e) early warning and evacuation, and (f) learning of lessons and strengthening of capacity at all levels.

[2] Climate change scenarios were considered by factoring the rainfall intensities by +10% (by year 2050) and +20% (by year 2100).

Section A1

Key European research projects on stress testing of critical infrastructure and management of the risk posed by rare, extreme events and by cascading hazards

The European Commission initiated a number of collaborative research projects to develop methods for stress testing of critical infrastructure and management of the risk posed by rare, extreme events and by cascading hazards or events. The key European research projects and their objectives are listed below. The project coordinator is mentioned in parenthesis after the project acronym. Some of these projects are still ongoing. The results of these and similar projects elsewhere in the world should improve the tools and methodologies available for managing the risk posed by extreme events to constructed facilities and critical infrastructure. For example, stress testing is a rapidly evolving field in engineering risk management. The new research initiatives in Europe and elsewhere are likely to provide the tools and methodologies needed for stress testing in the near future.

STREST: Harmonized approach to stress tests for critical infrastructures against natural hazards (ETHZ, Switzerland)

The aim of STREST is to develop appropriate stress tests for all classes of non-nuclear critical infrastructures (CIs). The project comprises the following activities:

- Establish a common and consistent taxonomy of non-nuclear CIs;
- Develop a rigorous, consistent approach to model hazard, vulnerability, risk and resilience assessment of Low Probability – High Consequence events;
- Design a stress test framework and specific applications to address the vulnerability, resilience and interdependence of CIs;
- Implement European policies for the systematic stress testing.

MATRIX: New Multi-Hazard and Multi-Risk Assessment Methods for Europe (GFZ, Germany)

The MATRIX project investigated multiple natural hazards and risks in a common theoretical framework. The methodology developed in MATRIX is particularly well suited for analysing hazard interactions and triggering or cascading effects in a holistic multi-risk analysis (Liu *et al.* 2015).

The main goals of MATRIX were:

- Determine and demonstrate under what conditions multi-type risk assessment provides (or not) better results compared with considering only single-type hazards;
- Provide tools for analysing multi-type risk problems within a European context;
- Establish a European knowledge base on multi-type risk in Europe;
- Disseminate multi-type risk concepts to potential end-users and other relevant members of the broader community;
- Provide support for the necessary decision-making by civil protection and disaster management authorities on the basis of probabilistic information.

INFRARISK: Novel Indicators for Identifying Critical Infrastructure (CI) at Risk from Natural Hazards (Roughan & O'Donovan Limited, Ireland)

The main goal of the INFRARISK project is similar to the one of STREST. Some of the main objectives of INFRARISK are:

- Identify low probability – high consequence events, which have a potential impact on CIs;
- Develop a stress test structure for specific natural hazards on CI networks and a framework for complex linear infrastructure systems (roads, highways and railroads);
- Develop an integrated approach to hazard assessment considering interdependences and cascading effects;
- Facilitate implementation through the development of GIS-based and web-based stress test algorithms for complex infrastructure networks;
- Test the framework developed through simulation of complex case studies;
- Develop exploitation strategies for disseminating the new knowledge.

INTACT: Impact of Extreme Weather on Critical Infrastructures (TNO, The Netherlands)

The INTACT project investigates the impact/resilience of extreme weather events (EWE) on critical infrastructures (CIs). The objectives of INTACT are:

- Assess regionally differentiated risk throughout Europe associated with EWEs;
- Identify and classify CIs and to assess the resilience of such CIs to the impact of EWEs;
- Raise awareness of decision-makers and CI operators on current and future EWE challenges for CIs;
- Identify potential measures and technologies to implement for planning, design and protection of CIs or for preparing for crisis response and recovery.

Country/Region Reports

Chapter 2

Technical report on landslide related disasters in Brazil

W.A. Lacerda

COPPE, PhD, Emeritus Professor, Federal University of Rio de Janeiro, Brazil

A.L. Coelho Netto

D.Sc., Full Professor, Head of GEOHECO-Laboratory of Geo-Hydro-ecology at Department of Geography, Institute of Geosciences, Federal University of Rio de Janeiro, Brazil

A.M. Sato

D.Sc., Associate Professor, Research Group on Socio-Natural Disasters at Department of Geography and Public Policies, Institute of Education of Angra dos Reis, Fluminense Federal University (IEAR/UFF), Brazil

2.1 INTRODUCTION

Most landslide related disasters are concentrated in the South and Southeast regions of Brazil, particularly in the mountainous domain that includes the Atlantic scarp of Serra do Mar ridge and coastal massifs (Figure 2.1). Such disasters may reach catastrophic proportions in response to extreme thunderstorms that may occur in the summer. Previous soil wetting tends to favor slope failures, but these interactions are not yet well understood from the literature. However, many efforts have been made to approach the relation between previous rainfall and critical rainfall conditions to trigger landslides, as summarized in this report.

Usually, landslides induced by extreme rainfall produce an excess of sediment yield that may partially converge to channels network, feeding the down-valley dense and turbulent debris flow that increases the rate of erosion due to bank scouring and streambed incision. As the rain stops, sediment storage tends to drown the surface flow through the local drainage system, favoring future large floods. Both landslides and related floods usually cause serious damage, as well as social, economic and ecological losses, including the destruction of houses, constructions, roads, oil and gas pipelines, tailings dams that spread toxic contamination throughout the soil-water system, etc.

The São Paulo Institute of Technological Research (IPT-SP) points out that landslides triggered by extreme rainfall in Brazil caused some 1,534 deaths from 1988 to 2004, greatly concentrated in two extreme events: in 1988, with 277 deaths, and 1996, with 228 deaths. However, the extreme landslide disaster that occurred in January 2011 involved a large area (above 450 sq. km) throughout the mountains of the state of Rio de Janeiro; official governmental numbers hit nearly one thousand deaths and around five hundred missing that should now be considered to have died as well. The latter is the worst disaster in Brazilian records. Coelho Netto *et al.* (2013) identified 3,623 landslide scars larger than 30 sq. m reaching partially the municipality of Nova Friburgo.

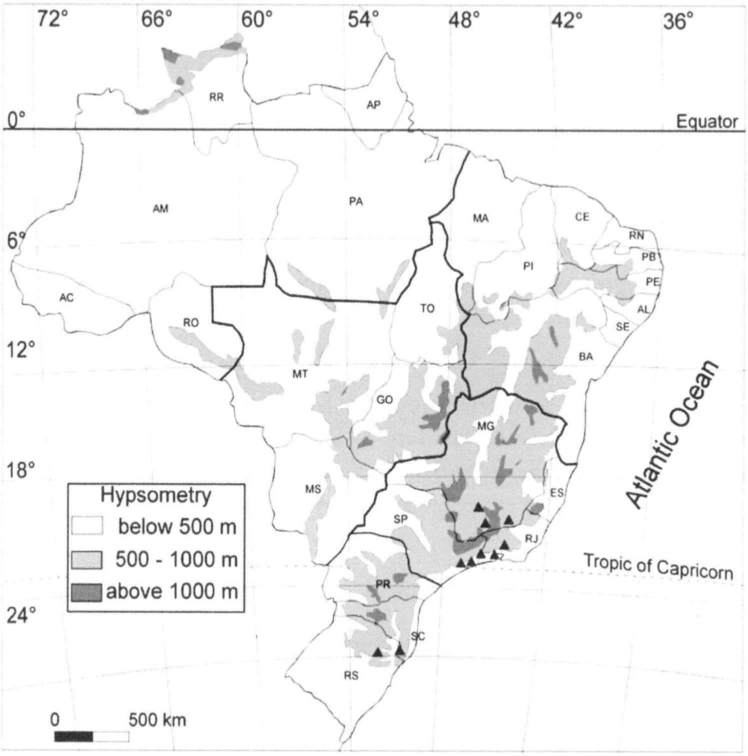

Figure 2.1 Hipsometry and areas most affected by landslide disasters in Brazil (black triangles). Source: Coelho Netto *et al.* (2009).

This report focuses on major landslide disasters in the South and Southeast Regions of Brazil, including the most extreme event ever recorded in the country, in January 2011.

2.2 RAINFALL VARIABILITY OVER SPACE AND TIME

Tropical humid climate predominates in most of the Brazilian coast, and extreme rainfall events occur in the summer in the Southern and Southeast Regions. The spatial variability of mean annual rainfall and the daily rainfall variability are shown in Figures 2.2 and 2.3 for the period between 1961 and 1990. Figure 2.3 represents the mean value yielded by three rain gauge stations in regional biomes in Brazil and their respective maximum and minimum monthly rainfall totals in the same period. Well-defined seasonal variability predominates in both Atlantic forest and Cerrado (a savanna-like biome) biomes, respectively in southeastern and central Brazil, and small parts of the North, Northeast and South regions. Monthly variability increases in the Atlantic biome in southeastern Brazil, in contrast with the well-distributed average

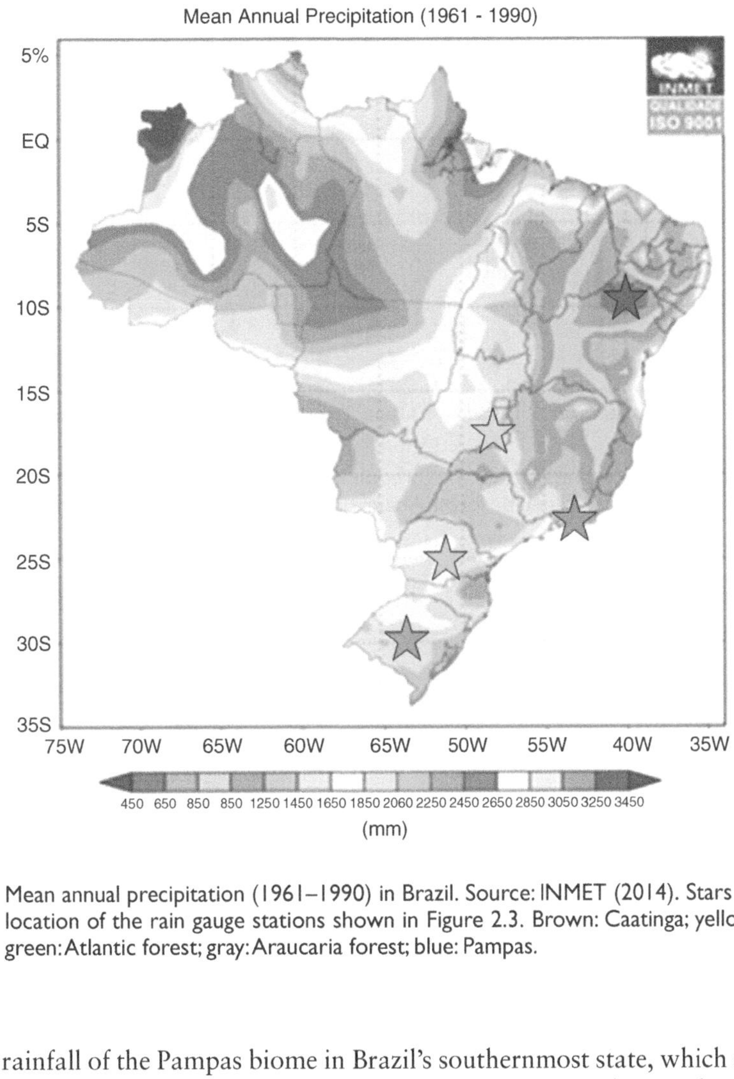

Figure 2.2 Mean annual precipitation (1961–1990) in Brazil. Source: INMET (2014). Stars indicate the location of the rain gauge stations shown in Figure 2.3. Brown: Caatinga; yellow: Cerrado; green: Atlantic forest; gray: Araucaria forest; blue: Pampas.

monthly rainfall of the Pampas biome in Brazil's southernmost state, which still shows considerable monthly variability, as indicated by the extreme values in the period.

A daily rainfall below 10 mm takes place in above 60% of rainy days in all regions and frequency decreases exponentially so that rainfall above 50 mm/day is lower than 4% (Figure 2.4). Intense daily rains (from 100 to 140 mm/day) are rare but attention should be called to the largest variability in the Atlantic southeastern Brazil and the Pampas in the south. The highest variability of daily rainfall (above 140 mm), in the southeastern Atlantic biome region, should also be considered.

Recent studies by Figueiró and Coelho Netto (2011) point out considerable variations in the distribution of rainfall frequency in the Southeast of Brazil. The authors observed, at the Resende station (in the state of Rio de Janeiro), throughout the 20th century, not only a decreasing trend in annual totals, but also an increasing frequency of daily rainfall totals above 100 mm, followed by a decreasing frequency of daily rains

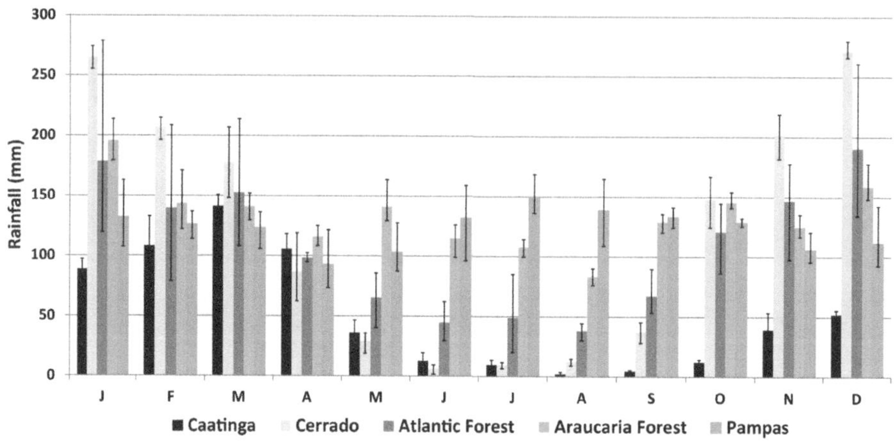

Figure 2.3 Histograms of average monthly precipitation in different bio-climatic regions in Brazil: Caatinga; Cerrado; Atlantic forest; Araucaria forest and Pampas (see location in Figure 2.2).

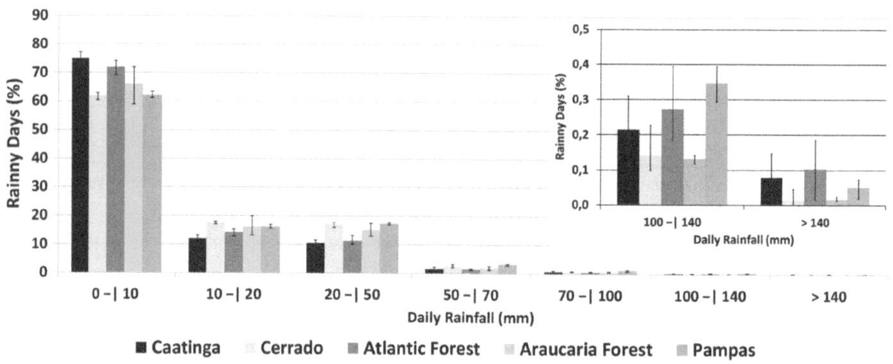

Figure 2.4 Histograms of average daily rainfall frequency in different bio-climatic regions of Brazil: Caatinga; Cerrado; Atlantic forest; Araucaria forest and Pampas (see location in Figure 2.2).

lower than 20 mm. The changing frequency distribution of daily rains points to two major concerns: 1- reducing frequency of smaller daily rain can disturb the soil-water-biota interactions; 2- the increasing number of days with extreme rain (>100 mm) will promote more landslides (see Figure 2.5).

2.3 RELATION BETWEEN RAINFALL AND LANDSLIDES

Table 2.1 shows the relationships between rainfall regime and landslides mechanisms in southern and southeastern Brazil as pointed out by Coelho Netto *et al.* (1999).

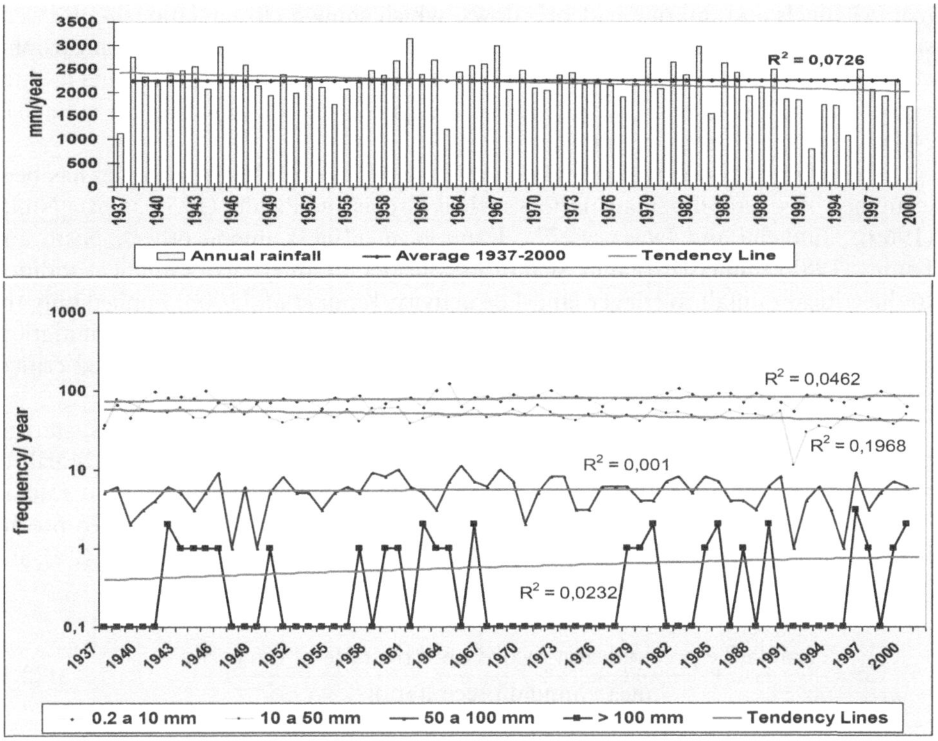

Figure 2.5 Above: Histograms of annual precipitation (mm) for the period 1937–2000; average annual rainfall line (blue) and tendency line (red). Below: Frequency of daily rainfall classes over the same time period and respective trend lines. Source: Figueiró & Coelho Netto (2011).

Table 2.1 Rainfall regimes and major landslide types at different locations. Source: Coelho Netto *et al.* (1999).

Regular rainfall regime		*Extreme rainfall regime*	
Location	*Landslide type*	*Location*	*Landslide type*
Valley bottom & thick colluvium	Creep	Rock escarpments	Rock falls
Steep slopes & Lithosol (thin soil)	Shallow Translational failures	Steep slopes, thin soil or shallow root zone	Shallow Translational failures
Channel banks & road cuts	Deep rotational failures	Steep slopes and river channels	Debris flow and Rock Debris Avalanches

Shallow translational failure is the dominant mechanism and commonly occurs at the mid-upper portion of mountain slopes on any slope geometry, as observed by Lima *et al.* (2014) regarding the January 2011 landslide disaster in the municipality of Nova Friburgo, in the state of Rio de Janeiro. It may be associated with shallow hydraulic discontinuities in soil profile. Large amounts of sediments moved down slopes into the main channels and ran out as debris flows, which show a close relationship between slide-flow transformations as another important cause of down valley catastrophic effects by flooding. Rotational slides may also occur mainly along road cuts in thick saprolitic soils (>5 m thick). The latter failure type is not common, but the volume of collapsed materials may be quite large.

The relationship between rainfall and landslides in southeastern Brazil has been examined by several Brazilian authors such as Vargas and Pichler (1957), Costa Nunes (1969), Guidicini and Iwasa (1977), Kanji *et al.* (2003) among others. Nieto and Barany (1988) emphasize that most authors consider the antecedent rainfall in addition to the critical rainfall to trigger landslide activity. Kanji *et al.* (2003) applied only the previous cumulative rainfall (using variable periods of time) for debris-flow initiation, as observed in many locations in Brazil and South America, and proposed critical values to produce landslides of different degrees of severity.

Guidicini and Iwasa (1977) considered the rainfall of previous months, starting the counting of cumulative rainfall in July (the dry season in southeastern Brazil) to preview critical rainfall to trigger shallow landslides. They highlighted that rainfall of more than 250–300 mm could trigger landslides under any condition. To predict landslide risk, they suggested the "cycle coefficient" (Cc), "event coefficient" (Ce), and "final coefficient" (Cf), as indicated by equations 2.1, 2.2 and 2.3 below:

$$Cc = \frac{\text{cumulative precipitation record upto the date of the event}}{\text{mean annual precipitation}} \tag{2.1}$$

$$Ce = \frac{\text{precipitation record of the event}}{\text{mean annual precipitation}} \tag{2.2}$$

$$Cf = Cc + Ce \tag{2.3}$$

For this report, the Guidicini-Iwasa coefficient was applied to a few critical disasters at Rio de Janeiro state (1967, 1988, 1996, 2010 and 2011) as shown in Figure 2.6. When the "cycle coefficient" (Cc) is plotted against the months of the year, it indicates four levels of landslide susceptibility: A, B, C and D, A being the highest. According to this graph, the 1967 event reached extreme rainfall conditions (above zone A), while the landslide events of 1996, 2010 and 2011 remained within zone A and the 1988 event at class B, which confirms that the Guidicini-Iwasa curves have a high success rate for forecasting these disasters.

Tatizana *et al.* (1987a, 1987b) analyzed the correlation between rainfall and landslides at Serra de Cubatão (state of São Paulo) and found a correlation between landslides and four-day cumulative rainfall between 80 and 120 mm as triggers for landslides. Other values are guiding the warning systems in the municipalities of Rio de Janeiro and Nova Friburgo. Futai (2014) highlights the importance of antecedent rainfall, since rainfall infiltration diminishes soil suction which, associated to intense rainfall, can initiate landslides.

Relationships between rainfall and landslide occurrences are difficult to generalize as variable terrain characteristics control soil moisture and, therefore, regulate slope

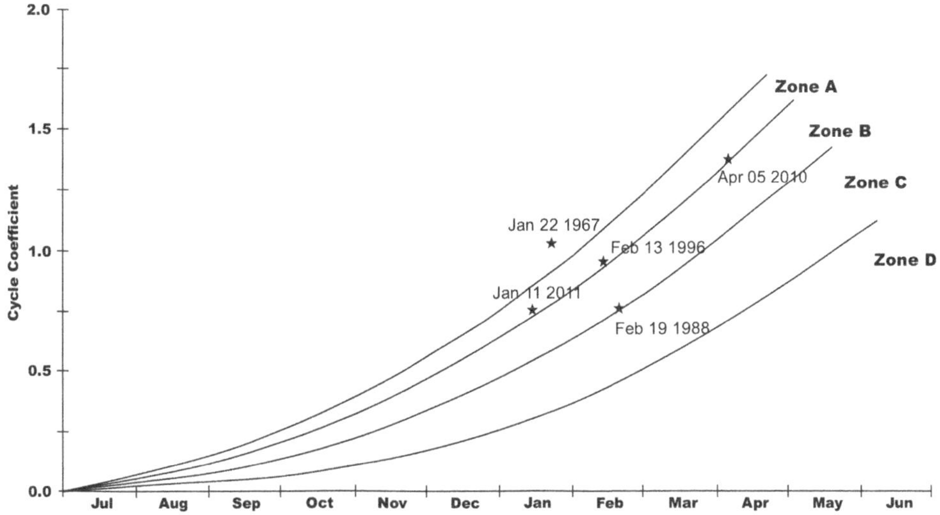

Figure 2.6 Guidicini-Iwasa rainfall curves applied to locations of landslide disasters in SE-Brazil: 1967 at Serra das Araras (RJ/state); 1988, 1996, 2010 at the Tijuca massif (RJ/city); and 2011 at Serra dos Órgãos (RJ/state).

failures. The most important terrain conditions include spatially non-uniform geo-bio-human aspects as indicated by Coelho Netto *et al.* (2012) regarding the Tijuca massif (in the municipality of Rio de Janeiro).

Coelho Netto *et al.* (2013) also observed this relation between rainfall and terrain at the Serra dos Órgãos region during the extreme, rainfall-induced landslides of January 2011. Cumulative rainfall from late December 2010 was followed by a dry period of about a week before the critical rainfall of January 11–12th, 2011. First, an electrical storm and cloudburst fell for 12 hours. Then, after a short time interval of eight hours, a second one fell for nine hours; the second storm reached the highest amount of rainfall (240 mm at some stations) which corresponds to the average monthly rainfall. More than 3,600 landslide scars (>30 sq. m) were mapped in a 420 sq. km area at the municipality of Nova Friburgo. The frequency of landslide scars decreased towards the less rainy isohyets, as might be expected. Only 2.7% occurred between 140 and 160 mm/day; 22.7% between 160 and 180 mm/day; 34.6% between 180 and 200 mm/day, and 40.0% in the area of rainfall of >200 mm/day (see Figure 2.7). These authors also indicate that landslides reached both urban and rural areas; but the density of landslides was relatively higher in the central zone of Nova Friburgo, despite the lower amount of rain under the lowest 30 minute-rainfall intensities as compared with other localities under similar geological-geomorphological conditions.

Coelho Netto *et al.* (2009) mention a list of geo-bio-human elements affecting the spatially non-uniform distribution of rainfall-induced landslides in southeastern Brazil. The January 2011 landslide event occurred at the municipality of Nova Friburgo also illustrates the spatial variability of landslide responses due to both rainfall characteristics and the net product of variable local environmental conditions that regulate

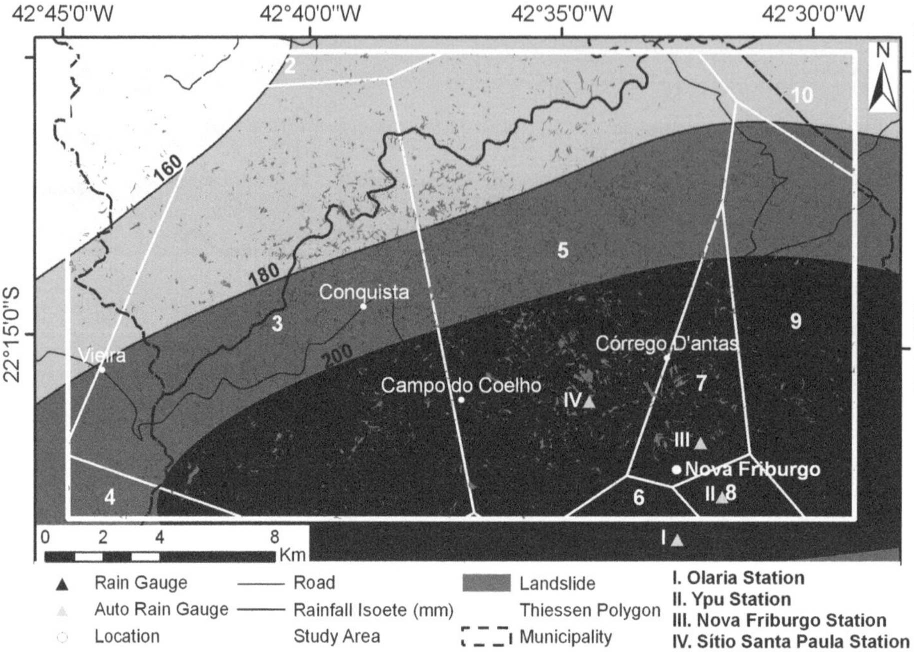

Figure 2.7 Map of landslide scars of January 2011 within the rainfall isohyets and Thiessen polygons: 1- Córrego Sujo, 2- Sumidouro, 3- Fazenda Mendes, 4- Friburgo/INMET, 5- Sitio Santa Paula, 6- Olaria, 7- Nova Friburgo, 8- Ypu, 9- Vargem Grande, 10- Bom Jardim. Source: Coelho Netto *et al.* (2013).

hillslope hydrology and soil properties, as highlighted by Coelho Netto *et al.* (2013). The authors emphasize that among these variable conditions are several landscape elements related to geological, geomorphological, geotechnical, vegetation and land use characteristics. It means that a thorough local investigation is required to define risk management. It also means that detailed thematic mapping of functional parameters or categories relevant to each type of landslide should fundamentally be taken into account in the development of a susceptibility map.

2.4 LANDSLIDE DISASTERS STUDY CASES

Landslides started being recorded in Brazil in the twentieth century. There is data for the municipality of Santos, in the state of São Paulo (March 1928, 1947, and 1956); southern Minas Gerais (December 1948); the metropolitan area of Rio de Janeiro and its neighboring sierras (January 1966, February 1967, and February 1988, February 1996, 2010); the Serra das Araras region, Rio de Janeiro (January 1967); Serra de Caraguatatuba, São Paulo (February 1967); the municipality of Tubarão, Santa Catarina (March 1974), the municipalities of Ouro Preto and Monlevade, Minas Gerais (March 1979), the Serra dos Órgãos region, Rio de Janeiro (December 1981, 2011); the municipality of Rio de Janeiro (February 1996) and the municipalities of

Timbé do Sul and Alto Feliz, Santa Catarina and Rio Grande do Sul (December 1995 and December 2000). Table 2.2 summarizes the main landslides disasters in Brazilian history.

Most landslide disasters of Rio de Janeiro city occurred in the mountainous Tijuca Massif (112 sq km), which is surrounded by a densely populated fluvio marine plain (the so-called "baixadas"). The years 1966 and 1967 were the first in which severe storms produced hundreds of landslides along the southeastern coast of Brazil, namely in the states of Rio de Janeiro and São Paulo. They were reported by Barata (1969), Costa Nunes (1969) and Jones (1973). The main occurrences were the 1966 disaster (Figure 2.8), the Laranjeiras slide (Figure 2.9), the Serra das Araras disaster (Figure 2.10), and the Caraguatatuba disaster (Figure 2.11), the latter three in 1967.

A well-studied landslide took place at the Soberbo slope in January 1966. This movement was primarily described by Meis and Silva (1968) using a geomorphological approach. Soon after, Fonseca (1969) produced a detailed geotechnical view of the case. His work and Meis and Silva's shared information that showed the importance of a geomorphological and geotechnical interface to better understand slope problems. In 1983, a new dislocation was observed on the same slope. Detailed geotechnical studies were carried out by Lacerda and Sandroni (1985) and later by Pedrosa *et al.* (1988). Complementary field and laboratory studies include Lacerda and Schilling (1992), Schilling *et al.* (1992), Silveira and Lacerda (1992), Barros *et al.* (1992), Avelar and Lacerda (1997), and Avelar *et al.* (2006).

According to Barros *et al.* (1992), the first sign of mass movement on the Soberbo slope was observed around 6 p.m., on January 14, 1966, and it was described as an avalanche of great proportions. It ran down about 800 m after a great rainfall, when a large rock moved from the highest part of the slope at an altitude of around 430 m, at the summit of the Soberbo Road. Figure 2.12, shows where this initial movement started in region A, with a partial rupture of the Soberbo Road. Barros *et al.* (1992) reports that 48 hours after the first avalanche, two new mass movements of distinct manner also occurred in regions B and C. In region B, the propagated translational movement was rising up, while in region C the terrain suffered degradation on a narrow strip about 4m wide that originated from loss of material.

Schilling (1993) attained artesian conditions in regions C and D for the piezometers installed in the soil and in the rock, which is explained by the blocking effect of transversal diabase dikes in these two regions. Lacerda (1997) proposed a model for landslide initiation by critical artesian flows provided by these dikes. To avoid a new rupture, an official alert was sent to the Geotechnical Bureau of the City of Rio de Janeiro stating that deep sub horizontal drainage was necessary. Since then, this slope has been stable.

Another good example of landslide in a colluvial slope is the Lagoa slide, in Rio de Janeiro, which took place in 1988. The main factor was the upward flow from the fractured gneiss rock underlying the colluvial mass, a few months after particularly intense summer rainstorms (Figure 2.13).

On February 13, 1996, there was another extreme rainfall in this mountainous area of the city of Rio de Janeiro. The rainiest area was located in the uppermost, western portion of the Tijuca massif. A local rain-gauge station recorded 380 mm in less than 24 hours, concentrated in two major storms: in the morning (reaching 150 mm); and around 7:50 p.m. and 8:50 p.m., reaching 230 mm, with a maximum rain intensity

Table 2.2 Historical landslides disasters in Brazil.

Date	*Local*	*State*	*Predominant Mechanisms*	*Deaths*	*Economic Damage*	*Rainfall*	*References*
Jan 10, 1966	Rio de Janeiro and Serrana Region	RJ	Rotational and translational landslides; Debris flow	1,000[a] 723[b]	Extremely diversified	287 mm (12 hours)[g] 487–675 mm (3 days)[a] *heavy electrical storm reported	Barata (1969) Costa Nunes (1969) Jones (1973) Guidicini and Iwasa (1977) S2ID (2014) EM-DAT (2014)
Jan 22, 1967	Serra das Araras	RJ	Translational landslides; Debris flow (> 10^7 m^3)[e]	1,700[a] 1,200[e]	Highway, > 100 houses, hydroelectric power plant flooded with sediments	218–275 mm (1 day)[a] *heavy electrical storm reported	Barata (1969) Costa Nunes (1969) Guidicini and Iwasa (1977) Jones (1973) Kanji *et al.* (2003)
Feb 20, 1967	Rio de Janeiro	RJ	Rotational and translational landslides; Debris flow	200[d]	2 residential buildings	137–198 mm (1 day)[h] 348–401 mm (3 days)[h]	Barata (1969) Costa Nunes (1969) Guidicini and Iwasa (1977) Jones (1973) S2ID (2014)
Mar 18, 1967	Caraguatatuba	SP	Translational landslides; Debris flow (7.6 Mton)[e]	120[e]	400 houses, roads	420 mm (1 day)[e] 586 mm (2days)[e]	Guidicini and Iwasa (1977) Kanji *et al.* (2003) Bressani (2009)
Dec 02, 1981	Petrópolis, Teresópolis and Guapimirim	RJ	Translational landslides	41[g]	Houses; gas station; highway	288 mm (1 day)[i]	Nunes *et al.* (2013)
Feb 27, 1985	Angra dos Reis (Piraquara)	RJ	Translational landslide (2.8×10^6 m^3)	0	220 m of highway, escape route and laboratory of nuclear power plant	129 mm (1 day)	Soares and Politano (1997)
Feb 05, 1988	Petrópolis	RJ	Translational landslides; Debris flow	171[e] 277[d]	5,000 houses affected[e]	145 mm (1 day)[e] 414 mm (6 days)[e]	S2ID (2014) Kanji *et al.* (2003)
Feb 19, 1988	Rio de Janeiro	RJ	Translational landslides; Debris flow	307[d]	US$ 935 Millions[d]	177 mm (1 day)[i] 272 mm (2 days)[i] 367 mm (3 days)[i]	S2ID (2014)

Feb 06, 1994	Cubatão	SP	Translational landslides; Debris flow (3×10^5 m^3)[e]	0	Oil refinery (US$ 44 Millions)[e]	60 mm (1 hour)[e] 250 mm (1 day)[f]	Kanji *et al.* (2003) Souza Jr. and Satyamurty (2008)
Dec 23, 1995	Timbé do Sul	SC	Translational landslides; Debris flow (3×10^6 m^3)[e]	29	20 houses, rural proprieties	500 mm (1 day)[e]	Kanji *et al.* (2003) Silveira (2003) Bressani (2009)
Feb 13, 1996	Rio de Janeiro (Tijuca Massif)	RJ	Rotational and translational landslides; Debris flow (13×10^4 m^3)[e]	1	Hundreds of houses destroyed	380 mm (1 day)	Coelho Netto (1996) Kanji *et al.* (2003) Fernandes *et al.* (2004)
Dec 08, 2002	Angra dos Reis	RJ	Rotational and translational landslides; Debris flow	35	One hundred houses destroyed	323 mm (1 day)	S2ID (2014)
Nov 2008	Vale do Itajaí	SC	Translational landslides; Debris flow	135	Hundreds of houses and gas pipeline	104 mm (1 day)[m] 225 mm (2 days)[m]	Lacerda (2009)
Dec 31, 2009	Angra dos Reis	RJ	Translational landslides	53	Hundreds of houses destroyed	170 mm (1 day)[j] 245 mm (2 days)[j] 415 mm (3 days)[j]	Silva and Spada (2013) Nunes *et al.* (2013)
Apr 05, 2010	Rio de Janeiro and Niterói	RJ	Rotational and translational landslides	222[d]	Hundreds of houses destroyed	425 mm (2days)[k]	Coelho Netto *et al.* (2012) Nunes *et al.* (2013) S2ID (2014)
Jan 11, 2011	Serrana Region	RJ	Translational landslides; Debris flow	900[c]	Extremely diversified (US$ 2.35 Billions)[c]	228 mm (1 day)[l] 300 mm (2 days)[l] *heavy electrical storm reported	Banco Mundial (2012) Dourado *et al.* (2012) Coelho Netto *et al.* (2013) Rodrigues *et al.* (2013) Nunes *et al.* (2013)

[a] Jones (1973).
[b] EM-DAT (2014).
[c] Banco Mundial (2012).
[d] S2ID (2014).
[e] Kanji *et al.* (2003).
[f] Souza Jr. and Satyamurty (2008).
[g] O Globo Archives (2014).
[h] Barata (1969).
[i] Hidroweb/ANA (2014).
[j] Civil Defence of Angra dos Reis (unpublished).
[k] Coelho Netto *et al.* (2012).
[l] Coelho Netto *et al.* (2013).
[m] BDMEP/INMET (2014).

DEZENAS DE MORTES NO MAIOR TEMPORAL DE TODOS OS TEMPOS

ANO XLI — Rio de Janeiro, 3.ª-feira, 11 de janeiro de 1966 — N.º 12 170

O GLOBO

FUNDAÇÃO DE IRINEU MARINHO

Diretor-Redator Chefe: ROBERTO MARINHO | Diretor-Tesoureiro: HERBERT MOSES
Diretor-Secretário: RICARDO MARINHO | Diretor-Substituto: ROGERIO MARINHO

Figure 2.8 Newspaper headline in January 11, 1966 – "Dozens of fatalities in the major storm ever recorded". The number of fatalities rose rapidly in the following days. Source: O Globo Archives (2014).

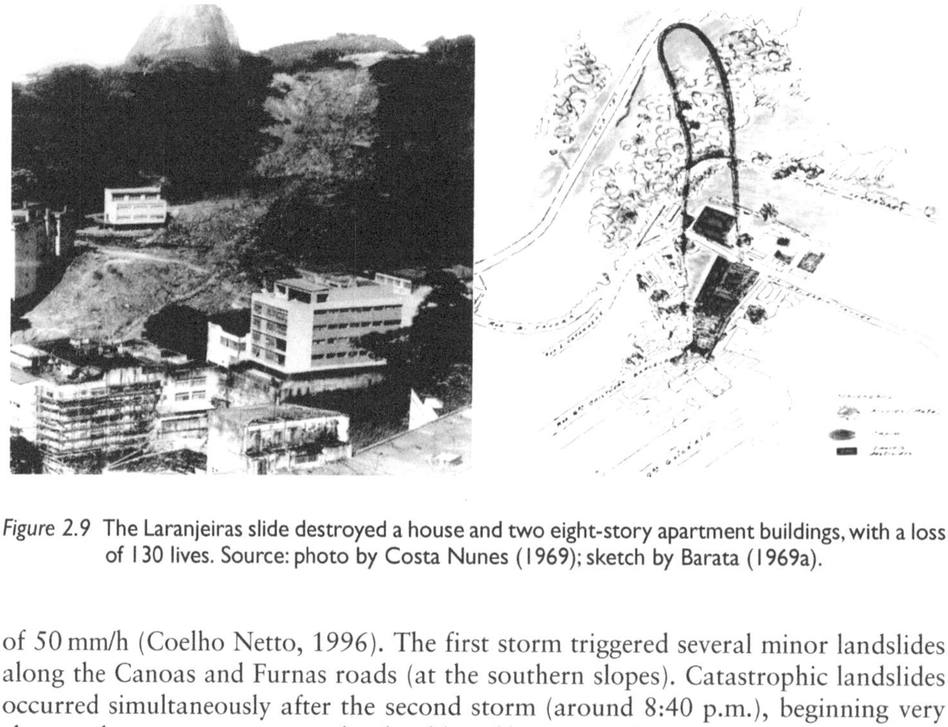

Figure 2.9 The Laranjeiras slide destroyed a house and two eight-story apartment buildings, with a loss of 130 lives. Source: photo by Costa Nunes (1969); sketch by Barata (1969a).

of 50 mm/h (Coelho Netto, 1996). The first storm triggered several minor landslides along the Canoas and Furnas roads (at the southern slopes). Catastrophic landslides occurred simultaneously after the second storm (around 8:40 p.m.), beginning very close to the crest-zone or near the shoulder of lower interfluves (Figure 2.14). Shallow landslides and debris flows propagated downslope into the main channels to feed extensive rock-debris avalanches (>4 km in length). Velocity reached 2.8 to 5.3 m/s as indicated by Vieira *et al.* (1997) and spread to the Jacarepaguá lowlands, causing serious environmental damage, property loss and the spread of diseases, in addition to 70 deaths.

Another extreme, rainfall-induced landslide took place at the Tijuca massif on April 5, 2010. Intense rainfall fell all over the Tijuca massif, but the highest intensity were concentrated in the eastern portion around the Sumaré/Corcovado area.

Figure 2.10 Debris flows in Serra das Araras in 1967. Source: Jones (1973).

Figure 2.11 Debris flows in Serra do Mar, which buried the city of Caragatuatuba, state of São Paulo, in 1967. Source: Bressani (2009).

Differently from the 1996 landslide disaster, scars did not propagate far from their trigger position. There were no debris flows, and only shallow landslides occurred, as indicated in Figure 2.15. The critical rainfall (total amount and rain intensity) were similar to the extreme event of February 1996, but the antecedent moisture conditions were lower as it followed a relatively dry summer period. Among the major

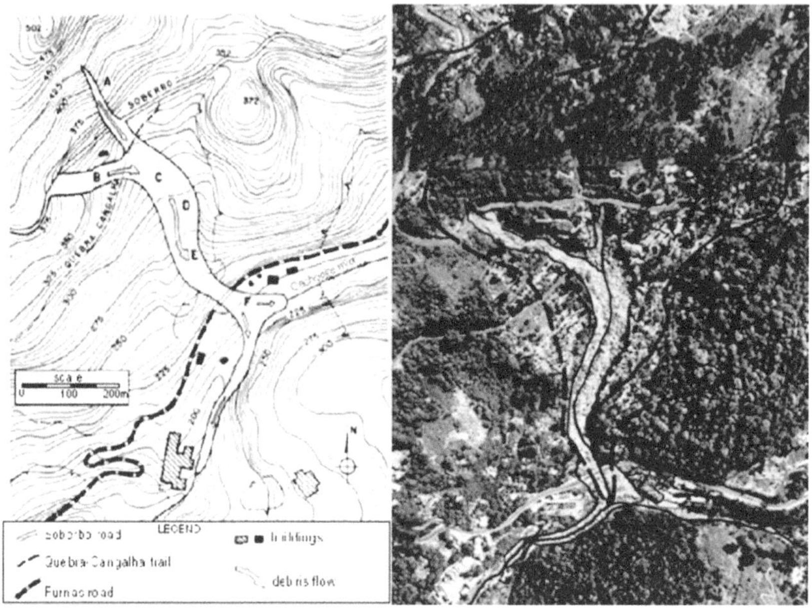

Figure 2.12 Topographic map (A) of the Soberbo mass movement in the southern slopes of Tijuca Massif, and buildings and roads destroyed by this event (Pedrosa *et al.*, 1988). The drainage area (dashed line) and land use around the landslide are shown in the adjacent aerial photograph (Coelho Netto *et al.*, 2009).

Figure 2.13 The Lagoa slide. Four pillars of the building were destroyed, but there was no damage to the main structure. No deaths, only a few cars destroyed (Photo by W. Lacerda).

Figure 2.14 1996 landslide scars in the Quitite river basin, western slopes of the Tijuca massif, Rio de Janeiro (Photo by A.L. Coelho Netto).

99 landslide scars (>30 sq m), 45 occurred along roadcuts, besides another 120 minor scars (<30 sq m). In some places, however, landslides reached a few houses and killed around 35 people, most of them in the Prazeres Hill (Figure 2.16).

One of the biggest landslides in Brazil, involving about 2.8 million cubic meters, took place in 1985 along the Rio-Santos highway near Angra dos Reis and its nuclear power plants (Figure 2.17). Soares and Politano (1997) described that the slide took place after a rainstorm that reached a peak of 120 mm of rainfall in one hour. The main cause was the inadequate drainage of surface water at the contact between the highway embankment and the natural slope, along a nearly horizontal 200 m stretch. The granitic rock was exposed, and the slide took place at the contact of the residual soil and rock. The landslide propagated to the sea, changing the coastal morphology and creating an 8-meter wave that produced secondary damages to boats and other constructions.

Angra dos Reis was affected again by an extreme rainfall event in December 2002 (about 300 mm in 24 hours). Dozens of landslides and floods affected 30 neighborhoods of the municipality. The main landslide occurred at Areal, with a combination of planar and debris flows mechanisms. Shallow and planar landslides occurred in the upper slope, and a temporary dam formed in the adjacent creek. Dam failure led to a debris flow of over 300 m amplitude, which hit many houses downstream and killed 40 people (Figure 2.18).

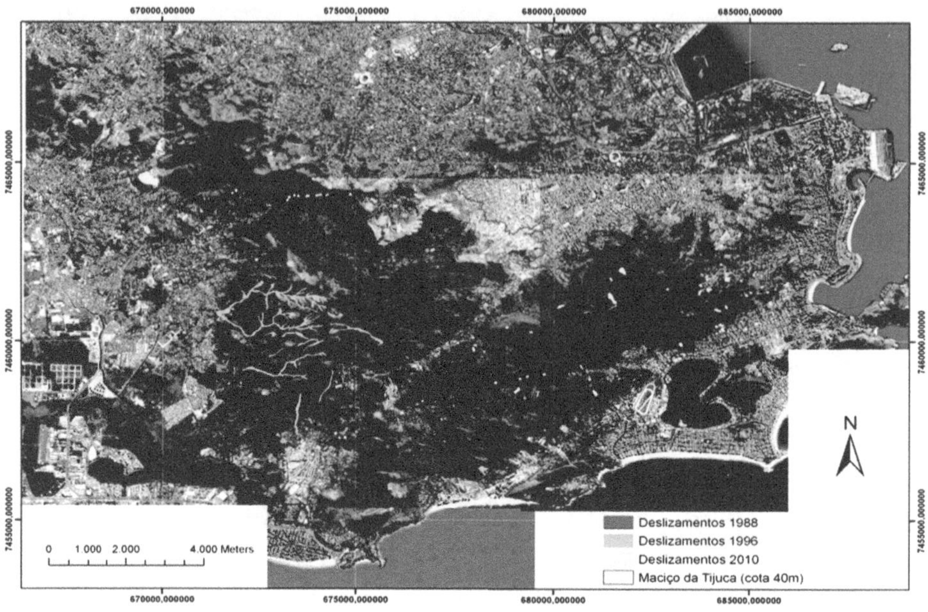

Figure 2.15 Landslide scars at the Tijuca massif, Rio de Janeiro: the 1996 landslides are in orange, the 2010 in yellow. Source: Coelho Netto *et al.* (2012).

Figure 2.16 Landslide adjacent to the Prazeres slum, Rio de Janeiro (Photo by G. Araújo, AFP).

Figure 2.17 The Piraquara slide of 1985 (Photo by R. Quinet).

Figure 2.18 A – Shallow landslide scar in the upper hanging valley at Areal, Angra dos Reis (Photo by A.L. Coelho Netto). B – Debris flows in the lower valley and destroyed houses (Photo by C. Coimbra, AP).

The most emblematic case of Angra dos Reis occurred in the New Year's Eve of 2009, when extreme rainfall (400 mm in 3 days) triggered hundreds of landslides that affected both the mainland and the islands of the city, causing 52 deaths. According to Nunes *et al.* (2013), 61 neighborhoods were affected, more than 70 large landslides were recorded by the Civil Defense and the BR-101 highway was interrupted in 37 places. At this event, the human influence on triggering landslides was made evident, as occurred at Morro da Carioca. The landslide was triggered by inappropriate water

Figure 2.19 Large shallow slide over rock destroyed houses and a small lodge, killing 32 people (Photo by W. Lacerda).

use in the houses, with pipe drains placed directly onto the upper, steep slopes. This water disposal triggered shallow translational landslide, destroying houses and killing 21 people in the foot slope. In the island portion of the city, the most serious case occurred at Bananal, where a shallow translational slide in the contact between soil and rock took place on the upper portion of a steep hillside that had no human intervention (Figure 2.19). The morphology of the adjacent slopes shows older landslide scars, which reinforces the recurrence of these movements in the area. Field observations showed a large amount of water exfiltration through fractures and joints in the predominantly granitic rock. The sliding scar begins 200 m above sea level and part of the material was deposited in the sea, killing a total of 32 people.

Angra dos Reis was a scenario of several landslide disasters in the last decade, resulting in dozens of deaths, thousands people affected and R$ 100 million in economic loss (S2ID, 2014). According to DRM-RJ (2014), Angra dos Reis has more than 200 areas of imminent risk. Studies conducted by Coelho Netto *et al.* (2014) showed that in the central area of Angra dos Reis there were 8,000 people living in areas of high and very high susceptibility to landslides (Figure 2.20 and 2.21).

In 1995 hundreds of shallow slides in the steeply slopes of a canyon were initiated by heavy rainstorm at Timbé do Sul, state of Santa Catarina, and formed a debris flows but with few fatalities (Figure 2.22).

The region around the city of Blumenau, also in Santa Catarina, suffered in 2008 the worst disaster known up to that date in that state. Nine cities decreed "state of

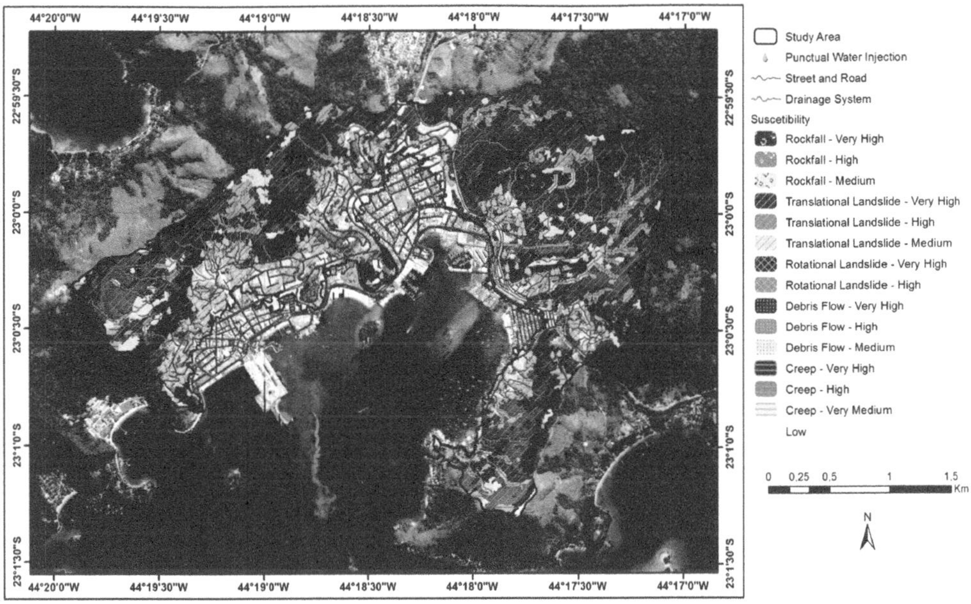

Figure 2.20 Landslide susceptibility map of Angra dos Reis considering dominant mechanisms. Source: Coelho Netto *et al.* (2014).

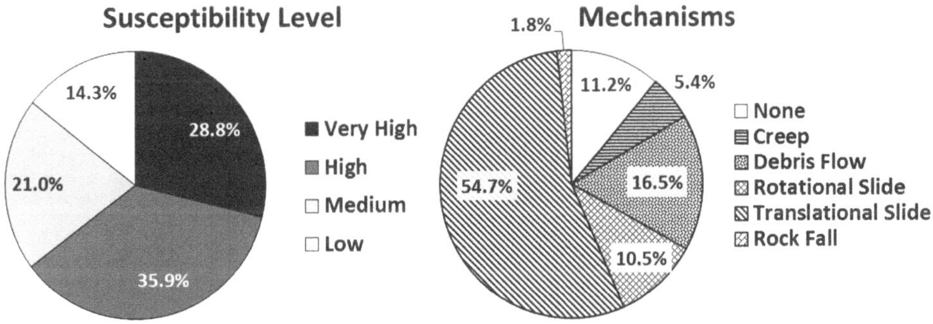

Figure 2.21 Areal cover (%) of landslide susceptibility level (left) and landslide mechanisms (right) in the area studied. Source: Coelho Netto *et al.* (2014).

public calamity" (Figure 2.23). More than 130 deaths occurred during that event. The amount of rainfall in the month of November 2008 was three times the average rainfall in the region. Figures 2.24 and 2.25 show the record of accumulated rainfall in 24 hours and in 30 days.

Flow slides initiated at the top of the hills, and destroyed houses, a gas duct and roads. Figure 2.26 illustrates just one of these debris flows.

In the summer of 2011, a tragedy occurred in the mountains of the state of Rio de Janeiro, with more than 1,500 deaths and a large loss of property. The region is part

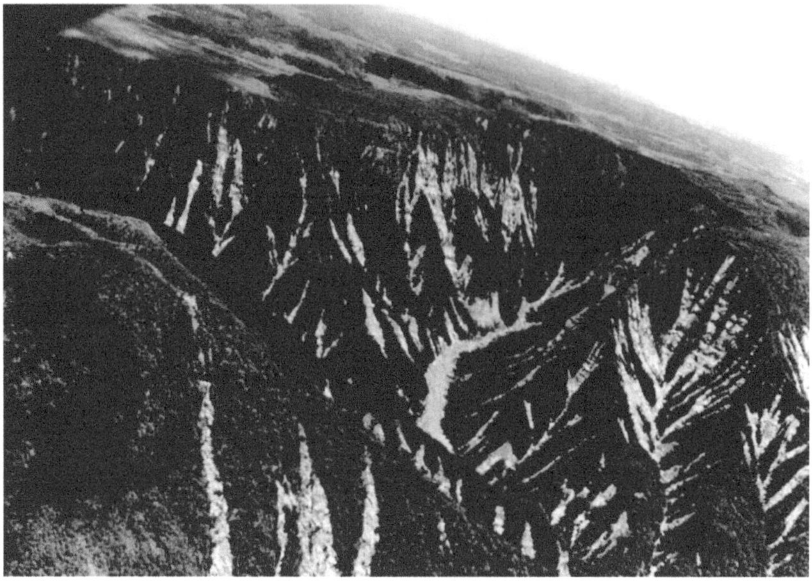

Figure 2.22 Debris flows in Canyon of Pinheirinho, Timbé do Sul, State of Santa Catarina, in 1995, after 400 mm of rain in one day. Source: Bressani (2009).

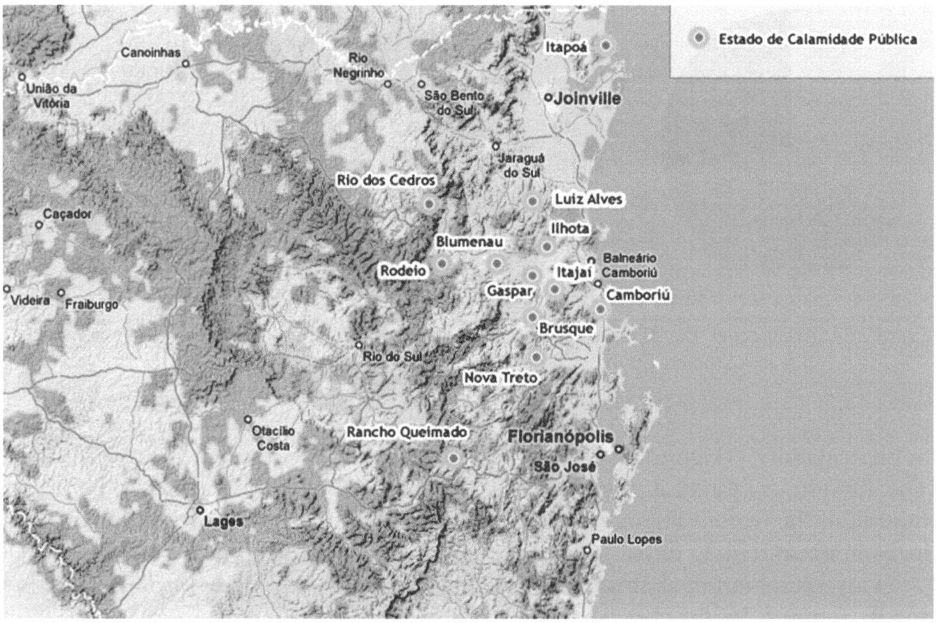

Figure 2.23 Cities affected by the 2008 debris flows and landslides in the state of Santa Catarina.

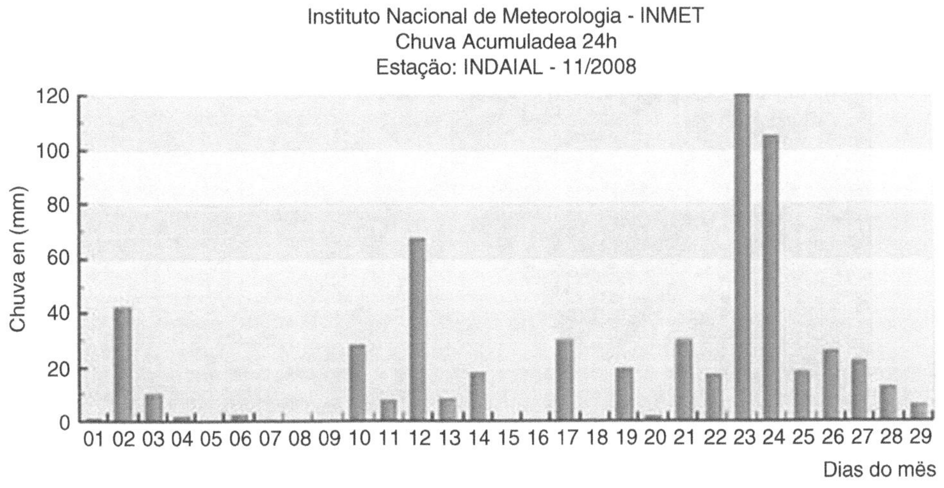

Figure 2.24 Daily rainfall of November, 2008. The days where debris flows occurred were 23 and 24, when the total amount of rainfall was 235 mm.

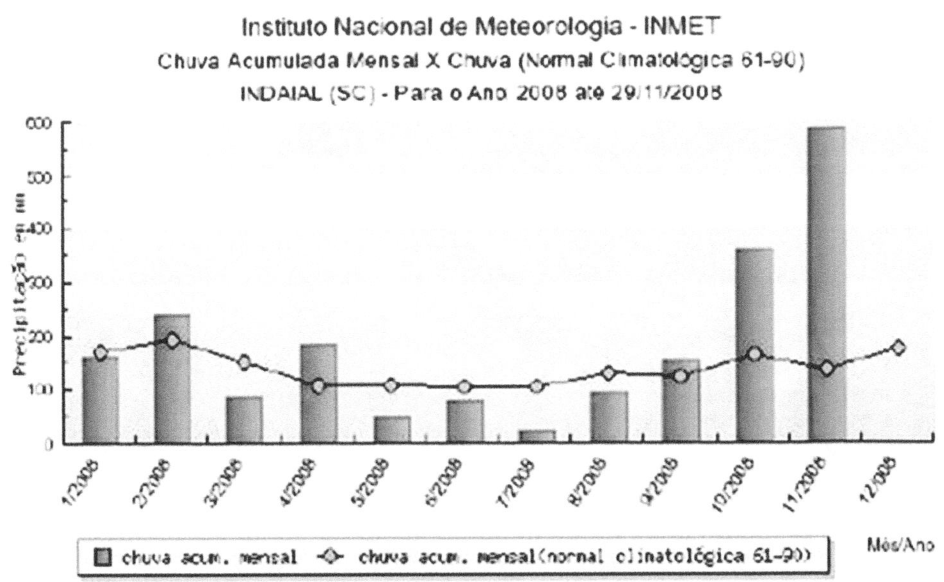

Figure 2.25 Monthly rainfall in 2008. October had 2 times the average precipitation and in November 3 times.

Figure 2.26 Debris flow near Blumenau in 2008 (Photo by W. Lacerda).

Figure 2.27 Shallow translational slides and deposition at the valley bottoms (responsible for most of the death toll) in Nova Friburgo (Photo by W. Lacerda).

Figure 2.28 Lateral spread of flow slides, Nova Friburgo (Photo by W. Lacerda).

Figure 2.29 Landslides in the mountainous region of state of Rio de Janeiro in 2011 (Photo by A.L. Coelho Netto).

of the Serra do Mar range, and the three main cities in the area are Nova Friburgo, Teresópolis and Petrópolis. Figure 2.27 shows a small part of the disaster at Córrego Dantas district, Nova Friburgo. Thousands of shallow slides occurred during the night, in a period of about 9 hours. Damming of rivers led to deadly flow slides, like the ones shown in figure 2.28 and 2.29. Coelho Netto *et al.* (2013) and Avelar *et al.* (2013) reported this event.

REFERENCES

Avelar, A.S. & Lacerda, W.A. (1997) Causas iniciais do movimento de massa da encosta do Soberbo, Rio de Janeiro, Brasil. In: *2a Conferência Brasileira sobre Estabilidade de Encostas*. Rio de Janeiro, ABMS, ABGE, ISSMGE. pp. 3–12.

Avelar, A.S., Lacerda, W.A. & Coelho Netto, A.L. (2006) Mecanismos de iniciação de fluxos detríticos no Maciço da Tijuca, Rio de Janeiro (RJ): O caso da encosta do Soberbo. *Revista Brasileira de Geomorfologia*, 7 (1), 73–83.

Avelar, A.S., Coelho Netto, A.L.C., Lacerda, W.A., Becker, L.B. & Mendonça, M.B. (2013) Mechanisms of the recent catastrophic landslides in the mountainous range of Rio de Janeiro, Brazil. In: *Landslide Science and Practice*. Berlin, Springer Berlin Heidelberg. pp. 265–270.

Banco Mundial (2012) *Avaliação de perdas e danos: inundações e deslizamentos na Região Serrana do Rio de Janeiro-Janeiro de 2011*. Brasília, BM.

Barata, F.E. (1969a) Landslides in the tropical region of Rio de Janeiro. In: *VII International Conference on Soil Mechanics and Foundation Engineering, Mexico*.

Brata, F.E. (1969b) Personal communication.

Barros, W.T., Bogossian, F. & Lopes, P.C.C. (1992) O escorregamento do Soberbo. In: *Publ. Esp. I Conf. Bras. Estab. de Encostas COBRAE, Rio de Janeiro*.

BDMEP/INMET (2014) *Banco de Dados Meteorológicos para Ensino e Pesquisa*. Available from: http://www.inmet.gov.br/portal/index.php?r=bdmep/bdmep [Accessed 14th November 2014].

Bressani, L.A. (2009) Escorregamento de encostas durante chuvas torrenciais: uma visão dos danos e possíveis causas. In: *Workshop Geológico-Geotécnico das Catástrofes Naturais de Santa Catarina*.

Coelho Netto, A.L. (1996) Produção de Sedimentos em Bacias Fluviais Florestadas do Maciço da Tijuca: respostas aos eventos extremos de fevereiro de 1996. In: *Anais do II Encontro nacional de Engenharia de Sedimentos, Rio de Janeiro, ABRH*, Vol. 1. pp. 209–217.

Coelho Netto, A.L., Avelar, A.S. & Cruz, E.S. (1999) Conditions and Problems for Landslides in Southeastern Brazil. In: *IDNDR Symposium on Mitigation of Water-Related Disasters, Nagoya – Japão Central. Proceedings, IDNDR – Symp. on Mitigation of Water-Related Disasters. Nagoya – Japan Central*, Vol. 1. pp. 105–110.

Coelho Netto, A.L., Avelar, A.S. & Lacerda, W.A. (2009a) Landslides and disasters in Southeastern and Southern Brazil. In: Latrubesse, E. (ed.) *Natural Hazards and Human-Exacerbated Disasters in Latin-America. Developments in Earth Surface Processes, 10 Special Volumes of Geomorphology*. pp. 223–243.

Coelho Netto, A.L., Avelar, A.S. & Lacerda, W.A. (2009b) Landslides and disasters in southeastern and southern Brazil. *Developments in Earth Surface Processes*, 13, 223–243.

Coelho Netto, A.L., Avelar, A.S., Sato, A.M., Dias, M.A., Schlee, M.B. & Negreiros, A.B. (2012) Vulnerabilidade em Geoecossistemas Montanhosos e Desastres Causados por Deslizamentos na interface Florestal-Urbana: Controles Geológicos, Geomorfológicos e Geoecológicos. In: *Desastres Naturais: Susceptibilidade e Riscos, Mitigação e Prevenção, Gestão e Ações Emergenciais*. Rio de Janeiro, COPPE/UFRJ. pp. 63–80.

Coelho Netto, A.L., Sato, A.M., Avelar, A.S, Vianna, L.G.G., Araújo, I.S., Ferreira, D.L. & Silva, R.P. (2013) January 2011: The extreme landslide disaster in Brazil. In: *Landslide Science and Practice*. Berlin, Springer Berlin Heidelberg. pp. 377–384.

Coelho Netto, A.L., Avelar, A.S., Sato, A.M., Fernandes, M.C., Oliveira, R.R., Costa, R.V.C., Barbosa, L.S., Lima, P.H.M. & Lacerda, W.A. (2014) Landslides susceptibility and risk zoning at Angra Dos Reis, Rio de Janeiro State, SE-Brazil: A quali-quantitative approach at 1:5,000 scale. In: *Extreme Rainfall Induced Landslides: An International Perspective, São Paulo, Oficina de Textos*, Vol. 1. pp. 262–296.

Costa Nunes, A.D. (1969) Landslides in soils of decomposed rock due to intense rainstorms, in *Proceedings of VII International Conference on Soil Mechanics and Foundation Engineering, Mexico*. pp. 547–554.

Dourado, F., Arraes, T.C. & Silva, M. (2012) O Megadesastre da Região Serrana do Rio de Janeiro: as causas do evento, os mecanismos dos movimentos de massa e a distribuição espacial dos investimentos de reconstrução no pós-desastre. *Anuário do Instituto de Geociências*, 35 (2), 43–54.

DRM-RJ (2014) *Plano de contingência do DRM-RJ* – período 12/2014 – 04/2015.

EM-DAT (2014) *The International Disaster Database*. Available from: http://www.emdat.be/ [Accessed 21st October 2014].

Fernandes, N.F., Guimarães, R.F., Gomes, R.A., Vieira, B.C., Montgomery, D.R. & Greenberg, H. (2004) Topographic controls of landslides in Rio de Janeiro: Field evidence and modeling. *Catena*, 55 (2), 163–181.

Figueiró, A.S. & Coelho Netto, A.L. (2011) Climatic variability and pluviometric trends in a humid tropical environment at Resende municipality-middle Paraíba do Sul river valley (Rio de Janeiro – Brazil). *Brazilian Geographical Journal: Geosciences and Humanities Research Medium*, 2 (2), 256–273.

Fonseca, A.M.M.C.C. (1969) Relato sobre causas e problemas das encostas da Guanabara. In: *1a Semana Paulista de Geologia Ap., São Paulo*, Vol. 1. pp. 1–10.

Futai, M.M. (2014) Landslide in unsaturated conditions caused by rainfall infiltration. In: *Extreme Rainfall Induced Landslides: An International Perspective*, Vol. 1, 11th edition. São Paulo, Oficina de Textos. pp. 42–69.

Guidicini, G. & Iwasa, O.Y. (1977) Tentative correlation between rainfall and landslides in a humid tropical environment. *Bulletin of the International Association of Engineering Geology-Bulletin de l'Association Internationale de Géologie de l'Ingénieur*, 16 (1), 13–20.

Hidroweb/ANA (2014) *Sistema de Informações Hidrológicas*. Available from: http://hidroweb.ana.gov.br/ [Accessed 7th December 2014].

INMET (2014) *Instituto Nacional de Meteorologia*. Available from: http://www.inmet.gov.br/ [Accessed 2nd October 2014].

Jones, F.O. (1973) *Landslides of Rio de Janeiro and the Serra das Araras escarpment, Brazil* (No. 697). USG, 42 pp.

Kanji, M.A., Massad, F. & Cruz, P.T. (2003) Debris flows in areas of residual soils: Occurrence and characteristics. In: *International Workshop on Occurrence and Mechanism of Flows in Natural Slopes and Earthfills*. pp. 1–13.

Lacerda, W.A. (1997) Stability of natural slopes along the tropical coast of Brazil. In: Almeida, M. (ed.) *Proceeding of the International Symposium on Recent Developments in Soil and Pavement Mechanics*. Balkema, Brookfield. pp. 17–39.

Lacerda, W.A. (2009) *Relatório sobre o acidente do gasoduto GASBOL ocorrido em 23 de Novembro de 2008, no Km 658+510, em Gaspar, Santa Catarina, Internal Report.*

Lacerda, W.A. & Sandroni, S. (1985) Movimentos de massas coluviais. In: *Mesa Redonda Sobre Aspectos Geotécnicos de Encostas. Clube de Engenharia, Rio de Janeiro*. pp. 1–19.

Lacerda, W.A. & Schilling, G.H. (1992) Rain induced creep-rupture of Soberbo road landslide. In: *Proceedings of the VIth International Symposium on Landslides*. pp. 10–14.

Lima, P.H.M., Coutinho, B.H., Gomes, G.B., Fernandes, M.C. & Coelho Netto, A.L. (2014) Parâmetros morfométricos relacionados às bacias de 1ª ordem e a ocorrência de deslizamentos rasos na bacia do Córrego Dantas: Nova Friburgo – RJ. Ver. *Geonorte*, 10 (1), 218–223.

Meis, M.R.M. & Silva, J.X. (1968) Considerações geomorfológicas a propósito dos movimentos demassa ocorridos no Rio de Janeiro. *Revista Brasileira de Geografia*, 30 (1), 55–73.

Nieto, A.S. & Barany, I. (1988) Interim report on catastrophic rain induced landslides in Rio de Janeiro and Petropolis, Brazil. In: *Report to Committee on Natural Disasters, National Research Council, Department of Geology, University of Illinois. Urbana, April.*

Nunes, A.L.L.S., Sayão, A.S.F.J., Rios Filho, M.G. & Dias, P.H.V. (2013) Desastres e Ações nas encostas do Rio de Janeiro. In: *Conferência Brasileira de Encostas, Angra dos Reis.* pp. 74–96.

O Globo Archives (2014) *Acervo O Globo.* Available from: http://acervo.oglobo.globo.com/ [Accessed 11th November 2014].

Pedrosa, M.G.A., Soares, M.M. & Lacerda, W.A. (1988) Mechanism of movements in colluvial slopes in Rio de Janeiro. In: *Proceedings of the 5th Int. Symp. Landslides, Lausane*, Vol. 2. pp. 1211–1216.

Rodrigues, J., Tupinambá, M. & Amaral, C. (2012) A corrida de massa do rio Vieira em Teresópolis, sudeste do Brasil: caracterização da área-fonte dos sedimentos transportados. *Anuário do Instituto de Geociências*, 35 (2), 152–164.

S2ID (2014) *Sistema integrado de informações sobre desastres.* Available from: http://s2id.mi.gov.br/ [Accessed 30th October 2014].

Schilling, G., Lacerda, W.A. & Barros, W.T. (1992) Instrumentação da Encosta do Soberbo. In: *Anais da I Conferencia Brasileira sobre Estabilidade de Encostas—I COBRAE, Rio de Janeiro*, Vol. 1. pp. 307–326.

Schilling, G.H. (1993) *Instrumentação e análise dos movimentos da encosta do Soberbo, Alto da Boa Vista, RJ.* MS dissertation. COPPE/UFRJ, Rio de Janeiro. 237 pp.

Silva, L.F.M. & Spada, J.L.G. (2013) A Tragédia em Angra dos Reis na Passagem de Ano de 2009 para 2010 – Projetos e Obras de Recuperação. In: *Conferência Brasileira de Encostas, Angra dos Reis.* pp. 96–104.

Silveira, G.C. & Lacerda, W.A. (1992) Caracterização mineralógica dos solos residuais e coluvionares do escorregamento do Soberbo. In: *I Conf. Bras. Estab. Encostas COBRAE, Rio de Janeiro*, Vol. 1. pp. 479–490.

Silveira, R.M. (2003) *Propriedades geotécnicas dos solos coluvionares do gasoduto Brasil-Bolívia em Timbé do Sul.* MS dissertation. PPGEC/UFRGS, Porto Alegre. 128 pp.

Soares, J.E.S. & Politano, C.F. (1997) Deslizamento de massa de solo de grandes dimensões, encosta de Piraquara. In: *Conferência Brasileira de Estabilidade de Encostas, Rio de Janeiro.* pp. 475–484.

Sousa Júnior, S.B. & Satyamurty, P. (2008) Dois Eventos Extremos de Chuva na Região da Serra do Mar. In: *XV Congresso Brasileiro de Meteorologia, São Paulo. A meteorologia e as cidades.*

Tatizana, C., Ogura, A.T., Cerri, L.E.S. & Rocha, M.D. (1987a) Análise de correlação entre chuvas e escorregamentos-Serra do Mar, município de Cubatão. In: *Congresso Brasileiro de Geologia de Engenharia*, Vol. 5. pp. 225–236.

Tatizana, C., Ogura, A.T., Cerri, L. & Rocha, M.D. (1987b) Modelamento numérico de análise de correlação entre chuvas e escorregamentos aplicado às encostas da Serra do Mar, município de Cubatão. In: *Congresso Brasileiro de Geologia de Engenharia*, Vol. 5. pp. 237–248.

Vargas, M. & Pichler, E. (1957) Residual soil and rock slides in Santos, Brazil. In: *Proceedings of the 4th Pan-American Confon SMFE*, Vol. 6 (27). pp. 394–398.

Vieira, B.C., Vieira, A.C.F., Fernandes, N.F. & Amaral, C.P. (1997) Estudo comparativo dos movimentos de massa ocorridos em Fevereiro de 1996 nas bacias do Quitite e Papagaio (RJ): uma abordagem geomorfológica. In: *2ª Conferência Brasileira sobre Estabilidade de Encostas, Rio de Janeiro.* pp. 165–174.

Chapter 3

Potential impacts of climate change on landslides occurrence in Canada

C. Cloutier and J. Locat
Université Laval, Québec, Canada

M. Geertsema
Government of British Columbia, British Columbia, Canada

M. Jakob
BGC Engineering, British Columbia, Canada

M. Schnorbus
Pacific Climate Impacts Consortium, University of Victoria, British Columbia, Canada

Presented at Joint Technical Committee JTC-1, TR3 Forum: Slope Safety Preparedness for Effects of Climate Change, November 17-18, 2015, Naples, Italy

3.1 INTRODUCTION

Canada's landmass is the second largest in the World and spans approximately 10 million km^2 from about latitude 41°40′ at Pelee Island in southern Ontario to the North Pole and from longitude 140°W in Yukon to about 53°W in Newfoundland. For these reasons, Canada consists of many climate regions (Fig. 3.1) including humid regions like Atlantic Canada and the Pacific Coast, but also arctic regions underlain by permafrost like the Arctic mountains and fjords in the north.

It is therefore unrealistic to represent climate change impact for the entire nation, and one must rely on regional and local analysis. Evidences of climate change since 1950 are shown in Figures 3.2 and 3.3 highlighting the increase in surface air temperature (Fig. 3.2) and increase in amount of precipitation (Fig. 3.3) in water equivalent averaged over Canada. Figure 3.4 represents the relative change in mean annual air surface temperature over Canada projected for the year 2100 relative to the period of 1981–2000. As expected, these changes are variable in space and time and show the most pronounced changes (up to +4.5°C) in central and northern Canada (Vincent *et al.*, 2015). In terms of annual average precipitation, projection to 2050 (Fig. 3.5) indicates that Canada is very likely to become wetter, except for south-west and south-central Canada where precipitation is projected to decrease in summer and fall (Bush 2015; Bush *et al.*, 2014). The highest increase is projected to occur in northern latitudes. While the confidence in the temperature projections is reasonably high, it is much lower for precipitation projections.

By the late 21st century, it is expected that globally, sea level may increase by about 0.25 to 0.82 m (Church *et al.*, 2013). In Canada, the impact of sea level change will vary according to whether or not, a given region has been under an ice cover and is still witnessing glacio-isostatic rebound. For example, in the Hudson Bay (located in Fig. 3.1), land is still rising at a rate of about 1 cm per year. In contrast, in some sectors

Figure 3.1 Canadian Climate regions. NF&L: Newfoundland and Labrador; N.B.:New-Brunswick; N.S.: Nova Scotia; P.E.I: Prince-Edward Island; Saskat.: Saskatchewan (modified from Environment Canada, 2014).

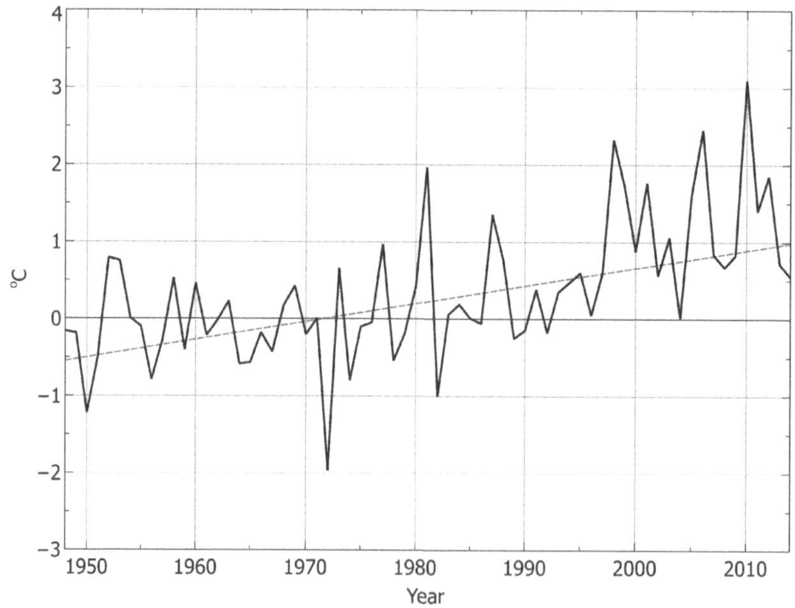

Figure 3.2 Yearly departures from the 1961–1990 average temperature averaged across Canada for the period from 1948 to 2014. The linear trend shows an increase of 1.6°C of the annual temperature (Data source: Environment Canada, 2015. For more information: http://ec.gc.ca/adsc-cmda/default.asp?lang=Fr&n=D48C5C94-1).

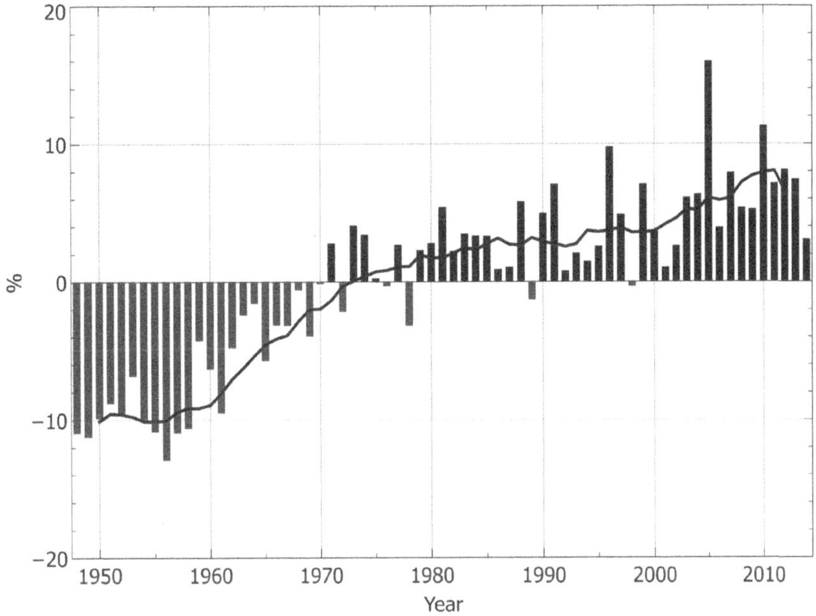

Figure 3.3 Yearly departures between 1948 and 2014 from the yearly precipitation averaged across Canada from the 1961–1990 average using a 9 year mobile mean (Data source: Environment Canada, 2015).

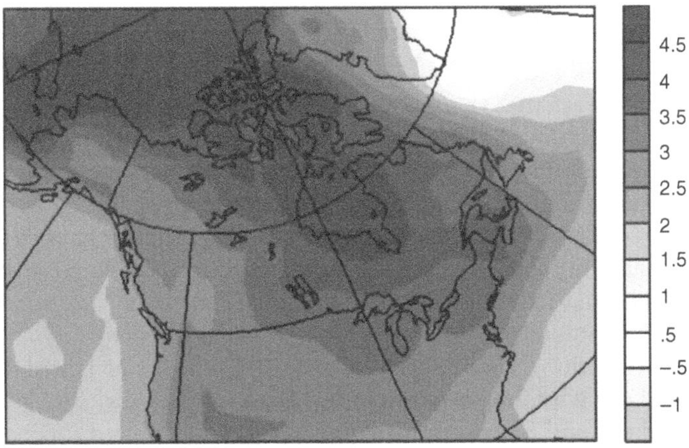

Figure 3.4 Projected change in the 5-year mean surface air temperature (°C) in 2100 relative to the 1981–2000 period as simulated by the CGCM3/T47 in the IPCC SRES A1B experiment (Source: Environment Canada 2010. For more information: http://www.cccma.ec.gc.ca/diagnostics/cgcm3/plot_st_glb_sresa1b_2090_vs_1990.shtml).

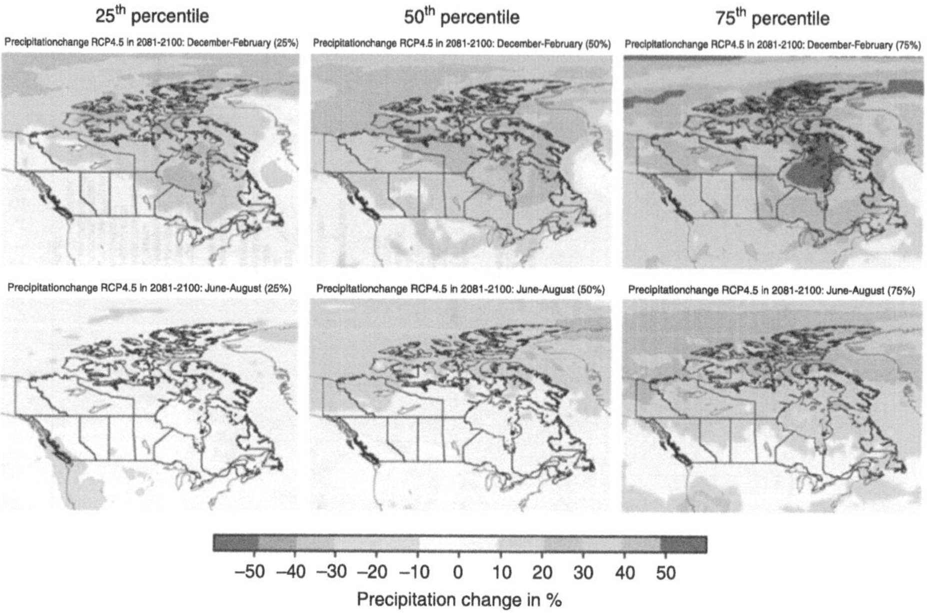

Figure 3.5 Maps of winter (DJF) (top) and summer (JJA) (bottom) precipitation change for the period 2081–2100 (relative to the 1986–2005 baseline period) projected by the CMIP5 multi-model ensemble for the RCP4.5 scenario, averaged over December–February and June–August, respectively (modified from Bush, 2015).

of the Atlantic region, the actual rate of sea level increase is twice the mean sea level increase due to the fact that this region is now subsiding (McCulloch *et al.,* 2002). Other factors, like tectonic activity may also introduce changes in sea level over time. The sensitivity of shoreline to erosion due to sea level changes, where applicable, is illustrated in Figure 3.6 for eastern Canada. In addition to sea level rise, the decrease in sea and lake ice cover and increase open water season (Paquette *et al.,* 2015; Vaughan *et al.,* 2013) results in increased impacts of wave action on the coast lines. Increase shoreline erosion can affect slope stability along the Canadian coasts, because erosion is a landslide-triggering factor.

As shown from the various examples below, the study of landslides in Canada has been carried out with greater efforts in western Canada (British Columbia (B.C.), Alberta and Yukon) and in eastern Canada around Ontario and Québec. In order to develop this report, we simplify the physiographic regions so that areas covered by B.C., Alberta and Yukon are referred to as western Canada, while Ontario, Québec and Atlantic provinces are included in eastern Canada and Northwest Territories and Nunavut in Northern Canada (provinces and territories are located in Fig. 3.1). Still, in some cases, examples will be provided on a more local basis.

Landslides have plagued many regions of Canada, not only in the mountainous regions of B.C., Alberta and the Yukon but also along major rivers in the Prairies, in

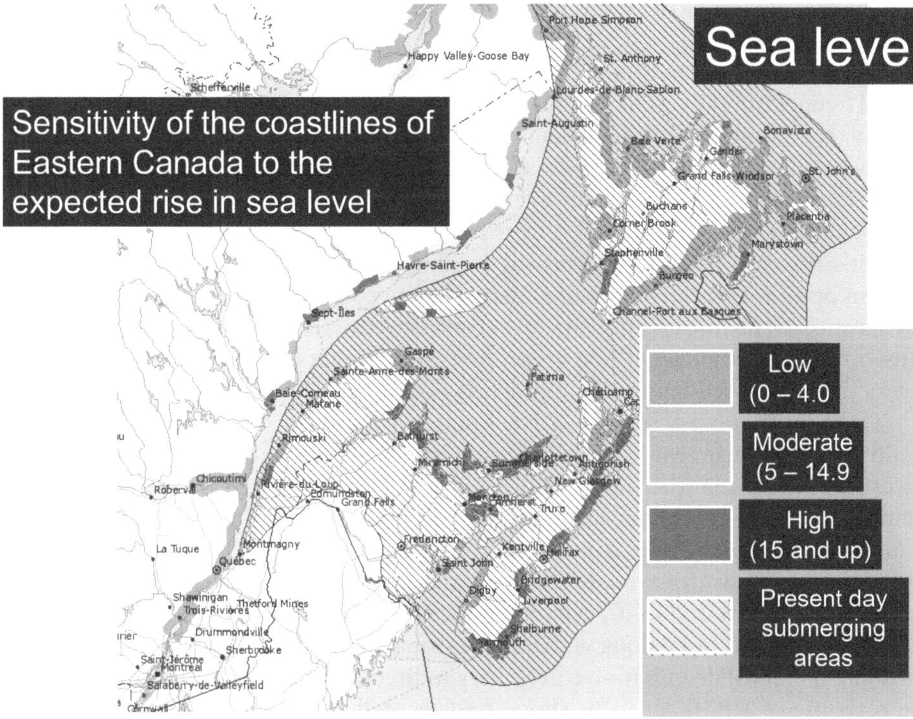

Figure 3.6 Sensitivity of shoreline erosion to sea level change (modified from Shaw *et al.,* 1994). Shoreline erosion is triggering factor of slope instability.

the St. Lawrence Lowlands and along the Mackenzie River. Excluding extreme events, landslides along the St. Lawrence Valley, mostly in quick clays, have led to significant loss of life (154 since 1870) and property with an average of about 200 slides per year including 80% in clayey soils. In Québec, about 40% of the landslides for which a governmental help is requested are related to human activities (Demers *et al.*, 2008). As described below, the July 1996 extreme storm event (total damage of CDA $ 1.5 billion and 10 fatalities) triggered about 1000 slides within one week. A high concentration of landslides is found in the mountainous regions of B.C. and Alberta. Both provinces have also witnessed significant growth over the past decade, urban and linear infrastructure development has encroached on previously undeveloped mountainous terrain. This has led to significant losses during times of heightened landslide activity, accentuated most noticeably during the June 2013 debris flows and debris floods near the town of Canmore, Alberta that led to approximately CDA $ 100 million in economic losses, though fortunately no fatalities occurred (Jakob *et al.*, 2014 and 2015).

With development continuing to penetrate hitherto untouched terrain, landslide prevention is critical to allow safe housing, transport of hydrocarbons, power, and people. In the past, analogues have formed the basis to build landslide frequency-magnitude relationships, which in turn, are the backbone of hazard assessments. Such analogues are based on reconstruction of landslide activity with tools such as radiocarbon analysis, dendrochronology, air photograph interpretation or varve chronologies, to name a few.

However, the past may no longer be an adequate surrogate for the future. Unprecedented temperatures in mountainous glacierized terrain may lead to increased landslide activity that has previously not been observed simply because portions of the landscape have been covered by ice. Progressive increases in the depth of the active layer in permafrost and associated landslides overlying frozen ground may increase in frequency and magnitude, particularly where forest fires will occur. Recent temporal clustering of debris flows in B.C. and Alberta is beginning to raise questions if this cluster is random or an earlier harbinger of climate change consistent with projections of more intensive rainfall. In eastern Canada, recent extreme storm events, like the one that caused the 1996 Saguenay flood event, have shown that areas considered stable were not as they witnessed landslides and debris flows.

This report summarizes the current state of knowledge on how climate change could affect the occurrence of landslides in Canada. This is a topic that is submerged in uncertainty given that the weather impact on landslide activity is complex and climate change is very complex, not uniform, and uncertain. Landslides will respond to the combined effect of various climatic parameters, for example, to study the effect of climate change on water availability in soil, which is important for landslide initiation, precipitation (snow and rain), evapotranspiration, infiltration, vegetation, etc. should all be taken into account along with their seasonal variability. Thus, the effect of global warming on landslide initiation can be very indirect. This report will investigate firstly the response of landslides to warming and secondly to changes in precipitations by illustrating the potential effect by giving examples of Canadian landslides and studies. The report starts with an overview of climate projections.

3.2 CLIMATE CHANGE PROJECTIONS IN THE CONTEXT OF SLOPE STABILITY

3.2.1 Climate change – background

The projections results presented in the following sections (Temperature, Precipitation and Hydrology) are derived by analysing output from ensemble-based modelling experiments. Climate projections are constructed from the output of an ensemble of global climate models (GCMs) driven by multiple emissions scenarios and spanning a projection horizon to 2100. Hydrologic projections are derived by downscaling the climate projections, which are then used to feed a hydrologic model. The results given in this section stem from two separate generations of climate modelling experiments known as the Coupled Model Intercomparison Project phase 3 (CMIP3; Meehl *et al.*, 2007) and the phase 5 (CMIP5; Taylor *et al.*, 2012). CMIP3 models are based on emissions scenarios documented in the Special Report on Emissions Scenarios (SRES, Nakicenovic *et al.*, 2000) whereas CMIP5 results are driven by the new Representative Concentration Pathways (RCPs; van Vuuren *et al.*, 2011). Although results from the two climate model experiments may differ due to differences in the complexity of GCMs and differences in emissions scenarios, future projection results are generally similar. No further distinction is made in the remainder of this section between

CMIP3- and CMIP5-derived results. Often, RCP8.5 scenario, the highest emission one, is used purposely to show the maximum projected change in climate.

A key step in projecting future climatic conditions for a specific location or region is the downscaling from GCMs to the area of interest. A downscaling algorithm depends, however, on a network of climate stations that record daily weather data (Wang *et al.*, 2012). In particular, for the northern regions there are very few climate stations that can be used in such a downscaling algorithm and their time series are short. The uncertainties increase when deriving local climate projection data from a GCM (Mbogga *et al.*, 2009; Wang *et al.*, 2012).

3.2.2 Temperature

With continuing emissions of anthropogenic greenhouse gases, temperature is projected to increase throughout Canada over the coming century (van Oldenborgh *et al.*, 2013). In western Canada, air temperature will increase in all seasons, although the largest increase is generally expected to occur over the winter months as a result of the snow albedo feedback. Temperature projections are very robust to emissions scenarios, inter-annual variability and inter-model variability, and changes are generally considered statistically significant throughout the projection period (van Oldenborgh *et al.*, 2013). Projected changes in temperature over western Canada range from 4°C to 7°C annually, 4°C to >11°C in winter, and 3°C to 7°C in summer by end-century (RCP8.5 scenario). Similarly, for eastern Canada, the projected rises are more pronounced in the north than in the south and are higher for the winter season than for summer with rises of 9°C to 15°C depending on the emissions scenario (Colombo *et al.*, 2007; Ouranos, 2014). In the Atlantic Provinces, increases of around 4°C are forecasted, except for Newfoundland-Labrador where the increase should be less due to the effect of the North American Oscilliation (Vasseur and Catto, 2008).

Temperature extremes are projected to increase in the future. Cold nights (minimum and annual daily minimum temperature) and hot days (maximum and annual daily maximum temperature) are both projected to get warmer over western Canada, with cold nights warming faster than hot days (Sillman *et al.*, 2013). Conversely, the number of frost days (daily minimum temperature below 0°C) will decrease, potentially by as much as 60 days or more by end-century (RCP8.5) (Sillmann *et al.*, 2013). The 20-year return period annual maximum and minimum temperatures are both projected to increase globally and throughout western Canada (Kharin *et al.*, 2013). Again, cold temperature extremes will warm substantially faster than the warm ones, where 20-year return period minimum and maximum temperatures will increase by as much as 6–10°C and 6–8°C, respectively, by end-century (RCP8.5 scenario). Similar projections are made for Eastern Canada: a strong reduction of the number of days below the freezing point is forecasted and the minimum temperature for the coldest day of the year will increase more than the hot extremes, with increases of 5°C to 7°C for the RCP 4.5 scenario and more than 10° C for RCP8.5 (Colombo *et al.*, 2007; Ouranos, 2014).

For the southern part of Canada, the quantity of freeze-thaw cycles, which are a rock fall preparatory factor, is unlikely to change much annually, but the timing of their occurrence will be shifting, with more events in winter and less in fall and spring (Logan *et al.*, 2011; Rapaic, 2012).

3.2.3 Precipitation

Annual precipitation amounts are projected to increase throughout Canada, although results are not as consistent as those for temperature and are not statistically significant over the region until roughly mid-century (van Oldenborgh *et al.*, 2013). Seasonally, precipitation will increase during the winter/spring half of the year (October–March) but precipitation changes are more uncertain during the summer/fall half of the year (April–September). Overall, western Canada will tend to become wetter in the north and drier in the south (van Oldenborgh *et al.*, 2013). In eastern Canada, the climatic models also agree on an increased amount of winter and spring precipitation (Ouranos, 2014; Bush *et al.*, 2014). In the Arctic, precipitation is projected to increase for every season (Fig. 3.5).

Climate change is expected to affect the frequency and intensity of extreme precipitation events. For example, heavy precipitation days (days where precipitation exceeds 10 mm), monthly or annual maximum 5-day accumulated precipitation and very wet day precipitation (precipitation amount on days when precipitation exceeds the historical 95th percentile threshold) are all projected to increase, whereas consecutive dry days (length of the longest period of consecutive dry days in a year) are projected to decrease throughout western North America over the coming century (Sillmann *et al.*, 2013). On a seasonal basis increases in winter (Precipitation deficits = DJF) precipitation extremes will be greater than those projected for the summer (Precipitation deficits = JJA). Low frequency-high magnitude precipitation events are projected to increase with continued greenhouse gas emissions. For instance, the 20-year annual maximum precipitation event is projected to increase everywhere across western North America (by as much as 20–30% by end-century for RCP8.5), with corresponding reductions in the return period of the historical event (to as low as 5 to 10 years by end-century for RCP8.5) (Kharin *et al.*, 2013). Relative changes in extreme precipitation events are expected to exceed changes in annual mean precipitation in western and eastern Canada (Kharin *et al.*, 2013; Sillmann *et al.*, 2013; Ouranos, 2014).

For example, in Quebec, an increase in the maximum annual quantity of precipitation for every duration and return period is projected. The 20-year return period of the annual maximum daily precipitation for the period between 1986 and 2005 could occur more frequently and have a 7 to 10-year return period around 2046–2065. Models agree strongly on the increase of high and extreme precipitations (Ouranos, 2014).

3.2.4 Hydrology

Changes in temperature and precipitation due to climate change are expected to impact hydrological processes. For instance, snowmelt-driven regimes will experience changes in the dynamics of snow accumulation and melt, which will result in increased winter runoff and an earlier spring freshet; hybrid systems (mixed-rain snow) will become increasingly rainfall driven, with runoff shifting from spring to winter; and rainfall-driven systems will respond predominantly to changes in precipitation, which will likely vary regionally and locally (Whitfield *et al.*, 2002; Elsner *et al.*, 2010; Shrestha *et al.*, 2012; Schnorbus *et al.*, 2014).

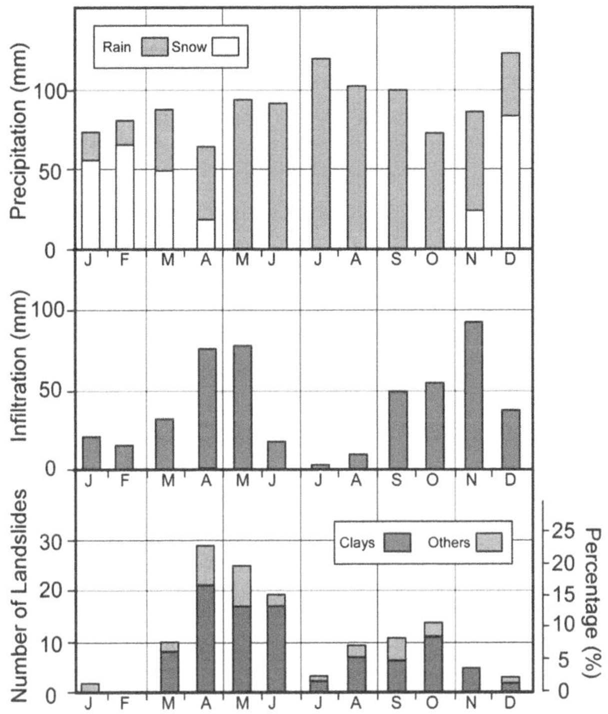

Figure 3.7 Relationship between landslide activity in the St. Lawrence Lowlands as a function of precipitation and evapotranspiration average monthly on the period 1971–1975 (modified after Lefebvre and Lafleur, 1978).

For eastern Canada and the St. Lawrence Lowlands in particular, the maximum occurrence of landslide activity is in the spring when the snowmelt is added to the spring rain and the evapotranspiration is at a minimum (Fig. 3.7). Hydrological changes could modify the timing of landslide events.

Although overall changes in hydrology and streamflow are informative for understanding impacts, changes in hydrologic extremes are most relevant for understanding potential changes in debris flow and debris flood activity. The tendency for flooding is affected by precipitation, temperature (snow and ice melt), and antecedent conditions, such as snow cover and soil storage. In large river basins in western Canada, flooding is typically a result of snowmelt, often in conjunction with long-lasting or intense rainfall. In smaller basins throughout the region, flooding is more often due to short-duration high-intensity rainfall events. Consequently, projected increases in both temperature and precipitation extremes imply that regional scale changes in the flood frequency distribution should be expected (Seneviratne *et al.*, 2012; Kundzewicz *et al.*, 2013). In small, steep creeks, debris floods and debris flows are the dominant hazardous process and an increase in debris flow frequency can be expected in coastal B.C. largely due to higher rainfall amounts and intensities, particularly during the late fall and winter season (Jakob and Lambert, 2009).

It seems reasonable to posit, where an increasing proportion of winter precipitation is expected to fall as rain, and where winter precipitation extremes are also expected to increase, the potential for localized rainfall-driven winter flooding will increase. This is supported by recent GCM experiments that indicate, as a result of increasing temperature and precipitation, extreme (considered as future events that exceed historical seasonal maximums) snowmelt runoff and total runoff during winter are projected to increase in western Canada (Diffenbaugh *et al.,* 2012). The future occurrence of large rain-on-snow flooding events may also change, although this will be complicated by counteracting trends of increasing precipitation extremes, an increasing proportion of rainfall, and decreasing snow occurrence and persistence (*i.e.* increasing snow line elevation). Under projected climate change conditions the rain-snow transition zone is expected to move up in altitude throughout the region (Shrestha *et al.,* 2012; Klos *et al.,* 2014; Schnorbus *et al.,* 2014). There is currently very little information on the potential changes to more localized summer rainfall-driven flooding in western Canada. Change in the magnitude and frequency of such events will depend primarily on changes in extreme summer precipitation, which is projected to be less pronounced and less certain than changes in winter (Sillmann *et al.,* 2013).

Annual maximum flood magnitudes in snow-dominated regimes in western North America (Yukon, Mackenzie, Columbia and Nelson rivers) are also expected to increase in the future (*e.g.* Hirabayashi *et al.,* 2013). Results from more regionally focused hydrologic modelling studies generally support this conclusion (*e.g.* Werner *et al.,* 2013).

While climate-related changes to flooding are not the focus of this study, flooding can affect landsliding due to undercutting of unstable banks, or due to rapid drawdown following floods. The latter process has been well-studied along the Thomson River (Porter *et al.,* 2002; Eshraghian *et al.,* 2007, 2008).

In southern Québec, a diminution of the snow cover along with a snow maximum reached in February should diminish spring freshet of about 10% (Ouranos, 2014). Throughout Québec, different hydroclimatic projections forecast an increase in winter river flow for the 2041–2070 period. On the other hand, in southeastern Canada, the mean flow is forecasted to decrease during summer, spring and fall. These changes will most likely influence riverbank erosion, which is a main landslide triggering factor. Therefore, more landslides are expected in the winter months.

The Atlantic region is affected each year by large storms passing through and producing powerful storm waves. The increase in sea level caused by global warming in conjunction with the projected larger and more frequent storms will likely increase the risk of erosion on soft sediment shores and of storm surge, putting low-lying coastal areas at risk of flooding (Vasseur and Catto, 2008). This effect is expected to be enhanced at the Arctic cost due to increase in open water season and enhanced impact from wave action on warming permafrost.

3.3 ILLUSTRATION OF POTENTIAL IMPACTS OF CLIMATE CHANGE ON SLOPE STABILITY

Slopes respond variably to changes in their environment and the result can be expressed, as suggested by Crozier and Preston (1999), as the landslide triggering threshold (LTT)

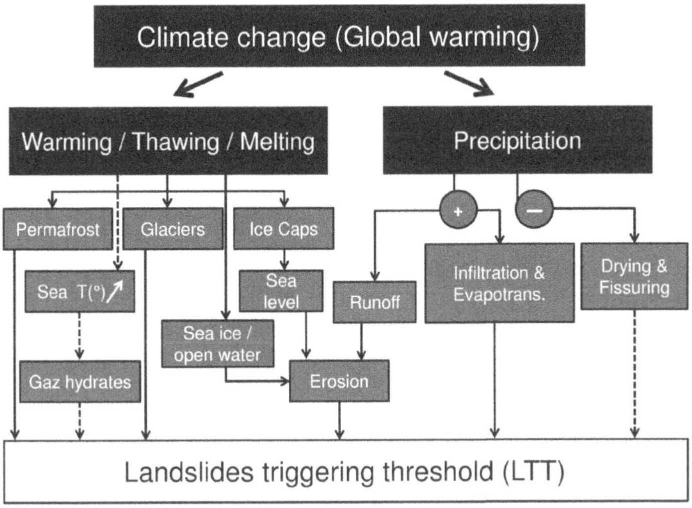

Figure 3.8 A conceptual view on how a warming climate can induce an increase in landslide triggering threshold. Arrows indicate the various flow of consequence with dashed arrows suggesting a less significant link.

which represents the stability condition of a natural slope at a given time (Fig. 3.8). The LTT is modified by various natural (*e.g.* climatic) or human-induced (*e.g.* deforestation) forcing. As illustrated in Figure 3.8, it is expected that an increase in temperature will have an impact on two major components of the climate: warming (degradation of permafrost, melting of glaciers and ice caps) and precipitation. Concerning precipitation, the effects are not always straightforward; for example, climate warming may in some regions increase the amount of precipitation, but associated with an increased evapotranspiration due to warming, it may lead to desiccation. The slope responses can also be directly linked, *e.g.* if climate changes lead to more frequent exceedances of hydroclimatic thresholds triggering debris flows. In the case of marine environments, an increase in sea water temperature with a small increase in sea level could lead to gas hydrate dissociation and unstable slopes (Lee, 2009). Slope can also respond to complex feedback mechanisms involving progressive destabilization, such as creep derived loading, or irreversible fracture widening.

In some cases, slopes are expected to become more stable under climate change, for example where morainal slopes become vegetated, or formerly outburst and debris-flow producing proglacial or subglacial lakes disappear in conjunction with glacial recession or glacial disappearance. For example, the 2012 ice avalanche and subsequent outburst flood of Ghost Glacier (Quinn *et al.*, 2014), which will be detailed later in this report, will likely not reoccur as almost the entire glacier detached and may not reform under projected climate conditions. In this case, the LTT would be very high for this type of mass movement. While such cases exist, they are the exception rather than the rule (Stoffel *et al.*, 2014).

Interestingly, in the few studies looking at landslide triggers and climate change, the role of human activity may often be seen as having a stronger effect on landslide activity than climate change alone. In support of this is the remark by Jakob and Lambert (2009) on debris flow occurrences along the Pacific coast where the debris flows may increase by a factor of about 1.3 over the next fifty years due to climate change while in areas of clear cutting it is already increasing by a factor of 6 to 9. A similar observation has also been made by Crozier (2010) in New Zealand. In Québec, it has been shown that about 40% of the reported landslides are triggered by human activity (Demers *et al.*, 2008). This may also help providing some perspective on the potential effect of climate change on landslide activity relative to other causes.

Aside from future anthropogenic changes, such as clear cutting and urban expansion, land cover changes (biospheric and cryospheric) associated with climate change are also occurring and expected to continue. For forests, these changes include an upward and northward migration of treeline in cold areas, and forest cover loss due to increased wildfires, and insect and disease outbreaks. Grasslands are expected to expand upward and northward at the expense of forests in warm and dry areas. Alpine glaciers are projected to continue thinning and retreating, while frozen ground is projected to continue its path of degradation as active layers are likely going to thicken. Both glaciers and vegetation can interact with permafrost in complex ways (Vitt *et al.*, 2000; Etzelmüller and Hagen, 2005; Fischer *et al.*, 2013). Biospheric and cryospheric changes in land cover can affect thresholds for landslide initiation. Those effects and how professionals ought to respond to them have been enshrined, for example, in the B.C. Guidelines for Flood Assessments in a Changing Climate (APEGBC, 2012).

3.4 LANDSLIDE RESPONSE TO WARMING

Air temperature increases ultimately result in two distinct morphodynamic responses in mountain environments (Deline *et al.*, 2015). The first of these is glacial thinning and glacier retreat. The second is difficult to recognize, because it describes a thermal state rather than a landform change: ground (permafrost) temperature warming. This warming causes permafrost degradation which can result in a change in landform.

3.4.1 Glacial conditioning

Glaciers play an important role in conditioning mountain slopes for mass movement (Geertsema and Chiarle, 2013). Slope stability in mountainous terrain is strongly influenced by the advance and retreat of glaciers (Evans and Clague, 1994; Haeberli *et al.*, 1997) given that glaciers effectively buttress soil and rock. Upon their retreat, the buttressing effect diminishes or disappears and slope relaxation and adjustments occur. Recently deglaciated areas are hotspots for a variety of landslide types in Canada (*e.g.* Bovis, 1982; Holm *et al.*, 2004; Geertsema *et al.*, 2006) including landslides in soil and in rock.

Glaciers erode, redistribute and deposit sediments. Upon retreat and thinning, glaciers expose both scoured bedrock and glaciogenic sediment (Fig. 3.9). Exposed soils can range from thin veneers over bedrock to complex sediment assemblages tens of metres thick. The unvegetated, and often still ice-cored sediments in the alpine belt, are subject to a variety of hillslope processes (Ballantyne, 2002) and can be readily

Figure 3.9 Thinning and retreat of Llewellyn Glacier in northern B.C. (Clague *et al.*, 2010) has exposed steep rock and soil-covered slopes, with two trim lines (arrows). Note the variable size of gullies, related in part to the age of surface exposure. The oversteepened slope below the lowest trimline is especially at risk for gully erosion, as is demonstrated in the inset photo (Geertsema and Chiarle, 2013) (Photos by M. Geertsema).

mobilized (Chiarle *et al.*, 2007). Where unconsolidated and cohesionless sediments are exposed on steep slopes, debris flows are the primary agents of sediment transport (Ballantyne, 2002). Bulking of sediment in gullies and along channel side walls often result in long runout debris flows. While intense rainstorms and rapid snowmelt are the principal triggers of debris flows, sediment supply on relatively young steep slopes, and the lack of vegetative cover govern the magnitude (volume and peak discharge) of debris flows. This characteristic differentiates debris flows from recently deglaciated areas apart from other landslide prone regions worldwide (Geertsema and Chiarle, 2013). For example, in a survey of 19 alpine basins in southern B.C. Coast Mountains, Holm *et al.* (2004) found that debris movements were concentrated along Little Ice Age (LIA) lateral moraines and trimlines. Once mature vegetation invades these slopes, their stability increases in response to a changing hillslope hydrology (canopy intercept and evapotranspiration changes) and increasing root soil strength.

Both glaciation and deglaciation condition rock slopes, making them prone to landslides (Figs. 3.10 and 3.11). This happens in four steps: (1) through deepening, widening and, importantly, oversteepening of valleys (glacial erosion), (2) ice loading

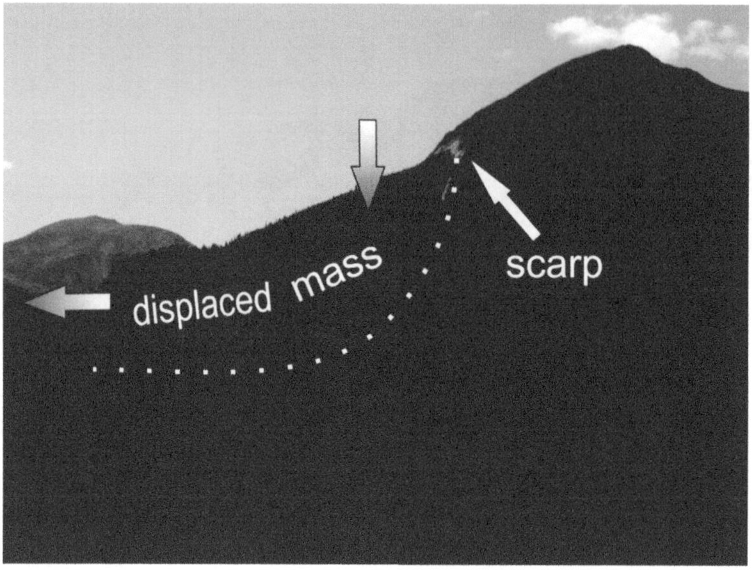

Figure 3.10 Rotational rock slide of mountain side as a result of late Pleistocene glacial debuttressing at Nevis Creek, northern Rocky Mountains, B.C. (Geertsema and Chiarle, 2013). Red arrows represent displacement vectors. View is to the South (Photo by M. Geertsema).

resulting in the development of stress fractures, (3) debuttressing (during deglaciation), and (4) stress release as manifested by joint expansion (Geertsema and Chiarle 2013). Landslides resulting from glacially conditioned bedrock include rock slides, topples, and falls, and slow deep-seated slope deformation. Both rock fall and slow rock slope deformation and sagging can be precursors of highly mobile rock avalanches (Fig. 3.11). The spatial and temporal association of rock landslides with previously glaciated areas suggests that glacial conditioning of rock slopes plays an important role in their occurrence. Many post LIA rock avalanches in Canada have occurred on rock slopes above glaciers (Evans and Clague, 1994; Holm *et al.*, 2004). Two thirds of the large rock avalanches in northern B.C. that occurred between 1973 and 2003 (Geertsema *et al.*, 2006) occurred on LIA cirque walls.

Landsliding due to increases in temperature and associated with glacial conditioning can be indirect as was evidenced by two recent events, the Ghost Glacier (Quinn *et al.*, 2014) and the Cathedral mountain ice avalanche and debris flood events (Arenson *et al.*, 2015) where warming temperature acts as a trigger of a natural hazard (ice avalanche) which in turn triggers a mass movement which causes consequences.

The first event was the ice avalanche that fell from Ghost Glacier on Mount Edith Cavell in Jasper National Park during the night of August 9th, 2012 (Quinn *et al.*, 2014). This event displaced water from the glacial lake at the foot of the rock face, which in turn caused an outburst flood that led to significant damages to the park's facilities (Fig. 3.12). Had the event occurred during day-time hours, numerous park visitors would have been exposed to the outburst floods and the trail paralleling the

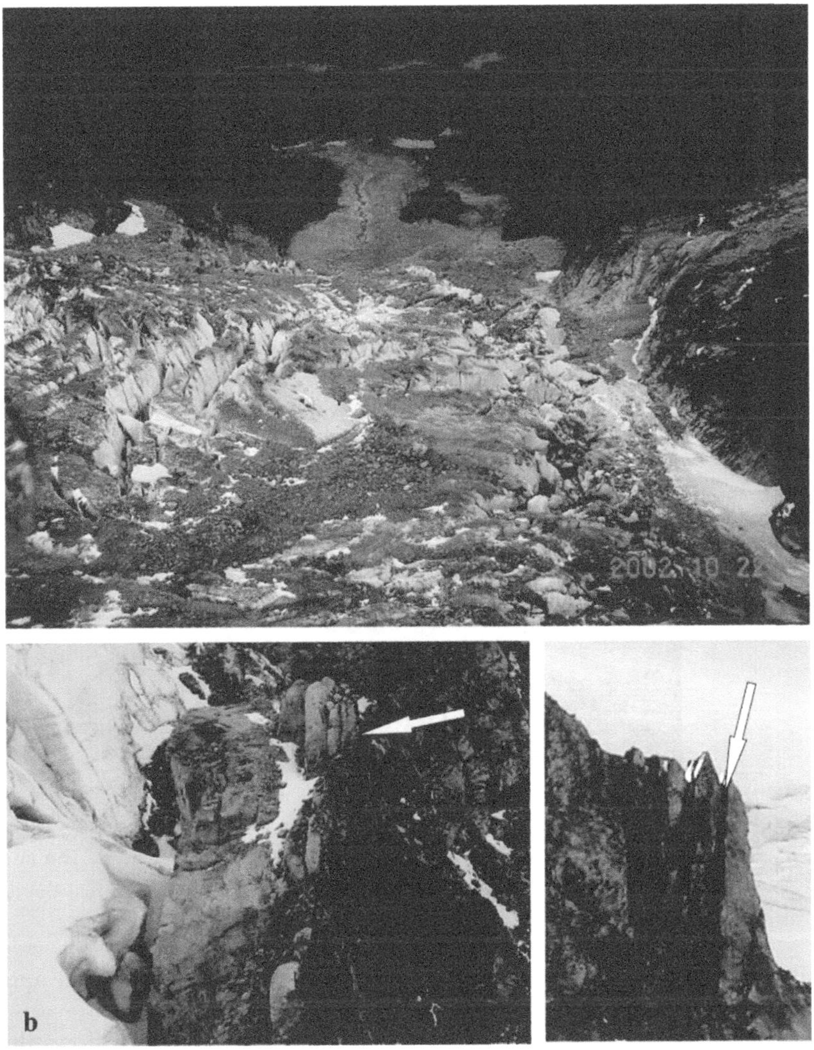

Figure 3.11 The two lower pictures (b, c) are showing joint expansion at the main scarp (arrows) of the 2.7 km runout 1999 Howson rock avalanche (a) in B.C.'s Telkwa Mountains (Geertsema and Chiarle, 2013) (Photos by M. Geertsema).

creek, probably leading to several fatalities. The majority of Ghost Glacier has disappeared and it is not known whether the glacier will re-build because of the projected increase in temperature and decrease in precipitation in this region of the Canadian Rocky Mountains.

The second event occurred at 06:30 on July 10, 2014. Approximately 100,000 m^3 of ice collapsed from Cathedral Glacier, south central B.C. allowing the supra-glacial lake to partially drain. The ice and water descended Cathedral Gulch and transformed

Figure 3.12 The western flanks of Mount Edith Cavell with the remnants of Ghost Glacier in the background near the top of the image. Arrow indicates failure zone. The inset image shows impacted park's facilities (Photo source: Parks Canada).

into a debris flow through entrainment of channel sediments that buried the uppermost CP tracks (Arenson *et al.*, 2015) (Fig. 3.13).

Historically, Cathedral Gulch has witnessed several large debris flows that have buried or eroded one or more levels of the CP railway during the summers of 1925, 1946, 1962, 1978, 1982, and twice in 1984. Of these seven events, five crossed at least two levels of CP track, and two crossed all three (Jackson *et al.*, 1989). In order to reduce the hazard from the supra-glacial lake creating a catastrophic glacier outburst flood, lake pumping and lake drainage began in 1985 but was stopped in the early 2000s, allowing for the most recent events to occur. Compared to similar small steep watersheds in the Rockies, the debris flows at Cathedral Gulch tended to be larger and more frequent. This increased activity is caused by the storage of water within the closed depression, the unstable condition of the Cathedral Glacier, and loose glacial and colluvial sediments.

It is yet unclear if the hazard of outburst floods will decrease with glacial shrinking and mass loss due to temperature increases. With a decrease in the elevation of the glacial ridge separating the supra-glacial lake from the gulch below, less water will

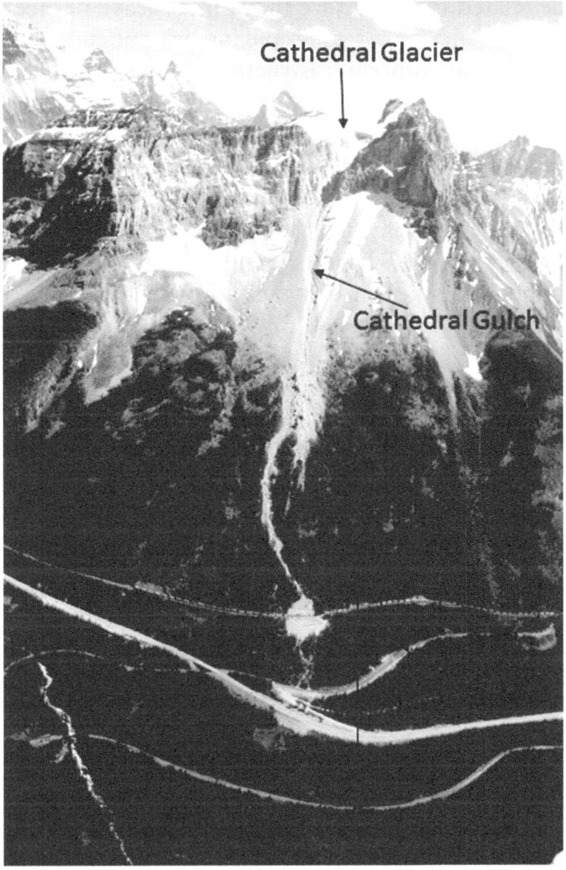

Figure 3.13 Overview of Cathedral Gulch showing the debris flow track and deposit, two of the three CPR tracks and the Trans-Canada Highway (1). The glacier is visible just above the Gulch (Photo by L. Arenson).

be able to accumulate in the lake thus, possibly decreasing the likelihood of repeated failure and high magnitude. However, the degrading permafrost in the source area of debris flows in the upper watershed may open the door to a new hazard as previously frozen talus slopes begin to thaw and lose their strength.

Nonetheless, from the cases of Cathedral Mountain and Ghost Glacier it appears that impacts of climate change may have contributed to the failure, but, over the long-term may decrease the prevailing hazard or even make it impossible to reoccur. In consequence, historic (non-)events may not be representative for future slope instabilities.

3.4.2 Permafrost degradation and landslides

The existence of perennially frozen ground (permafrost) can result in a unique suite of slope stability challenges. Slope stability issues associated with frozen ground are manifested in two main ways: on-site and off-site (Gruber *et al.*, 2015). On-site includes ground subsidence related to ground-ice thaw, frost heave or creep. In gentle terrain, on-site analysis is important for assessing the integrity of embankments, foundations, tailings dams, pipelines, and other infrastructure (Bommer *et al.*, 2010). This includes the effects of low temperatures and a permafrost foundation on waste rock piles (Pham *et al.*, 2013) or heap leach facilities (EBA 2011), amongst others.

Problems with off-site permafrost, such as process chains linked to the degradation of distant permafrost (Geertsema and Clague 2011), are more difficult to characterize and predict because more variables interact. This could be as simple as a rockfall debris cone (Gruber and Haeberli, 2007), or as complex as a rock slide transforming into a debris flow and impounding a river downstream, which in turn, could precipitate an outburst flood with further downstream consequences (Huggel *et al.*, 2012; Evans and Delaney, 2015).

Landslides in permafrost are becoming increasingly common. Two examples of talus moving on ice-rich permafrost at Pink Mountain (Geertsema and Foord, 2014), B.C. and in Nahanni National Park, Northwest Territories (Jermyn and Geertsema 2014) are shown in Figures 3.14a and b. At Pink Mountain, the landslide was a discrete event, but in Nahanni Park the movement is ongoing. In some cases, permafrost-related landslides retrogress upwards and impact infrastructure (Gruber *et al.*, 2015) as well sites which were impacted by retrogressive landslides along Temple Creek (Fig. 3.14c) in northern B.C., in an area where permafrost was present (Geertsema and Foord, 2014).

Sometimes there are synergies between disturbance agents in permafrost terrain. Forest fires can reduce the thickness of insulating mosses leading to active layer thickening, thaw consolidation, and additional water along the upper boundary of ice-rich permafrost leading to retrogressive thaw flows as shown in the Mackenzie River area in Figure 3.14d. Thaw slumps are now a dominant driver of landscape changes in the Arctic (Kokelj *et al.*, 2015). They occur along river, lake, coast and valley flank where ground ice is exposed. They are active during summer and retrogression rates ranging from 6 to 15 m/yr were measured (Wang *et al.*, 2016). As permafrost thaws, slices of material fall in the scar. Debris accumulation in the scarp can protect the slope from further thermal degradation and lead to stabilization. However, the failed material can flow out of the scarp permitting the process to retrogress. Increase in rainfall will cause the debris to flow out of the scarp and thus accelerates slump activity, which was shown by Kokelj *et al.* (2015).

Figure 3.15 shows the scar of a debris flow with exposed ground ice at the frontal area of a rock glacier in the south-eastern Rocky Mountains of Alberta (Arenson and Jakob 2014). Rock slides in northwestern B.C. (Fig. 3.16) transformed into far travelling debris flows and ruptured natural gas pipelines (Geertsema *et al.*, 2006).

Off-site permafrost-landslide hazard assessments are challenging because large areas must be assessed for geohazard identification, leading to important uncertainties (Gruber *et al.*, 2015) *e.g.* a permafrost-related landslide could come from a variety of locations in a catchment up to tens of square kilometres in size. Once a landslide occurs, other factors control whether it might reach infrastructure in the valley, whether it might be of sufficient magnitude and intensity to be damaging, and which several process chains might be involved.

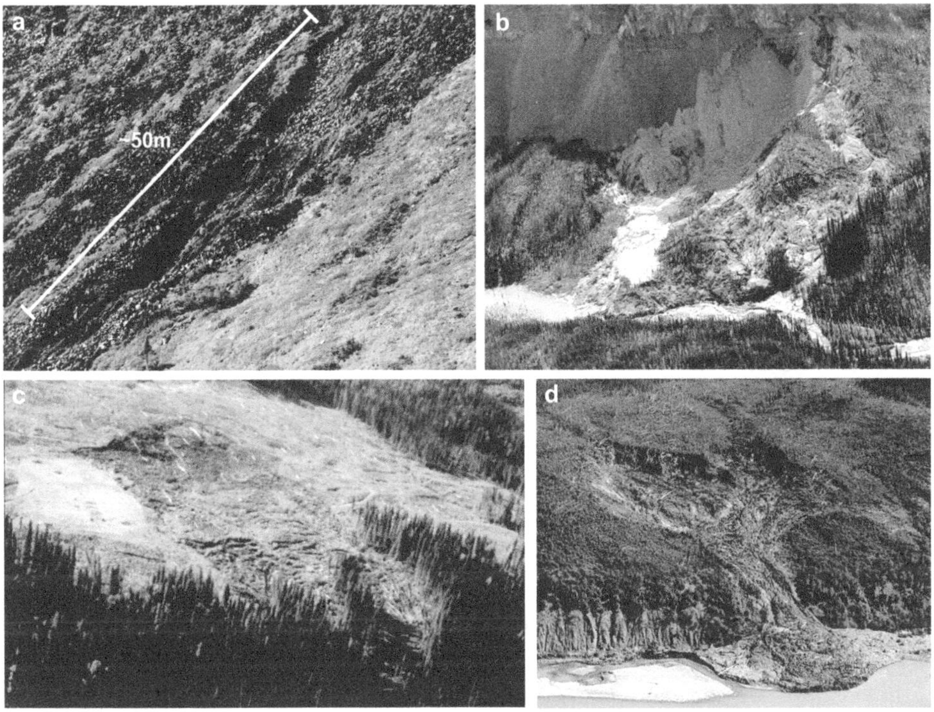

Figure 3.14 a) Active layer detachment slide moved on ice-rich permafrost, 3 m below the ground surface at Pink Mountain, B.C., fall 2008. b) Complex failure of an ice-rich talus apron, Nahanni National Park, Northwest Territories (NWT). c) Temple Creek landslide (> 12 ha) in northeastern BC. It occurred on a slope on 3.5° in clayey soil. Although not observed in the landslide, there is sporadic permafrost in the area. The landslide may have resulted as a hydrological response to timber harvesting. d) Retrogressive thaw flow in permafrost terrain, Wrigley area, NWT. This landslide occurred in an area burned by a large forest fire, which likely reduced the thickness of insulating mosses and in turn, would have allowed the active layer to thicken (Photos by M. Geertsema).

3.4.3 Ice cover reduction, coastal erosion and landsliding

Global warming impacts coastal dynamics by increasing the relative sea level, reducing the ice cover, increasing the amount of storm and accelerating cryogenic processes (Bernatchez and Dubois, 2008; Bernatchez *et al.*, 2011, Bernatchez *et al.*, 2015). In the St-Lawrence gulf and estuary, Senneville *et al.* (2014) projected that the ice cover will occur by a 10 to 20-day delay, while the river could be free of ice 20 to 30 days earlier for the 2041–2070 period, compared to the 1982–2011 period (Ouranos, 2014). Since the ice cover protects the coast from erosion associated with storms, its reduction could increase coastal erosion. Because toe erosion on coastal slopes is an important slope destabilizing factor, climate change has the potential to affect slope stability along Canadian coastlines in soils or weak rocks (*e.g.* Demers *et al.*, 2014).

Figure 3.15 Slump scar at frontal area of an unnamed rock glacier in the Cougar Creek watershed near Canmore, Alberta. Note the depth of active layer displayed above the wetted horizon indicating permafrost at depth (Photo by M. Jakob, July 2013).

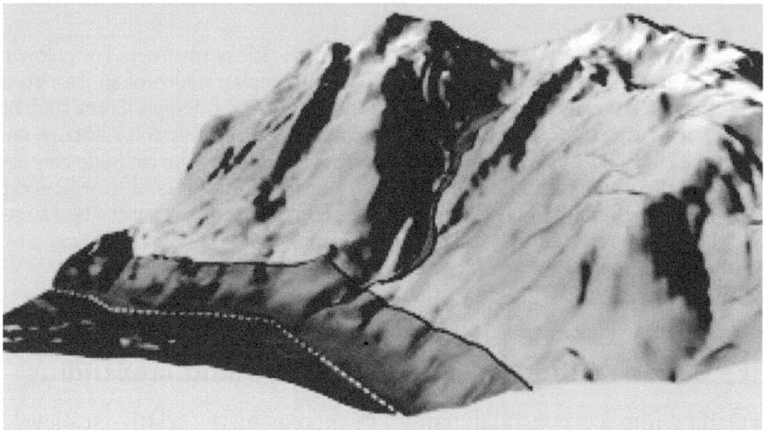

Figure 3.16 The 2006 Zymoetz landslide (red) was initiated as a rock slide in permafrost-underlain terrain, transformed into a 4 km long debris flow, and ruptured a pipeline (dotted yellow line) in the valley bottom (Geertsema and Clague, 2011).

One example for the above process is the Havre-St-Pierre landslide in sensitive clays that occurred in April 1980 and affected the only road reaching the town of Havre-St-Pierre and other communities of Québec's north-shore. The landslide involved more than 1 million m^3 of marine sensitive clay (quick clay), sand and peat originating

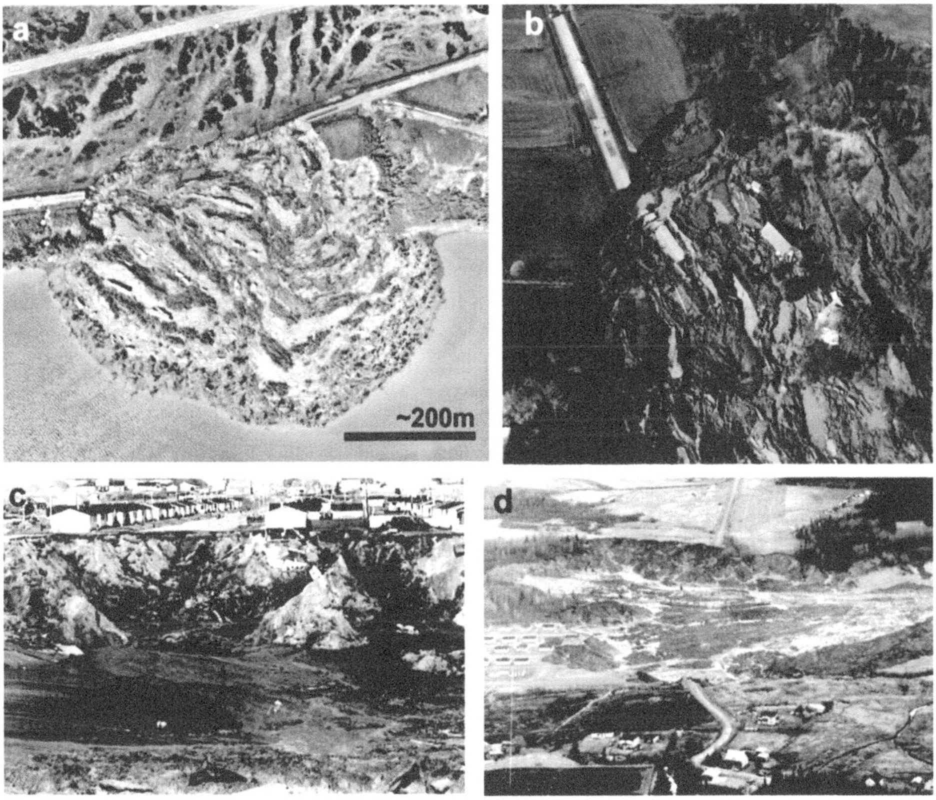

Figure 3.17 a) Aerial photo of Havre St-Pierre lateral spread that cut the national road (Locat *et al.*, 2014). b) Aerial photo of St-Jude lateral spread shortly after its occurrence. Note the house, garage, and road in the debris. c) (Photo source: Gouvernement du Québec) and d) St-Jean-Vianney 1971 earthflow (Photo source: Canadian Armed Forces). c) View of the 25 m high backscarp with the village atop where 200 houses were moved to safer ground in the months following the tragedy. d) Overview of the landslide scar. The landslide destroyed the road in the centre of the picture.

from a 23 m high coastal bluff (Locat *et al.*, 2014). The 410 m wide landslide had retrogressed up to 110 m and spread over the tidal flat (Fig. 3.17a).

In Canada, marine clays were deposited in post-glacial seas in B.C., Ontario, Quebec and Labrador-Newfound Land and are now above sea level (Hungr and Locat, 2015). When the salt is leached from the porewater in these soils, it increases their liquidity index, leading to low remolded undrained shear strength. Eighty-nine percent of Quebec's population is settled within post-glacial marine sedimentary basins (Demers *et al.*, 2014; Locat and St-Gelais, 2014) making this province particularly susceptible to quick clay landslides. Landslides in sensitive clays occur rapidly (within minutes) (Locat *et al.*, 2011), so they can be a threat to human life. Indeed, fatal large landslides in sensitive clays have marked in Canadian history, *e.g.*, the 2010 St-Jude lateral spread (Fig. 3.17b) (Transports Québec, 2011) which took the lives of a family of four and the 1971 St-Jean-Vianney earthflow (Fig. 3.17c, d) (Potvin *et al.*, 2001) which killed 31 people in less than 5 minutes.

These two landslides occurred on slopes along rivers, thus governed by a different hydrological dynamic than the coastal Havre-St-Pierre landslide. At St-Jude, the erosion of the slope is pointed out as one of the factors leading to the slope's destabilization, along with a high artesian groundwater pressure (Transports Québec, 2011). The projected climate change will increase slope erosion along marine environments, but the effect is not as clear for rivers. In Canada, the warming should increase winter flow in rivers (increasing erosion), but will reduce the spring flow (reducing erosion) (Boyer *et al.*, 2010; CEHQ, 2013). As not only erosion but also rapid drawdown after flood events can destabilize slopes (Eshraghian *et al.*, 2007), the relationship between climate change, hydrological cycle and landslides is difficult to establish. Moreover, river erosion is not a direct parameter of climate change and is very site-specific, and characterized by high spatial variability (Walling, 2009) making generalization very difficult.

3.5 LANDSLIDE RESPONSE TO INCREASED PRECIPITATION

Depending on the nature of the material, depth, and characteristics of weak zones in soil or rock, slope geometry, geological setting, and other factors, landslides can have rapid or delayed responses to changes in precipitation pattern. Threshold pore pressures develop more rapidly in shallow soils and low permeability substrate (high fines content) to result in a debris avalanche or debris slide than deeper materials with higher permeability. Both rapid and delayed landslide responses to climate occur in Canada given the vastly different climate zones and regional morphologic and geological differences.

Landslides that respond rapidly to changes in precipitation include shallow debris slides, debris flows, and rock fall. In this context "shallow" is defined as typically less than 2 metres thick. Landslides with delayed responses to changes in precipitation and temperature tend to be larger, deeper-seated rock and earth slides, slumps and earth flows.

3.5.1 Rapid response landslides

Shallow debris slides, flows and avalanches are typically associated with heavy and often prolonged rainfall, aided to a significant degree by high antecedent moisture conditions. Hogan and Schwab (1991) studied rainfall characteristics before and during verified shallow landslide events in north coastal B.C. They compared the temporal frequency of slope failures to precipitation trends for antecedent day to year time scales. Over decadal time scales, a positive correlation was found between annual moisture conditions and reported landslide frequency. Nonetheless, annual moisture had a lower statistical significance than short term precipitation (months or days) preceding a slope failure. On the shorter time scale, only the months immediately preceding landslides were important in conditioning the hillslope to failure. Similar analysis by Jakob and Weatherly (2003) and Jakob *et al.* (2011) on the B.C. south coast shows that 4-week antecedent rainfall amounts are critical in conditioning previously glaciated soils towards a state that facilitates shallow debris slides and debris avalanches. They also found that this corresponds roughly to some 120 mm of rain measured at low elevations and proportionally higher rainfall with increasing elevation.

Septer and Schwab (1995) and Schwab (1997; 1998) used tree ring dating to shed some light on the ages of shallow landslides along the B.C. north coast and Haida Gwaii back to the early 1800s. One of the conclusions of this work is that most landslides occurred during few very heavy regional storms. Four to six major storms since 1875 moved some 3/4 of the volume transported in various watersheds. Other recent landslide-triggering storms occurred in 2003 and again in 2004. The October 25, 2003, event triggered many large landslides, with 12 reported in the vicinity of Prince Rupert alone. Landslides on November 4, 2004, severed a natural gas pipeline serving Prince Rupert and closed the only highway link to the city. Similar storms have generated hundreds of landslides on the south coast of B.C. For example, on August 29, 2013 more than 260 landslides occurred in Knight Inlet after a convective weather system stalled over the area (Tom Millard, pers. comm.).

Under projected climate change scenarios, return periods of high precipitation events should be reduced and because extreme precipitation events are a landslide triggering factor, it could increase landslide frequency. In the Saguenay-Lac-St-Jean region in the province of Québec, such an event occurred in 1996 and caused tremendous damage. On July 19th and 20th between 170 and 200 mm of rain fell on this region in a 36 h period and 250 mm in 48 h. Groundwater levels were already high due to antecedent rainfall. The abnormal rise in water levels caused subsequent flooding in adjacent areas, several bridges were damaged or completely destroyed, and in some areas water bypassed dams and cut new adjacent channels (Fig. 3.18). The rains were

Figure 3.18 Damage and erosion caused by the river bypassing a dam during the 1996 Saguenay flood (Photo source: Natural Ressources Canada).

the primary trigger for more than a thousand landslides over 350 km^2, but the exact number of landslides is unknown and could be greater (Perret and Bégin, 1997). In fact, debris flows in the surrounding mountainous regions and the landslides that failed in the weeks following the precipitation are excluded from the inventory (Ministère des Transports Québec, 1999). The vast majority of landslides occurred in clayey soils and most were superficial (~90%) (Ministère des Transports Québec, 1999). Infiltration of water into slopes was the triggering factor of superficial landslides, because erosion was very minor or non-existent. These extraordinary floods were also responsible for active erosion, with some river beds being modified and sometimes completely shifting out of their beds, that triggered deep seated landslides (Ministère des Transports Québec, 1999).

These observations have very important implications with respect to the effects of climate change in this part of Canada. Previous studies have shown that an ideal combination of antecedent moisture conditions and heavy prolonged rain are the most likely trigger of abundant and widespread shallow landsliding. This motivates the question if one can expect (1) changes in the frequency of storms, (2) changes in the volume of rain associated with storms, and (3) changes in the intensity of storms or (4) a combination of those factors. The interaction of those variables will ultimately determine if landslides become more frequent and/or larger.

Extreme precipitation is also key in triggering debris flows, especially when antecedent moisture conditions are high. For example, in a study of climate change effects on landslides along the SW coast of B.C., Jakob and Lambert (2009) demonstrated that debris flows are expected to increase in frequency in accordance with a 10% increase in the 4-week antecedent moisture conditions and a 6% expected increase in short-term (24-hr) precipitation by the end of the century. This work was based on daily data from the Canadian Centre for Climate Modelling and Analysis (CCCma) model. The monthly mean precipitation data were extracted from an ensemble of 19 of the world's global climate models using three different levels of prescribed greenhouse gas concentrations during the 21st century.

Figure 3.19 shows the dependency of debris flow occurrences in the study area with antecedent and short-term precipitation and regional thresholds above which shallow landsliding and debris flows are more likely. The projected increase in antecedent and short-term rainfall conditions will shift more precipitation events above those thresholds thus leading to an increase in debris-flow activity. Note that since this work was completed in 2009, five more debris flows have occurred on Howe Sound and the North Shore Mountains.

3.5.2 Delayed response landslides

Deep-seated landslides tend to respond more slowly to increases in precipitation and snowmelt (Geertsema *et al.*, 2010) because of the time lag for water to infiltrate and create sufficiently high pore water pressures in potential slide masses. Geertsema *et al.* (2006, 2007) studied large landslides in northern B.C. where large landslides (greater than 0.5 Mm^3) between 1973 and 2003 have increased in frequency from 1.3 per year over the 30-year period to 2.3 per year in the last decade of that period. While a few of the soil landslides were triggered by individual storms, most responded to longer periods of above-average precipitation. Large sensitive clay landslides in the Mink Creek and Khyex River valleys (Schwab *et al.*, 2004) near Terrace, B.C. occurred

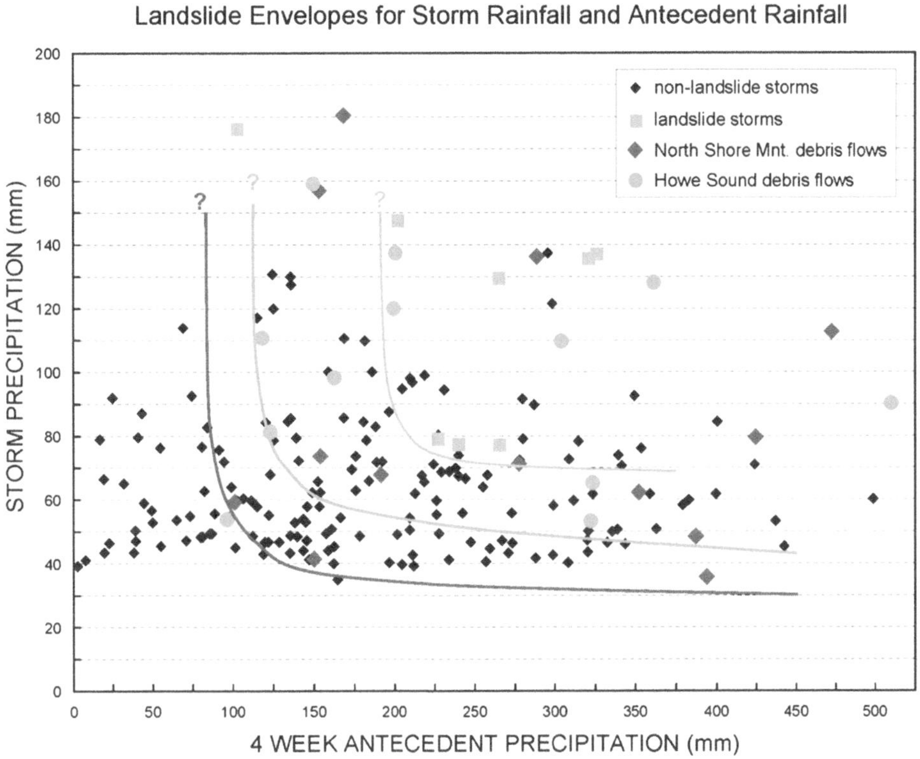

Figure 3.19 Landslide envelopes for 24-hour storm rainfall and antecedent rainfall for urban landslides on Howe Sound and the North Shore Mountains, located north of Vancouver, B.C. (modified from Jakob and Lambert, 2009).

after decade-long wet periods. More than one third of Holocene flowslides in the Terrace area fell within a wet millennium some 2–3 thousand years ago (Geertsema and Schwab, 1997). This suggests long-term climate trends may influence these deeper landslides.

3.6 CONCLUDING REMARKS

For western and Arctic Canada, the actual knowledge points towards some link between ongoing climate change and landslides, and in most instances, the link between landslide activity and climate has been made on an empirical basis. In eastern Canada, apart from extreme events, evaluation of the impact of climate change on slope stability still needs to be carried out. Still, in most instances, the link between landslide activity and climate has been made on an empirical basis. The main points about slope safety preparedness with regards to climate change are: (1) understanding the current situation, (2) downscaling the climate model projections, (3) integrating relative impacts of human activity and (4) mitigating and adapting to potential or existing consequences. The following does not present a detailed analysis, but ideas at a glance.

3.6.1 The current situation

Landslide investigation depends on the actual risk associated to the event, so the current knowledge between the triggering factors, the response of soil/rock, and water condition are limited to few cases compared to the actual extent of landsliding in Canada. For example, how many sites have pore pressure readings at a rate high enough to record detail changes linked with precipitation? Linking antecedent precipitation and storm precipitation with a given slide is an empirical link, and does not provide the actual geotechnical information on how the mass reached its instability. To move forward along these lines, we need to set up a series of slope observatories that will help understanding the link with the actual climate to slope instability and the evolution of the landslide triggering threshold (LTT) in various geological and physiographic settings. This may be a very expansive operation and might be an opportunity for international cooperation.

It is also very relevant to understand how environmental and ground conditions evolve with landslide development. For example, if a glacier disappear it may no longer be a threat. If a till layer covering a glaciated bedrock surface is removed by a debris flow, the actual failure surface will not be replenished until there is another glaciation. In a sense, as mentioned by Crozier and Preston (1999), one must understand how landslide conditions develop and evolve with time, in addition to potential impacts from climate change. Such a condition is illustrated in Figure 3.20 showing that once the debris flow activity is increased by an increase in precipitation (lowering of the LTT shown by the dashed line), the slide intensity could be reduced because there should be a decrease in material availability in supply-limited channels (rise of the LTT shown by the hard line) (Bovis and Jakob, 1999).

Figure 3.20 may also be used to underline the very large amount of uncertainty related to such a schematic view, including: the lack of knowledge about the timing of the amount of precipitation providing the maximum debris flow activity, the removal rate, and actual value of the precipitation model itself. This LTT curve may in fact vary with time in such a way that the actual variability in the precipitation may in fact bring the LTT up again in some sort of cycle.

3.6.2 Downscaling the climatic models

As it may be the case elsewhere in the World, one of the main difficulty with projecting the impact of climate change on slope stability is to develop the capacity to project climate change at the scale of the slide of interest. In most cases, the actual scale of interest would require climatic 'cells' of less than 1 km^2. The downscaling may be more crucial with regard to precipitation than temperature since temperature is a more global parameter at a given site and can be adjusted to terrain conditions.

In addition, how can we project the future rate of climate change and what are the possibilities for the terrain to adjust to these changes thus preventing catastrophic events?

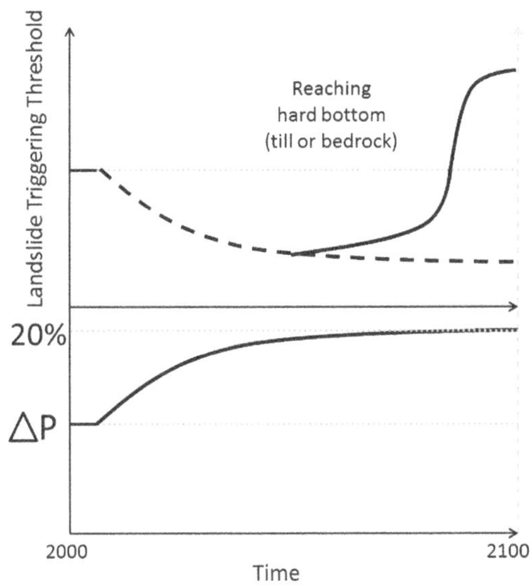

Figure 3.20 Schematic view of the possible evolution of the LTT for debris flows until year 2100 as a result of an increase in precipitation (ΔP) and as a function of removal rate of source material. Here it is assumed that the increase in precipitation would be about 20%.

3.6.3 Relative impacts of human activities

As it has been shown above and elsewhere, in inhabited areas, the direct impacts of human activity on landsliding may be more significant that what can be projected from climate change alone. This is a very important consideration that shall not prevent us from understanding how climate influences landslide activity but may help looking at adaptation technologies or practices that will reduce landslide hazard overall. Besides, in some areas, earthquake triggered landslides may be dominant but climate change may modify the sliding intensity and magnitude.

3.6.4 Adaptation and mitigation

Developing adaptation practices or technologies to reduce the potential impact of climate change on landslide activity will be a very challenging issue. In Canada, apart from the study of the impacts of warming of permafrost (Benkert *et al.*, 2012), there has been no real attempt to produce zoning maps which will take climate change into account. This suggests that adaptation technology may be easier to plan for counteracting warming effects of climate change on slope stability than adapting to an increase in precipitation. The latter will require much better and refined models of the link between landslide susceptibility and climate leading to changes in the design of drainage systems and slope protection measures. Adaptation efforts will likely be a function of the potential risk associated with landsliding so that active, but isolated areas in high altitude may not require the same efforts than in lowlands where the hazard may be less but the consequences higher.

About hazard mapping in quick clays, the Québec program considers landslide susceptibility and the retrogression potential for creating the maps which are not directly linked to climatic conditions. However, further research is needed to evaluate if this approach needs to be adapted for the impact of climate changes.

ACKNOWLEDGEMENTS

The authors would like to thank Andrée Blais-Stevens, Lukas Arenson, Dominique Turmel and Serge Leroueil for their comments on the report.

REFERENCES

APEGBC (2012) *Professional Practice Guidelines: Legislated Flood Assessments in a Changing Climate in BC*. 144 pp.

Arenson, L.U. & Jakob, M. (2014) Periglacial geohazard risks and ground temperature increases. *Engineering Geology for Society and Territory*, Vol. 1. Cham, Springer International Publishing. pp. 233–237.

Arenson, L.U., Bunce, C.M., Gauthier, D., Guthrie, R.H., Hutchinson, D.J., Jakob, M. & McGregor, C. (2015) Cathedral mountain 2014 ice fall and debris flow. In: *GEOQuébec 2015. 68th Canadian Geotechnical Conference, 7th Canadian Permafrost Conference*. September 20–23, 2015, Québec-City, QC, Canada.

Ballantyne, C.K. (2002) Paraglacial geomorphology. *Quaternary Science Reviews*, 21, 1935–2017.

Benkert, B., Kennedy, K., NCE Researchers & Kluane First Nation (2012) *Landscape Hazards in Yukon: Geoscience Mapping for Climate Change Adaptation Planning*. Available from: http://yukoncollege.yk.ca/downloads/2012_geoscience_poster.pdf.

Bernatchez, P. & Dubois, J.M.M. (2008) Seasonal quantification of coastal processes and cliff erosion on fine sediments shoreline in a cold temperate climate, Ragueneau Region, Quebec. *Journal of Coastal Research*, 24, 169–180.

Bernatchez, P., Jolivet, Y. & Corriveau, M. (2011) Development of an automated method for continuous detection and quantification of cliff erosion events. *Earth Surface Processes and Landforms*, 36, 347–362.

Bernatchez, P., Dugas, S., Fraser, C. & Da Silva, L. (2015) *Évaluation économique des impacts potentiels de l'érosion des côtes du Québec maritime dans un contexte de changements climatiques*. Laboratoire de dynamique et de gestion intégrée des zones côtières, Université du Québec à Rimouski. Rapport remis à Ouranos. 45 pp.

Bommer, C., Phillips, M. & Arenson, L.U. (2010) Practical recommendations for planning, constructing and maintaining infrastructure in mountain permafrost. *Permafrost and Periglacial Processes*, 21 (1), 97–104.

Bovis, M.J. (1982) Uphill-facing (antislope) scarps in the Coast Mountains, southwest British Columbia. *Geological Society of America Bulletin*, 93, 804–812.

Bovis, M.J. & Jakob, M. (1999) The role of debris supply to determine debris flow activity in southwestern BC. *Earth Surface Processes and Landforms*, 24, 1039–1054.

Boyer, C., Chaumont, D., Chartier, I. & Roy, A.G. (2010) Impact of climate change on the hydrology of St. Lawrence tributaries. *Journal of Hydrology*, 384 (1), 65–83.

Bush, E.J. (2015) Plenary 2: Climate change in Canada: Observations and projected changes. In: *The Proceedings of Mine Water Solutions in Extreme Environments, Vancouver, Canada*. pp. 6–19.

Bush, E.J., Loder, J.W., James, T.S., Mortsch, L.D. & Cohen, S.J. (2014) An overview of Canada's changing climate. In: Warren, F.J. & Lemmen, D.S. (eds.) *Canada in a Changing Climate: Sector Perspectives on Impacts and Adaptation*. Ottawa, ON, Government of Canada. pp. 23–64.

Centre d'expertise hydrique du Québec (CEHQ) (2013) *Atlas hydroclimatique du Québec méridional – Impact des changements climatiques sur les régimes de crue, d'étiage et d'hydraulicité à l'horizon 2050*. Québec. 51 pp.

Chiarle, M., Iannotti, S., Mortara, G. & Deline, P. (2007) Recent debris flow occurrences associated with glaciers in the Alps. *Global and Planetary Change*, 56 (1), 123–136.

Church, J.A., Clark, P.U., Cazenave, A., Gregory, J.M., Jevrejeva, S., Levermann, A., Merrifield, M.A., Milne, G.A., Nerem, R.S., Nunn, P.D., Payne, A.J., Pfeffer, W.T., Stammer, D. & Unnikrishnan, A.S. (2013) Sea level change. In: Stocker, T.F., Qin, D., Plattner, G.-K., Tignor, M., Allen, S.K., Boschung, J., Nauels, A., Xia, Y., Bex, V. & Midgley, P.M. (eds.) *Climate Change 2013: The Physical Science Basis. Contribution of Working Group I to the Fifth Assessment Report of the Intergovernmental Panel on Climate Change*. Cambridge, and New York, NY, Cambridge University Press.

Clague, J.J., Koch, J. & Geertsema, M. (2010) Expansion of outlet glaciers of the Juneau Icefield before and during the last two millenia. *The Holocene*. Available from: 10.1177/00959683609353433.

Colombo, S.J., McKenney, D.W., Lawrence, K.M. & Gray, P.A. (2007) *Climate Change Projections for Ontario: Practical Information for Policymakers and Planners*. Ontario Ministry of Natural Resources. 49 pp. ISBN: 978-1-4249-2126-3.

Crozier, M.J. (2010) Deciphering the effect of climate change on landslide activity: A review. *Geomorphology*, 124 (3), 260–267.

Crozier, M.J. & Preston, N.J. (1999) Modelling changes in terrain resistance as a component of landform evolution in unstable hill country. In: *Process Modelling and Landform Evolution*. Springer Berlin Heidelberg. pp. 267–284.

Deline, P., Gruber, S., Delaloye, R., Fischer, L., Geertsema, M., Giardino, M., Hasler, A., Kirkbride, M., Krautblatter, M. & Magnin, F. (2015) Ice loss and slope stability in high-mountain regions. In: Haeberli, W. & Whiteman, C. (eds.) *Snow and Ice-Related Hazards, Risks, and Disasters*. Amsterdam, Elsevier B.V. pp. 521–561.

Demers, D., Robitaille, D., Potvin, J., Bilodeau, C. & Dupuis, C. (2008) La gestion des risques de glissements de terrain dans les sols argileux au Québec. In: Locat, J., Perret, D., Turmel, D., Demers, D. & Leroueil, S. (eds.) *Proceedings of the 4th Canadian Conference on Geohazards: From Causes to Management*. Québec, Presse de l'Université Laval. 594 pp.

Demers, D., Robitaille, D., Locat, P. & Potvin, J. (2014) Inventory of large landslides in sensitive clays in the Province of Québec, Canada: Preliminary analysis. In: L'Heureux, J.-S. (eds.) *Landslides in Sensitive Clays: From Geosciences to Risk Management, Advances in Natural and Technological Hazards Research*. Dordrecht, © Springer Science+Business Media. pp. 77–89. Available from: 10.1007/978-94-007-7079-9_11.

Diffenbaugh, N.S., Scherer, M. & Ashfaq, M. (2012) Response of snow-dependent hydrologic extremes to continued global warming. *Nature Climate Change*. Available from: 10.1038/nclimate1732. http://www.nature.com/nclimate/journal/vaop/ncurrent/full/nclimate1732.html (Accessed 20th November 2012).

EBA, A Tetra Tech Company (2011) *Case Study of Brewery Creek*. EBA File V15101033, Presented to Government of Yukon – Energy, Mines and Resources. 29 pp.

Elsner, M.M., Cuo, L., Voisin, N., Deems, J.S., Hamlet, A.F., Vano, J.A., Mickelson, K.E.B., Lee, S.-Y. & Lettenmaier, D.P. (2010) Implications of 21st century climate change for the hydrology of Washington State. *Climatic Change*, 102, 225–260.

Environment Canada (2010) Available from: http://www.cccma.ec.gc.ca/diagnostics/cgcm3/plot_st_glb_sresa1b_2090_vs_1990.shtml.

Environment Canada (2014) *Climate Trends and Variations Bulletin – Annual 2013*. Available from: https://www.ec.gc.ca/adsc-cmda/default.asp?lang=En&n=8C7AB86B-1.

Environment Canada (2015) *Climate Trends and Variations Bulletin – Annual for 2014, Her Majesty the Queen of Right of Canada*. 4 pp. ISSN: 2367-9808.

Eshraghian, A., Martin, C.D. & Cruden, D.M. (2007) Complex earth slides in the Thompson River valley, Ashcroft, British Columbia. *Environmental & Engineering Geoscience*, 13, 161–181.

Eshraghian, A., Martin, C.D. & Morgenstern, N.R. (2008) Movement triggers and mechanisms of two earth slides in the Thompson River valley, British Columbia, Canada. *Canadian Geotechnical Journal*, 45, 1189–1209.

Etzelmüller, B. & Hagen, J.O. (2005) Glacier-permafrost interaction in Arctic and alpine mountain environments with examples from southern Norway and Svalbard. *Geological Society, London, Special Publications*, 242, 11–27.

Evans, S.G. & Clague, J.J. (1994) Recent climatic change and catastrophic geomorphic processes in mountain environments. *Geomorphology*, 10, 107–128.

Evans, S.G. & Delaney, K.B. (2015) Catastrophic mass flows in the mountain glacial environment. In: Haeberli, W. & Whiteman, C. (eds.) *Snow and Ice-Related Hazards, Risks, and Disasters*. Amsterdam, Elsevier B.V. pp. 563–606.

Fischer, L., Huggel, C., Kääb, A. & Haeberli, W. (2013) Slope failures and erosion rates on a glacierized high-mountain face under climatic changes. *Earth Surface Processes and Landforms*. Available from: 10.1002/esp.3355.

Geertsema, M. & Schwab, J.W. (1997) Retrogressive flowslides in the Terrace-Kitimat, British Columbia area: From early post-deglaciation to present and implications for future slides. In: *The Proceedings of the 11th Vancouver Geotechnical. Society Symposium*. Vancouver, B.C. pp. 115–133.

Geertsema, M. & Clague, J.J. (2011) Pipeline routing in landslide prone terrain. Innovation. *Journal of the Association of Professional Engineers and Geoscientists of BC*, July/August, 17–21.

Geertsema, M. & Chiarle, M. (2013) Mass-movement causes: glacier thinning. In: Shroder, J.F., Marston, R.A. & Stoffel, M. (eds.) *Treatise on Geomorphology, Vol. 7, Mountain and Hillslope Geomorphology*. San Diego, Academic Press. pp. 217–222.

Geertsema, M. & Foord, V.N. (2014) Landslides in the isolated patches permafrost zone, northeastern British Columbia (NTS mapsheet 94G east half). In: *Proceedings of the 3rd World Landslide Forum*. Beijing, China.

Geertsema, M., Clague, J.J., Schwab, J.W. & Evans, S.G. (2006) An overview of recent large catastrophic landslides in northern British Columbia, Canada. *Engineering Geology*, 83, 120–143.

Geertsema, M., Egginton, V.N., Schwab, J.W. & Clague, J.J. (2007) Landslides and historic climate in northern British Columbia. In: McInnes, R., Jakeways, J., Fairbank, H. & Mathie, E. (eds.) *Landslides and Climate Change: Challenges and Solutions*. London, Taylor and Francis. pp. 9–16.

Geertsema, M., Schwab, J.W., Jordan, P., Millard, T.H. & Rollerson, T.P. (2010) Chapter 8 – Hillslope processes. In: Pike, R.G., Redding, T.E., Moore, R.D., Winkler, R.D. & Bladon, K.D. (eds.) *Compendium of Forest Hydrology and Geomorphology in British Columbia*. B.C. Min. For. and Range, For. Sci. Prog., Victoria, B.C. and FORREX Forum for Research and Extension in Natural Resources, Kamloops, B.C. *Land Manag. Handb*. 66. pp. 213–273. Available from: http://www.for.gov.bc.ca/hfd/pubs/Docs/Lmh/Lmh66.htm.

Gruber, S. & Haeberli, W. (2007) Permafrost in steep bedrock slopes and its temperature-related destabilization following climate change. *Journal of Geophysical Research: Earth Surface (2003–2012)*, 112 (F2), S18.

Gruber, S., Burn, C.R., Arenson, L., Geertsema, M., Harris, S., Smith, S.L., Bonnaventure, P. & Benkert, B. (2015) Permafrost in mountainous regions of Canada. In: *The Proceedings of the 68th Canadian Geotechnical Conference, 7th Canadian Permafrost Conference. September 20–23, 2015, Québec-City, Qc, Canada.*

Haeberli, W., Wegmann, M. & Vonder Mühll, D. (1997) Slope stability problems related to glacier shrinkage and permafrost degradation in the Alps. *Eclogae Geologicae Helvetiae*, 90, 407–414.

Hirabayashi, Y., Mahendran, R., Koirala, S., Konoshima, L., Yamazaki, D., Watanabe, S., Kim, H. & Kanae, S. (2013) Global flood risk under climate change. *Nature Climate Change*, 3, 816–821.

Hogan, D.L. & Schwab, J.W. (1991) *Meteorological Conditions Associated with Hillslope Failures on the Queen Charlotte Islands*. B.C. Min. For., Victoria, B.C. Land Manag. Rep. No. 73.

Holm, K., Bovis, M.J. & Jakob, M. (2004) The landslide response of alpine basins to post-Little Ice Age glacial thinning and retreat in southwestern British Columbia. *Geomorphology*, 57, 201–216.

Huggel, C., Clague, J.J. & Korup, O. (2012) Is climate change responsible for changing landslide activity in high mountains? Earth Surface Processes and Landforms, 37:77–91.

Hungr, O. & Locat, J. (2015) *Examples of Common Landslide Types in Canada – Canadian Technical Guidelines and Best Practices Related to Landslides: A National Initiative for Loss Reduction*. Geological Survey of Canada, Open File 7897, 90 pp.

Jackson Jr., L.E., Hungr, O., Gardner, J.S. & Mackay, C. (1989) Cathedral mountain debris flows, Canada. *Bulletin of the International Association of Engineering Geology-Bulletin de l'Association Internationale de Géologie de l'Ingénieur*, 40 (1), 35–54.

Jakob, M. & Weatherly, H. (2003) A hydroclimatic threshold for landslide initiation on the North Shore Mountains of Vancouver, British Columbia. *Geomorphology*, 54 (3), 137–156.

Jakob, M. & Lambert, S. (2009) Climate change effects on landslides along the southwest coast of British Columbia. *Geomorphology*, 107 (3), 275–284.

Jakob, M., Owen, T. & Simpson, T. (2011) A regional real-time debris-flow warning system for the District of North Vancouver, Canada. *Landslides*, 9 (2), 165–178.

Jakob, M., Weatherly, H., Perkins, A., Bale, S. & Sorensen, J. (2014) Hydro-geomorphic response to the June 19-21, 2013 storm in Canmore and the Municipal District of Bighorn, Alberta. In: *Geohazards 6, Canadian Geotechnical Society Conference, Kingston, ON, June 15–18, 2014.*

Jakob, M., Clague, J.J. & Church, M. (2015) Rare and dangerous: Recognizing extraordinary events in stream channels. *Canadian Water Resources Journal/Revue canadienne des ressources hydriques*, 41, 1–13.

Jermyn, C. & Geertsema, M. (2014) An overview of some recent large landslide types in Nahanni National Park, Northwest Territories, Canada. In: *The Proceeding of the IAEG Conference, Torino, Italy.*

Kharin, V.V., Zwiers, F.W., Zhang, X. & Wehner, M. (2013) Changes in temperature and precipitation extremes in the CMIP5 ensemble. *Climatic Change*, 119, 345–357.

Klos, P.Z., Link, T.E. & Abatzoglou, J.T. (2014) Extent of the rain-snow transition zone in the western U.S. under historic and projected climate. *Geophysical Research Letters*, 41. Available from: 10.1002/2014GL060500.

Kokelj, S.V., Tunnicliff, D., Lacelle, D., Lantz., T.C. & Fraser, R.F. (2015) Retrogressive thaw slumps: From slope process to the landscape sensitivity of northwestern Canada, in GEOQuébec 2015. 68th Canadian Geotechnical Conference, 7th Canadian Permafrost Conference. September 20–23, 2015, Québec-City, QC, Canada, 7p.

Kundzewicz, Z.W., Kanae, S., Seneviratne, S.I., Handmer, J., Nicholls, N., Peduzzi, P., *et al.* (2013) Flood risk and climate change: Global and regional perspectives. *Hydrological Sciences Journal*, 59, 1–28.

Lafleur, J. & Lefebvre, G. (1978) *Influence des écoulements souterrains sur la stabilité des pentes naturelles d'argile.* Ministère des richesses naturelles, Direction générale des mines, Direction de la géologie, Service de géotechnique.

Lee, H.J. (2009) Timing and occurrence of large submarine landslides on the Atlantic Ocean margin. *Marine Geology*, 264, 53–64.

Locat, A., Leroueil, S., Bernander, S., Demers, D., Jostad, H.P. & Ouehb, L. (2011) Progressive failures in eastern Canadian and Scandinavian sensitive clays. *Canadian Geotechnical Journal*, 48 (11), 1696–1712.

Locat, J. & St-Gelais, D. (2014) Nature of sensitive clays from Québec. In: L'Heureux, J.-S., Locat, A., Leroueil, S., Demers, D. & Locat, J. (eds.) *Landslides in Sensitive Clays: From Geosciences to Risk Management, Advances in Natural and Technological Hazards Research.* Dordrecht, © Springer Science+Business Media. pp. 25–37.

Locat, P., Leroueil, S., Locat, J. & Demers, D. (2014) Chapter 11: Characterization and post-failure analysis of the 1980 landslide in sensitive clays at Havre-St_Pierre, Québec, Canada. In: L'Heureux, J.-S., Locat, A., Leroueil, S., Demers, D. & Locat, J. *Landslides in Sensitive Clays: From Geosciences to Risk Management, Advances in Natural and Technological Hazards Research.* Dordrecht, © Springer Science+Business Media. pp. 133–144.

Logan, T., Charron, I., Chaumont, D. & Houle, D. (2011) *Atlas de scénarios climatiques pour la forêt québécoise. Montréal.* Ouranos; Ministère de ressources naturelles et de la faune du Québec (MRNF). Available from: http://www.ouranos.ca/media/publication/162_AtlasForet2011-Sans-Annexes.pdf.

McCulloch, M.M., Forbes, D.L., Shaw, R.W. & the CCAF A041 Scientific Team. (2002) Coastal impacts of climate change and sea-level rise on Prince Edward Island. Forbes, D.L. & Shaw, R.W. (eds.) Geological Survey of Canada, Open File 4261. 62 pp.

Mbogga, M.S., Hamann, A. & Wang, T. (2009) Historical and projected climate data for natural resource management in western Canada. *Agricultural and Forest Meteorology*, 149 (5), 881–890.

Meehl, G.A., Covey, C., Taylor, K.E., Delworth, T., Stouffer, R.J., Latif, M., McAvaney B. & Mitchell, J.F.B. (2007) THE WCRP CMIP3 multimodel dataset: A new era in climate change research. *Bulletin of American Meteorological Society*, 88, 1383–1394.

Ministère des Transports Québec (1999) *Gestion des risques de glissement de terrain liés aux pluies des 19 et 20 juillet 1996 au Saguenay – Lac-St-Jean.* Rapport remis au Bureau de reconstruction et de relance du Saguenay – Lac-St-Jean. 58 pp. ISBN: 2-550-36816-9

Nakicenovic, N., Coauthors (2000) *Emissions Scenarios.* Nakicenovic, N. & Swart, R. (eds.) Cambridge, Cambridge University Press. 570 pp. Available from: http://www.ipcc.ch/ipccreports/sres/emission/index.php?idp=0.

Ouranos (2014) *Vers l'adaptation. Synthèse des connaissances sur les changements climatiques au Quebec. Partie 1: Évolution climatique au Québec.* Édition 2014. Montréal, Quebec, Ouranos. 79 pp.

Paquette, M., Fortier, D., Mueller, D.R., Sarrazin, D. & Vincent, W.F. (2015) Rapid disappearance of perennial ice on Canada's most northern lake. *Geophysical Research Letters*, 42, 1–8.

Perret, D. & Bégin, C. (1997) *Inventaire documenté des glissements de terrain associés aux fortes pluies de la mi-juillet 1996, région du Saguenay-Lac Saint-Jean.* Institut national de la recherche scientifique (INRS-Géoressources). Report presented to the Bureau de reconstruction et de relance du Saguenay-Lac Saint-Jean.

Pham, N.H., Sego, D.C., Arenson, L.U., Blowes, D.W., Amos, R.T. & Smith, L. (2013) The Diavik Waste Rock Project: Measurement of the thermal regime of a waste-rock test pile in a permafrost environment. *Applied Geochemistry*, 36, 234–245.

Porter, M., Savigny, K.W., Keegan, T.R., Bunce, C.M. & MacKay, C. (2002) Controls on stability of the Thompson River landslides. In: *Proceedings of the 55th Canadian Geotechnical Conference.* Niagara Falls, Ontario, Canada, October 20–23, 2002. pp. 1393–1400.

Potvin, J., Pellerin, F., Demers, D., Robitaille, D., La Rochelle, P. & Chagnon, J.-Y. (2001) Revue et investigation supplémentaire du site du glissement de Saint-Jean-Vianney. In: *Proceedings of the 54th Canadian Geotechnical Conference, Calgary*, Vol. 2. pp. 792–800.

Quinn, P., Jakob, M., McDougall, S., Strouth, A., Danielson, J., Wainstein, P., Arenson, L., Jones, A. & Hungr, O. (2014) Ghost-busting: Risk response following the Ghost Glacier failure, Mt. Edith Cavell, Jasper National Park. In: *Proceedings of the Canadian Conference on Geohazards*. Paper 202.

Rapaic, M. (2012) *Atlas de la Biodiversité du Québec nordique. Analyses et scénarios climatiques*. Montréal, Québec, Ouranos. Available from: http://www.ouranos.ca/media/publication/271_RapportRapaic2013red.pdf.

Schnorbus, M., Werner, A. & Bennett, K. (2014) Impacts of climate change in three hydrologic regimes in British Columbia, Canada. *Hydrological Processes*, 28, 1170–1189.

Schwab, J.W. (1997) Historical debris flows, British Columbia north coast. In: *The Proceedings of Forestry Geotechnique and Resource Engineering, Richmond, B.C.* Vancouver, B.C., BiTech Publishers.

Schwab, J.W. (1998) Landslides on the Queen Charlotte Islands: Processes, rates, and climatic events. In: Hogan, D.L., Tschaplinski, P.J. & Chatwin, S. (eds.) *Carnation Creek and Queen Charlotte Islands Fish/Forestry Workshop: Applying 20 Years of Coast Research to Management Solutions*. B.C. Min. For., Victoria, B.C. *Land Manag. Handb.* Vol. 41. pp. 41–48.

Schwab, J.W., Geertsema, M. & Blais-Stevens, A. (2004) The Khyex River landslide of November 28, 2003, Prince Rupert British Columbia, Canada. *Landslides*, 1 (3), 243–246.

Seneviratne, S.I., Coauthors (2012) Changes in climate extremes and their impacts on the natural physical environment. Managing the risks of extreme events and disasters to advance climate change adaptation. In: Field, C.B., *et al.* (eds.) Special Report of Working Groups I and II of the Intergovernmental Panel on Climate Change (IPCC), Cambridge, and New York, NY, Cambridge University Press. pp. 109–230.

Senneville S., St-Onge-Drouin S., Dumont D., Bihan-Poudec A.-C., Belemaalem Z., Corriveau M., Bernatchez P., Bélanger, S., Tolszczuk-Leclerc, S. & Villeneuve R. (2014) *Modélisation des glaces dans l'estuaire et le golfe du Saint-Laurent dans la perspective des changements climatiques (Rapport final)*. Rimouski Quebec. Institut de sciences de la mer de Rimouski.

Septer, D. & Schwab, J.W. (1995) *Rainstorm and Flood Damage: Northwest British Columbia 1891–1991*. B.C. Min. For., Victoria, B.C. Land Manag. Rep. No. 31.

Shaw, J., Taylor, R.B., Forbes, D.L., Ruz, M.-H. & Solomon, S. (1994) *Sensitivity of the Coasts of Canada to Sea-Level Rise*. Map M-270. Open File 2825. Geological Survey of Canada. Available from: http://ftp2.cits.rncan.gc.ca/pub/geott/ess_pubs/194/194054/of_2825.pdf.

Shrestha, R.R., Schnorbus, M.A., Werner, A.T. & Berland, A.J. (2012) Modelling spatial and temporal variability of hydrologic impacts of climate change in the Fraser River basin, British Columbia, Canada. *Hydrological Processes*, 26, 1840–1860.

Sillmann, J., Kharin, V.V., Zwiers, F.W., Zhang, X. & Bronaugh, D. (2013) Climate extremes indices in the CMIP5 multimodel ensemble: Part 2. Future climate projections. *Journal of Geophysical Research Atmospheres*, 118, 2473–2493.

Stoffel, M., Tiranti, D. & Huggel, C. (2014) Climate change impacts on mass movements – Case studies from the European Alps. *Science of the Total Environment*, 493, 1255–1266.

Taylor, K.E., Stouffer, R.J. & Meehl, G.A. (2012) An overview of CMIP5 and the experiment design. *Bulletin of the American Meteorological Society*, 93, 485–498.

Transports Québec (2011) *Rapport MT11-01: Glissement de terrain du 10 mai 2010 à St-Jude, Montérégie. Rapport sur les caractéristiques et les causes*. Bibliothèque et Archives Nationales du Québec, Gouvernement du Québec. 101 pp. ISBN: 978-2-550-63733-2.

van Oldenborgh, G.J., Collins, M., Arblaster, J.M., Christensen, J.H., Marotzke, J., Power, S.B., Rummukainen, M. & Zhou, T. (2013) Annex I: Atlas of global and regional climate projections. In: Stocker, T.F., Qin, D., Plattner, G.-K., Tignor, M., Allen, S.K., Boschung, J.,

Nauels, A., Xia, Y., Bex, V. & Midgley, P.M. (eds.) *Climate Change 2013: The Physical Science Basis. Contribution of Working Group I to the Fifth Assessment Report of the Intergovernmental Panel on Climate Change*. Cambridge, and New York, NY, Cambridge University Press.

van Vuuren, D.P., Edmonds, J., Kainuma, M., Riahi, K., Thomson, A., Hibbard, K., *et al.* (2011) The representative concentration pathways: An overview. *Climatic Change*, 109, 5–31.

Vasseur, L. & Catto, N. (2008) Atlantic Canada. In: Lemmen, D., Warren, F., Lacroix, J. & Bush, E. (eds.) *Impacts to Adaptation: Canada in a Changing Climate* 2007. Ottawa, ON, Government of Canada. pp. 119–170.

Vaughan, D.G., Comiso, J.C., Allison, I., Carrasco, J., Kaser, G., Kwok, R., Mote, P., Murray, T., Paul, F., Ren, J., Rignot, E., Solomina, O., Steffen, K. & Zhang, T. (2013) Observations: Cryosphere. In: Stocker, T.F., Qin, D., Plattner, G.-K., Tignor, M., Allen, S.K., Boschung, J., Nauels, A., Xia, Y., Bex, V. & Midgley, P.M. (eds.) *Climate Change 2013: The Physical Science Basis. Contribution of Working Group I to the Fifth Assessment Report of the Intergovernmental Panel on Climate Change*. Cambridge and New York, NY, Cambridge University Press.

Vincent, L.A., Zhang, X., Brown, R.D., Feng, Y., Mekis, E., Milewska, E.J., Wan, H. & Wang, X.L. (2015) Observed trends in Canada's climate and influence of low frequency variability modes. *Journal of Climate*, 28, 4545–4560.

Vitt, D.H., Viereck, L.A. & Zoltai, S.C. (2000) The changing landscape of Canada's western boreal forest: The current dynamics of permafrost. *Canadian Journal of Forest Research*, 30, 283–287.

Walling, D.E. (2009) *The Impact of Global Change on Erosion and Sediment Transport by Rivers: Current Progress and Future Challenges*. UNESCO. 26 pp. ISBN: 9231041355, 9789231041358.

Wang, T., Hamann, A., Spittlehouse, D.L. & Murdock, T.Q. (2012) ClimateWNA-high-resolution spatial climate data for western North America. *Journal of Applied Meteorology and Climatology*, 51 (1), 16–29.

Wang, B., Paudel, B. & Haoqiang, Li. (2016) Behaviour of retrogressive thaw slumps in northen Canada – three-year monitoring results from 18 sites, Landslides, 13(1):1–8.

Werner, A.T., Schnorbus, M.A., Shrestha, R.R., Eckstrand, H.D. (2013) Spatial and temporal change in the hydro-climatology of the Canadian portion of the Columbia River Basin under multiple emissions scenarios. *Atmosphere-Ocean*, 51, 357–379.

Whitfield, P.H., Cannon, A.J. & Reynolds, C.J. (2002) Modelling streamflow in present and future climates: Examples from the Georgia Basin, British Columbia. *Canadian Water Resources Journal*, 27, 427–456.

Chapter 4

Enhancing slope safety preparedness for extreme rainfall and potential climate change impacts in Hong Kong

K.K.S. Ho, H.W. Sun, A.C.W. Wong and C.F. Yam
Geotechnical Engineering Office, Department of Civil Engineering and Development, Hong Kong SAR Government

S.M. Lee
Hong Kong Observatory, Hong Kong SAR Government

ABSTRACT

A comprehensive Slope Safety System has been implemented by the Geotechnical Engineering Office (GEO) to manage the landslide risk in Hong Kong. The system has generally proved to be effective in coping with the prevailing landslide problems. However, the more frequent occurrences of extreme rainfall causing serious landslides and major casualties in different parts of the world in recent years have highlighted the impact of extreme rainfall on the community and concerns about the possible effect of climate change on slope safety. The impact of extreme rainfall poses new challenges which need to be addressed from a strategy, policy and technical perspective, as the number, scale and mobility of the corresponding landslides could be unprecedented. To manage the new challenges, a systematic approach was adopted by the GEO to (a) take stock of the relevant climate change studies, (b) identify the nature and scale of the credible extreme landslide events, (c) assess the severity of landslide consequences, (d) evaluate the capacity of emergency management, and (e) examine the scope for improving crisis preparedness as well as system and community resilience. The preliminary findings are presented in this report.

4.1 INTRODUCTION

Hong Kong has a population of 7 million and a small land area of 1,100 km^2, only 15% of which is developed land. The terrain is hilly (with 75% of the land steeper than 15° and 30% steeper than 30°) (See Figures 4.1 and 4.2). The vulnerable setting, comprising dense urban development, a legacy of inadequate geotechnical control before the 1970s, coupled with high seasonal rainfall (24-hour intensity exceeding 300 mm and 1-hour intensity exceeding 70 mm are not uncommon) has given rise to acute landslide problems in Hong Kong.

The Geotechnical Engineering Office (GEO), which was established by the Hong Kong Government in 1977, has implemented a comprehensive Slope Safety System. This embraces a range of initiatives covering policy, strategic, legislative, administrative, technical and procedural frameworks that serve to manage landslide risk in a

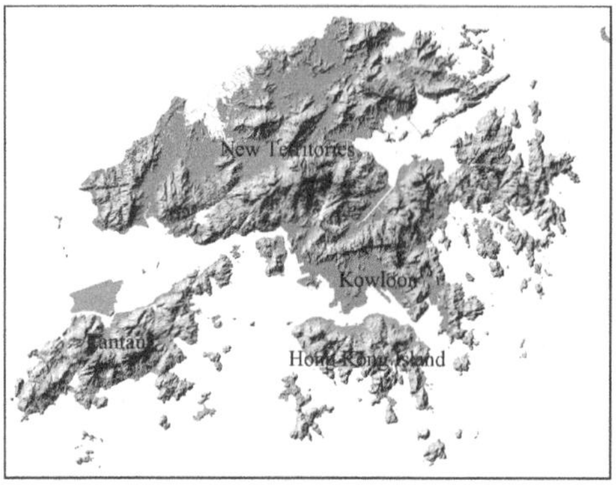

Figure 4.1 Topography of Hong Kong.

Figure 4.2 Aerial view of Hong Kong Island.

holistic manner. Risk management is done through an explicit risk-based approach and strategy. The goals are: (a) to reduce landslide risk to the community through a policy of priority and partnership, and (b) to address public perception and tolerability of landslide risk in order to avoid unrealistic expectations.

The key components of the Slope Safety System are shown in Table 4.1. The risk-based methodology embraces a holistic consideration of the likelihood of landslide hazards and their adverse consequences to reduce the landslide risk posed to existing developments, and to contain the increase in landslide risk associated with new developments. It is a multi-pronged approach involving hazard avoidance and prevention, study and mitigation of risk, public education and warning, and emergency management.

The overall landslide risk in Hong Kong has been substantially reduced over the years, as reflected by Quantitative Risk Assessment (QRA) and significant reduction in landslide fatalities (see Figure 4.3).

4.2 EVOLVEMENT OF SLOPE SAFETY SYSTEM UP TO 2008

The Slope Safety System has been evolving with time to incorporate enhancements arising from improved knowledge and practice.

Hong Kong is not immune from heavy rainfall events, which occur from time to time, either in association with low pressure troughs or in connection with tropical cyclones. The following is a brief account of the severe rainstorms experienced in Hong Kong since the 1960s, up to the late 1990s:

(a) Rainstorm of June 1966 – the rainstorm primarily struck Hong Kong Island and triggered hundreds of landslides and flooding which resulted in 64 fatalities, >2,000 people homeless, >8,600 people temporarily evacuated and >400 houses damaged (Chen, 1969).

(b) Rainstorm of June 1972 – the severe rainstorm caused many significant landslides in various places of Hong Kong, including the landslide disasters at Sau Mau Ping and Po Shan Road which killed more than 130 people.

Table 4.1 Key components of the Hong Kong slope safety system.

Slope Safety System components	*Contribution by each component* — *reduce landslip risk*: *hazard*	*reduce landslip risk*: *consequence*	*address public attitude & perception*
Policing			
– cataloguing, safety screening and statutory repair orders for slopes	✓		
– checking new works	✓	✓	
– slope maintenance audit	✓		
– inspecting squatter areas and recommending safety clearance		✓	
– input to land use planning	✓	✓	
Safety requirements, technical standards and research			
[e.g. *natural terrain hazard study and mitigation, debris mobility, etc.*]	✓	✓	✓
Works projects			
– upgrading existing Government man-made slopes	✓		
– mitigating natural terrain landslide hazards	✓	✓	
Regular slope maintenance			
– routine and preventive slope maintenance	✓		
Education and information			
– slope maintenance campaign	✓		✓
– personal precautions campaign		✓	✓
– slope safety awareness programme	✓	✓	✓
– information services	✓	✓	✓
– landslide warning and emergency services	✓	✓	✓

Note: Regular maintenance of registered Government man-made slopes and natural terrain defence/stabilisation measures are carried out by the responsible Government departments.

(c) Rainstorms of May and August 1982 – over 600 mm of rainfall was recorded within the four days from 28 to 31 May 1982 and more than 520 mm of rainfall from 15 to 19 August 1982. The two intense rainstorms triggered over 1,400 natural terrain landslides and caused serious damage particularly to squatter areas (with over 20 fatalities) (Hudson, 1993; Lee, 1983; Wong *et al.*, 2006).

(d) Rainstorm of November 1993 – over 700 mm of rainfall was recorded in 24 hours over Lantau Island triggering >800 natural terrain landslides and >300 man-made slope failures, which resulted in closure of roads and evacuation of houses (Wong *et al.*, 1997; Wong & Ho, 1995).

(e) Rainstorm of July 1994 – during the heavy rainfall episode from 22 to 24 July, over 600 mm rainfall was recorded at Hong Kong Observatory, together with record-breaking maximum 1-hour, and 24-hour rainfall of 185.5 mm and 954 mm respectively recorded at Tai Mo Shan. This resulted in about reported 200 landslides, 5 fatalities, 4 injuries, closure of roads and evacuations of houses (Chan, 1996).

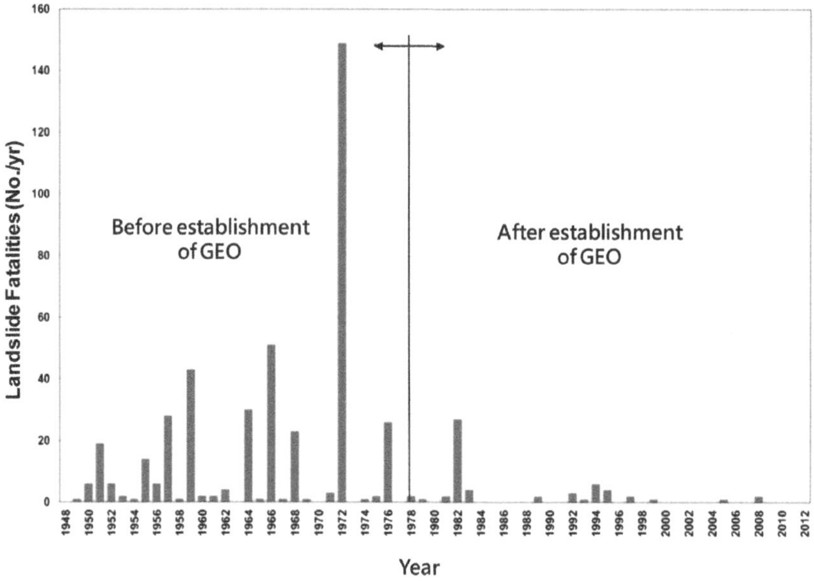

Figure 4.3 Landslide fatalities in Hong Kong since 1948.

(f) Rainstorm of August 1999 – over 500 mm of rainfall was recorded in 24 hours over the southern part of the New Territories triggering >200 man-made slope failures, which resulted in 1 fatality and permanent evacuation of 3 public housing blocks (Ho *et al.*, 2002).

It is noteworthy that many of the past major rainstorms in Hong Kong were not centred at the most densely populated locations, or that the intensity of the rainstorms that struck the more densely populated locations was not particularly severe. Otherwise, the number of landslides and the casualty rate would have been much higher.

The GEO has been maintaining comprehensive records on reported landslides since the late 1970s. In addition, recent and relict natural terrain landslides have been systematically mapped using the available stock of territory-wide, high-resolution aerial photographs dating back to the 1920s (e.g. to produce a natural terrain landslide inventory that contains >100,000 landslides), together with the use of air-borne Light Detection and Ranging (LiDAR) technology. In addition, Hong Kong has implemented a comprehensive network of automatic raingauges (>100 nos.) that collect and transmit rainfall data at 5-minute interval round the clock since the mid-1980s.

The above data provide essential information for evaluating slope and hillside responses to severe rainfall and form a basis of GEO's Landslip Warning System, landslide QRA, back analysis of landslide debris mobility, rainfall-landslide correlations for man-made slopes and natural terrain respectively, etc.

The high quality landslide data, together with lessons learnt from the study of serious landslides under GEO's systematic landslide investigation programme, has

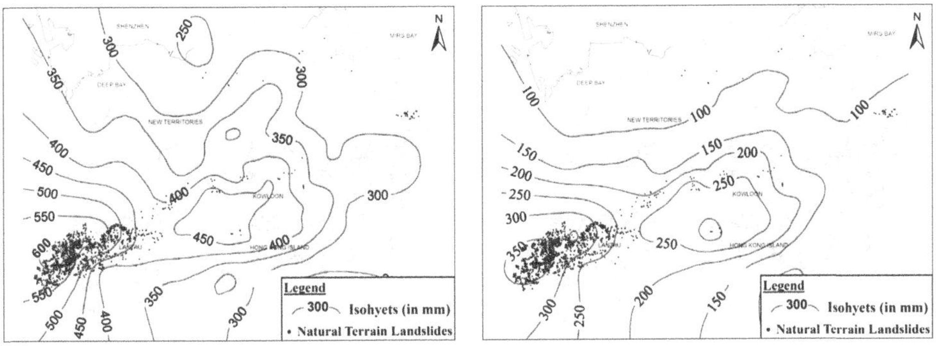

(a) Maximum rolling 24-hour rainfall distribution (b) Maximum rolling 4-hour rainfall distribution

Figure 4.4 Maximum rolling rainfall distributions for the 7 June 2008 rainstorm.

contributed to enhancing the technical understanding of actual slope and hillside performance.

The impact of extreme rainfall was previously assessed by the GEO, albeit in a rather rudimentary manner based on the knowledge and data available at the time. In 2000, GEO undertook a preliminary examination of its landslide emergency preparedness by reference to a severe rainstorm corresponding to 80% of the Probable Maximum Precipitation (PMP). The corresponding landslide scenario was taken to be 650 reported landslides, comprising 430 man-made slope failures and 220 natural terrain landslides affecting existing developed areas (Sun *et al.*, 2001). In 2006, the risk posed by natural hillsides was quantified using QRA techniques, under which the risk scenario of an extreme rainstorm was considered by reference to the prevailing rainfall-landslide relationship at the time. It was assessed that the risk associated with such extreme rainfall would contribute about 30% of the overall risk posed by natural hillsides (Wong *et al.*, 2006).

Focused R&D work on natural terrain landslides, together with the development and application of digital technology, has culminated in the Hong Kong Government launching an expanded retrofitting programme for slopes affecting existing buildings and infrastructure (Development Bureau, 2007). This expanded programme, known as the Landslip Prevention and Mitigation Programme, involves mitigation of the risk posed by vulnerable natural terrain catchments pursuant to the 'react to known hazard' policy, upgrading of substandard Government man-made slopes, and safety screening of private man-made slopes.

4.3 THE SEVERE RAINSTORM IN JUNE 2008

Hong Kong experienced a very severe rainstorm in June 2008, which primarily struck Lantau Island. It resulted in widespread natural terrain landslides, with many large scale and mobile failures.

The maximum rolling 4-hour and 24-hour rainfall recorded for the rainstorm were 384 mm and 623 mm respectively (see Figure 4.4). An estimate using the available rainfall records of the corresponding raingauges in that area in the last two decades reveals

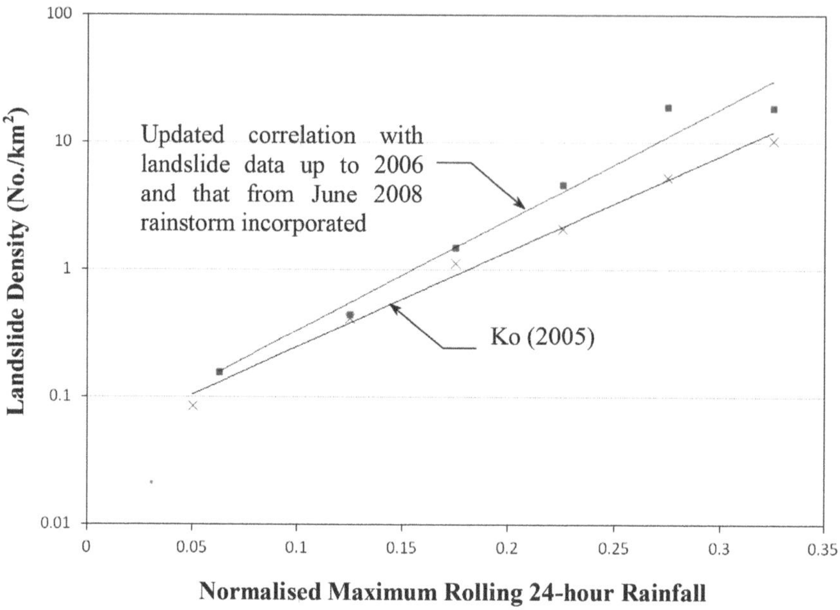

Figure 4.5 Year-based rainfall-natural terrain landslide correlation based on normalised maximum rolling 24-hour rainfall (Chan *et al.*, 2012).

that the statistical return period of such 4-hour and 24-hour rainfall may be of the order of 1,000 and 200 years respectively. This was probably the most severe rainstorm since rainfall record began in Hong Kong in 1885, as corroborated by the significant response of natural hillsides in terms of occurrence of rain-induced landslides.

The June 2008 rainstorm resulted in widespread natural terrain landslides (some 2,400 nos. on Lantau Island). The landslides caused serious disruptions to the communities (e.g. sole road access cut off, water supply to some local villages was cut off, large scale evacuation needed, the trunk road to the Hong Kong International Airport was shut down for several hours by debris flows and flooding, etc.), and major difficulties in respect of the response and recovery operations rendering Government's emergency response system on the verge of breakdown.

Due to the significant residual risk that might exist following natural terrain landslides (e.g. presence of tension cracks, overhanging boulders, loose debris or debris dams on drainage lines), the ability to deploy sufficient experienced staff to undertake timely inspections of failed hillsides that are close to developed areas, in order to establish the hillside condition and the necessary follow-up actions, has proved to be an important element of the landslide emergency system.

The landslides that occurred in 2008 were studied extensively and the technical insights were summarised by Wong (2009). In essence, the number, scale and mobility of natural terrain landslides under such intense rainfall conditions were unprecedented.

The rainfall-landslide response reinforces the key observation made in the past studies that the density of natural terrain landslides in Hong Kong increases exponentially with normalised rainfall intensity (Figure 4.5). It is noteworthy that the landslide

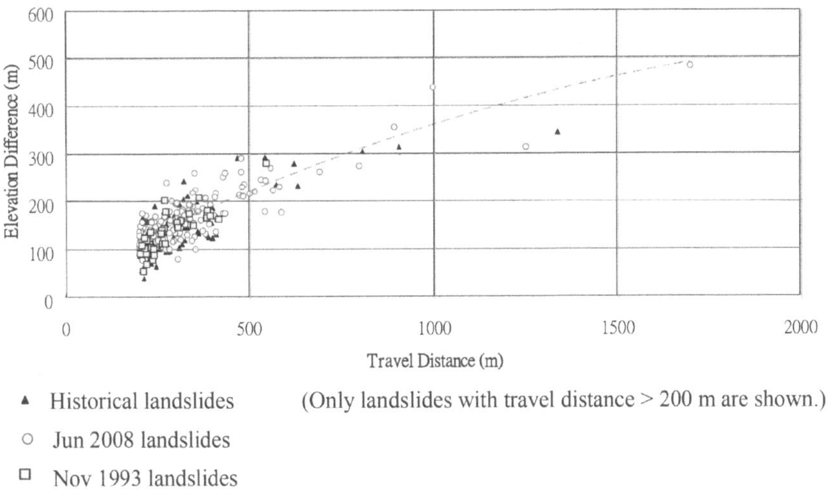

Figure 4.6 Mobility of June 2008 channelised debris flows and historical long-runout landslides.

Figure 4.7 Example of long runout June 2008 debris flow (Shek Pik Reservoir).

density in the June 2008 rainstorm was higher than that projected from the correlation with normalised 24-hour rolling rainfall based on the previous data from several decades of 'recent' landslide data. The mobility of landslide debris (Figures 4.6 and 4.7) and the scale (i.e. volume) of the 2008 landslides were also observed to have increased significantly as compared with that in the landslide inventory up to that time.

All in all, the data from a relatively short time window (i.e. of several tens of years for natural terrain landslides) are liable to under-represent the effects of extreme rainfall. In light of the severe rainstorm of June 2008, the corresponding rainfall-landslide correlation was duly updated, as presented below.

The torrential rain in June 2008 fell largely on Lantau Island, the largest outlying island and one of the sparsely populated areas in Hong Kong, hence it could be regarded as a near-miss event. The consequences could have been much more serious had the 2008 rainstorm hit the more densely developed urban areas of Hong Kong.

4.4 CONSIDERATION OF EXTREME RAINFALL EVENTS

According to the Intergovernmental Panel on Climate Change (IPCC), extreme weather refers to "the occurrence of a value of a weather or climate variable above (or below) a threshold value near the upper (or lower) ends of the range of observed values of the variable" (IPCC, 2012). The US National Oceanic and Atmospheric Administration (NOAA) echoes that by the following: "... charts the occurrence of specific extreme events over time since 1910. In most cases, extreme events are defined as lying in the outermost ("most unusual") ten percent" of a place's historical data (where analyses are done at national and regional scales, and "Climatic extremes are an important component of a location's climatology and are used for quality controlling meteorological observations, setting engineering limits, and helping authorities to develop climate-related safety plans, among other things" (NOAA, 2014). In essence, 'extreme event' may be taken to refer to rare and adverse condition at the extremities of the relevant historical or statistical distribution at national/regional scale.

The above highlights the important observation that an extreme event could be a natural phenomenon given the natural and intrinsic variability of the weather system. As such, extreme events would need to be considered irrespective of whether climate change prevails or not. Climate change is liable to render extreme rainfall events more frequent, as well as more intense.

In so far as landslide risk management is concerned, the extreme events that need to be considered are those which would bring about many sizeable and/or mobile landslides (either due to rainfall or earthquake). Such extreme landslide scenarios are characterised by widespread landslides which could impact on developed areas and lead to serious loss of life and damage to buildings, infrastructure and essential services. Around the globe, catastrophic landslides induced by extreme rainfall causing serious casualties have been reported increasingly in recent years. For example, the landslide at Xiaolin village during the passage of Typhoon Morakot in Taiwan in 2009 killed more than 400 people; the rain-induced landslides in Brazil in 2011 caused more than 900 fatalities; the debris flow in Gansu in 2010 led to more than 1,400 fatalities; the torrential rain in South Korea in 2011 brought about more than 30 landslide-related fatalities; Tropical Storm Washi brought flash floods and severe landslides, resulting in 1,010 fatalities in the Philippines in 2011; Severe Typhoon Wipha struck eastern Japan in 2013 and caused more than 40 landslide-related fatalities, etc. Hong Kong was not immune from extreme rainfall events. The occurrence of widespread natural terrain landslides during the June 2008 rainstorm, which involved many large-scale and mobile failures, has emphasised the critical need to address extreme rainfall events in a more rigorous manner.

The more frequent occurrences of extreme rainfall causing serious landslides and major casualties around the world in recent years have highlighted the real threat of the potential impact of extreme rainfall on community and concerns about the possible effect of climate change on slope safety. The impact of extreme rainfall poses new challenges which need to be addressed from a strategic, policy and technical perspective, given that the number, scale and mobility of the corresponding landslides could be unprecedented.

To address the acute challenges of more frequent occurrences of extreme rainfall, a systematic approach has been adopted by the GEO as follows:

(a) to take stock of the relevant climate change and rainfall studies,
(b) to identify the key component of risk and the nature and scale of credible extreme landslide events,
(c) to assess the severity of the landslide consequences,
(d) to evaluate the capacity of emergency management, and
(e) to examine the scope for improving crisis preparedness, as well as system and community resilience in respect of slope safety.

4.4.1 Climate change and rainfall studies

4.4.1.1 Globally observed changes in extreme rainfall and future projections

The United Nations Intergovernmental Panel on Climate Change (IPCC) released its Fifth Assessment Report of Working Group I (WGI AR5) in 2013, reaffirming the observed and unequivocal warming of the Earth's climate due to the human-caused increases in atmospheric concentration of greenhouse gases (IPCC, 2013). Apart from the warming atmosphere, one of the many responses of the climate system to the enhanced greenhouse warming is an increase in tropospheric water vapour (Santer *et al.*, 2007; Wentz *et al.*, 2007), and hence an increase in the availability of precipitable water for extreme rainfall events. There is also evidence of an intensifying water cycle over the last 50 years (Durack *et al.*, 2012). Hence, a warmer atmosphere holding more moisture can lead to notable changes in the frequency and intensity of extreme precipitation events (Kharin *et al.*, 2007; O'Gorman *et al.*, 2009; Lenderink & Meijgaard, 2010). Although precipitation changes are less spatially coherent than temperature changes, studies have shown that where sufficient data are available, there have been statistically significant increases in the number of heavy precipitation events in more regions than there have been statistically significant decreases (Figures 4.8 (a) and (b)).

IPCC's latest projections of future climate are based on the climate model data contributed to the Coupled Model Intercomparison Project Phase 5 (CMIP5) from various modelling centres/groups around the world. Although the science of climate modelling has advanced significantly in the last few decades, limitations in climate models still exist. For instance, there are small-scale physical, biological and chemical processes which cannot be fully described by mathematical equations. These processes are approximated by the so-called "parameterisations" within the climate models. Different climate models have different but equally plausible numerical representations of the climate system, with different strengths and weaknesses. Such a diversity is considered a healthy aspect by the climate modelling community. Although the collection of

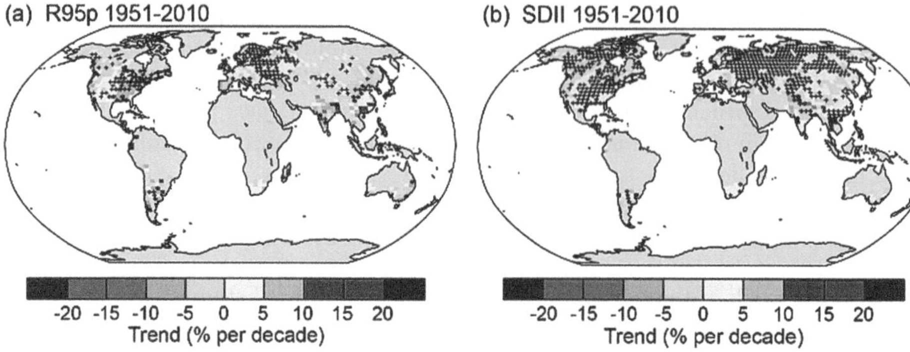

Figure 4.8 Trend in (a) the annual amount of precipitation from days above the 95th percentile (R95p) and (b) the daily precipitation intensity (SDII) during 1951–2010. Grey areas indicate incomplete or missing data. Black plus signs indicate statistically significant trends (IPCC, 2013).

CMIP5 model data is not meant to be a systematic or comprehensive sampling of possible future climate, it does provide a range of plausible climate projections as well as a basis for assessing the uncertainties in the projections. The "multi-model" approach is now a standard technique used by the climate science community to assess projection of a specific climate variable.

In WGI AR5, the climate projections are computed according to four new greenhouse gases concentration scenarios with different target radiative forcing at the top of the atmosphere at the end of the 21st century, namely Representative Concentration Pathway RCP2.6, RCP4.5, RCP6.0 and RCP8.5. A higher radiative forcing corresponds to a higher trajectory of greenhouse gases concentration in the 21st century and hence stronger warming on Earth. Although all these scenarios are plausible in principle, the latest observations have a strong indication that the world is likely to be tracking along the trajectory of RCP8.5 scenario in terms of carbon dioxide emission and that the RCP2.6 scenario seems rather unlikely (Figure 4.9). Under the RCP8.5 scenario (i.e. high concentration scenario), most land regions on Earth would experience an increase in extreme rainfall for the rest of this century (Figures 4.10 (a) and (b)).

4.4.2 Observed changes in extreme rainfall over China and future projections

A number of studies detected changes in annual precipitation as well as extreme precipitation frequency and intensity in China in the last few decades (e.g. Zhai *et al.*, 1999; Gong & Wang, 2000; Wang & Zhou, 2005; Zhai *et al.*, 2005; Zhang *et al.*, 2009; Lu *et al.*, 2010). Zhai *et al.* (2005) reported that, for the period of 1951–2000, the total precipitation in China as a whole had little trend but there were distinctive regional and seasonal patterns of trends. For instance, the annual precipitation as well as the frequency of extreme precipitation events has increased significantly over western China, the Yangtze River valley, and the southeastern coast. Over Guangdong, there was an increasing trend in annual precipitation during 1961–2006, though statistically insignificant (Composing team for Assessment Report on Climate Change of

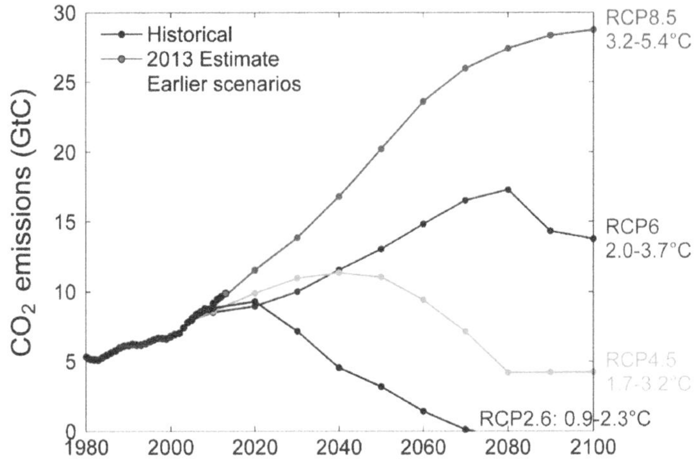

Figure 4.9 Observed emissions and emissions scenarios (image source: Global Carbon Project).

Guangdong, 2007). The number of rain days over southern China had a decreasing trend but the precipitation intensity increased significantly during 1961–2008 (Wu *et al.*, 2011). The frequency of heavy precipitation (40 mm per hour) increased from 1.3 times per year in the 1960s to 1.8 times per year in recent years (Composing team for Assessment Report on Climate Change of Guangdong, 2007).

Using 19 climate models, Sun & Ding (2009) projected an increase in summer (June–August) rainfall over southern China due to the increase in atmospheric water vapour under a warming climate and the presence of an anticyclone over western Pacific and South China Sea which enhances water vapour transport. Kitoh *et al.* (2013) found that, over the East Asian region, the daily precipitation intensity and seasonal (May–September) maximum 5-day precipitation total would increase under the RCP4.5 and RCP8.5 scenarios. Chen (2013) and Zhou *et al.* (2014) also found an intensification of precipitation over southern China under these two scenarios. Zhou *et al.* (2014) projected that the proportion of heavy precipitation in annual precipitation and maximum daily rainfall would increase over southern China. These changes can be attributed to intensification of the East Asian monsoon circulation (Kitoh *et al.*, 2013), and more unstable atmosphere over southern China (Chen, 2013).

4.4.3 Observed changes in extreme rainfall in Hong Kong

The rainfall trend in Hong Kong is generally consistent with that in southern China. In the last 60 years or so (1947–2013), Hong Kong's annual precipitation exhibited an increasing trend of 36 mm/decade, though statistically insignificant. On the analysis of the past trend in extreme rainfall in Hong Kong, Wong & Mok (2009) found that the return periods of short-duration extreme rainfall events had decreased significantly from the year 1885 to 2009. For example, the return period of hourly rainfall of 100 mm or more shortened from 37 years in 1900 to 18 years in 2000, indicating that such heavy rainfall events have become more frequent.

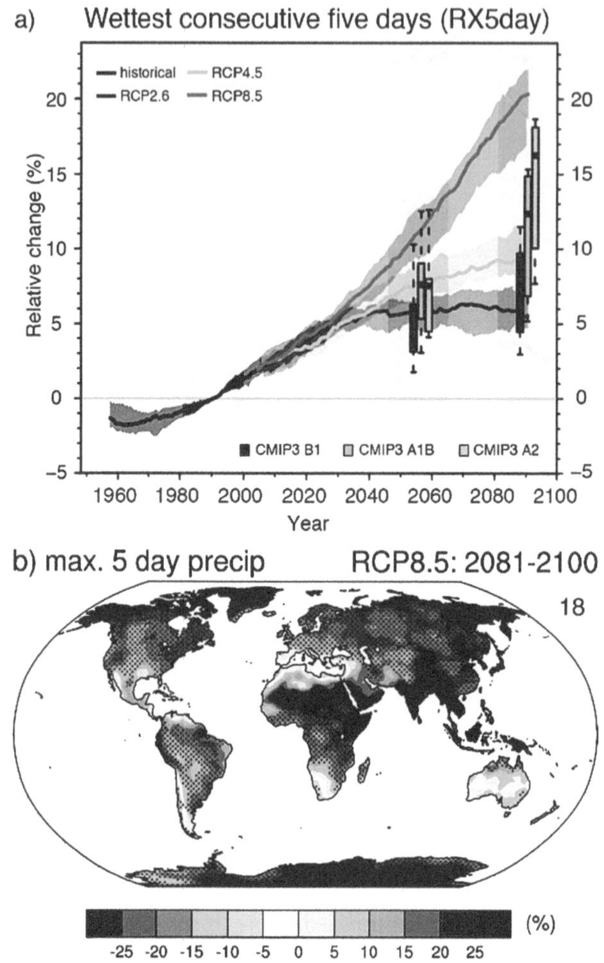

Figure 4.10 (a) Projected global average percent changes (relative to 1981–2000) over land regions in the annual maximum five-day precipitation accumulation (RX5day) for the RCP2.6, RCP4.5 and RCP8.5 scenarios. Shading in the time series represents the interquartile ensemble spread (25th and 75th quantiles). The box-and-whisker plots show the interquartile ensemble spread (box) and outliers (whiskers) of projections from 11 CMIP3 model simulations of the SRES scenarios A2 (orange), A1B (cyan) and B1 (purple) averaged over the periods 2046–2065 and 2081–2100. (b) Projected percent change of RX5day over the 2081–2100 period (relative to 1981–2000) in the RCP8.5 scenario. Stippling indicates statistical significance at the 5% level (IPCC, 2013).

The maximum hourly rainfall record of the Hong Kong Observatory Headquarter has been broken several times since 1885 and the time interval between new records is getting shorter. In particular, the record was broken three times since 1966. The latest record of 145.5 mm was set on 7 June 2008, breaking the previous one by a wide margin of 30 mm (Figure 4.11).

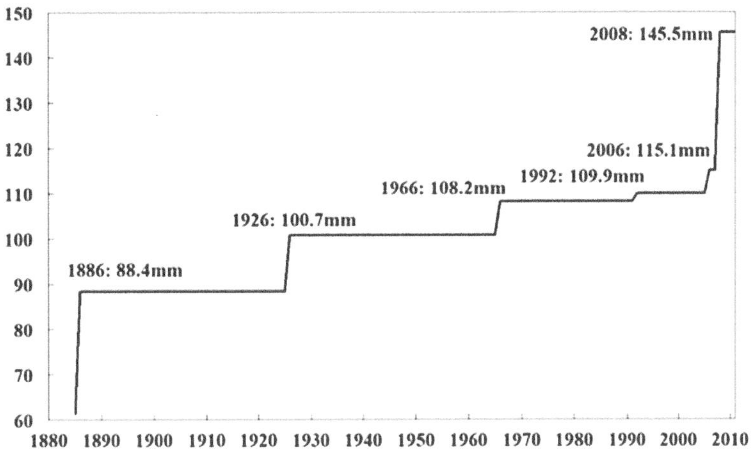

Figure 4.11 Maximum hourly rainfall recorded at Hong Kong Observatory.

4.4.4 Downscaling strategy in climate projection

To obtain rainfall projection for a small area like Hong Kong, either dynamical or statistical downscaling of global climate models (GCM), which are normally of coarse spatial resolution (e.g. 200 km × 200 km), should be performed. Dynamical downscaling involves a regional climate model (RCM) of a finer resolution (typically 50 km), which uses GCM output as boundary conditions to produce projections at a finer scale. This approach is physics-based and its results are dynamically consistent. Since this technique is very computationally demanding, it is not uncommon to see RCM using only one GCM as input (Gao *et al.*, 2013), rendering the uncertainty of the projection difficult to assess. Generating a large ensemble of projections based on a single GCM and a single RCM to assess the uncertainty is possible but could be very expensive (e.g. the UKCP09 project producing climate projections of a scale down to 25 km grid squares for the UK cost about £11 million, http://ukclimateprojections.metoffice.gov.uk/22681). The inclusion of an RCM inevitably introduces further model uncertainty to the projection (Rowell, 2006). In addition, the precipitation output from an RCM is still subjected to bias and usually cannot be used to support hydrological model without bias correction (Piani *et al.*, 2010; Wang & Chen, 2014). Even if bias-correction is carried out, the resolution of an RCM may still be inadequate to capture local forcings of the scale of a few kilometres (Murphy, 1999).

Statistical downscaling develops statistical relationships between large-scale predictors and local-scale predictands, and assumes that these relationships will still hold in future climate. Climate projection for a specific area can then be obtained using GCM output as predictors. This technique is far less computationally demanding but its performance is comparable to dynamical downscaling (Murphy, 1999). The same downscaling procedures can be applied to an ensemble of GCMs and hence a range of possible climate projections could be obtained.

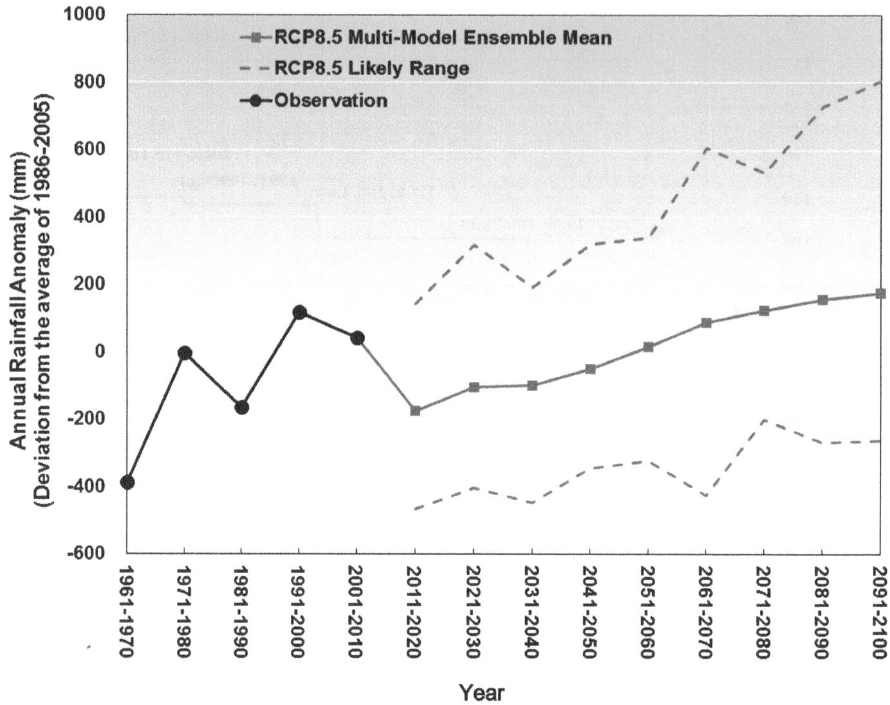

Figure 4.12 Past and projected annual rainfall anomaly of Hong Kong under the RCP8.5 scenario.Likely range refers to the region embraced by the 5th and 95th percentiles of the multi-model ensemble.

4.4.5 Rainfall projections for Hong Kong in the 21st century

Since the publication of the Third Assessment Report of IPCC, the Hong Kong Observatory (HKO) has conducted climate projection studies using the technique of statistical downscaling. Subsequent to the release of WGI AR5, the HKO is now working on the update of the climate projections for Hong Kong. The HKO utilised monthly data of 25 CMIP5 models and employed multiple linear regression to project the rainfall changes in Hong Kong in the 21st century. Under the RCP8.5 scenario, the number of extremely wet years (annual rainfall >3,168 mm) is expected to increase from three in 1885–2005 to about 12 in 2006–2100 (with a likely range of 5 to 20). The corresponding annual rainfall in late 21st century is expected to increase by about 180 mm compared to the 1986–2005 average (with a likely range of −250 mm to +800 mm) (Figure 4.12).

4.4.6 Recent update of Probable Maximum Precipitation (PMP) estimate for Hong Kong

PMP refers to the theoretical upper limit of the amount of rain that can fall according to hydro-meteorological principles in a given time duration over a given area, with no

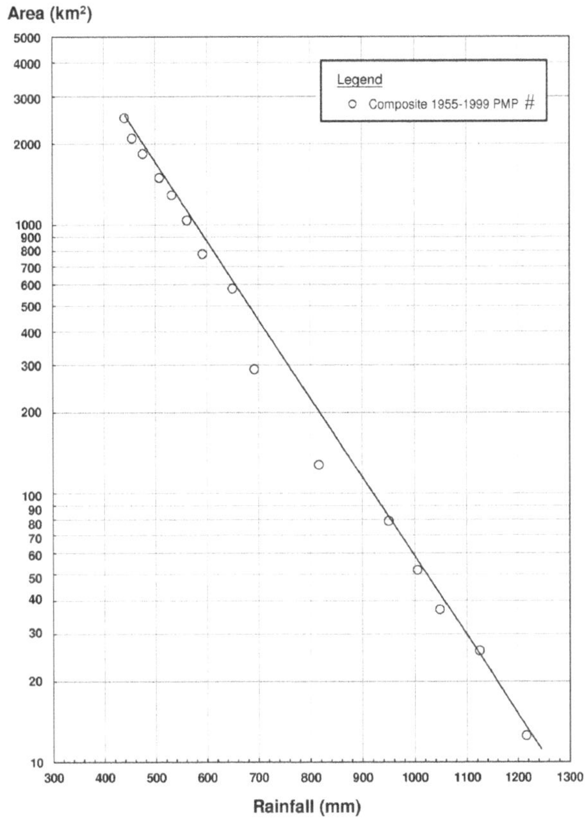

"Composite 1955-1999 PMP" refers to composition of the controlling points on the PMP which come from different rainstorms in the period of 1955 to 1999

Figure 4.13 Previous 24-hour PMP estimates for Hong Kong (HKO, 1999).

allowance made for long-term climatic trends (WMO, 2009). PMP is commonly used in reservoir and dam engineering and is hence familiar to the engineering profession. It may be adapted in the present context to serve as a yardstick or index for benchmarking the severity of extreme rainfall events to be considered for landslide impact assessment and emergency preparedness planning purposes. The use of the storm transposition technique (see below), which considers all the severe rainstorms actually observed in the neighbouring meteorologically similar regions, can add some realism and confidence to the assessed PMP.

The PMP for Hong Kong was first derived in 1968 for waterworks developments by the HKO (Bell & Chin, 1968). In 1998, the GEO and HKO collaborated on a study to update the 24-hour PMP using the moisture maximisation approach, see Figure 4.13 (HKO, 1999). This was used by the GEO in the early 2000s to develop possible extreme landslide event scenario to serve as a benchmark for reviewing the adequacy of the Government's landslide emergency preparedness measures (Sun *et al.*, 2001).

The GEO, with technical support by the HKO and various other Government departments and tertiary institutes, commissioned a study in 2011 to further update the 24-hour PMP estimates. In this third-generation update, the storm separation technique (viz. the Step Duration Orographic Intensification Factors (SDOIF) Method) was adopted. The rainfalls for each selected rainstorm are decomposed into two components (i.e. convergence component and local orographic component respectively) to facilitate the transposition analysis. A statistical approach was adopted to estimate the orographic factors (denoted as orographic intensification factors) by examining the geographical variation of historical annual maximum rainfalls of the corresponding duration. The associated rainstorm survey covered various meteorologically similar areas in the region, including Taiwan, Guangdong, Guangxi, Hainan, Fujian and Zhejiang. Following a detailed review of the severity of the storms and data availability and quality, four major rainstorms in Taiwan (viz. Herb in 1996, Aere in 2004, Haitang in 2005 and Morakot in 2009) were selected for transposition analysis.

By accounting for the orographic intensification factors of Taiwan (to eliminate the orographic effect), the isohyets of the above four rainstorms were converted into the convergence components. Then the convergence components of these four rainstorms were generalised to elliptical-shaped isohyets with an aspect ratio of 1.6 to establish the so-called "generalised convergence component". The generalised convergence component was then multiplied by orographic intensification factors of Hong Kong to take into account the local orographic effect. The results were further adjusted by a moisture maximisation factor (WMO, 2009) to produce the PMP estimate. The merging of the generalised convergence component and the orographic intensification factors was subjected to different orientations of the generalised convergence component and different relative positions of the convergence component to Hong Kong in order to ascertain the worst combination (see Figure 4.14).

The findings are given in AECOM & Lin (2014). The Depth-Area-Duration curve of the updated PMP estimate is shown in Figure 4.15, while the isohyets of the updated PMP estimates are given in Figure 4.16. The updated PMP for 10 km^2 has increased from 1,250 to 1,510 mm (i.e. 21% increase), whilst for 100 km^2 the updated PMP has increased from 930 mm to 1,350 mm (i.e. 45% increase), as compared with the 1999 PMP (denoted as PMP_{1999}).

By definition, the PMP estimates do not explicitly consider or fully account for the potential impact of climate change. AECOM & Lin (2014) have attempted to investigate the impact of climate change on the PMP estimates for Hong Kong through a literature review (e.g. studies on impact of climate change on PMP, extreme rainfalls and tropical cyclone activities in North Pacific, etc.), and a trend analysis of the related climate parameters that affect the PMP estimates. Based on the trend analysis, the time series of monthly average sea surface temperature in July and August in the relevant areas (i.e. the sea to the south-east of Hong Kong) shows a statistically significant increasing trend from 1961 to 2010 with a confidence level of 95%. However, the time series of annual maximum persisting 12-hour dew point temperature at the HKO does not show an increasing trend from 1961 to 2010. AECOM & Lin (op. cit.) opined that the uncertainties involved are too high to quantify the projection of extreme rainfalls, potential changes in the amount of precipitable water and the relationship between extreme rainfalls and PMP. As a result, they concluded that it was not credible to quantify the impact of climate change on the PMP in Hong Kong in

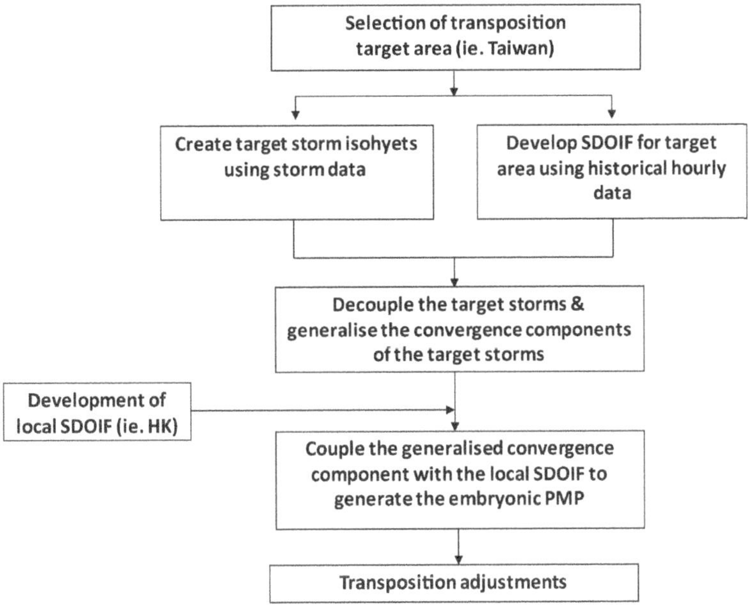

Figure 4.14 Storm transposition via SDOIF Method.

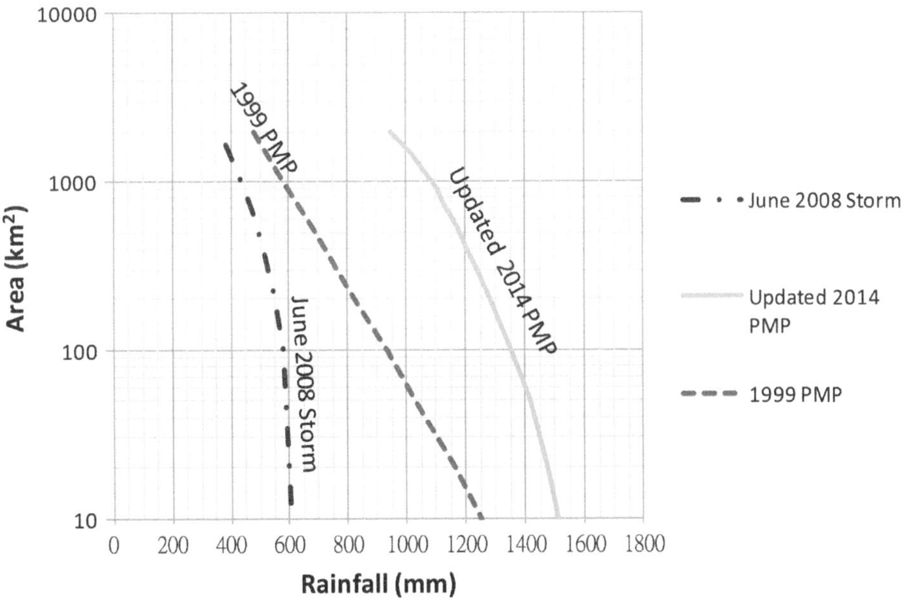

Figure 4.15 Depth-Area-Duration curves for updated 24-hour PMP for Hong Kong.

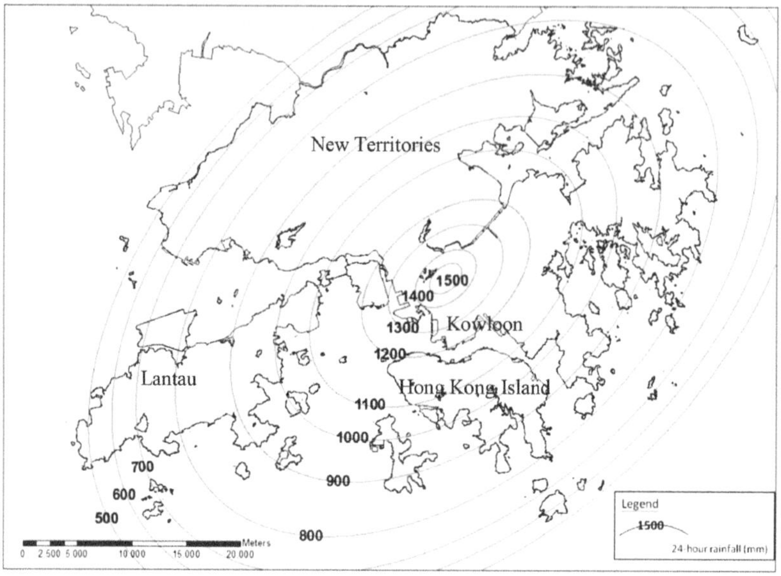

Figure 4.16 Isohyets of 24-hour rainfall of a hypothetical rainstorm corresponding to 2014 PMP estimate.

the 21st century with sufficient confidence based on the currently available data and technical-knowhow.

4.4.7 Geological evidence on relationship between landslide activity and climate change

Apart from the empirical correlation between rainfall and landslides, a more scientific approach may be adopted to study the potential effect of climate change by reference to the geological record which contains evidence of how the Earth's climate has changed over hundreds of millions of years. This evidence includes direct and proxy data for the ocean and atmosphere. The GEO has embarked on a programme of systematically dating ancient debris fan deposits by quantitative means with a view to studying the potential relationship between landslide activity and climate change.

The approach entails (a) reviewing past and present studies on palaeo-climate relevant to the Pearl River Delta region, (b) acquiring numerical age data (using a combination of techniques comprising optically stimulated luminescence (OSL), radiocarbon dating and cosmogenic nuclide surface exposure) on ancient coastal debris fan complexes, together with their characterisation and interpretation, and (c) integrating these data with the results of previous dating studies by GEO on ancient deep-seated sizeable landslides (volumes ranging from 700 m^3 to 850,000 m^3, most of which had ages of the order of tens of thousands of years) in Hong Kong. This is intended to provide a basis for projecting the possible response of the landscape to the estimated climatic and environmental changes, including the probable landslide characteristics under these scenarios.

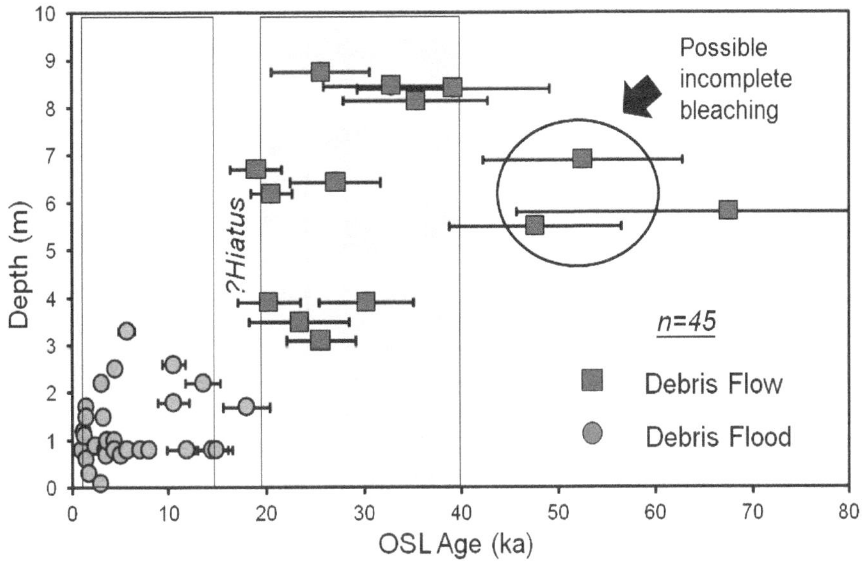

Figure 4.17 Age-depth relationship of debris samples taken from sites at West Lantau (Sewell *et al.*, 2013).

Studies of Hong Kong's palaeo-climate have made reference to various proxies including microfossil diatom evidence, pollen evidence, speleothem data, heavy metal contamination, carbon isotopes in vegetation, pulses of organic carbon, high resolution sea surface temperature, together with response of geochemical proxies to monsoon intensification. The dating of the ancient debris fans revealed that the fan complexes on Lantau Island typically consisted of multiple overlapping debris deposits. It was found that clast-supported debris flood deposits generally overlie older, matrix-supported debris flow deposits. The ages for the debris flood deposits range from 14,000 to 1,000 years old, with an average thickness of 1 to 2 m, whereas the ages for the debris flow deposits range from 20,000 and 40,000 years old, with an average thickness of 2 to 3 m. There appears to be an apparent hiatus at around 14,000 yr BP, as suggested in the boreholes by a change in decomposition grade and debris characteristics (see Figure 4.17).

The available age dating data from sizeable relict landslides suggest the following pulses of deep-seated landslides: 3.5–8 ka, 11–17 ka, 26–31 ka and 47–59 ka, with an apparent increased frequency of sizeable landslides post-Last Glacial Maximum (LGM) corresponding to sea level rise and monsoon intensification (see Figure 4.18). There is no obvious neotectonic connection, but this cannot be entirely ruled out.

The available information points to the possibility of a period of intense landslide activity during the early Holocene, coincident with the rapid rise in sea level and elevated temperatures (Sewell & Campbell, 2005; Sewell *et al.*, 2006). A post-glacial sea level of approximately 2 m above present has been recorded in Hong Kong around 5,500 yr BP (Fyfe, 2000), corresponding to the peak of the period known as Holocene Climatic Optimum (HCO).

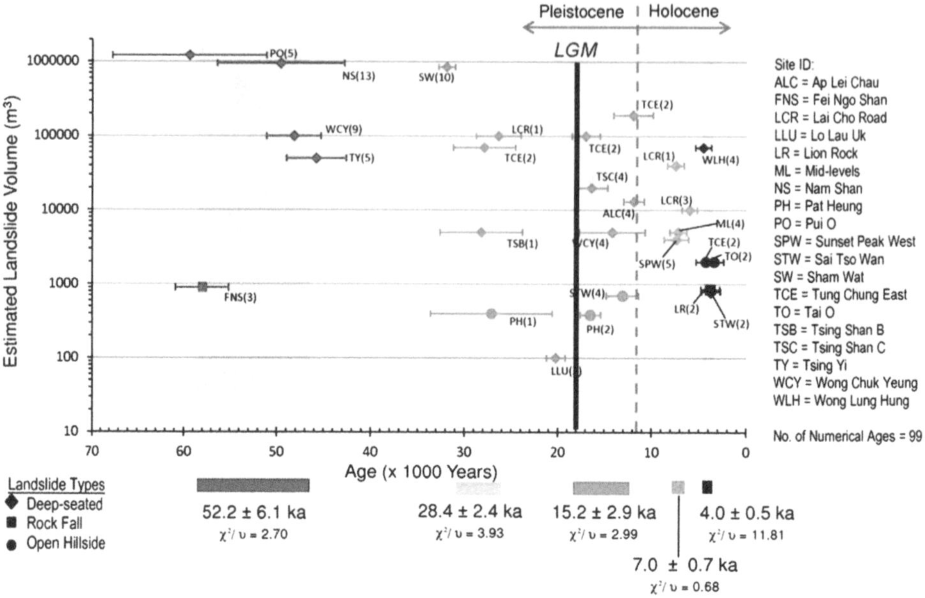

Figure 4.18 Ancient deep-seated landslide studies (Sewell *et al.*, 2013).

The mode and scale of landslide activity in Hong Kong appear to have changed over the past 50,000 years. Debris flow deposits seem to dominate during cool/dry periods whereas debris flood deposits seem to be more frequent during warm/moist periods.

The projected intensification of the East Asia Monsoon is likely to trigger more frequent sizeable landslide events and deep-seated landslides may be liable to become more common as the sea level rises significantly.

4.4.8 Landslide patterns corresponding to extreme rainfall

To obtain an estimate of the scale of the problem, the overall landslide patterns corresponding to extreme rainfall are extrapolated by projecting from the available information using the prevailing rainfall-landslide correlation. Hong Kong is data-rich in terms of high-quality information on rainfall and landslides, hence there is a reasonable basis for establishing the correlations between rainfall and different types of landslides by empirical means.

4.4.9 Rainfall-landslide correlations for man-made slopes

Rainfall-landslide correlations are practical tools to predict the response of man-made slopes and natural hillsides to rainfall. The GEO has been operating in conjunction with the HKO a territory-wide Landslip Warning System for over 35 years to alert the general public of possible landslide risk during periods of heavy rainfall. As part of the Landslip Warning System, the GEO operates an extensive network of automatic

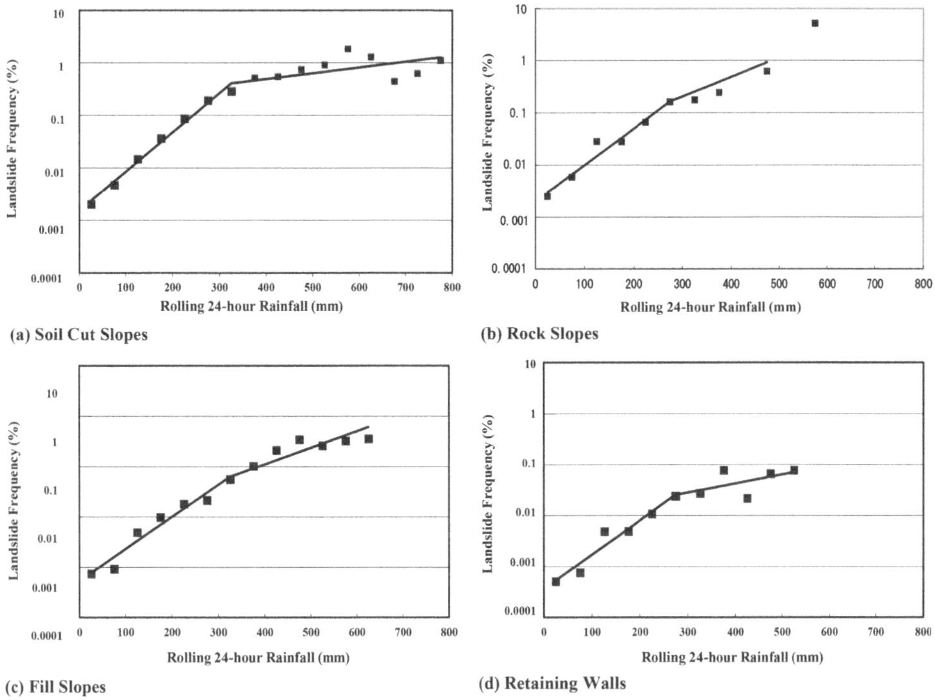

Figure 4.19 Correlations between landslide frequency and rolling 24-hour rainfall (Wong *et al.*, 2014).

raingauges which provide real-time data that can be combined with a short-term rainfall forecast undertaken by the HKO to predict, in a GIS platform, the number of landslides that may occur. This prediction is used for decision-making on whether a Landslip Warning would need to be issued or not.

The wealth of data on rainfall and landslides collected by the GEO over the past few decades has allowed the establishment of rainfall-landslide correlations for man-made slopes and natural hillsides respectively. Over the years, the GEO has been updating the rainfall-landslide correlations for man-made slope failures (Yu *et al.*, 2004). The correlations were based on the comprehensive data available on the spatial and temporal distribution of rainfall intensity and landslide occurrence.

The correlation between man-made slope failures and rainfall adopts a probability-based framework, which has been integrated with the spatial distribution of the registered man-made slopes together with the spatial and temporal variation of rainfall, for the prediction of the number and distribution of landslides. The model allows a realistic projection of the landslide pattern in real time as the rainfall pattern develops, which is essential for the effective operation of the Landslip Warning System. From past experience, different types of man-made slopes have different responses to the same rainfall intensities. Rainfall-landslide correlations have been developed for soil cut slopes, rock cut slopes, fill slopes and retaining walls respectively (Yu *et al.*, 2004). The latest correlations are based on rainfall and landslide data from 1996 to 2010 (see Figure 4.19).

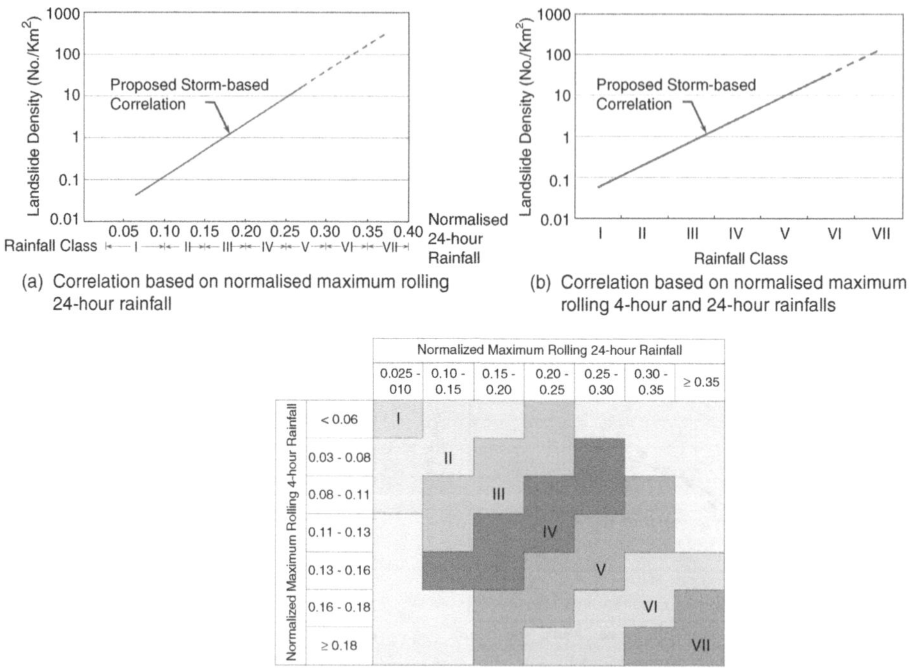

(a) Correlation based on normalised maximum rolling 24-hour rainfall

(b) Correlation based on normalised maximum rolling 4-hour and 24-hour rainfalls

(c) Rainfall classes referred in (b)

Figure 4.20 Rainfall-storm-based natural terrain landslide correlations for Hong Kong (Chan *et al.*, 2012).

4.4.10 Rainfall-landslide correlations for natural hillsides

As part of the R&D initiatives on the study and mitigation of natural terrain landslide hazards, the GEO has been developing relationship between rainfall and landslides on natural hillsides. Ko (2005) established a correlation between 24-hour rainfall and landslide density (viz. number of natural terrain landslides per km^2) on natural terrain based on normalised maximum rolling 24-hour rainfall intensities (i.e. dividing the maximum rolling 24-hour rainfall by the mean annual rainfall at the same location) and year-based natural terrain landslide data in the Natural Terrain Landslide Inventory (NTLI) from 1985 to 2000. The normalised maximum rolling 24-hour rainfall was found to be a better indicator of the severity of rainfall compared with maximum rolling 24-hour rainfall, because of the significant uneven rainfall distribution in Hong Kong. The year-based correlation was converted to a storm-based correlation based on rigorous spatial analysis of rainfall and natural terrain landslide data (Chan *et al.*, 2012). This approach is considered sufficiently reliable and useful for practical application, e.g. for formulating a natural terrain landslide warning system.

Following the severe rainfall in June 2008, the GEO updated the storm-based rainfall-landslide correlations for natural hillsides based on rainfall and landslide data from 1985 to 2008, see Figure 4.20(a) (Chan *et al.*, 2012). In addition to updating the

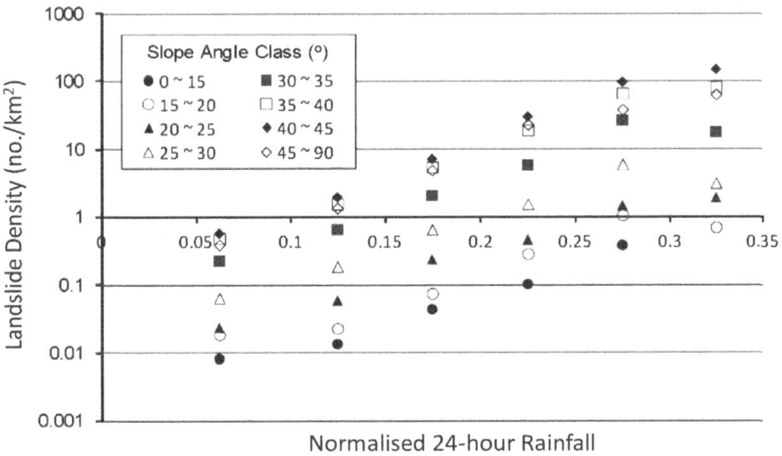

Figure 4.21 Year-based rainfall-landslide correlation.

correlation with normalised maximum rolling 24-hour rainfall, another correlation based on the combined consideration of both normalised maximum rolling 4-hour and 24-hour rainfalls respectively was also developed (Figure 4.20(b)).

The study indicates that the natural terrain landslide density tends to increase exponentially with increase in rainfall intensity. For an extreme rainfall event, significant implications on emergency preparedness and response planning are expected as widespread natural terrain landslides, probably with increased scale and mobility, will occur, which are expected to become the predominant landslide hazards. However, it should be noted that only limited data are available on landslides under extreme rainfall conditions for the establishment of the rainfall-landslide correlation. Hence, the forward prediction of extreme landslide scenarios would require considerable data extrapolation and should therefore be interpreted with caution.

The above rainfall-landslide correlations are being further refined by investigating the correlations for various slope angle classes in a territory-wide rainfall-based landslide susceptibility analysis for natural terrain in Hong Kong (Lo *et al.*, 2015 and Lo & Ko, 2015). Some of the results are shown in Figure 4.21.

4.4.11 Extreme rainfall scenarios

By nature, PMP is similar to the 'Maximum Credible Earthquake' concept in seismic hazard assessment and evaluation of extreme seismic event. In practice, realistic and credible extreme rainfall scenarios can be established by reference to the PMP estimates for a 'stress test' on the prevailing landslide emergency system (Wong, 2013).

For this purpose, the GEO opted to consider two extreme rainfall scenarios separately, namely (a) Level-1 extreme rainstorm involving an actual or near-miss event that has occurred in the past, which is unusual, severe and at the extremity of the local historical rainfall record. This is defined by transposing the actual June 2008

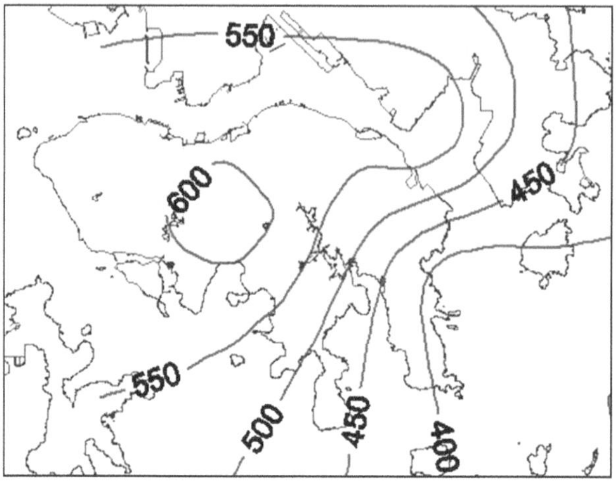

Figure 4.22 Level-1 extreme rainstorm: 24-hour rainfall distribution of June 2008 transposed from Lantau to Hong Kong Island.

rainstorm (which has a return period in the order of 1,000 years and corresponds to 60% PMP_{1999}) spatially from Lantau Island to Hong Kong Island (which is a densely urbanised area); and (b) Level-2 extreme rainstorm, which is a more severe, and rare but credible rainfall scenario as compared with the worst observed rainstorm. This is taken to correspond to 85% to 90% PMP_{1999} (with a projected return period in the order of 10,000 years) striking Hong Kong Island. This is similar to the range of notional return period mandated for earthquake emergency preparedness planning of major cities in Mainland China by the Chinese Authority.

The characteristics of the above two extreme rainfalls in terms of their spatial rainfall distributions are shown in Figures 4.22 and 4.23 respectively.

Based on the expert advice by HKO, GEO has also separately established a simulated rainstorm as an alternative approach to characterising the Level-2 extreme rainfall scenario for use in examining potential landslide impact, with due consideration of the effect of climate change. Rainfall isohyets of Typhoon Morakot that hit Taiwan in 2009 are transposed to Hong Kong taking into account the effect of local topography and the possible increase in tropical cyclone rainfall due to the effect of climate change projected to the end of the 21st century. It is estimated that the maximum 24-hour rainfall would be about 1,040 mm. In effect, this simulated rainstorm corresponds to about 70% PMP_{2014}. It should be noted that the corresponding notional return period (of about 10,000 years) would be reducing with time given the effect of climate change. This alternative approach allows a sensitivity check to be made on the extreme landslide scenario.

Recently, more refined assessments have been carried out making use of the newly developed rainfall-landslide correlation model developed in the territory-wide rainfall based landslide susceptibility analysis (Lo *et al.* 2015). The results of this latest study are generally in agreement with the findings of the assessments reported above.

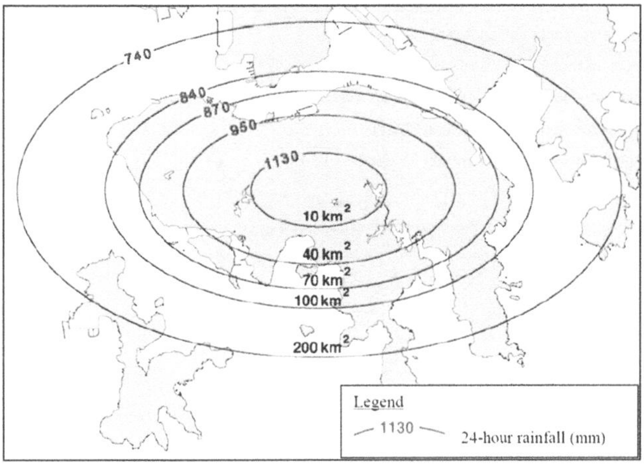

Figure 4.23 Level-2 extreme rainstorm: 24-hour rainfall distribution of a rainstorm corresponding to 85%/90% PMP_{1999} occurring over Hong Kong Island (after Leung & Lo, 2013).

4.4.12 Change in landslide portfolio with development of extreme rainfall

The June 2008 rainstorm was the most severe rainstorm hitting Hong Kong in the past few decades and resulted in widespread natural terrain landslides. Using the updated rainfall-landslide correlations for man-made slopes and natural hillsides, the landslide pattern under an extreme rainfall can be projected.

Figure 4.24 shows the curves that depict the estimated cumulative number of landslides for natural hillsides and man-made slopes respectively during the June 2008 rainstorm based on the rainfall-landslide correlations. It can be observed from Figure 4.24 that when the area was subjected to a 24-hour rolling rainfall greater than about 300 mm (at around 07:30 on 7 June 2008), the predicted number of landslides for natural terrain would start to increase rapidly resulting in a significant increase in the total landslide risk as well as the corresponding risk proportion. This is largely a result of the nature of the rainfall-landslide correlation for natural hillsides, whereby the landslide density increases exponentially with the normalised rainfall.

The projections therefore suggest that under extreme rainfall scenarios, natural terrain landslides would become the dominant landslide hazard.

4.4.13 Assessment of extreme landslide scenarios

Combining the probabilistic rainfall-landslide correlations for natural hillsides and man-made slopes, the rainfall characteristics corresponding to the above two extreme event scenarios for the stress test, together with the information in the Slope Catalogue and natural hillside inventory, the corresponding landslide scenarios can be projected. It should be reiterated that only limited data are available on landslides under extreme rainfall conditions. The forward prediction of extreme landslide scenarios therefore requires considerable data extrapolation and hence carries significant uncertainty.

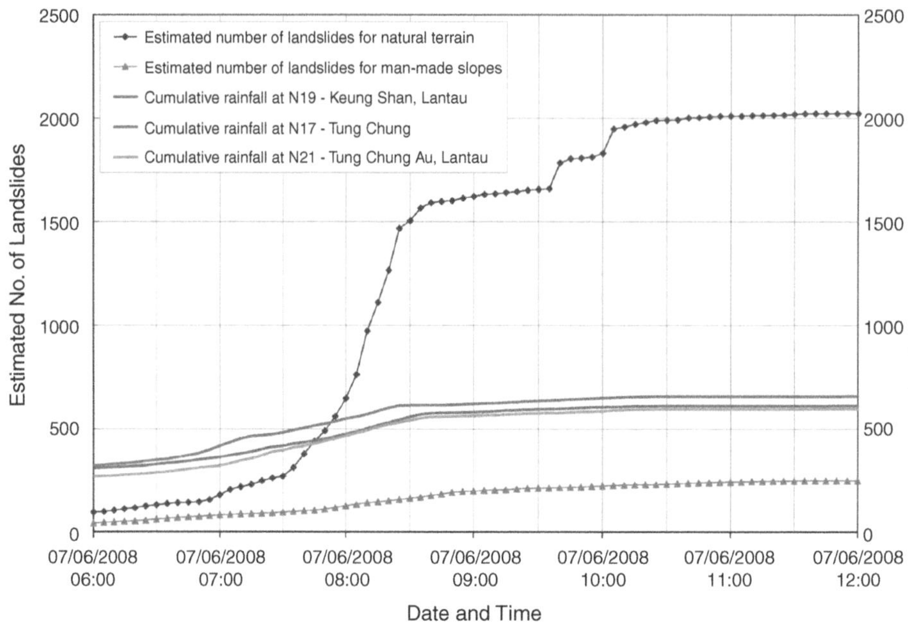

Figure 4.24 Comparison of estimated number of landslides from man-made slopes and natural terrain (based on the June 2008 rainstorm).

The historical data from the worst observed rainstorm in Hong Kong (i.e. the June 2008 rainstorm) since records began in 1885 provided a means to project the scale of natural terrain failure and debris mobility under extreme rainfall intensities in order to minimise data extrapolation as far as possible. Nonetheless, caution is needed when using historical data as a basis for assessing the scale of landslides and mobility of landslide debris under extreme rainfall events. Extreme rainfalls are liable to give rise to more watery debris, which tends to travel further and faster and is liable to overshoot from a drainage line. The increased surface runoff may overwhelm surface drainage provisions turning road pavement into drainage channels (or catchwater channels can become blocked and breached by landslides), overspilling a large volume of water onto the downhill slope at road bends and leading to washout type failures as well as flooding of the downhill areas. The possibility of major knock-on effects, such as the scenario of a service water reservoir being breached by major landslides releasing a colossal amount of water running violently downhill could also lead to major washout failures or debris floods, and grave consequences.

In essence, the possibility of large scale and mobile landslides being under-represented through data extrapolation needs to be borne in mind. Case histories of sizeable landslides triggered by world record or near world record rainfall intensities have indicated four main types of landslide hazards under extreme rainfall, namely mobile debris flows, mobile rockslides (or rock avalanches), large scale deep-seated landslides, and formation and subsequent collapse of landslide debris dams. The occurrence of the above hazards would depend on the site-specific setting (e.g. existence of

Table 4.2 Landslide consequences with Hong Kong Island struck by 60% PMP_{1999} and 85%/90% PMP_{1999} rainstorms respectively.

		Landslide scenarios					
		Level-1 extreme rainfall		*Level-2 extreme rainfall*			
				Approach A		*Approach B*	
		Hong Kong Island struck by 60% of PMP_{1999} (i.e. transposing the June 2008 rainstorm)		Hong Kong Island struck by *85%/90% PMP_{1999}*		*transposed to Hong Kong Island, with allowance for the projected effect of climate change (≈70% PMP_{2014})*	
Types of landslides and consequences		*No. of landslides*	*No. of landslides affecting buildings/ roads*	*No. of landslides*	*No. of landslides affecting buildings/ roads*	*No. of landslides*	*No. of landslides affecting buildings/ roads*
Man-made slope failures		105	65	745	465	1,000	620
Natural terrain landslides	All	1,900	150–250	50,000	3,000–10,000	49,000	3,800–8,000
	Volume > 1,000 m^3	20	2	500	30–100	490	38–80
	Volume > 10,000 m^3 (0.1%)	2	<1	50	3–10	49	4–8

fault zones or preferential sliding planes). The possibility of under-representing certain failure modes due to limited historical data under severe rainfall conditions cannot be overly emphasised.

4.4.14 Impact of Level-1 extreme rainfall

The Level-1 extreme rainfall scenario-based assessment for 60% PMP_{1999} rainfall, is based on transposition of the actual June 2008 rainstorm profile from Lantau Island (corresponds to 60% PMP_{1999}) to centre on the Mid-levels area of Hong Kong Island (see Figure 4.22).

The number of landslides is projected using the latest rainfall-landslide correlations. In assessing the landslide impact, reference was made to QRA models developed previously by the GEO for the quantification of landslide consequence (Ho *et al.*, 2000; Wong, 2005) through superimposing the landslide runout model onto the developed areas in a probabilistic framework. Details of the assessments are given by Lau *et al.* (2012). The results of the assessments are summarised in Table 4.2.

As explained above, natural terrain landslides would likely dominate under the extreme rainfall scenarios. The assessment indicated that about 1,900 natural terrain landslides could occur, some 150 to 250 of which would impact on buildings or roads. The assessment also showed that about 100 man-made slope failures (of which some 65 could affect buildings or roads) could occur. Using quantitative risk assessment, the corresponding Potential Loss of Life (PLL) given the occurrence of the events was also evaluated. For the 60% PMP_{1999} scenario, the calculated PLL for the event is about 10, which is about two orders of magnitude higher than the annual average PLL for the area.

The above assessment was also benchmarked against the outcome of the 1966 rainstorm, which was the most severe rainstorm that hit Hong Kong Island (corresponding to about 40% to 45% PMP_{1999} with the normalised maximum rolling 24-hour rainfall in the range of 0.1 to 0.25). The present assessment took into consideration that about 15% of the natural terrain landslides triggered by the June 1966 rainstorm reached the present-day development areas and allowed for the increased mobility of landslides under extreme rainfall, given that the June 2008 rainstorm at Lantau was about 40–50% more severe than the June 1966 rainstorm in terms of normalised maximum rolling 24-hour rainfall.

The 1966 rainstorm resulted in much damage and social disruption, including complete closure of many roads and 64 fatalities as a result of landslides and flooding. In comparison, the present assessment appears to be credible and the assessed consequences seem realistic and are broadly consistent with the past observations.

4.4.15 Impact of Level-2 extreme rainfall based on 85%/90% PMP_{1999}

A similar analysis was conducted to assess the extreme landslide scenario on Hong Kong Island in the event of the Level-2 extreme rainfall based on 85%/90% PMP_{1999} (Leung & Lo, 2013).

The assessment indicated that about 50,000 natural terrain landslides could occur, some 3,000 to 10,000 of which could impact on buildings or roads. This corresponds to approximately one landslide per 1,000 m^2 of natural terrain area, which is about 20 times the estimate obtained in the assessment by Lau *et al.* (2012) for the scenario involving the June 2008 rainstorm hitting Hong Kong Island. The landslides would result in the detachment of about 10% of the natural terrain area.

The assessment also showed that about 750 man-made slope failures (of which some 450 cases could affect buildings or roads) could occur (see Table 4.2).

It is expected that PLL of this event would be much greater than that associated with the scenario of 60% PMP_{1999} (roughly by an order of magnitude). However, there is much uncertainty in the expected PLL values, given that the collapse of a number of high-rise buildings due to landslide debris impact cannot be ruled out in adverse combination of circumstances.

4.4.16 Impact of Level-2 extreme rainfall based on transposition of Typhoon Morakot to Hong Kong with climate change effect projected to the end of 21st century

Additional analyses were also carried out to assess the extreme landslide scenario on Hong Kong Island under the Level-2 extreme rainfall scenario of a transposed Typhoon Morakot rainstorm, with allowance made for the effect of climate change projected to the end of 21st century. This assessment indicated that about 49,000 natural terrain landslides could occur, some 4,000 to 8,000 of which could impact on

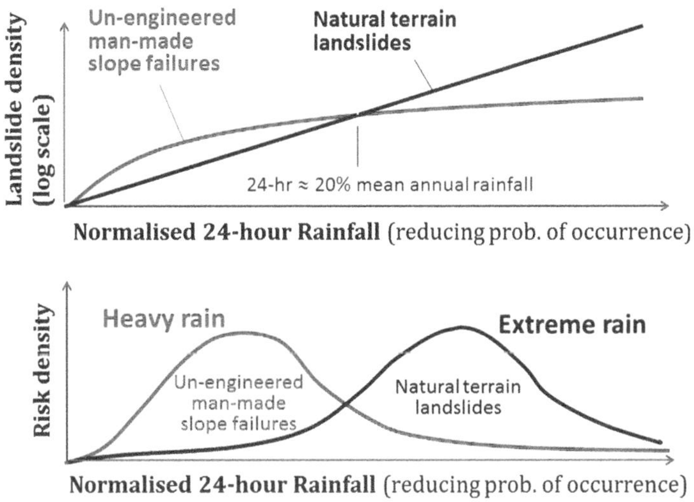

Figure 4.25 Distribution of landslide and risk density in response to extreme rainfall and climate change.

buildings or roads. The assessment also showed that about 1,000 man-made slope failures (of which some 620 cases affecting buildings or roads) could occur (see Table 4.2). It can be seen from Table 4.2 that the outcomes of the assessments using the above two different approaches in establishing the Level-2 extreme rainfall scenario were in the same ballpark.

4.4.17 Landslide risk due to extreme rainfall and climate change

Figure 4.25 illustrates diagrammatically the relationship between landslide density of man-made slopes and natural terrain respectively and increasing rainfall (with reducing chance of occurrence) based on the available data in Hong Kong.

When an area is subjected to severe rainfall that remains below the threshold (rolling 24-hour corresponding to about 20% of the mean annual rainfall), the dominant landslide problem is associated with man-made slopes, in particular non-engineered substandard man-made slopes. However, when the above rainfall threshold is exceeded, the number of natural terrain landslides would increase substantially based on the rainfall-landslide correlations and overtake that on man-made slopes, along with a notable increase in the volume of failure and the mobility of landslide debris. The corresponding shift in the risk density associated with the occurrence of extreme rainfall has major implications, as natural terrain landslides would then become the dominant problem.

The overall societal risk posed by natural terrain landslides was previously quantified by Cheng & Ko (2010) and the annual Potential Loss of Life (PLL) was assessed without the consideration of extreme rainfall events. With the latest projected extreme landslide scenarios, there are notable implications to the risk profile due to extreme rainfall events that are expected to become more frequent and more intense due to possible climate change. Making reference to the updated PMP_{2014} by AECOM and

Table 4.3 Summary of GEO's landslide emergency service from 1994 to 2013.

No. of Landslip Warning issued	*No. of reported landslide incidents*	*No. of building units evacuated**			*No. of road sections closed*
		Block	*House*	*Flat/Unit*	
62	5,274	21	119	1,003	2,434

Note: *A 'block' is a multi-storey building, which may comprise up to several dozens of flats/units. A 'house' is typically within 3 storeys, which comprises several flats/units.

Lin (2014), consideration of natural terrain landslide risk arising from extreme rainfall will bring about 50% increase in the overall risk in terms of annual PLL (Ko & Sun, 2016). However, one must not lose sight of the major uncertainties involved in assessing the risk of extreme landslide scenarios. The above risk quantification would at best give a very rough order of the possible risk level of extreme rainfall events and one must not be misled by the apparent precision of the estimate. Notwithstanding the above, the present assessment findings reinforce the notion that extreme rainfall would constitute an increasing concern and hence expanded efforts are warranted to enhance the resilience of Hong Kong's Slope Safety System in order to better manage the risk of extreme weather conditions.

4.4.18 Evaluating the capacity of landslide emergency management system

The findings from the above scenario-based assessments were used as a form of 'stress test' to review the adequacy of the GEO's prevailing landslide emergency preparedness.

Since 1978, the GEO has been operating a 24-hour year round landslide warning system and landslide emergency system to provide warning and emergency services and give professional advice to Government departments on actions to be taken in case of landslide danger. The aim is to minimise casualty and damage to property under heavy rainfall conditions, and facilitate recovery.

When it is predicted that numerous landslides will occur based on recorded and forecasted rainfall, the HKO in consultation with the GEO will issue the Landslip Warning to alert the general public of the potential landslide danger. As shown in Table 4.3 above, the landslide emergency service has played a significant role in minimising the exposure of the general public to landslide danger, thereby reducing landslide risk.

The landslide emergency system in Hong Kong has proved to be effective in dealing with emergency management under heavy rainfall conditions, such as those typically encountered when Landslide Warning is issued to the general public. However, the system has not been specifically designed to cope with, and tested for, extreme events, particularly in respect of preparedness, response and recovery.

An assessment has been made of the likely capacity of the landslide emergency management system, based on consideration of the available human resources for emergency inspection of landslide incidents that require GEO's input in emergency response (Wong, 2013).

The findings indicate that the current system is able to handle about 210 reported natural terrain landslides, or about 360 reported man-made slope failures.

It is noteworthy that the above is the existing capacity following the current practice of dealing with reported landslides and making recommendations on emergency actions to be taken, such as making recommendation on evacuation of buildings, closure of affected areas and urgent works needed to remove the immediate landslide danger.

All in all, it is projected that a Level-1 extreme rainstorm (viz. a 60% PMP_{1999} hitting a densely populated area on Hong Kong Island) would stretch the existing system to the limit. In comparison, about one-third of the system capacity was mobilised in June 2008 when the rainfall event hit the relatively sparsely populated and less accessible western part of Lantau Island.

In case a Level-2 extreme rainstorm (viz. an extreme rainstorm corresponding to 85%/90% PMP_{1999} or the transposed Morakot rainstorm allowing for the effect of climate change) were to hit a densely populated area on Hong Kong Island, the very large number of predicted natural terrain landslides would completely overwhelm the capacity of the current landslide emergency management system. Both the response and recovery phases would face acute challenges and the landslide consequences are likely to be very serious.

Other bottlenecks will almost certainly exist in emergency management for extreme events, such as transport arrangements, communication facilities, and provision for prompt and safe settlement of the affected community, which are constrained by the capacity of other Government agencies or non-government organisations involved in emergency management. All these issues need to be addressed in a holistic manner as part of the emergency preparedness for extreme landslide scenarios.

4.5 ADAPTATION STRATEGIES FOR MANAGING EXTREME LANDSLIDE EVENTS

In the context of risk management, emergency management is the discipline that deals with or mitigates the risk of weather-related extreme events or natural disasters. In the context of protecting the community from the adverse consequences of landslide disasters, the following adaptation strategies are pertinent:

(a) Prevention: This involves preventive activities to provide protection from disasters, e.g. setting up and enforcing the use of suitable slope investigation, design, construction, supervision and maintenance standards.
(b) Mitigation: This involves implementation of engineering measures to lessen the impact of landslides, e.g. retrofitting substandard slopes to reduce the chance of landslide, or providing mitigation measures (e.g. barriers or check dams) to reduce adverse landslide consequences.
(c) Preparedness: This focuses on formulating procedures, managing human resources, setting up emergency systems, preparing equipment, etc. for prompt mobilisation when a disaster occurs. It also involves preparing the vulnerable community to respond promptly and effectively to the landslide disaster, e.g. taking suitable personal precautionary actions during Landslip Warning.
(d) Response: The response phase of an emergency may involve search and rescue, as well as evacuation and fulfilling the basic humanitarian needs of the affected population. Emergency inspections for identification of any imminent danger

and advice on necessary response actions, safe settlement of those people who have been evacuated and provision of relief measures, are examples of response activities.

(e) Recovery: The recovery phase starts after the immediate threat to life has been dealt with. The goal is to bring the affected area back to normal, e.g. carrying out landslide repair or mitigation works.

Items (a) and (b) may be construed as broad risk reduction strategies whereas emergency management relates mainly to items (c) to (e). Resilience embraces items (a) to (e) and reflects the ability of a system to withstand shocks and stresses whilst maintaining its essential functions. Resilient systems are also more amenable to repair and recovery afterwards. Whilst slope safety systems commonly include provisions for emergency management, the existing systems are rarely designed for, nor tested against, extreme events.

Other options may include protection of the built environment by means of engineering works, allowing for 'safe' failure, embedding redundancy, better land-use planning, possible relocation of existing development, etc.

The various adaptation strategies and measures that are being implemented by the GEO to manage the risk of extreme rainfall events are outlined in the following.

4.5.1 Promulgate enhanced technical guidance and promote more robust mitigation measures

Following the June 2008 rainstorm, enhanced technical guidance was promulgated on the delineation of adverse site settings for the potential development of highly mobile channelised debris flows associated with major drainage lines fed by sizeable catchments. In addition, guidance was issued on the appropriate parameters for landslide debris runout modelling based on back analyses of actual mobile debris flows in 2008. Appropriate assessment of the mobility of natural terrain landslide debris is of the essence, as this will affect decision-making on whether or not mitigation measures are needed to protect a given facility and the detailed design of the corresponding mitigation measures. Furthermore, enhanced technical guidance on natural terrain hazard studies, together with empirical design of standard barriers for open hillside landslides, was promulgated in 2013 following consolidation of experience in the past few years and a probabilistic assessment of the scale and mobility of open hillside landslides triggered by the June 2008 rainstorm.

Designing slopes to a higher safety standard against possible extreme weather conditions (e.g. by adopting a higher factor of safety) may not be practicable or cost effective. Experience from disaster management of other natural hazards (e.g. earthquakes) indicates that it is generally more effective and appropriate to improve emergency management and enhance both system resilience and community resilience with a view to minimising landslide damage and protecting the community from the adverse consequences of landslide disasters.

Lessons learnt from systematic landslide studies in Hong Kong have emphasised the need to give due attention to improve the robustness of slope design and detailing by means of measures that will make the slopes less vulnerable to uncertainties (Ho & Lau, 2008). These include the use of more robust slope stabilisation schemes (e.g. soil nailing, retaining wall, etc.), improved design of slope surface drainage system (GEO, 2004; Tang & Cheung, 2010), application of prescriptive slope surface protection and drainage measures as contingency provisions (Wong *et al.*, 1999), as well as improved detailing of and enhanced redundancy in drainage provisions.

4.5.2 Low-frequency, large-magnitude landslides

In addition to the need to address extreme landslide scenarios involving widespread natural terrain landslides, the increase in the severity and frequency of rainfall also calls for attention to the challenges of low-frequency, large-magnitude landslides impacting on densely populated areas (viz. several tens of thousands m^3 or more, which can be highly damaging given Hong Kong's vulnerable setting).

Some low-frequency, large-magnitude landslides have occurred in Hong Kong within the last few decades. For example, the 1990 Tsing Shan landslide was triggered by a relatively moderate rainstorm, which involved a 400 m^3 failure at the source area and the adverse site setting resulted in a large and mobile debris flow due to significant entrainment along the runout path. The total active volume amounted to about 20,000 m^3 (King, 1996).

Another example is the 1995 Shum Wan Road failure, which involved a large-scale and mobile open hillside landslide. The drainage provisions along Shum Wan Road above the landslide source area were substandard and became blocked due to lack of maintenance, which resulted in the uncontrolled discharge of surface water flow onto the hillside below during severe rainfall. This led to concentrated water ingress in addition to rainwater infiltration arising from the severe rainstorm, triggering a fairly deep-seated and long runout open hillside landslide by exploiting the adverse geological features in the saprolite with weak clay infill. Two persons were killed in the incident. The additional source of concentrated water ingress arising from uncontrolled overflow of surface water that contributed to the large-scale failure illustrates the potential impact of extreme rainfall on the mode of landslide.

A further example is the 2008 Shek Pik reservoir landslide which comprised a sizeable and mobile channelised debris flow involving watery debris and a total debris deposition volume of about 46,000 m^3. The severe rainstorm in November 1993 only triggered a relatively small failure in the same catchment with limited mobility (travel distance of about 400 m), whereas the more severe June 2008 rainstorm triggered a very mobile failure (travel distance of about 1.3 km).

The identification of potential 'hot-spots' that are liable to give rise to low-frequency, large-magnitude landslide events is being examined along several fronts. These include (a) examination of large relict landslides and known distressed hillsides, (b) identification of vulnerable site settings based on an engineering geomorphological approach, and (c) delineation and mapping of major drainage lines fed by sizeable catchments.

Systematic assessment of digital elevation model (DEM) generated from territory-wide LiDAR data, together with field mapping, will assist with the identification of notable morphological features such as pre-historic deep-seated rockslides, relict scars of large landslides, sizeable colluvial lobes or fans, terrain with signs of disturbed drainage patterns, hillside with geomorphological signs of possible past large-scale slope instability, concentration of talluvium, etc.

To address the risk of potential debris flows or debris floods with significant entrainment and high mobility under extreme rainfall events, the need to give special attention to major drainage lines fed by sizeable catchments was highlighted by the June 2008 rainstorm, as these are potential locations of sizeable and mobile debris flows with watery debris. Catchment size, channelisation ratio, drainage line morphology, drainage order, length and straightness, together with presence of potentially entrainable material, are some of the important factors that govern the mobility of channelised debris flows. A GIS methodology has recently been developed by the GEO for the systematic delineation of major drainage lines and the corresponding catchment boundaries on a territory-wide basis (Lo & Ko, 2014). This was used to identify major drainage catchments (MDC), which are defined as hillside catchments with any drainage line of a length greater than 500 m that is originated from a watershed with a plan area of at least 10,000 m^2. The above definition has been benchmarked against the past large-magnitude channelised debris flows in Hong Kong.

4.5.3 Systematic implementation of mitigation measures on existing features posing highest risks to community

'Hardware' resilience may be enhanced through engineering measures in accordance with a risk-based approach to upgrade or retrofit the city infrastructure. The GEO has been systematically upgrading substandard man-made slopes since 1977 under the Landslip Preventive Measures (LPM) Programme. Over the last three decades, some 7,000 substandard man-made slopes were dealt with at a total expenditure of US$1.5B, resulting in about 75% reduction in overall landslide risk.

To dovetail the LPM Programme, the GEO launched in 2010 the Landslip Prevention and Mitigation (LPMit) Programme with an expanded scope to systematically mitigate natural terrain landslide hazards and retrofit the remaining substandard man-made slopes, in accordance with risk-based priority ranking systems. The pledged annual outputs approximately correspond to dealing with the worst 1% of the vulnerable hillside catchments and substandard man-made slopes every year on a rolling basis (Development Bureau, 2007).

The annual output in terms of number of catchments and slopes dealt with is not significant in comparison with the slope population and the extensive area of steep natural terrain in Hong Kong. Thus, the LPMitP cannot be the sole answer in addressing the impact of extreme rainfall events. However, as this systematic retrofitting programme systematically mitigates the risks of those hillside catchments and man-made slopes that are posing the highest risk to life, it will contribute to contain the overall landslide risk to an "as low as reasonably practicable" (ALARP) level and help to reduce the adverse consequences under extreme events, particularly in respect of the chances of multiple-fatality landslide events.

4.5.4 Improving crisis preparedness

The available resources of the existing landslide emergency system have a limited capacity in handling the reported landslide incidents under an extreme event. To further enhance the resilience of the Hong Kong Slope Safety System, crisis preparedness needs to be improved, albeit with different strategies and emphasis for the two extreme rainfall scenarios under consideration.

(a) Up to Level-1 extreme rainstorm scenario (viz. 60% PMP_{1999})
The main focus is to ensure effective and efficient emergency management when the system is stretched to the limit. This calls for streamlining the emergency response with due priority given to more critical cases, and enhancing preparedness by identifying and dealing with possible bottlenecks in the operation of the system.

(b) Exceeding Level-1 extreme rainstorm (viz. 60% PMP_{1999}) and up to Level-2 extreme rainstorm scenario (viz. 85%/90% PMP_{1999})
The capacity of the existing system will be completely overwhelmed. The current mode of emergency management operation including deploying professional geotechnical engineers to landslide sites to give advice on emergency actions will become impracticable. The development and implementation of a new strategy and mode of operation in managing the landslide emergency is therefore of the essence.

4.5.5 Enhancing and streamlining landslide emergency services

Based on the findings of the stress tests using the scenario-based assessments, the Hong Kong's landslide emergency system could in principle just about cope with the Level-1 extreme rainfall scenario (notional return period of 1,000 years) by stretching to its limits. There is scope, however, for further streamlining and continuous improvement and development of suitable contingency plans to enhance robustness. The constraints and bottlenecks associated with other relevant Government departments also require attention. The current landslide emergency system will break down under the Level-2 extreme rainfall scenario (notional return period of 10,000 years under the current climate condition).

The above scenario-based assessments and stress tests emphasise that the GEO needs to gear up its slope safety crisis preparedness under extreme rainfall conditions in respect of preparedness, response and recovery. As an illustration, the following measures have been undertaken recently by the GEO to strengthen and streamline the emergency service:

- Draw up a list of more experienced geologists/geotechnical engineers as a contingency provision for prompt deployment to the field to assist in assessing the residual risk of major natural terrain landslides.
- Put in place arrangements for GEO works contractors to be mobilised, if and when needed, to undertake emergency repair or risk mitigation works on large-scale natural terrain landslides. Flexible barrier components have been stockpiled for possible use in landslide emergency works.
- Use of modern information technology for sharing of key information and enhanced communication in order to facilitate prompt emergency response. A new, web-based landslide information management system was developed. This system is equipped with a mobile map and geo-location service that enables users to identify and record landslide locations conveniently. The timely reporting of

Table 4.4 Natural terrain alert criteria framework.

Equivalent Number of Predicted Natural Terrain Landslides	*Alert Level*	*Scenarios*
≥500 to <1000	1	Alert of possible widespread natural terrain landslides
≥1000 to <2000	2	Warning of widespread natural terrain landslides
≥2000	3	Warning of very widespread of natural terrain landslides

Note: The equivalent number of predicted natural terrain landslides is calculated based on the rainfall-landslide correlations and scaling factors (which reflect the relative hazard levels for different regions in Hong Kong).

key landslide information to aid identification and classification of landslides is particularly important under an extreme landslide scenario, as it will support the Emergency Manager in decision-making, action planning and setting priorities for the deployment of emergency teams.

- Put in place an Emergency Command System to handle situations that individual infrastructure departments might be unable to cope with the widespread damage in order to enhance the coordination of repair and recovery works.

4.5.6 Enhancing landslide warning system

The GEO has recently developed a separate set of landslip alert criteria for natural terrain (see Table 4.4) to supplement the current Landslip Warning system (Chan *et al.*, 2012). These have been in use since 2012 for internal reference by the GEO Emergency Manager so that a timely alert on possible widespread occurrence of natural terrain landslides can be provided in order to facilitate enhanced emergency preparedness and response.

4.5.7 Enhancing coordination of Government emergency services

The Hong Kong Government has put in place a Contingency Plan for Natural Disasters to ensure that all departments concerned will respond quickly and effectively in a coordinated manner to deal with emergency situations. The Government conducted an inter-departmental exercise in April 2012 to test its response and capabilities in the event of a serious nuclear incident at the Daya Bay Nuclear Power Station. The exercise included a drill on emergency evacuation of about 120 people caused by a landslide scenario that might occur during prolonged period of heavy rain incidental to the nuclear accident. The landslide drill provided an opportunity to test the Government's response at various critical stages of emergency evacuation, including evacuation of occupants, leading them out of the danger zone, taking stock of the evacuees and transporting them to temporary shelters.

The GEO has recently organised a pilot desktop exercise with 30 key departments/agencies to 'walk through' together several hypothetical extreme landslide scenarios and assess the likely major constraints in respect of emergency management. In addition, the GEO has met with the relevant stakeholders to brainstorm the possible cascading events with adverse knock-on effects that may be triggered by large-scale landslides.

4.6 FURTHER WORK

4.6.1 Strengthening and streamlining existing landslide emergency system

Potential areas for further improvement or streamlining of the prevailing landslide emergency system include the following:

- Sourcing additional and contingency resources to narrow the gap between the demand for emergency services and the existing emergency management capacity (e.g. develop plans for mobilisation of resources from the private sector and quasi-government organisations to assist in emergency management under extreme landslide events).
- Streamlining the emergency management procedures and practice for dealing with extreme landslide events, focusing on cases where priority attention is required in order to facilitate provision of the essential emergency response actions.
- Revamping the emergency management strategy to make it more pragmatic and effective to implement under the constraints of extreme landslide events. For example, geotechnical engineers may be deployed in district police/fire service command centres for more effective provision of geotechnical advice on emergency rescue actions.
- Improving public awareness of possible extreme landslide scenarios and equipping the general public with the knowledge and skills required for emergency management, e.g. launch appropriate publicity campaigns and public education programmes.
- Utilising the recently formulated natural terrain landslip alert criteria to enhance forewarning and landslide emergency mobilisation in the event that widespread natural terrain landslides are predicted under critical rainfall conditions.
- Conducting forums, workshops and drills to engage the vulnerable community and relevant stakeholders in emergency management and response (e.g. development of evacuation plans in advance).
- Revamping the emergency information system to facilitate sharing of information and communication among the different Government bureaux/departments/agencies responsible for different types of hazards via a common GIS operating platform.
- Improving communication with the public during a crisis, such as pre-written press releases and clear protocols in terms of the officers that will engage with the media.
- Providing enhanced advice and training to the Police and Fire Services Department on appropriate actions to be taken by their front-line staff on emergency duties for typical landslide incidents.
- Developing an emergency service continuity plan to react efficiently to unexpected situations, e.g. prolonged interruption of utility services (including power grids and communication networks).

Other potential bottlenecks and constraints still exist in the emergency management system, such as the provision and arrangements of emergency transport and communication facilities. All the potential bottlenecks need to be addressed in a holistic manner and contingency plans developed in order to further enhance system resilience.

4.6.2 Need to consider concurrent multiple hazard scenarios

In September 2013, when Super Typhoon Usagi was still travelling over the Pacific, it was recognised to have the potential to bring about significant storm surge to Hong Kong. In the event, the typhoon made landfall just to the east of Hong Kong at Shanwei, Guangdong Province, where at least 15,000 homes were destroyed and more than 30 fatalities were reported. Based on storm surge modelling, the HKO found that should the typhoon have travelled along a track to the west of Hong Kong, the highest sea level at Victoria Harbour would reach 5 m above Chart Datum, instead of 2.8 m as recorded during the typhoon (Ng *et al.*, 2014). In this event, serious and widespread flooding would have occurred in the densely populated urban areas adjoining the harbour, with major impact to the infrastructure including both above-ground and below-ground facilities, and at the same time heavy downpour would probably have triggered extensive landslides.

By virtue of the vulnerable and densely urbanised settings, it is perceivable that an extreme weather event hitting Hong Kong could result in multiple hazards, such as landslides, flooding, storm surges, tree falls, damages to squatter huts and buildings by wind, breach of reservoirs and catchwaters, damage to port facilities by waves, etc., which call for different forms of emergency operations.

To effectively handle the impacts of such multiple hazards that could happen concurrently, the adoption of a holistic approach is of the essence in assessing the severity of the plausible combination of extreme events and development of suitable crisis preparedness plans. The GEO in conjunction with the relevant departments that manage other weather-related hazards is conducting scenario-based assessments of multiple hazard scenarios under extreme weather conditions, including the consideration of potential cascading events. In this regard, the extreme weather events to be adopted in emergency preparedness will be refined, and the projected landslide scenarios and consequences will be updated.

4.6.3 New strategy for landslide emergency management under extreme rainfall condition

Apart from the need to further strengthen and streamline the landslide emergency system to make it more robust and resilient, a new strategy and mode of operation are needed in enhancing the preparedness for extreme rainfall events. This new strategy would primarily involve key elements that include (a) enhancing the coordination and collaboration of emergency services provided by different Government departments, (b) enhancing community resilience by promoting the public to take personal precautionary measures, and (c) empowering and facilitating the public to take appropriate emergency response actions to protect themselves and minimise exposure to landslide hazards, by providing them with the necessary knowledge and information in a timely manner for their decision-making on emergency response.

Under the current system in dealing with normal emergency situations, the GEO will endeavour to attend to each significant or serious landslide incident in order to undertake an inspection and give advice on the immediate precautionary measures (e.g. evacuation or road closure) to safeguard public safety, and the necessary emergency works to remove the immediate or impending danger.

Such a mode of emergency service will become impractical during extreme rainfall and landslide conditions. Apart from the very large number of landslides that will be reported, many of them will likely become inaccessible as a large number of roads and footpaths are expected to be blocked by landslide debris, or flooding. A new strategy for managing the landslide emergency is therefore needed.

Enhancing the resilience of the vulnerable population and the relevant stakeholders in facing up to the extreme natural events is of the essence. Maintaining vigilance of the extreme landslide scenarios and awareness of emergency management constraints is critical to crisis preparedness. Hence, public education and communication would be as important as technical assessments, formulation of emergency plans and procedures, and preventive and mitigation works.

The key is to enhance the resilience of the community, without causing undue alarm or generating misunderstanding by the public on the Government's motives. Members of the public will need to be empowered and encouraged to take appropriate emergency response actions in order to protect themselves and their families, and to minimise their exposure to landslide hazards. This may be done by providing them with the necessary knowledge and information on the latest landslide and rainfall situations in a timely manner so as to assist them to make their own decisions on emergency response, before the arrival of GEO staff or other emergency staff. An analogy will be the general public taking appropriate emergency response actions in the event of a fire before the arrival of the firemen.

An account of GEO's expanded efforts on public education, information and communication under the Hong Kong Slope Safety System is given by Tam & Lui (2013). Focused and enhanced publicity and public education efforts are warranted to improve the community resilience against extreme landslide events.

4.7 OTHER INITIATIVES

In the 2008 Policy Address, the Chief Executive announced that the Hong Kong Government will make early preparations to meet the challenge of climate change, in particular in enhancing fuel efficiency, using clean fuels and relying less on fossil fuels to reduce carbon footprints of our city. It was proposed to adopt a voluntary carbon intensity reduction target of 50%–60% by 2020 as compared with 2005 level. (Environment Bureau, 2010).

The GEO has continued to promulgate environmental-friendly construction techniques in slope works. Soil nailing is now widely used by the local geotechnical profession to upgrade soil cut slopes, fill slopes and masonry walls. Apart from being robust, this technique is also more environmentally friendly and sustainable than other alternative solutions (such as trimming back a cut slope or re-compaction of loose fill) in terms of carbon emission. In comparison, carbon emission of soil nailing is some 40% lower than that of cutting back of a slope and 50% lower than that of fill re-compaction. Vegetation covers are also provided to slopes upgraded under the LPMP/LPMitP as far as practicable. This will not only enhance slope appearance but also contribute to negative carbon emission.

The GEO has recently completed a study on the carbon footprint of slope construction works under the LPMitP and concluded that the carbon emission intensity (i.e. carbon emission/project value) of LPMit works was lower than that of general civil and infrastructure works by about 15%. Also, the total carbon footprint of LPMit works is about 2% of the construction sector's total carbon emission in Hong Kong. Notwithstanding this, GEO will continue with the efforts to enhance green and sustainable construction practices in LPMit works.

4.8 DISCUSSION AND CONCLUSIONS

All the climate scenarios projected for Hong Kong so far, based on various approaches, are consistent in that they suggest increasing precipitation with greater extremes, as well as increasing spread (i.e. variability) with time. These are generally consistent with the projected regional climate change patterns in South China. In essence, both extreme changes and chronic changes need to be considered.

It is pertinent to note that there are major uncertainties associated with the various climate change projection models (Ho, 2014). The resolution of the predictions is not high, particularly for extreme rainfalls, because of the constraints of data availability and the current technical know-how. The approaches are mostly not amenable to verification or calibration. The physical processes involved (e.g. cloud microphysics, small-scale turbulence, etc.) are generally not well understood and often not modelled in detail. Thus, both model uncertainty and parameter uncertainty can be significant despite the use of state-of-the-art climate models, not to mention there may also be unknown unknowns in addition to the known unknowns. Given the many uncertainties involved, an adaptive management approach is promoted (e.g. Hall & O-Connell, 2013). Decision-making in public policy and risk management should be risk-based as far as possible.

A key observation is that the occurrence of extreme rainfall could be within the natural variability of the weather system, irrespective of whether or not there will be climate change. If climate change does lead to notable impacts, it is likely to lead to an increase in the frequency and severity of extreme rainfall, apart from other adverse consequences such as increase in sea water level, storm surge, droughts, etc. Hence, the risk of extreme rainfall would need to be managed. In view of the potential dire consequences to densely urbanised areas that are vulnerable to rain-induced landslide hazards, the impact of extreme rainfall should be duly considered from a strategy, policy and technical perspective, in order to assess the scale of the problem and the necessary risk management actions.

A 'stress test' of GEO's landslide emergency management system in a desktop exercise using scenario-based assessments, as described above, has provided considerable insights into the nature and scale of extreme landslide scenarios, as well as the gaps in emergency preparedness. Based on the study findings, the current landslide emergency system is currently not properly geared up to deal with extreme rainfall events in terms of preparedness, response and recovery. Both system resilience and community resilience to landslides under extreme rainfall conditions need to be enhanced. The general public generally expect or believe that the responsibility for slope safety lies with the Government. Most people tend to have a short memory of past landslide disasters. Landslide hazards are, by and large, not perceived to be of major concern in Hong Kong and that there is a general perception that the landslide problem is now under good control by the Government. At the same time, the public tend to have a very low tolerance level of fatal landslides (especially multiple-fatality landslides). Also, there appears to be an increasing 'blame culture' by the media in pinpointing which senior Government officials are to be accountable for major mishaps or incidents.

It is imperative that the community maintains highly vigilant about serious landslides and be better prepared to deal with extreme rainfall events. It is neither practical nor cost-effective to rely solely on engineering solutions to manage the risk of extreme rainfall. The engineering approach needs to be supplemented by non-works, 'software' measures involving enhanced emergency preparedness, response and recovery.

Consideration should be given to the possible concurrent occurrence of multiple hazards in a holistic manner and the development of integrated strategies using a systems approach. This would require concerted efforts by various Government bureaux and departments, and engaging the community through public education and communication, together with public awareness and publicity programmes, in order to enhance community resilience and response capabilities. The challenges of and difficulties in enhancing community resilience in a truly effective and sustained manner should, however, not be under-estimated, given the public (mis)perception or indifference to slope safety in Hong Kong.

A holistic and systems view is called for, including the consideration of the system of systems, possible interdependencies amongst different systems, and local distributed capacity. This is pertinent in the context of multi-hazards and the corresponding emergency responses. In addition, a more rigorous consideration of the resilience framework is warranted (e.g. Arup, 2014). Such a framework may include key elements such as robustness, redundancy, responsiveness, integration, inclusive, reflective and diversity. For example, a review is in progress by the Hong Kong Government with a view to establishing a 'Common Operation Picture' GIS platform for all emergency works agencies to share emergency information in real-time. This would bring about better integration and coordination, which would also enhance the responsiveness to emergency and is conducive to better planning of the emergency response. In addition, the adoption of suitable smart technologies (e.g. a city-wide Global Positioning System) and information technology and communication tools could promote better coordination and communication amongst the different emergency departments.

In terms of enhancing slope safety preparedness for extreme rainfall, further work is needed in respect of exploring the practicability of developing landslide evacuation plans for vulnerable communities, large-scale drills of emergency evacuations, consideration of the use of landslide detection and alert systems to supplement landslide risk mitigation works, or the possible use of such system as a risk management tool in lieu of mitigation works, etc.

A recent report by the Institution of Mechanical Engineers (2013) calls for a much greater focus on preparing people for the possibility of an extreme natural event occurring and building disaster resilience into communities. In particular, it estimates that every $1 spent on building preparedness and resilience can save as much as $4 in relief, recovery and reconstruction later.

For enhanced slope safety preparedness and resilience to extreme rainfall events to take root, it has to be embraced by the community. It is of the essence to engage the public for community support, education and preparedness. Soft skills such as capacity building need further development in Hong Kong so that the population could be better informed about how to avoid the dangers of extreme weather and the community can recover more quickly afterwards.

Well coordinated open dialogues and honest discussions with the general public and stakeholders regarding uncertainties and risks are likely to be needed in engaging the community and promoting buy-in by the public of the need to enhance community resilience.

The judicious use of suitable social science approaches and modern social media tools in public education and engagement merits further consideration. Creative ideas for suitable risk communication are needed in helping the community to build up a new culture of safety, i.e. one that acknowledges uncertainties, system limitations and the likely occurrence of failures with potentially very serious consequences, as well as securing community buy-in, and promoting self-protection and neighbourhood support. It is possible that different approaches may be needed for different sectors of the community.

ACKNOWLEDGEMENTS

This paper is published with the permission of the Head of the Geotechnical Engineering Office and the Director of Civil Engineering and Development, Hong Kong SAR Government.

The authors thank Dr Roderick Sewell and Dr Sam Ng of the GEO for their advice on related geological aspects presented in this paper. All other contributions from GEO and HKO colleagues are also gratefully acknowledged.

REFERENCES

AECOM Asia Co. Ltd & Lin, B.Z. (2014) *24-Hour PMP Updating Study*. Hong Kong, Geotechnical Engineering Office. 97 pp. (Vol. 1) + 200 pp. (Vol. 2).

Arup (2014) *City Resilience Framework* Report prepared for the Rockefeller Foundation. 21 pp.

Bell, G.J. & Chin, P.C. (1968) *The Probable Maximum Rainfall in Hong Kong*. Hong Kong, Royal Observatory. 145 pp.

Chan, C.H.W., Ting, S.M. & Wong, A.C.W. (2012) *Development of Natural Terrain Landslip Alert Criteria (Special Project Report SPR 1/2012)*. Hong Kong, Geotechnical Engineering Office. 68 pp.

Chan, W.L. (1996) *Hong Kong Rainfall and Landslides in 1994 (GEO Report No. 54)*. Hong Kong, Geotechnical Engineering Office. 161 pp. plus 1 drawing.

Chen, H.P. (2013) Projected change in extreme rainfall events in China by the end of the 21st century using CMIP5 models. *Chinese Science Bulletin* 58 (12), 1462–1472.

Chen, T.Y. (1969) *Supplement to Meteorological Results 1966 – The Severe Rainstorms in Hong Kong During June 1966* Hong Kong, Royal Observatory. 79 pp.

Cheng, P.F.K. & Ko, F.W.Y. (2010) *An Updated Assessment of Landslide Risk Posed by Man-Made Slopes and Natural Hillsides in Hong Kong (GEO Report No.252)*. Hong Kong, Geotechnical Engineering Office. 46 pp.

Composing Team for Assessment Report on Climate Change of Guangdong (2007) Assessment report on climate change of Guangdong (selection). *Guangdong Meteorology* 29 (3), 1–14.

Development Bureau (2007) *Legislative Council Brief – Post-2010 Landslip Prevention and Mitigation Programme*. Hong Kong, Development Bureau. 11 pp.

Durack, P.J., Wijffels, S.E. & Matear, R.J. (2012) Ocean salinities reveal strong global water cycle intensification during 1950 to 2000. *Science*, 336, 455–458.

Environment Bureau (2010) *Hong Kong's Climate Change Strategy and Action Agenda – Consultant Document* Hong Kong, Environment Bureau. 67 pp.

Fyfe, J.A., Shaw, R., Campbell, S.D.G., Lai, K.W. & Kirk, P.A. (2000) *The Quaternary Geology of Hong Kong*. Hong Kong, Hong Kong Geological Survey, Geotechnical Engineering Office. 209 pp. plus 6 maps.

Gao, X.J., Wang, M.L. & Giorgi, F. (2013) Climate change over China in the 21st century as simulated by BCC_CSM1.1-RegCM4.0. *Atmospheric and Oceanic Science Letters*, 6, 381–386.

GEO (2004) *Enhancing the Reliability and Robustness of Engineered Soil Cut Slopes (GEO Technical Guidance Note No. 11)* Hong Kong, Geotechnical Engineering Office. 8 pp.

Gong, D.Y. & Wang, S.W. (2000) Severe summer rainfall in China associated with enhanced global warming. *Climatic Research*, 16 51–59.

Hall, J. & O'Connell, E. (2013) Earth systems engineering: Turning vision into action. *New Civil Engineers International*, Feb 2013 issue, 23–31.

HKO (1999) *The 24-Hour Probable Maximum Precipitation Updating Study for Hong Kong*. Hong Kong, Hong Kong Observatory. 32 pp.

Ho, K.K.S. (2014) Managing the uncertainties of natural terrain landslides and extreme rainfall in Hong Kong. In: *Landslide Science and Practice Volume 6: Risk Assessment, Management and Mitigation (Proceedings of the Second World Landslide Forum)*. pp. 285–302.

Ho, K.K.S. & Lau, T.M.F. (2008) The systematic landslide investigation programme in Hong Kong. In: *Landslides and Engineered Slopes: From the Past to the Future*, Vol. 1, pp. 243–248.

Ho, K.K.S., Leroi, E. & Roberds, W.J. (2000) Quantitative risk assessment: Application, myths and future direction. In: *Proceedings of the International Conference on Geotechnical & Geological Engineering, GeoEng 2000, Melbourne, Australia*, Vol. 1. pp. 269–312.

Ho, K.K.S., Chan, H.W. & Lam, T.M. (2002) *Review of 1999 Landslides (GEO Report No. 127)*. Hong Kong, Geotechnical Engineering Office. 70 pp.

Hudson, R.R. (1993) *Report on the Rainstorm of August 1982 (GEO Report No. 26)*. Hong Kong, Geotechnical Engineering Office. 93 pp. plus 1 drg.

Institution of Mechanical Engineers (2013) *Natural Disasters Saving Lives Today, Building Resilience for Tomorrow*. IMechE, 47 pp.

IPCC (2012) Managing the risks of extreme events and disasters to advance climate change adaptation. In: Field, C.B., Barros, V., Stocker, T.F., Qin, D., Dokken, D.J., Ebi, K.L., Mastrandrea, M.D., Mach, K.J., Plattner, G.-K., Allen, S.K., Tignor, M. & Midgley, P.M. (eds.) *A Special Report of Working Groups I and II of the Intergovernmental Panel on Climate Change*. Cambridge, and New York, NY, Cambridge University Press. 582 pp.

IPCC (2013) Climate change 2013: The physical science basis. In: Stocker, T.F., Qin, D., Plattner, G.-K., Tignor, M., Allen, S.K., Boschung, J., Nauels, A., Xia, Y., Bex, V. & Midgley, P.M. (eds.) *Contribution of Working Group I to the Fifth Assessment Report of the Intergovernmental Panel on Climate Change*. Cambridge and New York, NY, Cambridge University Press. 1535 pp.

Kharin, V.V., Zwiers, F.W., Zhang, X.B. & Hegerl, G.C. (2007) Changes in temperature and precipitation extremes in the IPCC ensemble of global coupled model simulations. *Journal of Climate*, 20 (8), 1419–1444.

King, J.P. (1996) *The Tsing Shan Debris Flow Volume 1 (Special Project Report No. SPR 6/96)*. Hong Kong, Geotechnical Engineering Office 133 pp. plus 7 drawings.

Kitoh, A., Endo, H., Kumar, K.K., Cavalcanti, I.F.A., Goswami, P. & Zhou, T. (2013) Monsoons in a changing world: A regional perspective in a global context. *Journal of Geophysical Research Atmospheres* 118, 3053–3065.

Ko, F.W.Y. (2005) *Correlation Between Rainfall and Natural Terrain Landslide Occurrence in Hong Kong (GEO Report No. 168)* Hong Kong, Geotechnical Engineering Office. 77 pp.

Ko, F.W.Y. & Sun, H.W. (2016) The role of landslide risk assessment in slope safety management in Hong Kong. *ASCE Geotechnical Special Publication in Memory and Honor of the Late Professor Wilson Tang: Risk Assessment and Management in Geotechnical Engineering: From Theory to Practice*, in press.

Lau, J.W.C., Li, A.C.O. & Pun, W.K. (2012) *Assessment of Possible Natural Terrain Landslide Event Scenarios Under Severe Rainstorm Events (Technical Note 3/2012)*. Hong Kong, Geotechnical Engineering Office. 37 pp.

Lee, B.Y. (1983) *Hydrometeorological Aspects of the Rainstorms in May 1982. (HKO Technical Note (Local) no. 68)*. Hong Kong, Royal Observatory. 52 pp.

Lenderink, G. & van Meijgaard, E. (2010) Linking increases in hourly precipitation extremes to atmospheric temperature and moisture changes. *Environmental Research Letters*, 5, 025208.

Leung, J.C.W. & Lo, D.O.K. (2013) *Assessment of an Extreme Natural Terrain Landslide Event Scenario Under a Rainstorm of 0.9 Times Probable Maximum Precipitation Estimate (Technical Note TN 2/2013)*. Hong Kong, Geotechnical Engineering Office. 43 pp.

Lo, F.L.C. & Ko, F.W.Y. (2014) *Review of Selection Criteria of Historical Landslide Catchments and Sizeable Catchments with Major Drainage Lines (TN 5/2014)*. Hong Kong, Geotechnical Engineering Office. 28 pp.

Lo, F.L.C. & Ko, F.W.Y. (2015) Rainfall-based landslide susceptibility analysis for natural terrain in Hong Kong. In: *15th Asian Regional Conference on Soil Mechanics and Geotechnical Engineering, 9-13 November 2015, Fukuoka, Japan*

Lo, F.L.C., Law, R.P.H. & Ko, F.W.Y. (2015) *Territory-wide Rainfall-based Landslide Susceptibility Analysis (Special Project Report No. SPR 1/2015)* Hong Kong, Geotechnical Engineering Office. 27 pp.

Lu, H., He, H. & Chen, S. (2010) Spatiotemporal variation of extreme precipitation frequency in summer over South China in 1961–2008. *Chinese Journal of Ecology*, 29 (6), 1213–1220.

Murphy, J.M. (1999) An evaluation of statistical and dynamical techniques for downscaling local climate. *Journal of Climate*, 112, 2256–2284.

Ng, M.C., Lui, W.H. Yeung W.L. & Mok, H.Y. (2014) Storm surge risk assessment in Hong Kong during the passage of super typhoon usagi. In: 28*th* *Guangdong-Hong Kong-Macao Seminar on Meteorological Science and Technology, 13-15 January 2014, Hong Kong*. HKO reprint no. 1096.

NOAA (2014) *Extreme Events*. USA, National Oceanic and Atmospheric Administration. Available from: http://www.ncdc.noaa.gov/climate-information/extreme-events [Last accessed 14th October 2014].

O'Gorman, P.A. & Schneider T. (2009) Scaling of precipitation extremes over a wide range of climates simulated with an idealized GCM. *Journal of Climate*, 22, 5676–5685.

Piani, C., Haerter J.O. & Coppola, E. (2010) Statistical biase correction for daily precipitation in regional climate models over Europe. *Theoretical and Applied Climatology*, 99, 187–192.

Rowell, D.P. (2006) A demonstration of the uncertainty in projections of UK climate change resulting from regional model formulation. *Climatic Change*, 79, 243–257.

Santer, B.D., Mears, C., Wentz, F.J., Taylor, K.E., Gleckler, P.J., Wigley, T.M.L., Barnett, T.P., Boyle, J.S., Brüggemann, W., Gillett, N.P., Klein, S.A., Meehl, G.A., Nozawa, T., Pierce, D.W., Stott, P.A., Washington, W.M. & Wehner, M.F. (2007) Identification of human-induced changes in atmospheric moisture content. *Proceedings of the National Academy of Science of the United States of America*, 104, 15248–15253.

Sewell, R.J. & Campbell., S.D.G. (2005) *Report on the Dating of Natural Terrain Landslides in Hong Kong (GEO Report No. 170)*. Hong Kong, Geotechnical Engineering Office. 154 pp.

Sewell, R.J., Barrows, T.T., Campbell, S.D.G. & Fifield, L.K. (2006) Exposure dating (^{10}Be, ^{26}Al) of natural terrain landslides in Hong Kong, China. *Geological Society of America Special Papers 2006*, 415, 131–146.

Sewell, R.J., Wong, J.C.F. & Wang N.S. (2013) The potential relationship between landslide activity and climate change – Dating of Ancient debris fan complexes, Lantau Island, Hong Kong. In: *American Geophysical Union Fall Meeting, San Francisco, 9–13 December 2013 – Poster Session*

Sun, H.W., Evans, N.C. & Pang, P.L.R. (2001) Possible extreme landslide event scenarios for Hong Kong. In: Ho, S. & Li, K.S. (eds.) *Proceedings of the Fourteenth Southeast Asian Geotechnical Conference Hong Kong/10-14 December 2001, Geotechnical Engineering Meeting Society's Needs* Vol. 3. pp. 225–230.

Sun, Y. & Ding, Y.H. (2009) A projection of future changes in summer precipitation and monsoon in East Asia. *Science China Earth Sciences*, 53 (2), 284–300.

Tam, R.C.K. & Lui, B.L.S. (2013) Public education and engagement in landslip prevention and hazard mitigation In: *Paper Submitted to the International Conference on Post-disaster Reconstruction – Sichuan 5.12* Hong Kong, The Hong Kong Institution of Engineers. 9 pp.

Tang, C.S.C. & Cheung, S.P.Y. (2011) Frequency analysis of extreme rainfall values. In: *Proceedings of the HKIE Geotechnical Division Annual Seminar on Landslide Risk Reduction through Works: Thirty-Five Years of Landslip Preventive Measures Programme and Beyond* Hong Kong, Hong Kong Institution of Engineers pp. 204–209

Wang, L. & Chen, W. (2014) Equiratio cumulative distribution function matching as an improvement to the equidistant approach in bias correction of precipitation. *Atmospheric Science Letters*, 15, 1–6.

Wang, Y. & Zhou, L. (2005) Observed trends in extreme precipitation events in China during 1961–2001 and the associated changes in large-scale circulation. *Geophysical Research Letters*, 32, L09707.

Wentz, F.J., Ricciardulli, L. Hilburn, K. & Mears, C. (2007) How much more rain will global warming bring? *Science*, 317 233–235.

WMO (2009) *Manual on Estimation of Probable Maximum Precipitation (PMP) (WMO-No. 1045)*. Geneva, World Meteorological Organisation. 257 pp.

Wong, A.C.W., Ting, S.M., Shiu, Y.K. & Ho, K.K.S. (2014) Latest developments of Hong Kong's landslip warning system. In: *Proceedings of World Landslide Forum 3, 2–6 June 2014, Beijing*. 6 pp.

Wong, H.N. (2005) Landslide risk assessment for individual facilities. In: *Proceedings of the International Conference on Landslide Risk Management, Vancouver, Canada*
break pp. 237–296.

Wong, H.N. (2009) Rising to the challenges of natural terrain landslides. Natural hillsides: Study and risk management measures. In: *Proceedings of the 29th Annual Seminar of the HKIE Geotechnical Division, Hong Kong, 17 April 2009*. Hong Kong, Hong Kong Institution of Engineers. pp. 15–53.

Wong, H.N. (2013) Enhancing the resilience of slope safety system against extreme events. In: *The International Conference on Sichuan Reconstruction – The Review and Prospects of Sichuan (5.12) Post-Disaster Reconstruction*. Hong Kong. 11 pp.

Wong, H.N. & Ho, K.K.S. (1995) *General Report on Landslips on 5 November 1993 at Man-Made Features in Lantau (GEO Report No. 44)*. Hong Kong, Geotechnical Engineering Office. 78 pp.

Wong, H.N., Chen, Y.M. & Lam, K.C. (1997) *Factual Report on the November 1993 Natural Terrain Landslides in Three Study Areas on Lantau Island (GEO Report No. 61)*. Hong Kong, Geotechnical Engineering Office. 42 pp.

Wong, H.N., Pang, L.S., Wong, A.C.W., Pun, W.K. & Yu, Y.F. (1999) *Application of Prescriptive Measures to Slopes and Retaining Walls (GEO Report No. 56)*. 2nd edition Hong Kong, Geotechnical Engineering Office. 73 pp.

Wong, H.N., Ko, F.W.Y. & Hui, T.H.H. (2006) *Assessment of Landslide Risk of Natural Hill-sides in Hong Kong (GEO Report No. 191)*. Hong Kong, Geotechnical Engineering Office. 117 pp.

Wong, M.C. & Mok, H.Y. (2009) Trends in Hong Kong climate parameters relevant to engineering design. In: *Presented in HKIE Civil Division Conference 2009, January 2009*. HKO reprint no. 832.

Wu, H.Y., Du, Y.D., Chen, Z.H., Pan, W.J. & Duan, H.L. (2011) Climatic variations of wet days and rain intensity in South China. *Journal of Tropical Meteorology* 27, 877–888.

Yu, Y.F. (2004) *Correlations Between Rainfall, Landslide Frequency and Slope Information for Registered Man-Made Slopes (GEO Report No. 144)*. Hong Kong, Geotechnical Engineering Office. 109 pp.

Zhai, P., Ren, F. & Zhang, Q. (1999) Detection of trends in China's precipitation extremes. *Acta Meteorologica Sinica*, 57, 208–216.

Zhai, P., Zhang, X., Wan, H. & Pan, X. (2005) Trends in total precipitation and frequency of daily precipitation extremes over China. *Journal of Climate* 18, 1096–1108.

Zhang, Q., Xu, C.Y., Becker, S., Zhang, Z.X., Chen, Y.D. & Coulibaly, M. (2009) Trends and abrupt changes of precipitation maxima in the Pearl River Basin, China. *Atmospheric Science Letters*, 10 (2), 132–144.

Zhou, B., Wen, Q.H., Xu, Y., Song, L. & Zhang, X. (2014) Projected changes in temperature and precipitation extremes in China by the CMIP5 multimodel ensembles. *Journal of Climate*, 27, 6591–6611.

Chapter 5

Potential climate changes in Italy and consequences for land stability

L. Picarelli and L. Comegna
DIcDEA, Seconda Università di Napoli, Aversa, Italy

S.L. Gariano
Istituto di Ricerca per la Protezione Idrogeologica, Consiglio Nazionale delle Ricerche, Perugia, Italy
Università degli Studi di Perugia, Dipartimento di Fisica e Geologia, Perugia, Italy

F. Guzzetti
Istituto di Ricerca per la Protezione Idrogeologica, Consiglio Nazionale delle Ricerche, Perugia, Italy

P. Mercogliano and G. Rianna
CMCC Foundation, Euromediterranean Center on Climate Change, Regional Models and geo-Hydrological Impacts Division, Capua, Italy

M. Santini
CMCC Foundation Euromediterranean Center on Climate Change, Impacts on Agriculture, Forests and Ecosystem Services Division, Viterbo, Italy

P. Tommasi
Istituto di Geologia Ambientale e Geoingegneria, Consiglio Nazionale delle Ricerche, Roma, Italy

ABSTRACT

Changes in the main weather features are becoming perceivable in Italy. In the last years, the average temperature and the intensity of rainstorms are increasing. Such phenomena could lead to changes in the earth-atmosphere interaction. Accounting for the high vulnerability of the Italian territory, an obvious consequence would concern the effects on the hydro-geological hazards, including landslides and floods. As a matter of fact, the higher frequency of high magnitude flood and landslide events represent an alarm bell. It is necessary to explore with all available instruments the consequences of the potential climate changes. Since quantitative predictions are not possible, the only way to run is depicting rational scenarios for the most vulnerable contexts adopting a multidisciplinary approach. Based on projections of potential climate changes in the 21st century, the report examines their potential impact in the context of the Italian peninsula. After a general overview about typical slope responses to weather forcing some quantitative scenarios have been depicted for representative geomorphological contexts. The last part of the paper examines changes in the expected risk, based on the characteristics of the Italian territory and its vulnerability, looking at the initiatives to undertake for its mitigation.

5.1 FOREWORD

In the last years, the average temperature in Italy has increased. The frequency and the intensity of high intensity rainstorms is also increasing, although the average yearly cumulated precipitation is not showing significant changes. High magnitude flood and landslide events are becoming more frequent; an alarm bell for the population, enhanced – and often exaggerated – by the mass media. Naturally, losses due to geo-hydrological hazards are not a sign of climate change, necessarily, as they also depend of the significant increase in vulnerability occurred in the last seventy years (i.e., mainly after the Second World War). However, politicians, administrators and institutions in charge of land management have the duty to focus on the consequences of the potential climate changes in the near future.

The expected changes in climate will modify the earth-atmosphere interactions, with special reference to humidity exchanges. This will lead to changes in the soil moisture and pore pressure regime, with relevant consequences on the mechanical behaviour of soils. A key question then concerns the potential consequences on geo-hydrological hazards, and the related risk. Since a quantitative forecast is not yet possible, the only way forward is depicting rational scenarios looking for the most vulnerable geomorphological contexts. This analysis needs to be carried out adopting a multidisciplinary approach, exploiting the consolidated knowledge in the domain of the concerned disciplines, including, atmospheric physics, hydrology, geology, hydraulics, geotechnical engineering, geomorphology, and finally economics and social sciences.

Based on recent projections of the potential climatic changes on a global and local scale, the report examines some potential consequences on the expected changes of the main atmospheric variables in Italy. To design realistic scenarios, the projection is limited to the end of this century. Longer term phenomena such as the effects of ice melting on the stability of valley slopes, or the consequences of eustatism on coastal cliffs, are ignored. After a general overview on the present knowledge about the incoming climate changes in Italy and the relationships between weather and slope instability, quantitative scenarios are presented for typical geomorphological settings. The final chapter discusses expected variations in landslide risk, considering the vulnerability of the Italian territory, and puts forwards considerations for landslide risk mitigation in a changing climate.

5.2 PRESENT AND FUTURE CLIMATE TRENDS IN ITALY

5.2.1 Weather in Italy

Italy extends for about 1300 km between the 37°N and 45°N. Flat areas, hills and mountains (above 700 m) cover 23.2%, 41.6% and 35.2% of the total surface, respectively. The two main mountain ranges are the Alps (1200 km in length, with an average elevation of 1250 m and the highest peak at the Monte Bianco, reaching 4810 m), which mark the Northern boundary, and the Apennines (1200 km in length, with an average elevation of 500 m, and the highest peak at the Gran Sasso, reaching 2912 m), which form the backbone of the Italian peninsula. This configuration is characterised by the coexistence of areas with different precipitation and thermometric regimes, and where landslides are abundant and frequent.

Figure 5.1a shows the mean annual temperature in Italy, in the 30-year period from 1961–1990 (used here reference time interval). As shown, this ranges from about 4° to about 18°, with variations governed by elevation and latitude. The role of orography is evident from the maps showing the mean annual cumulated precipitation (Figure 5.1b) and the mean daily maximum precipitation (Figure 5.1c). Inspection of the maps reveals a lower cumulated annual value (<1000 mm per year) along the coasts, in the main islands and in the Po River valley, whereas in hilly and mountain areas, on average, the cumulated precipitation exceeds 1500 mm, with peaks of 2500 mm in the Eastern Alps. The geographical distribution of the mean daily maximum precipitation shows quite a different geographical pattern, with the highest values (around 150 mm/day) in the Eastern Alps and in the areas (as Liguria and Eastern Calabria) where mountains are located near the coasts, most probably an effect of wet fronts from the sea. Lower values of the mean daily maximum precipitation are recorded in the other areas, with a minimum of about 80 mm/day in the inner zones.

A useful summary of the main climatic features for Italy was provided by Kottek *et al.* (2006) through the Köppen-Geiger classification, which outlines different climate zones based on annual and seasonal values of precipitation and temperature. Using available datasets (CRU, Climatic Research Unit [http://www.cru.uea.ac.uk/data] and GPCC, Global Precipitation Climatic Centre [http://www.esrl.noaa.gov/psd/data/gridded/data.gpcc.html]) with a ground resolution of 0.5° × 0.5° for the 1951–2000 period, they provided the map shown in Figure 5.2, where a number of different macro areas can be distinguished. The Alps are associated with snow/tundra climate conditions, and the remaining part of the territory is ranked as warm temperate climate with differences based on precipitation and temperature summer regimes.

5.2.2 Evidence of climate changes in atmospheric patterns in Italy

Different studies carried out by the Intergovernmental Panel on Climate Change (IPCC) consider Southern Europe as a highly vulnerable area to climate change (Giorgi, 2006). This peculiarity is well illustrated by Brunetti *et al.* (2006), who studied the temperature and precipitation evolution in Italy between 1750 and 2000, revealing negative temperature anomalies in the first part of records (until 1860), followed by a positive trend until 1950 and by a slight decrease from 1950 to 1970 (Fig. 5.3). Beginning in 1970, a relevant positive anomaly was observed until 2003, the warmest year in the records. A comparison with global trends suggests that the positive trend in Italy is more pronounced than in other geographical areas.

A detailed analysis of the available climate data reveals significant differences among the seasons. In particular, the warming trend is more evident in the spring and the summer than in the fall and the winter. Concerning minimum (T_{min}) and maximum (T_{max}) annual and seasonal temperatures, Brunetti *et al.* (2000) reported a significant positive trend in the period 1865–1996, greater in Central and Southern Italy than in

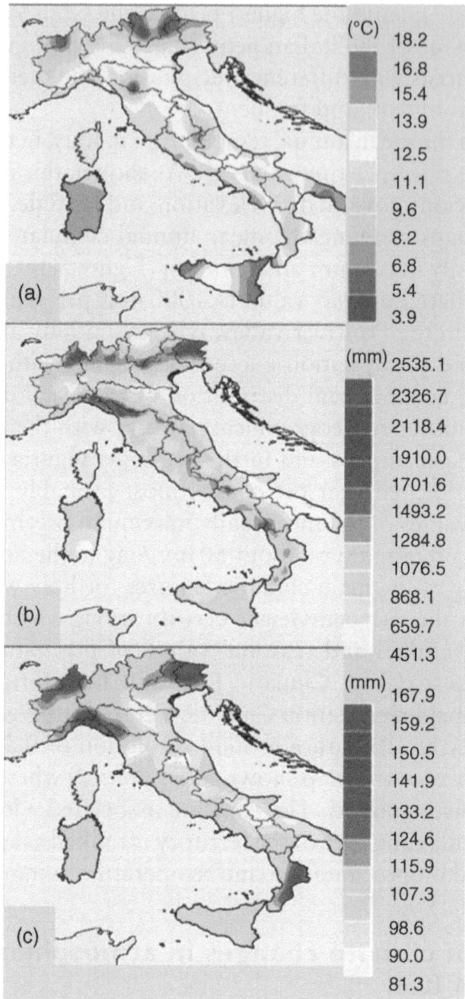

Figure 5.1 Precipitation and temperature regimes in Italy from 1961 to 1990 (source: http://scia.ispra.it): a) mean annual temperature; b) mean cumulated annual precipitation; c) mean daily maximum precipitation.

Northern Italy. The positive trend concentrates in the last 20 years in Northern Italy, and in the last 50 years in Southern Italy. The Diurnal Temperature Range (DTR) also shows a significant trend, in the same observation period. The increase is larger in Northern Italy (except in the winter, when it displays a decreasing trend) and less relevant in Southern Italy (except in the spring and summer, when the variation is not statistically significant). It is worth remembering that DTR represents an important proxy for climate change (Braganza *et al.*, 2004) due to its susceptibility to a variety of environmental effects, including water vapour, cloudiness, and urban influences (IPCC 2007). The observed temperature increase leads to greater evaporation, and thus

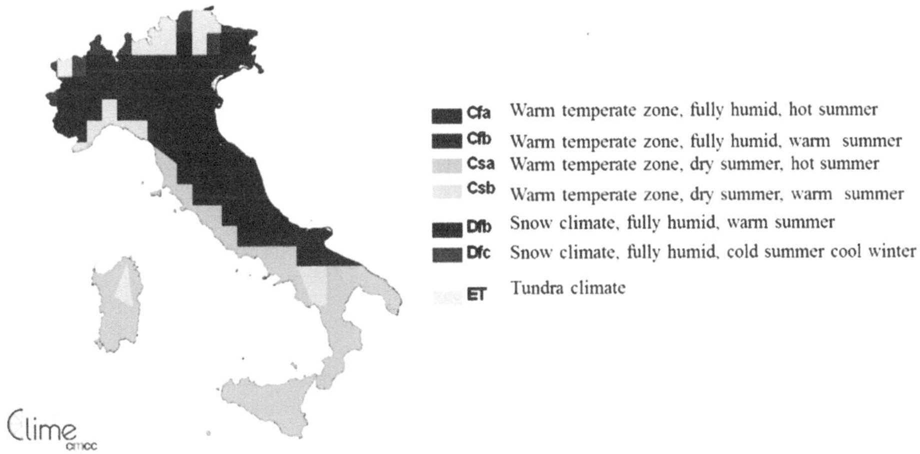

Figure 5.2 Map of Köppen-Geiger climate classification updated with mean monthly temperature and precipitation data for the period 1951–2000 (Kottek *et al.*, 2006).

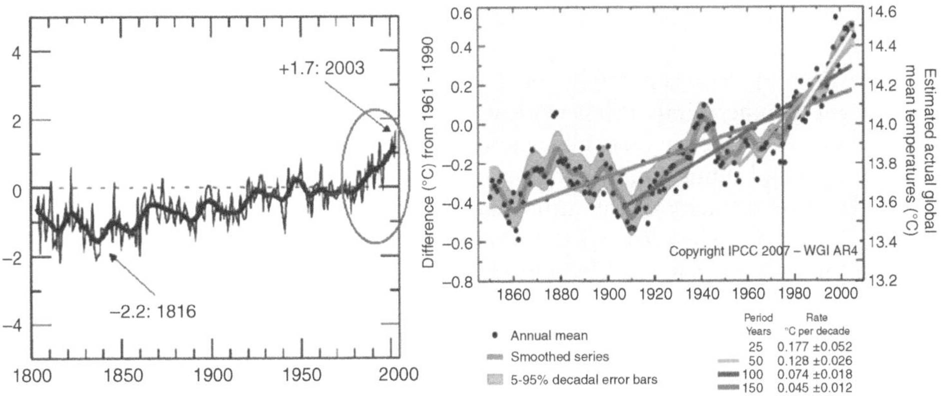

Figure 5.3 Left, pattern of mean temperature anomalies (calculated with respect to the reference period 1961–1990) from 1800 up to 2003 (Brunetti *et al.*, 2006): the coldest (1806: −2.2°C) and the hottest year (2003: +1.7°C) are shown; right, trend of the global mean temperature from 1850 to 2005 (IPCC 2007 –WG1 AR4): the linear trends for the entire 150-year period and for the last 100, 50 and 25 years, reveal that the temperature has been increasing significantly.

surface drying, causing an increase in the intensity and duration of drought periods. Brunetti *et al.* (2002, 2004a) found an increase in dry spells in the period 1951–2000.

Different studies suggest a slight increase in rainfall trends in Western Italy, which represents a local anomaly given the general agreement about the opposite trend in the rest of Italy towards dryer conditions (Piervitali *et al.*, 1998; Brunetti *et al.*, 2004b; Blasi *et al.*, 2007). The decrease in total precipitation is associated with a significant decrease in the number of rainy days. Considering the amount of precipitation during

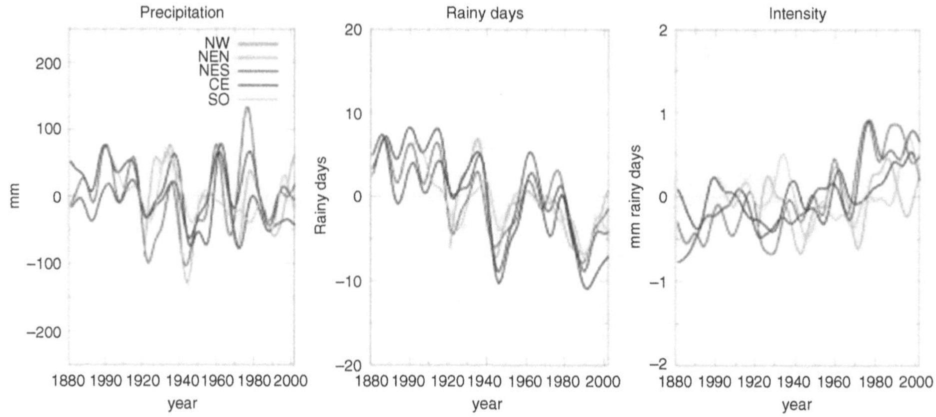

Figure 5.4 Cumulated yearly precipitation, annual number of rainy days, and rainfall intensity from 1880 to 2005. NW (North West), NEN (North East, northern part), NES (North East, southern part), CE (central part), SO (South). The series are represented by a Gaussian filter with a 5-year standard deviation.

rainy days, an increase is noted in some areas in Northern Italy, whereas the trend in Central and Southern Italy is less evident (Brunetti *et al.,* 2004b, Fig. 5.4). For heavy precipitations, Trenberth (2011) has highlighted the role of feedbacks associated with global warming leading to an increase of such events, first of all the increase in the water retention capacity of the atmosphere under the effect of temperature growth (about 7% per °C). The IPCC (2007) reports that an increase in heavy events could be inferred for wide areas in the Mediterranean (Alpert *et al.,* 2002). Similar findings for Italy are given by Kostopoulou & Jones (2005).

Warming temperatures and a generalized tendency to less precipitation have led to consistent changes also in snowfall, and in the extent of the snowpack in mountainous areas. In the Italian Alps less snowfall and snow accumulation was observed by Valt *et al.* (2005). Shorter snow cover periods have been associated with earlier melting, despite the later occurrence of the first snowfall (Laternser & Scheebeli, 2003).

Climate represents a complex of inter-related variables. On average, through a set of regulatory mechanisms, a smooth change in one variable triggers smooth changes in other variables. In particular, changes in temperature have effects on precipitation, but also on water vapour pressure and potential evapotranspiration. Considering these correlations, a likely effect of climate changes is the intensification of the hydrological cycle. Evidences of this effect have been observed in the Mediterranean area (Mariotti *et al.,* 2008). However, it is difficult to identify long-term trends for the components of hydrological systems, as they are often masked by a significant natural variability, on time-scales from inter-annual to decadal. This variability, together with limitations in the spatial and temporal coverage of existing monitoring networks, can explain the current uncertainty in the definition of the general effects of climate change over the hydrological cycle (Bates *et al.,* 2008).

5.2.3 Climate change scenarios in Italy

An assessment of the future evolution of weather forcing is usually carried out by exploiting a modelling chain which includes three main modules.

First, based on assumptions on the future economic development and population change at the global and regional scales, Integrated Assessment Models (IAM) provide an evaluation of future concentrations of greenhouse gases (GHG), aerosols, chemically active gases (climate-alterant gases), and changes in land-use until 2100. In this regard, the IPCC has selected four reference standard pathways (commonly known as RCPs, Representative Concentration Pathways) for subsequent analyses by means of Climate Models (CMs), allowing for comparisons among different climate projections. The four pathways estimate an increase in radiative forcing levels of 8.5, 6, 4.5 and 2.6 W/m^2, by the end of the present century, compared to the pre-industrial era (1765). Every RCP covers the 2006–2100 period, but extensions have been formulated until 2300. Concerning the main greenhouse gas, CO_2, the RCP8.5 follows the upper range suggested by the literature (i.e., rapidly increasing concentrations) associated with huge demographic increase and continued exploitation of resources. RCP6 and RCP4.5, representing a stable concentration of CO_2, are considered medium-range concentration or high mitigation scenarios. Finally, RCP2.6 assumes a peak in CO_2 concentrations around 2050, followed by some decline by the end of the century (low-range mitigation scenario). Concerning CH_4 and N_2O, respectively the second and third prevalent greenhouse gases produced by human activities, their future evolutions are strictly associated with climate policy (van Vuuren *et al.*, 2011). Practically, while RCP2.6 assumes that, through drastic policy intervention, greenhouse gas emissions will be reduced almost immediately leading to a slight reduction of present levels by 2100, the worst case scenario, RCP8.5, assumes more or less unabated emissions.

Second, such pathways are then used as forcing for Global Climate Models (GCMs). These are numerical models adopted to assess the impact on the climate system of variations of climate-alterant gases. Nevertheless, due to their coarse horizontal resolution (currently hardly exceeding 70–80 km), GCMs can simulate only large-scale atmospheric states (Von Storch, 1993). Several studies (Breugem, 2007; Cubasch *et al.*, 2013) have shown that these models can reproduce the climate and the global response to the changes in climate-alterant gases with higher reliability for some variables (temperature) and lower for others (precipitation). However, despite significant recent developments, the available GCMs are inadequate for estimates of trends and impacts at local/regional scales, for which some areal features (e.g., distance from the sea, topography) are crucial.

Lastly, different downscaling techniques have been developed in the last years. They are classifiable as "statistical" or "dynamical" approaches, and largely differ for their computational costs, prerequisites, and limitations. The first approach adopts frameworks based on empirical statistical relationships between large-scale "predictors" and the "predictand" local climate variables, calibrated and validated on observed data, then applied to GCM variables. These statistical models require limited computational resources and allow for analyses at the single station scale, but need long time series of observed data for the definition of the statistical relationships. The dynamical approach requires the use of climate models at limited area and highest resolution (RCM, Regional Climate Model) nested on the global models from which they can draw suitable boundary conditions. Currently adopted resolutions, in the order of 10 × 10 km, allow for a better resolution of the orography solving local atmospheric phenomena. Different experiments have proven the capability of the dynamic models to reproduce the regional climate variability due to the effects of local orography, land-sea contrast and small-scale atmospheric features such as convective cells (Feser *et al.*, 2011).

In the last years, several consortia within different international projects have carried out "ensemble" simulations on fixed domains in order to size the biases in estimating the main weather variables and to assess the potential future variations. The adoption of ensemble approaches is currently highly recommended by the climatological community and is considered the main tool for assessing the current uncertainties in the evaluation of ongoing and future weather patterns. For these reasons, in more recent years, the WCRP Coordinated Regional Downscaling Experiment (CORDEX) project was established (Giorgi *et al.*, 2009). CORDEX provides a global coordination for Regional Climate Downscaling aimed at improving climate change adaptation measures and related impact assessments. Within this consortium, different partners provide climate simulations on current and future time spans on fixed geographical domains and established horizontal resolution. For Italy, a high

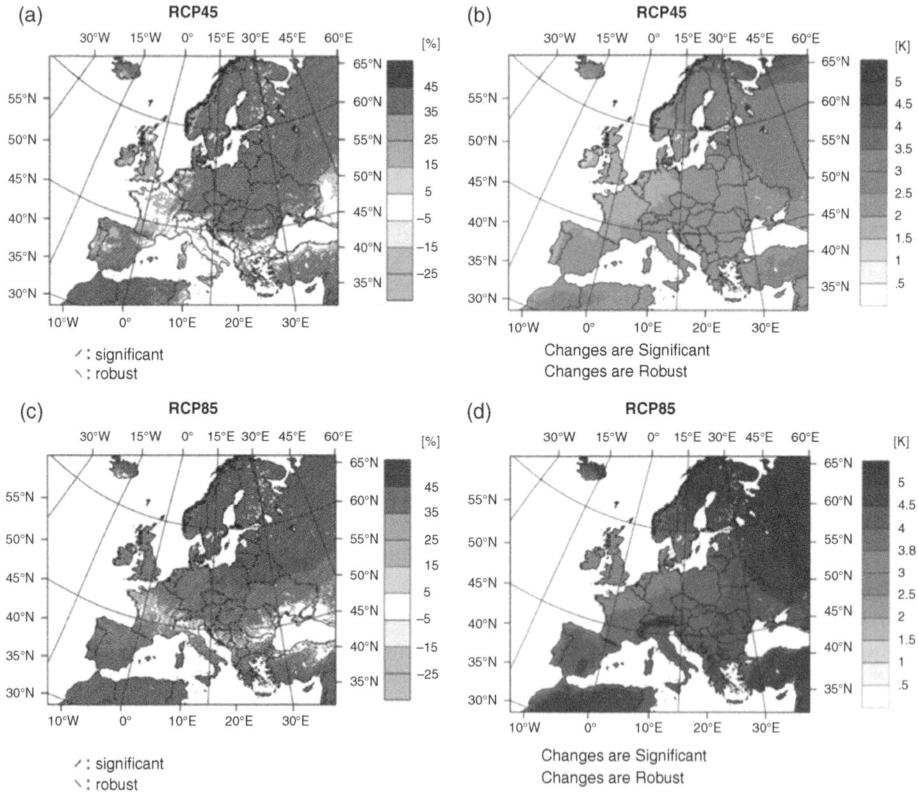

Figure 5.5 Projected changes of annual precipitation (%) and annual mean temperature [K] for 2071–2100 compared to 1971–2000 under RCP8.5 (c, d) and RCP4.5 (a, b) scenarios.

resolution RCM ensemble (EURO-CORDEX) has been completed at different spatial resolutions, the best of which is 0.11° × 0.11° (about 11 × 11 km). Simulations employ the IPCC RCP2.6, RCP4.5 and RCP8.5 concentration scenarios.

The available climate projection analyses are based on nine simulations adopting RCP4.5, and ten simulations using RCP8.5. The main findings are presented by Jacob *et al.* (2014) in terms of mean values and climate indices useful for impact studies. The results show a robust and statistically significant warming all over Italy, in the range of 2.0–3.5°C for RCP4.5 (Fig. 5.5b) and of 4.0–5.5°C for RCP8.5 (Fig. 5.5d). Associated with the strong increase in temperature for RCP8.5, there exist significant changes in annual precipitation (Fig. 5.5c) characterized by a statistically significant decrease in Southern Europe and an increase in the North-Eastern area. Seasonal mean variations for the heavy precipitations (defined as the 95th percentile of wet days) are shown in Figure 5.6 (for RCP8.5) and reveal a generalized increase for a large part of the domain. The largest increases characterise again North-Eastern Europe. For Italy, an increase is expected in all seasons except June, July and August (JJA) (higher in September, October, and November, SON). The change in pattern is very similar

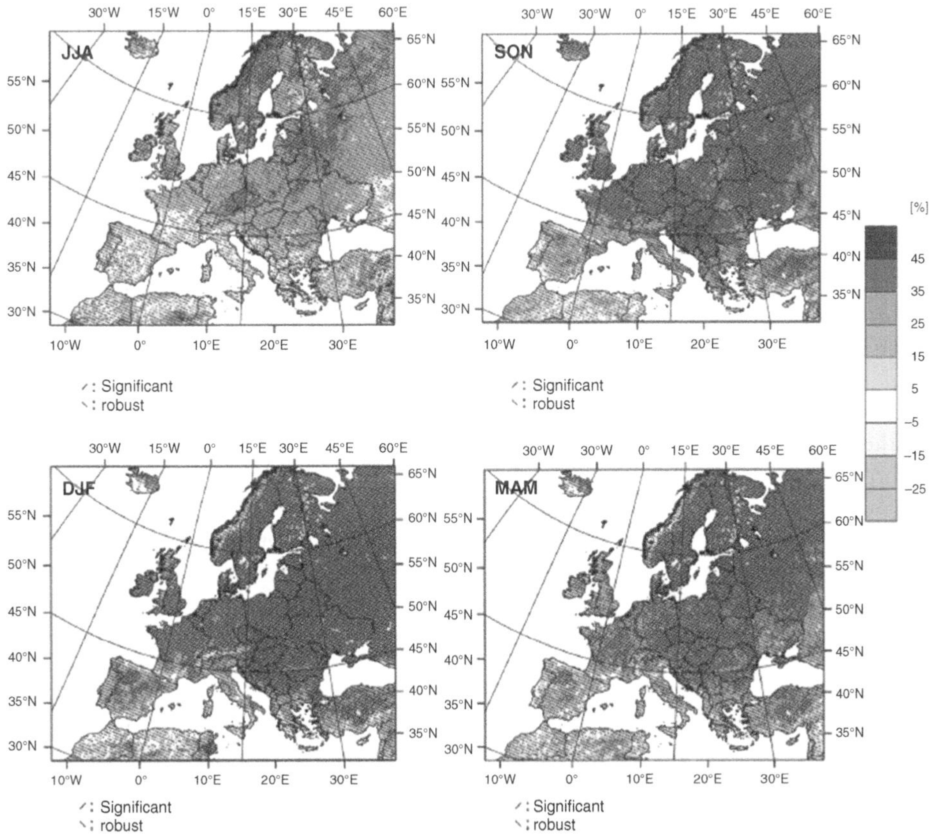

Figure 5.6 Projected seasonal changes of heavy precipitation (%) for the RCP8.5 scenario in 2071–2100 compared to 1971–2000. Hatched areas indicate regions with robust and/or significant change.

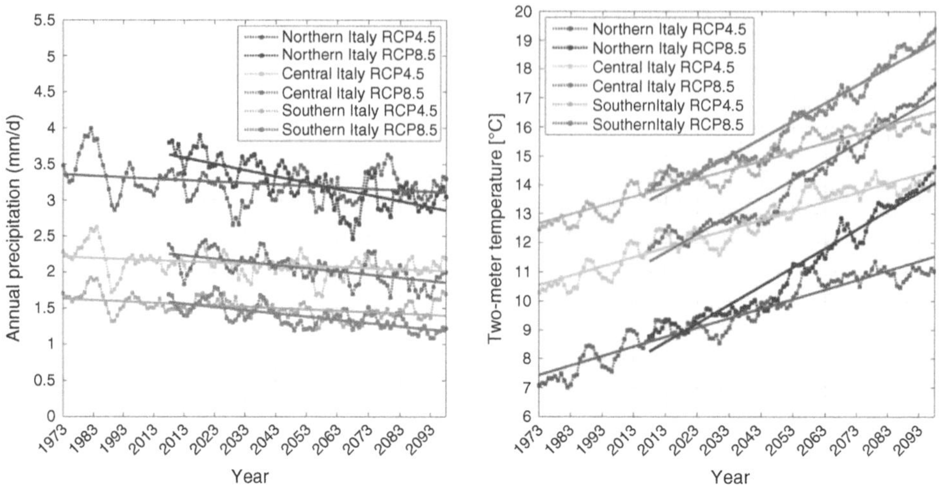

Figure 5.7 Left, time series (and trend lines) for annual precipitation over Northern, Central and Southern Italy, based on the RCP4.5 and RCP8.5 scenarios; right, time series (and trend lines) for the 2-meter temperature over Northern, Central and Southern Italy, for the same scenarios.

for RCP4.5, but less pronounced either for the annual (Fig. 5.5a) and for the heavy precipitation (not shown in the figures). The results also allow testing the robustness and significance of the climate simulations, confirming that the different RCMs are significantly consistent.

With special reference to the Italian domain, recent climate simulations adopting a resolution up to 8 × 8 km (Montesarchio *et al.*, 2014; Bucchignani *et al.*, 2015; Zollo *et al.*, 2015), confirm previous findings regarding a reduction of cumulated annual precipitation and an increase in temperature and heavy precipitation. Trends provided by the simulations in the period 1971–2100 for RCP4.5 and RCP8.5 are shown in Figure 5.7, that portrays separate values for Northern, Central, and Southern Italy. The warming expected in Italy at the end of the 21st century is more evident using the RCP8.5 scenario (Fig. 5.7b), whereas a slight decrease in precipitations for RCP4.5, and more evident for RCP8.5, is given (Fig. 5.7a). Specifically, the projections under RCP8.5 suggest a significant reduction of precipitation in the spring and the summer and an increase of the winter precipitations in Northern and Central Italy (not shown in Figure 5.7).

5.3 INFLUENCE OF WEATHER ON LANDSLIDING IN ITALY

The geologic, orographic and weather conditions in Italy lead to a high landslide susceptibility, produced by landslides of different types. Triggering (or acceleration) of almost all types of landslides is frequently governed by weather, in the form of soil moisture and/or pore pressure changes. This relationship can be relatively simple or

Table 5.1 Main landslide types susceptible to climate changes in Italy.

Landslide type	*Varnes (1978)*	*Hungr et al. (2014)*
Rock failure on cliffs and steep slopes	Fall, Topple	Topple, Rock wedge slide, Rock irregular slide, Rock planar slide, Rock fall
Slow rock slide	Slide	Rock rotational slide, Rock compound slide
Rapid shallow slide ("soil slip") in coarse grained soils, possibly generating debris flow or flowslide	Slide, Flow	Gravel/sand/debris slide, Debris flow, Flowslide
Rapid shallow slide in cohesive soils ("soil slip")	Slide	Clay-silt rotational slide, Clay/silt planar slide
Slow to very slow translational slide in cohesive soils	Slide	Clay/silt planar and compound slide
Earthflow	Flow	Earthflow

complex depending on the "hydraulic stratigraphy" of the slope, which conditions both infiltration and flow paths. Pore pressure changes are predominant in shallow layers, but open discontinuities or other hydraulic connections allow infiltrated water to reach deeper layers. Weather can also exert other actions, including concentrated erosion due to runoff or debonding caused by ice melting. As an example, Piccarreta *et al.* (2012) stressed the effects of changes in the rainfall regime on erosion processes in the Fossa Bradanica (Southern Italy). By analysing half-century rainfall data (1949–2000), they found an increase in erosion linked to the extension of the dry topsoil layer and the concentration of high intensity rainfall in macro-events of 3–4 consecutive days.

On the one hand, extreme weather conditions can trigger first-time landslides, reactivate quiescent slope movements or accelerate active phenomena. On the other hand, they can slow down or arrest landslide movements. Notice that the term "extreme" refers to both short-term and long-term conditions. Short-term extreme conditions include sharp increases in the rainfall amount within hours or a few days (increase in intensity, duration or both) or sharp rises in temperature. Long-term extreme conditions include the increase of daily rainfall and the number of rainy days over long periods, as well as prolonged dry and/or warm periods. Since perturbations of rainfall and temperature regimes in the short- and the long-term act differently on the different types of slope movements, and the expected climate changes should exacerbate in a different way short- and long-term perturbations, a description of the landslide types that are more sensitive to weather forcing in Italy can contribute to understand the potential impact of the expected climate changes. Such types, consistent with the Varnes classification (1978), are listed in Table 5.1.

5.3.1 Rock failures on cliffs and steep rock slopes

Climate changes have a major impact on steep rock slopes, where the change in temperature increases ice melting and the altitude of the frost line. Fractured rock slopes are subjected to sudden failures involving small volumes of material. Increase in temperature can act through an augmented frequency of freeze and thaw cycles that weaken the rock mass, and through ice melting which causes a strength loss in intensely fractured rocks and an increase of water pressures in joints.

Referring to an inventory including several mountain ranges in Italy, Fischer *et al.* (2012) noticed an increase in rock failures during a period of global temperature rise in zones located in permafrost areas, especially along the lower permafrost boundary for small-volume events. Huggel *et al.* (2012) examined slope failures in the mountains of Europe, America and Caucasus distinguishing between rock and ice avalanches and debris flows from de-glaciated areas. In particular, they examined the influence of air temperature on landslide activity (involving both ice and rock) at Monte Rosa, in the Italian Alps, and hypothesized that changes in permafrost could have had a major role in the 2004 and 2007 failures occurred on the Eastern flank of the mountain (4000 m a.s.l.) and on the Southern flank of Punta Thurvieser (3250–3600 m a.s.l.). They further noticed that some of the failures occurred at the end of the 1980s and early 1990s, following a strong increase in local temperature. The consequent large mass loss produced significant changes in the local geomorphological conditions, which influenced subsequent failures. They concluded that slope instability in that area may be initially triggered by climate changes, but can evolve independently from additional climatic stimuli. Duca (2013) identified ice segregation as a mechanism leading to rock mass degradation and instability at Matterhorn peak (NW Italian Alps). Ice-rock interaction has also been considered among the causes of the 40000 m^3 failure on the northern cliff of Cima Una in the Eastern Dolomites (Viero *et al.*, 2013).

5.3.2 Slow rock slides

Even though large slow moving slides on mountain slopes are widespread over the entire Alpine range, and especially where metamorphic rocks crop out, quantitative data are available only where the movements interact with civil engineering structures. Rockslides along shear surfaces located in the uppermost fractured and loosened part of the bedrock are often superimposed on deep-seated deformations with more complex kinematics. Selected monitoring-based case studies testify a clear relationship between the rainfall regime and movement accelerations. Documented examples include the Spriana (Belloni & Gandolfo, 1997), Rosone (Pisani *et al.*, 2010) and Ruinon landslides (Crosta & Agliardi 2003). In some cases, the presence of a thick shear band of fine-grained materials at the slide bottom strongly limits groundwater circulation to the shallowest part of the slope, favouring the increase of pore pressures within the slide mass. This scheme is consistent with monitoring data at the Rosone and the Spriana landslides. In the former case it is supported also by numerical seepage analysis in transient conditions. In the Ruinon landslide rainfall had a significant influence both on the short-term velocity in the wet season (Del Ventisette *et al.*, 2011) and on the medium-term velocity (general seasonal trend) (Crosta & Agliardi, 2003). At Spriana, impulsive accelerations (correlated to short and intense rainfall) and long-lasting accelerations (correlated to prolonged rainfall) were observed (Agliardi & Crosta, 2014). However, when only surface displacements are measured is not easy to recognize whether the short-term accelerations are due to local movements of the debris cover or to deep-seated movements.

In all the mentioned rock slides snow melting represents a significant source of infiltration. Snow can cover wide areas with considerable thickness especially in the upper part of the slope (up to 1800 m, 1600 m and 2002 m a.s.l., for the Spriana,

Rosone and Ruinon landslides, respectively). Another action of snow/ice melting is suggested by Crosta *et al.* (2004), who pointed out that melting of the snow cover and glacier (0°C line raised to high elevation for a long-period), before the 1987 Val Pola landslide, increased considerably the surface runoff and hence erosion of the gullies bordering the unstable slope.

Similarly, quantitative data showing a striking link between snow melting, piezometer levels, and slope displacements were outlined by Weidner *et al.* (1998) for the Gradenbach deep-seated slope movement (Germany) involving debris and fractured metamorphic bedrock. Snow melting provided a slow but remarkable groundwater recharge which produced an increase of pore pressures within the whole slope. Consequences can be recognized in both short-term accelerations (in conjunction with short and intense rainfall) and the long-term slope behaviour.

The expected climate changes could have moderate effects on these movements. In fact, groundwater pressure seems to depend on long-term rainfall, which should not increase significantly in the near future. Similarly, the incidence of snow melting could be only translated during the year as an effect of the increase of temperature in the cold season, but the total snow height is not expected to increase. Only the increased melting rate of glaciers could raise groundwater levels in slopes located at high elevations or in those which are hydraulically connected to the glacier areas.

5.3.3 Rapid shallow slides and flowslides in coarse-grained soils

Shallow landslides in Italy typically involve moraines, debris and pyroclastic soils. These materials are often only partially saturated. Their behaviour is then governed by changes in suction due to short-term rainfalls which make the recurrence of these events extremely prone to climate variations (Picarelli *et al.*, 2012; Damiano & Mercogliano, 2013). Slope collapse may occur as a general shear failure and possibly evolve into a flow as a consequence of static liquefaction (flowslide). The latter mechanism requires soil saturation and can affect even relatively gentle slopes (Picarelli *et al.*, 2004). Flow conditions establish immediately after failure as an effect of pore pressure raising. Climate variations can also have an influence on debris flows initiated by mechanisms related to the erosion due to surface flow (riling of "firehouse" effect), which are favoured by concentration of rainfall in very short time intervals.

The relationship between rainfall and pore pressure/flow regime, and hence landslide initiation, depends on the hydraulic properties of the materials, their stratigraphy and the hydraulic boundary conditions, including bedrock depth and the presence of springs close to the landslide source area (Cascini *et al.*, 2003). The possible influence of stratigraphy on the pore pressure distribution in pyroclastic slopes is illustrated in Figure 5.8.

For the pyroclastic covers that mantle parts of the slopes in the Campania Region (Southern Italy), the relationships have been investigated through numerical and physical modelling based on laboratory and *in situ* testing, which were subsequently validated again *in situ* by monitoring data (Olivares & Damiano, 2007; Damiano *et al.*, 2012; Pirone *et al.*, 2012). For the debris covers of the Alpine and Apennine mountain areas, the relationships between rainfall and landslide initiation have been mainly investigated through analyses of the rainfall regime (intensity and duration) following

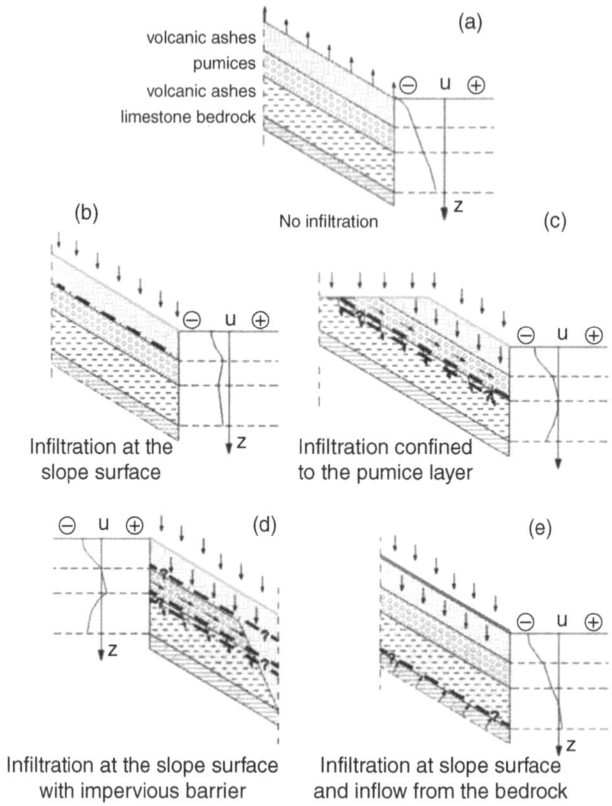

Figure 5.8 Triggering scenarios and possible evolution of pore pressure vs. depth for different hydraulic stratigraphy and boundary conditions (modified after Picarelli & Olivares, 2001).

major landslide and rainfall events. Based on numerical models and *in situ* monitoring, Pirone (2010) highlighted that duration-intensity curves for the same stratigraphy and boundary conditions are to be referred to a specific value of the initial suction, explaining the ineffectiveness of some apparently severe rainfall time series. In this respect, Damiano & Mercogliano (2013) pointed out that short storms during dry periods may influence the stability of only very shallow layers, whilst intense storms after long rainy periods may lead to failure the whole pyroclastic cover, depending on the initial saturation.

Rapid shallow landslides involving colluvial and debris covers formed by coarse-grained materials ("soil slips") are known to be very sensitive to the short-term rainfall regime, as it has been evidenced by intensity-duration rainfall thresholds.

Extensive investigations on soil slips were conducted in the Central Alps (Crosta & Frattini, 2009), Piedmont (North-Western Italy) (Aleotti, 2004), Northern Tuscany (Garfagnana) (Giannecchini *et al.*, 2012) and Calabria (Southern Italy) (Antronico *et al.*, 2004). General studies on rainfall regime-occurrence relationship have also been conducted (Guzzetti *et al.*, 2007). These phenomena are all related to short and very intense rainfall events thus assuming a particular importance in the light of the observed changes in rainfall regime.

In the Alps, debris flows are very frequent and locally associated with glacier outbursts (Chiarle *et al.*, 2007). Stoffel *et al.* (2014) analysed the changes in frequency, seasonal distribution and number of shallow landslide occurrences in the Piedmont Region from 1960 to 2011 and found that until 2002 the scenario was dominated by shallow landslides during prolonged and often extreme autumn rainfall events. After 2002, landslides occurred more frequently in early spring and are commonly triggered by moderate rainfalls, whereas in autumn events characterized by a large spatial density of landslides are less abundant. The authors linked the change in landslide activity to the change in the seasonal distribution of precipitation that changed from a bi-modal distribution with maxima during spring and summer, to an increase in winter precipitation and in dry conditions in spring and summer. In the decade 2002–2011, a significant increase in landslide frequency and density was noticed. Also, the Authors related the changes in landslide activity to elevation. Above 1500 m a.s.l. the expected decrease in snowpack and duration in future winters and springs will probably affect the frequency and seasonality of landslide activations.

In the central Apennines, visual inspection of historic stereoscopic aerial photographs revealed that debris flow activity was largest in the first half of the 20th century (Guzzetti & Cardinali, 1991). In the period from about 1920 to 1950, significant and widespread afforestation was conducted in mountain areas in the Apennines and elsewhere in Italy. The new forests, coupled with the construction of extensive systems of check dams, have reduced the frequency and magnitude of the debris flow/debris torrent events and their destructive consequences.

5.3.4 Rapid shallow slides in cohesive soils

This type of landslide includes chiefly small translational and roto-translational slides occurring primarily on pre-existing weak layers in overconsolidated clays or in weak rocks. A typical example are the planar slides along clay horizons (up to 20 m deep) of a regularly stratified rock-like marly formations in the Langhe region, North-Western Italy. On the occasion of prolonged intense rainfall, several events were more or less simultaneously triggered over a vast area (Govi *et al.*, 1985) producing damage and casualties. Tiranti *et al.* (2013) indicated that these planar slides may be significantly influenced by snow melting, when occurring at low elevations.

Large roto-translational landslides are less abundant. A well documented case is the failure of a 50 m thick moraine deposit composed of silty clays of low plasticity, overlying a marly flysch (Giau Pass, Dolomites) (Angeli, 1992). Also in this large landslide snow melting was recognized to influence groundwater circulation both at the site (local) scale and at the catchment scale, and to exert a major influence on the acceleration of the displacement.

5.3.5 Slow to very slow translational slides in cohesive soils

This type of landslide occurs in overconsolidated soils and in structurally complex formations (especially tectonized clay shales and marly flysch). Case histories in colluvial-eluvial covers of crystalline formations have also been described. Movements are localized along pre-existing slip surfaces in slopes that are close to limit equilibrium conditions. The switch from stable to unstable conditions is due to water infiltration, which benefits from prolonged rainfall over long periods and reduced evapotranspiration. In this view, the expected climate changes seem to have an overall positive effect on the stability of slopes affected by this type of failure. In fact, the expected climatic changes should lead to some increase in the average temperature and to longer dry periods.

Due to the large diffusion of marine and lacustrine overconsolidated deposits and of structurally complex fine-grained materials (Esu, 1977), this type of movement is relatively frequent in Italy. Two different types can be singled out that locally coexist in the same slope (e.g., multi-storied landslide of Skempton & Hutchinson, 1969): (a) slow to moderate shallow movements which occur in the colluvial cover of slopes carved in overconsolidated clays or in the colluvial-eluvial cover of structurally complex formations and (b) very slow to extremely slow deep-seated movements. In overconsolidated clays the slip surface is located in the uppermost softened, at times weathered, part of the formation. In structurally complex formations it is located at the base of landslide debris deposits or in the very first meters of the underlying bedrock. Depths usually range between 20 and 40 m excepting for a few cases, such as the Ancona landslide (Cotecchia, 2006), where exceptional geological features make the movements extend to larger depth.

Type (a) and type (b) movements exhibit a different response to variations in the rainfall regime, and hence to the expected climate changes. The former movements are shallower, and hence have shorter infiltration paths. Hydraulic conductivity depends on the different response to "degradation" (e.g., slaking, weathering, softening) and remoulding processes of the parent formation. Conversely, the remarkable depth of type (b) landslides entails long infiltration and seepage paths. Some structural features (conductive layers, as sand layers in overconsolidated clays or fractured limestone/sandstone layers in structurally complex formations) can have significant influence on hydraulic conductivity.

Type (a) landslides are widespread in the colluvial cover of many overconsolidated clays and of structurally complex formations with a clay or marly component. Values of v_{max} are high, and movements are much more affected by variations in rainfall patterns than those of the type (b) landslides.

In Orvieto, Central Italy, a 10 m thick landslide debris cover slides at $v_{av} = 40$–50 mm/year and reactivates typically more than once every hydrological year. As a consequence of the reduced infiltration paths, pore pressure raises and hence accelerations correlate well to increments of rainfall accumulated over 15 to 30 days (Tommasi *et al.*, 2013). A similar behaviour is exhibited by movements in the colluvial cover of the Adriatic overconsolidated clays (Central-Eastern Italy), whose periods of activity are dictated by the suction changes in the upper few metres (Alonso *et al.*, 2003).

Well documented examples of type (b) landslides in marly-clays of the Adriatic area are described by Bertini *et al.* (1984, 1992) and by Angeli (1992). The former (18 to 30 m thick) move at an average rate v_{av} of 1 to 2 mm/year, along 1 to 5 m thick shear bands located within the uppermost softened part of the basal formation. The latter display values of v one order of magnitude larger. Piezometer levels in the uppermost part of the basal formation are more uniform than in the overlying cover and accelerations (up to 35–60 mm/year) correlate well with the occurrence of long rainy periods. Seasonal pore pressures change and multiannual trends of deep piezometers correlate with medium-term and long-term variations of the rainfall regime (Bertini *et al.*, 1987). Similar movements occur also in the Orvieto slope; their kinematical features and potential response to climate changes will be analysed in detail in the following section.

Slopes in structurally complex formations with prevailing lithic component present locally a similar behaviour. As an example, the large Montemartano slide (Grana & Tommasi, 2014) mobilizes a strength close to residual values of marly-clayey interbeds of the unit forming the bedrock, along a slip surface located at a depth of 38 m. Movements show rainfall-induced seasonal accelerations with a maximum displacement rate of 75 mm/year and do not stop completely even in the dry season when v_{max} drops to a few millimetres per year.

5.3.6 Earthflows

Earthflows are widespread in the piedmont areas of the Apennines and in many areas of the Alps. They mostly involve clay-marly structurally complex formations (Esu, 1977). Composition of the parent rock is important because the higher the marly-clay fraction, the higher is the attitude to soil degradation and earthflow triggering (Picarelli & Di Maio, 2010). These movements are characterized by cyclic reactivations (Giusti *et al.*, 1996) which make displacement rates to vary widely (more than one order of magnitude). Based also on the movement rate, Guida & Iaccarino (1991) recognized four stages in the evolution of the earthflows (Table 5.2).

In stage 1, reactivation is triggered by the build up of excess pore pressures induced by a change in the state of stress. This usually results from the re-mobilization of the entire head of the earthflow mass or from the accumulation, over the upper part of the earthflow, of debris produced by failures at the slide scarp or along the lateral sides of the track. Both mechanisms, leading to an undrained loading process, occur after periods of intense and prolonged rainfall (Picarelli *et al.*, 2007; Bertolini & Pizziolo, 2008) and even of snow melt at higher elevations. "Feeding" can also be provided by lateral landslides. This is the case of the Tessina earthflow (Cola *et al.*, 2014) where a large (6×10^5 to 2×10^6 m^3) roto-translational landslide could easily connect to the

Table 5.2 Stages of earthflow evolution (modified from Guida & Iaccarino, 1991).

Stage	*Kinematics and morphological features*
1	Quick mobilisation of a stable slope or reactivation or of a quiescent or slow landslide, and consequent flow at v_{av} ranging from very rapid to moderate ("surging" phase according to Hungr *et al.*, 2014)
2	Flow within well defined and easily recognisable lateral shears with v_{av} decreasing down to slow, due to progressive dissipation of excess pore pressures
3	Slow to very slow steady motion during which the geomorphological features of the landslide are smoothed by erosion
4	Steady motion (very or extremely slow) during which the landslide morphology is not easily recognisable

body of the main landslide, initiating surging. Movement response to rainfall peaks is less than 4 days, with displacement rates increasing up to 0.5 mm/day. At this site, snow melting has a significant influence on pore pressures rise and movement acceleration.

In many cases reactivation of old quiescent landslides in the form of an earthflow can be relatively sudden and followed by long periods of high average displacement rates. The Corniglio earthflow (Northern Apennines) attained a peak velocity of 2 m/hour (Bertolini & Pizziolo, 2008). The Covatta earthflow (Southern Apennines) reached a rate of some meters per hour (Picarelli & Napoli, 2003) and the Montaguto earthflow (Southern Apennines) four years after the reactivation exhibited an acceleration with a peak velocity of 5 m/day (Giordan *et al.*, 2013).

Long-term behaviour is governed by drained conditions (Picarelli *et al.*, 2005; Berti & Simoni, 2010). Two well-monitored earthflows in an advanced stage of evolution are the Alverà earthflow in the Dolomites, North-Eastern Italy, the subject of a pioneering work on the effects of climatic changes on landslides (Dehn *et al.*, 2000) and the Costa della Gaveta landslide in Basilicata, Southern Italy (Di Maio *et al.*, 2010, 2013). The Alverà landslide involves clayey soils deriving from degradation/weathering of marls. It displays v_{av} of 80 mm/year and exhibits sharp responses of piezometer levels, and hence of displacements, to any rainfall intensification (Angeli *et al.*, 1998). In such a case, peaks of the groundwater levels have longer duration in the snow melting period. The Costa della Gaveta landslide involves a structurally complex formation with a prevailing clay component down to a maximum depth of 35 m. At present, the landslide moves continuously, even during dry periods, with very low displacement rates (11 to 16 mm/year).

For this category of movements, the expected climate changes could increase the incidence of re-activations due to intense precipitations and it could further reduce the displacement rate in the steady-state stage because of the decrease in the cumulated yearly rainfall. Based on investigations of the Costa della Gaveta landslide, in the Basento River valley, Comegna *et al.* (2013) proposed a methodology based on the coupling of climatic scenarios and geotechnical analyses with the goal to predict the future behaviour of the earthflow. The approach includes: (i) the calibration of a model linking weather parameters and pore pressure regime, (ii) the definition of a relationship between pore-water pressure and landslide movement, and (iii) the assessment of the long-time landslide behaviour based on the available climate scenarios. Studying historical daily rainfall and temperature records for the Costa della Gaveta landslide landslide, they found a negative trend in rainfall (−2.4% per decade) and a slightly positive trend in temperature (+0.04% per decade). Through the regional climate model COSMO-CLM, a simulation of the climate in the Mediterranean area for the period 1965–2100 was then performed. As a result of the expected future decrease in precipitation and increase in temperature, the Authors obtained a reduction in the groundwater level of about 8 mm per decade. By means of the relationship between pore pressure and landslide velocity, they calculated a decrease in the displacement rate of 1.5–3.0 mm/decade, leading to a maximum total displacement of 77–86 cm over next 50 years. In the considered case, the climatic changes are not expected to play a relevant role on the behaviour of the earthflow, providing useful information for planning purposes.

5.4 EFFECTS OF THE EXPECTED CLIMATE CHANGE

Assessing the consequences of expected climate changes is not an easy task, due to the difficulty in evaluating the variations of the weather parameters that more closely affect slope stability/instability conditions. However, rational scenarios can be depicted based on available knowledge. An example is provided by Polemio & Petrucci (2010), who proposed a statistical approach for the characterization of the relationship between climatic data and landslides in Calabria (Southern Italy). The analysis of monthly rainfall and temperature records highlighted the major role played by rainfall in the month before single landslide events. In contrast, the role of temperature appears negligible. However, despite some decrease in monthly rainfall, available data suggest that the frequency of landslide occurrence is not decreasing. The Authors attribute this trend to a slight underestimation in the number of landslides in the oldest part of their record and, indirectly, to the amplification of induced damage (increasing density of vulnerable elements).

Using the same landslide catalogue and daily rainfall series recorded by a network of 318 rain gauges in the same region, Gariano *et al.* (2015) reconstructed a record of 1466 rainfall events that triggered landslides in Calabria between 1921 and 2010. They found that the geographical and the temporal distributions of the rainfall-induced landslides, and in particular their monthly distribution, have changed in the observation period. In the most recent period 1980–2010 rainfall-induced landslides were more frequent in the SE part of the region and were concentrated mainly in the winter. Also, less cumulated rainfall was necessary to trigger landslides in the time interval 1981–2010 than in the preceding period 1951–1980, probably due to increasing vulnerability.

A different approach is based on the use of modelling chains through: (i) the analysis of the correlations between variations in atmospheric forcing and changes in frequency and magnitude of landslides, based on data regarding the past; and (ii) the forecast of potential trends through climate projections and quantitative analysis of their effects on the slope-atmosphere system.

An approach of the first type, based on paleo-evidences on landslide occurrence was adopted by Soldati *et al.* (2004) and by Borgatti & Soldati (2010), who investigated the relationship between climatic changes and hillslope evolution in the Dolomites, Eastern Alps, since the Late Glacial. They analysed paleo-environmental records relative to mobilisation of different types of landslides during the Late Quaternary in two study areas. The landslides were dated by stratigraphic data and radiocarbon analyses. They found a high number of events in the Late glacial – Holocene transition and related this cluster to the definitive permafrost melting at high elevations. Moreover, they associated the increase in landslide activity found in the Upper Holocene with deforestation due to the human impact. Generally, the periods of enhanced slope instability are quite in a good correlation with the indicators of cold and humid climate, suggesting that a positive moisture balance could have played a major role on landslide activity.

Further considerations and examples about the potentialities of modelling chains are reported below.

5.4.1 Modelling chains

When dealing with the relationships between climate and landsliding, it should be considered that the slope response mainly depends on the characteristics of the physical

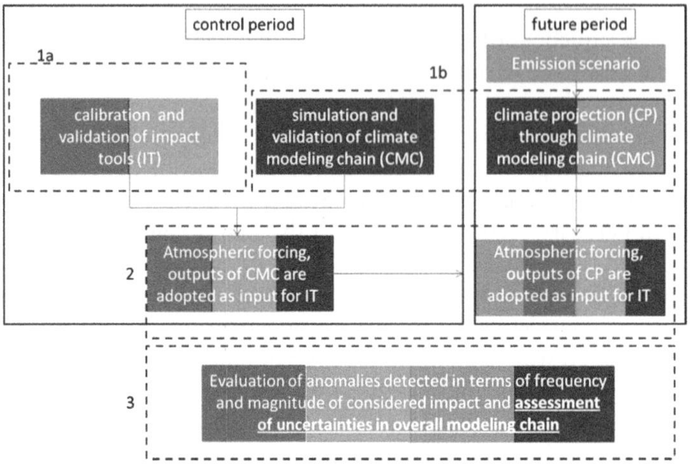

Figure 5.9 Flow chart of the adopted simulation chain. Colours identify the main skill required to address each step (red: geo-hydrological expertise; green: geotechnical; blue: climatic; orange: socio-economic and demographic).

system in which it is located, including the local weather input and the geomorphological features of the slope. Dixon & Brook (2007) notice that "landslides with a shorter antecedent period, or those that exhibit a strong response to relatively short term rainfall events, could be more vulnerable to the predicted increase variability of winter rainfall and so could experience a shorter return period for slope movements. However, slope instability caused by longer periods of antecedent rainfall may occur less often in response to drier summers with increased evapotranspiration". The case of the Costa della Gaveta earthflow discussed above (Comegna *et al.*, 2013) suggests that the displacement rate of slow landslides will be governed by the longest dry periods that are expected in the area. In contrast, slopes in fractured rocks, or in nearly saturated coarse grained soils, could be sensitive to short and intense rainstorms that are expected to be more frequent in the next decades (Damiano & Mercogliano, 2013). Such opposite effects can be supported by quantitative analyses carried out with the help of projections of the climate trends obtained by RCMs. Two well-documented case studies will be presented. The analyses will be carried out exploiting the simulation chain shown in Figure 5.9, which highlights with different colours the main skills required in any step of the simulation. We acknowledge that uncertainties increase with each step of the modelling (Wilby & Dessai, 2010).

5.4.2 Examples

5.4.2.1 Considered case histories

The first case study concerns the Orvieto area. Orvieto is an historical town located 100 km North of Rome built on a 50-m thick tuff slab with sub-vertical lateral cliffs, overlying overconsolidated, stiff clays. The slopes surrounding the volcanic platform are blanketed by an irregular debris cover produced by disruption of volcanic materials

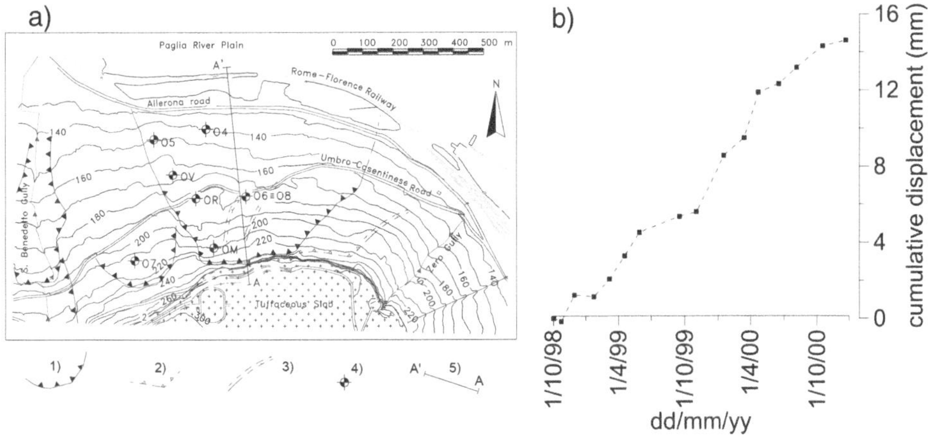

Figure 5.10 Northern side of the Orvieto hill; legend: (1) slide scarp, (2) boundary of the 1900 Porta Cassia landslide, (3) boundary of cinematically independent portions of the slide, (4) instrumented borehole, (5) geotechnical section; b) cumulated displacement at a depth of 18 m (Tommasi *et al.*, 2006).

and by slope movements (Tommasi *et al.*, 2012). A large number of slides affect the northern slope of the hill (Lembo Fazio *et al.*, 1984). The main historic landslide in this area was the Porta Cassia landslide triggered in 1900 by human activities. Further active slow movements, widespread in the Apennines chain, are directly related to the soil-atmosphere interaction.

Since 1982, the Northern side of the Orvieto hill is being monitored through seven instrumented stations. Monitoring allowed to identify deep movements along slip surfaces located in the softened part of the basal clayey formation, which display displacement rates ranging between 2 and 6 mm/year and shallower movements involving the cover, whose rates attain 7–12 mm/month. Previous studies have shown that any acceleration is strictly correlated to cumulated precipitations in time windows ranging from 15 to 30 days for the shallowest movements, and the entire wet season for the deepest movements (Tommasi *et al.*, 2006).

In order to investigate the influence of weather forcing on groundwater and displacement trends, the active slide monitored by station O7 (Fig. 5.10) has been used as a reference case. The same figure shows the displacement trend for the period 1998–2000. The slip surface of the landslide is located at a depth of 18 m, at the interface between the stiff and the softened clay. Pore water pressure readings are available at depths of 6 m and 15 m (Fig. 5.11).

The response of the groundwater regime to atmospheric forcing has been investigated via a 1D numerical simulation using the SEEP/W FEM code. The influx at the top has been calculated as the algebraic sum of daily precipitation and actual evaporation E_a obtained through the FAO dual approach (Allen *et al.*, 1998). The soil surface water balance required for a proper evaluation of water availability has been obtained following Boldini *et al.* (2014). Since only temperature values were available, the other forcing

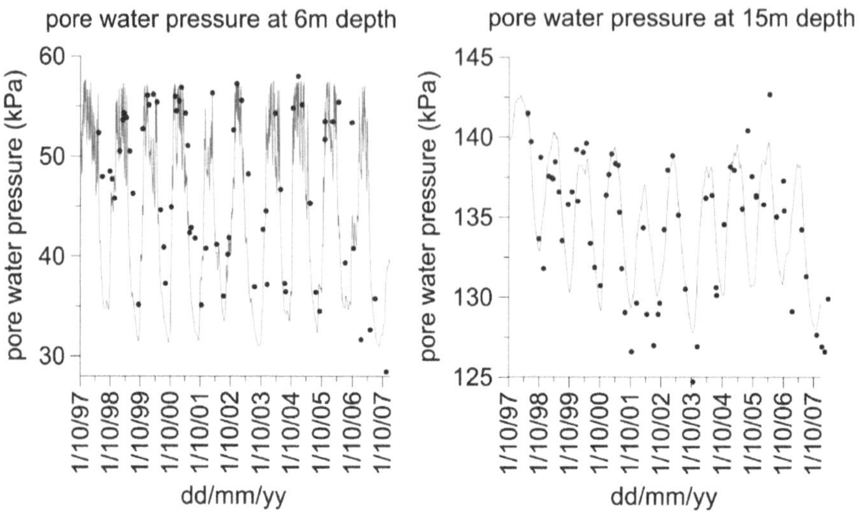

Figure 5.11 Orvieto monitoring station: estimated (red line) and measured (blue dots) pore water pressures at depths of 6 and 15 m.

factors have been retrieved using the empirical relationships suggested by the FAO guidelines. Other distinctive features of the slope (i.e., exposure, direction) have been also taken into account. In particular, natural grassland has been assumed to cover the entire soil surface. The stratigraphy has been simplified adopting two equivalent layers at 6 and 24 m depth (Tommasi *et al.*, 2006). Finally, impervious lateral and bottom hydraulic conditions have been adopted at the boundaries of the control volume.

For the calibration period 1998–2008 the best matching between the estimated and the observed piezometer levels (Fig. 5.11) has been obtained assuming $k_s = 10^{-7}$ m/s and $k_s = 10^{-8}$ m/s, for the uppermost and lowermost layers, respectively (both soils have been characterised by $m_{w,s} = 10^{-5}$ m^2/kN). The rather high hydraulic conductivity required to match the analysis results and readings (two order of magnitude higher than the values measured through laboratory tests) accounts for the effects of opened fissures and joints. From the data shown in Figure 5.11 it can be argued that the piezometer level ranges between 3 m at 6 m and between 1.5 m at 15 m, depending on depth. The differences between the calculated and the measured values, especially at higher depth, can be explained by the simplifying assumptions (in special way, 1D seepage conditions) adopted for the analysis.

The second case study concerns the 40° Cervinara slope (Fig. 5.12), located 50 km North-East of Naples, where on 16 December 1999 a rapid flowslide triggered by a total precipitation of 320 mm in about 50 hours caused huge damage and five fatalities (Olivares & Picarelli, 2003). The area is occupied by highly fractured limestone blanketed by a few meters thick unsaturated loose pyroclastic cover. Due to the effects of suction, such typically steep silty-sandy covers are stable. Precipitations cause the increase of water content and the decrease of matric suction and of the associated

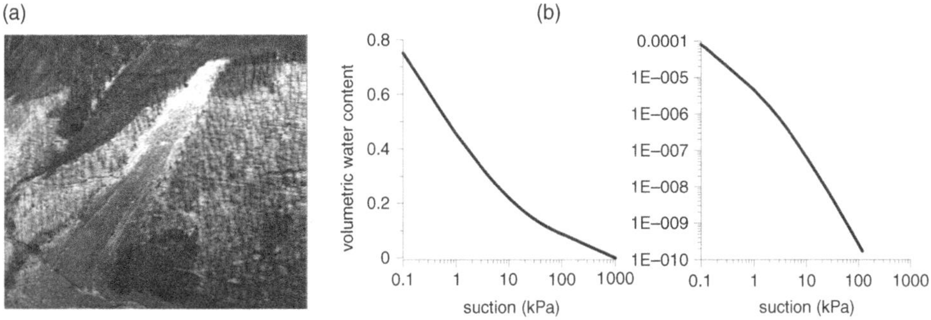

Figure 5.12 The Cervinara flowslide; b) the hydraulic properties of the cover (Greco *et al.*, 2013).

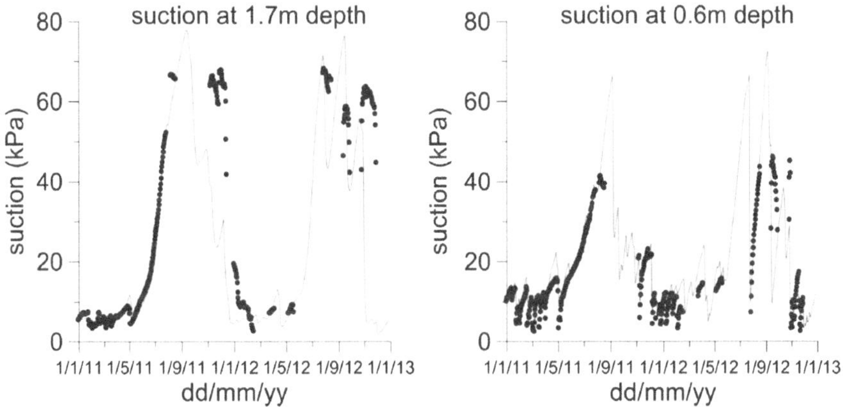

Figure 5.13 Observed (blue dots) and estimated suction (red line) at Cervinara in the calibration time window 2011–2012.

apparent cohesion. In extreme cases, this can lead to slope failure, soil liquefaction and generation of a flowslide (Olivares & Picarelli, 2003).

Based on the results of monitoring and laboratory tests (Damiano *et al.*, 2012), numerical simulations under simplified 1D conditions have been carried out assuming a homogeneous cover thickness of 1.9 m and a set of soil hydraulic properties based on available data (Fig. 5.12) (Greco *et al.*, 2013). At the ground surface, the algebraic sum of daily precipitation and actual evaporation has been imposed (only maximum and minimum atmospheric temperature measurements were available to estimate the potential evaporation). Based on monitoring data, an outgoing water flow depending on local suction has been imposed at the bottom of the cover. Estimated and observed suction values at depths of 0.6 and 1.7 m in the time interval 2011–2012 are shown in Figure 5.13. Numerical analysis provides reliable estimates especially during the wet season. During the dry season and autumn, the agreement between readings and calculated values is less effective, this especially during the first phase of suction decrease (September–November). The biases are due to the assumptions adopted regarding flux characteristics (1D) and boundary conditions at the uppermost and lowermost interfaces.

5.4.2.2 Climate modelling performance and potential future slope behaviour

The assessment of the effects of potential climate changes requires the adoption of weather forcing parameters provided by climate models. The reliability of the adopted simulation chain would be proven if, for the control period, it could reproduce at least the values of the atmospheric variables used for calibration i.e., precipitations and temperature.

In this work, atmospheric forcing for the two investigated areas has been retrieved through a simulation performed by means of the CMCC_CM+COSMO_CLM models with a horizontal resolution of 8 km for both the control (1981–2010) and future (2071–2100) time window periods. For the first time interval, the comparison in terms of mean monthly cumulative precipitation (Fig. 5.14a) shows that climate simulations (continuous magenta line) lead to significant errors (observations reported in black line). Also, a significant overestimation of the mean monthly number of wet days (Fig. 5.14b), and an underestimation of the peak temperature (Fig. 5.15) have been obtained in both cases. Similar biases, common in climate simulations over the European domain, are essentially due to coarse horizontal resolutions inducing imperfect representations of the atmospheric physics (e.g., diurnal convective cycle) and preventing a direct use of atmospheric variables in impact models (Maraun *et al.*, 2010). To this aim, in recent years numerous bias correction techniques have been proposed to adjust the model output towards observations in a post processing step (Ehret *et al.*, 2012; Teutschbein & Seibert, 2012). Among them, distribution mapping approaches have been shown to outperform the others methods (Lafon *et al.*, 2013). To explain how this approach works, the exemplary procedure proposed by Teutschbein & Seibert (2012) can be considered (Fig. 5.16). If F_{RCM} and F_{OBS} are the cumulative distribution function (CDF) of simulated and observed precipitation for day d, the bias corrected value $P^*(d)$ of RCM precipitation $P(d)$ can be obtained using the equation $P^*(d) = F_{OBS}^{-1}(F_{RCM}(P(d)) = h(P(d))$. A similar approach can be used for temperature. The main differences between the distribution mapping methods are due to h transformation. To this aim, theoretical distribution, parametric relationships quantile-quantile or directly the empirical quantiles are generally adopted.

As shown in Figures 5.14 and 5.15, for the cases in consideration the distribution quantile mapping approach (Gudmundsson *et al.*, 2012) leads to a good matching between observed and bias corrected data (continuous green line) allowing to confidently employ the obtained atmospheric variables in impact tools for evaluation of future trends. Figures 5.14 and 5.15 show the bias corrected trends for the time interval 2071–2100, under the emission scenario RCP 8.5 (dashed green line), in addition to trends over the control period. For both locations, a substantial decrease in cumulated precipitation is estimated for a large part of the year (April–September for Orvieto, and March–October for Cervinara). For the other months, roughly coincident with the wet season, an increase in monthly cumulated rainfall values has been obtained. A reduction in the mean monthly number of rainy days has been estimated during the entire year. In the drier periods the result could be partly due to the reduction of total

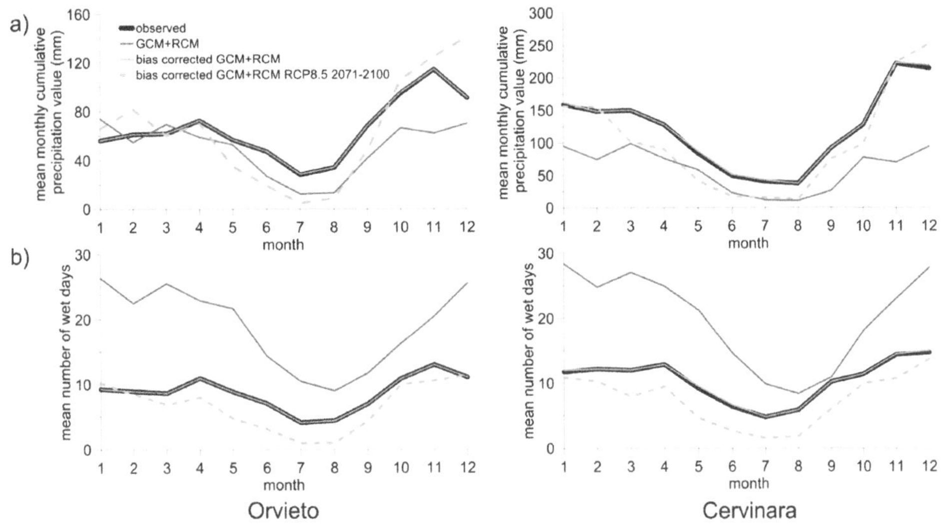

Figure 5.14 (a) Monthly cumulated precipitations; (b) mean monthly number of wet days. Black lines are observed values, magenta lines are simulated values through climate models GCM+RCM and green lines are simulated values through bias corrected (BC) climate models GCM+RCM.

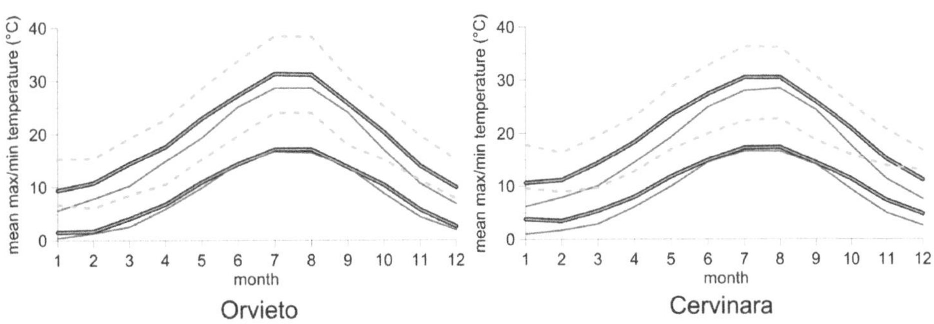

Figure 5.15 Maximum and minimum temperature observed (black line) and simulated through climate models GCM+RCM (magenta) and GCM+RCM+BC (green).

values. In the wet season, obtained data could indicate an increase of the inter-arrival time between precipitation events leading to a higher mean intensity for each event. The result is consistent with simulated temperature growths (Fig. 5.15) causing a substantial increase in the atmospheric water retention capacity and then a harder matching of the conditions bringing precipitations. While the evaluation of the average trends of atmospheric variables allows for understanding the general features of potential future climate conditions, for the assessment of the slope response, the analysis of trends of the extreme values of the major variables represents the crucial issue.

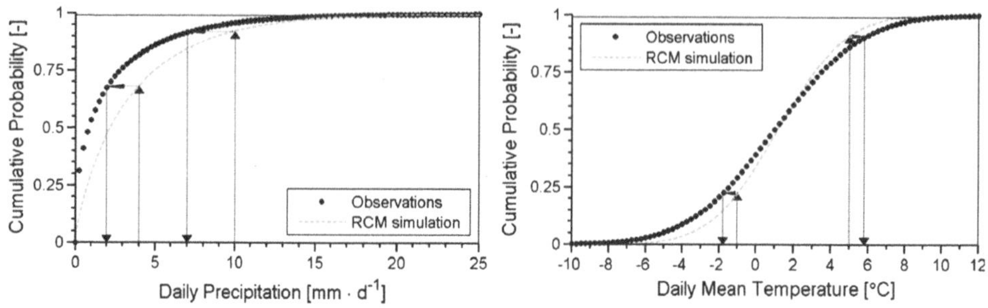

Figure 5.16 Procedure of the distribution mapping for precipitation and temperature (Teutschbein & Seibert 2012).

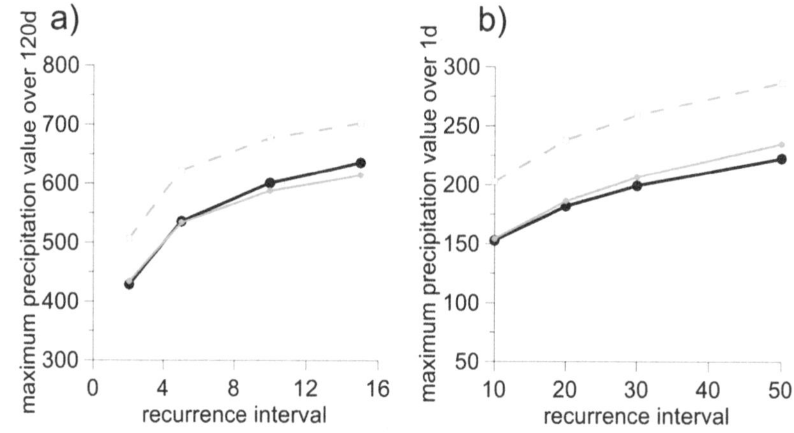

Figure 5.17 Recurrence interval for maximum values of precipitation (GEV statistical distribution): (a) Orvieto, cumulated precipitation over 120 days; (b) Cervinara, cumulated daily precipitation. Black lines are observed values, continuous green lines are simulated values over the same time interval, and dashed green lines are simulated values for the 2071–2100 period.

To account for the different characteristics of the considered landslides, following Tommasi *et al.* (2006), here the annual maximum cumulative precipitation over 120 days (P_{120d}) has been assumed as the reference variable for Orvieto. In contrast, for Cervinara the reference variable is the maximum daily precipitation P_{1d}. The time series of such maxima (P_{120d} and P_{1d}.) have been fitted through the General Extreme Value (GEV) statistical distribution. In Figure 5.17, the observed values are indicated by the black line. The same procedure has been repeated for the simulated values for the same time interval and for the 2071–2100 time window, represented by the continuous green line and the dashed green line, respectively. Through this way, the expected recurrence interval can be assessed, which suggests a significant increase.

The recurrence intervals shown in Figure 5.17 have different meanings. Whereas landslide acceleration in clay (as in Orvieto) is quite a frequent phenomenon, flowslide

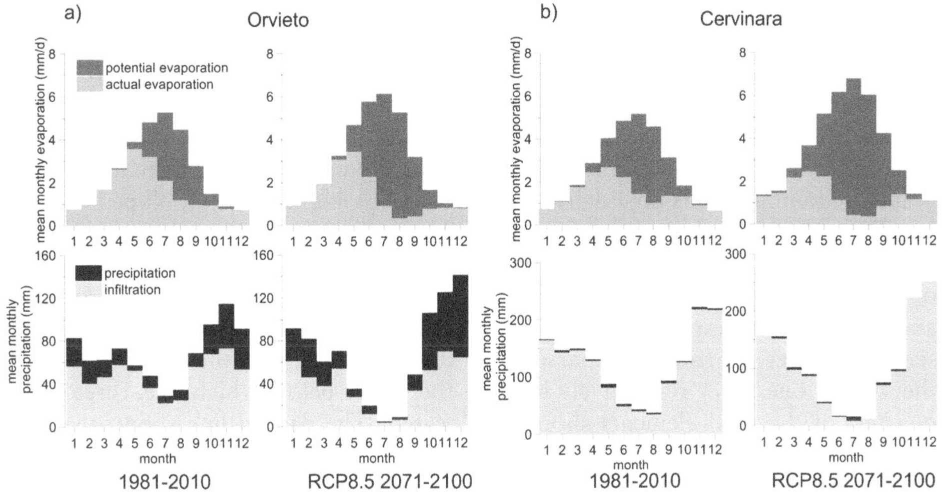

Figure 5.18 Orvieto (a) and Cervinara (b) sites. Comparison between monthly potential and actual evaporation, and monthly total and infiltrated rainfall for the control period and the 2071–2100 time window (RCP 8.5 emission scenario).

triggering in pyroclastic covers (as in Cervinara) is a less common event. It is worth to mention that the analysis is based on simplified assumptions that, for example, neglect the effects of evaporative processes.

5.4.2.3 Evaluation of the potential effects of climate changes in Orvieto and Cervinara areas

The hydro-mechanical response of both Orvieto and Cervinara slopes to potential climate changes has been investigated through the same numerical models described above, accounting for the atmospheric forcing supplied by bias corrected models. The reference atmospheric evaporative demand has been calculated using as unique input the simulated temperatures, despite climate simulations could provide estimates of all variables governing the evaporative process. For both examined sites, Figure 5.18 shows the simulated monthly average values of potential and actual evapotranspiration and of total and infiltrated precipitation, for the control period and the reference interval 2071–2100 (RCP 8.5 scenario). It is worth mentioning that the total precipitation and the potential evaporation represent the maximum imposed fluxes at the soil-atmosphere interface. When considering the outward and inward water fluxes, the hydraulic behaviour of the soil should be accounted for, introducing an additional governing factor.

In Orvieto (Fig. 5.18a), the estimated increase in potential evaporation for 2071–2100 could turn into slightly higher actual values only in the spring and winter, thanks to water availability due to precipitations. In contrast, in the summer and partly in the autumn, the larger evaporative demand should not lead to a greater water loss because of the reduction in cumulated precipitation during the warm months. Due to the increase in the mean intensity of precipitation, and to the low permeability of the fine-grained materials that crop out in the landslide area, the increase in winter precipitation might not result in a growth of the amount of infiltrated water, but only in an increase of the run-off. Figure 5.18b shows the main components of soil surface water balance for the Cervinara landslide. The evaporative water loss is similar to the Orvieto case. However, due to the higher hydraulic conductivity of the soil, rainfall could infiltrate also during intense precipitation events. Then, the estimated increase in intense rainfall events on a daily scale (Fig. 5.18b) might affect the slope behaviour.

The potential effects of soil water balance changes in Orvieto are shown in Figure 5.19a that portrays the estimated piezometer level at a depth of 18 m (i.e. the one of the sliding surface). In blue are shown the simulated values in the control period and in red the corresponding values in the future time span. The thinner line represents the lower quartile; the coarser line, the upper quartile; the intermediate line, the median one. They respectively indicate the values below which one, three or two quarters of the total elements should be located. The dashed line shows the water level for which, in the simplified infinite slope model, the safety factor reaches unity. Such a value has been calculated adopting an average slope angle of 11.5° and a friction angle of 20° (Tommasi *et al.*, 1997). It is worth mentioning that some aspects of the soil behaviour disregarded in the analysis could induce opposite effects. For example, increasing temperature could cause surface cracking favouring higher infiltration capacity. Despite the limitations of the adopted approach, the following general remarks can be made. On the average (median value), the threshold should be exceeded over shorter time intervals than at present. In fact, while for the period 1981–2010 this should occur for about 200 days, for the interval 2071–2100 the same phenomenon is estimated to occur over about 120 days. Moreover, in the examined future time span, for 8 years the threshold might be exceeded for less than 60 days. According to these results, the effects of the expected increase in the maximum precipitation cumulated over 120 consecutive days (P_{120d}, Fig. 5.18a) would be completely cancelled by the increase in evaporation and the changing rainfall distribution during the wet season that are explicitly taken into account in the analysis. The final result is a moderate improvement in the slope stability conditions, just as shown in the study carried out by Comegna *et al.* (2013).

Moving to Cervinara, a landslide might take place when suction at the base of the cover (i.e., at 1.9 m) practically vanishes. In fact, the slope angle (40°), that is quite constant (infinite slope), is very close to the friction angle (38°). An assessment of the potential critical precipitation events is reported in Figure 5.19b. Based on a numerical simulation of the suction profile performed by 1D infiltration analyses based on the rainfall history, here are shown the combinations $P_{1d} - P_{29d}$ that could trigger a landslide (P_{29d} is the cumulated precipitation over 29 consecutive days before the present one). The red dots represent the critical events during the control period obtained from simulated precipitation data; the green dots, critical events over 2071–2100. Finally, the blue dots indicate critical events estimated through observed data in the control period. The circled dot corresponds to the 16 December 1999 flowslide and the blue dots should indicate missed landslide events. Concerning this last point, it is worth mentioning that the corresponding recorded precipitation triggered landslides in nearby areas characterised by a similar geomorphological context (Palma Campania, in 1986; Alfaterna, in 2003; the same Cervinara area, in 2010). A comparison between

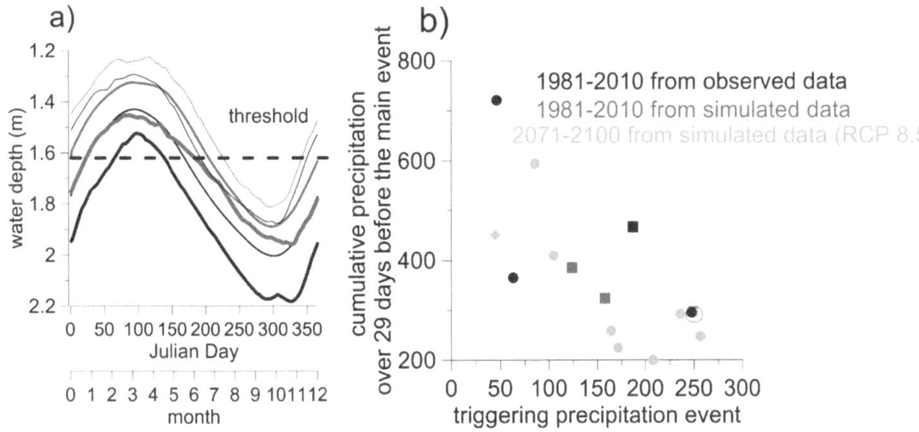

Figure 5.19 a) Orvieto, calculated piezometer level at the depth of the sliding surface for the control period (red lines) and the 2071–2100 time window (blue lines); thinner line, lower quartile; thicker line, upper quartile; intermediate line, median value; b) Cervinara, rainfall history governing potential landslide triggering. Blue dots, critical rainfall from observed data; red dots, critical rainfalls from simulated data in the control period; green dots, critical rainfalls from simulated data in the future time window.

the clusters of red and green dots suggest that future extreme precipitation events could produce an increase in the probability of landslide occurrence. An additional interesting remark is the potential use of such data for early warning. In fact, the envelope of blue, red and green dots represents sort of threshold. Based on that, in the examined geomorphological context cumulated precipitations P_{29d} less than 200 mm, and P_{1d} less than 50 mm should not result in slope failures. For higher values, the critical conditions should be governed by the combination $P_{29d} - P_{1d}$.

As a final consideration, it should be observed that the simulations of the effects of climate changes can capture only statistical distributions of precipitations at daily scale (Maraun *et al.*, 2010), not allowing the analysis of the effects of sub-daily precipitation patterns. A number of authors (Westra *et al.*, 2013; Lenderink *et al.*, 2011) have shown that the effects of atmospheric warming on extreme values of hourly precipitation could be significantly higher than those determined by daily values. As a consequence, soils presenting a high hydraulic conductivity might be even more severely affected by such intense short-duration events.

5.4.3 Further considerations on the potential effects of climate changes on the hydrological slope response

Figure 5.19 shows that the effects of similar climate changes can differ substantially, depending on the geomorphological context. In order to more closely examine this point, additional numerical simulations have been carried out in which, on a daily

Table 5.3 Main hydraulic properties of typical "standard" soils used in the analysis.

	Porosity (–)	*Saturated hydraulic conductivity (m/s)*
Clay	0.48	6.80E–09
Silt loam	0.50	1.00E–07
Loamy fine sand	0.46	5.00E–06
Coarse sand	0.42	1.00E–04

scale, the main components of the soil surface water balance and the soil water content resulting from the slope-atmosphere interaction, which affect the mechanical slope response, are estimated for different "standard" soils (Rianna *et al.*, 2016). The analyses have been performed by the code HELP (Hydrologic Evaluation of Landfill Performance) (Schroeder *et al.*, 1994), which allows the assessment of the effects of simplified surface processes (interception of rainfall by vegetation, surface runoff and surface evaporation) and of subsurface processes (evaporation, plant transpiration, unsaturated vertical drainage).

The analyses have been carried out assuming an infinite soil layer, covered by natural vegetation, subjected only to gravitational gradients (1 D flux conditions). The daily temperature and precipitation data are those obtained from the corrected climate simulations for both control period (1981–2010) and time window 2071–2100 under the two emission scenarios RCP 4.5 and RCP 8.5. For sake of simplicity the adopted weather forcing parameters are those calculated for Cervinara (Figs. 5.14 and 5.15). The soil properties (Table 5.3) have been retrieved by the USDA soil catalogue (1985). Naturally, for analysis lasting 30 years the initial conditions play a minor role on the solution of the problem. In any case, average values of the water content corresponding to the month of January have been imposed.

The mean and peak values of daily precipitation and the average and peak values of the saturation degree in the shallowest part of the deposit (4 meters) are reported in the left column in Figure 5.20. During the dry-warm season the estimated average trend of the saturation degree for the 2071–2100 time span, deeply affected by the temperature increase, is below the analogous curve over the current period. In the wet-cold seasons, the effects of evapotranspiration play a minor role and the differences are less pronounced. However, the nature of soil plays a fundamental role. In fact, the expected increase of inter-arrival times of rainfall and the higher rainfall intensity will cause a reduction in the average water content mainly in soils characterised by lower hydraulic conductivity, while in coarse soils the saturation degree should maintain values comparable to those estimated for the present. With regard to the peaks, the trends shown in the figure indicate that the increase in maximum rainfall height for both considered concentration scenarios does not automatically result in an increase of the highest yearly values of the saturation degree. The intense rainfall events estimated for future periods should be characterised by almost full infiltration in the most pervious soils leading to much higher peaks especially in the drier season, while this should not occur in fine grained soils because of their lower hydraulic conductivity. In any

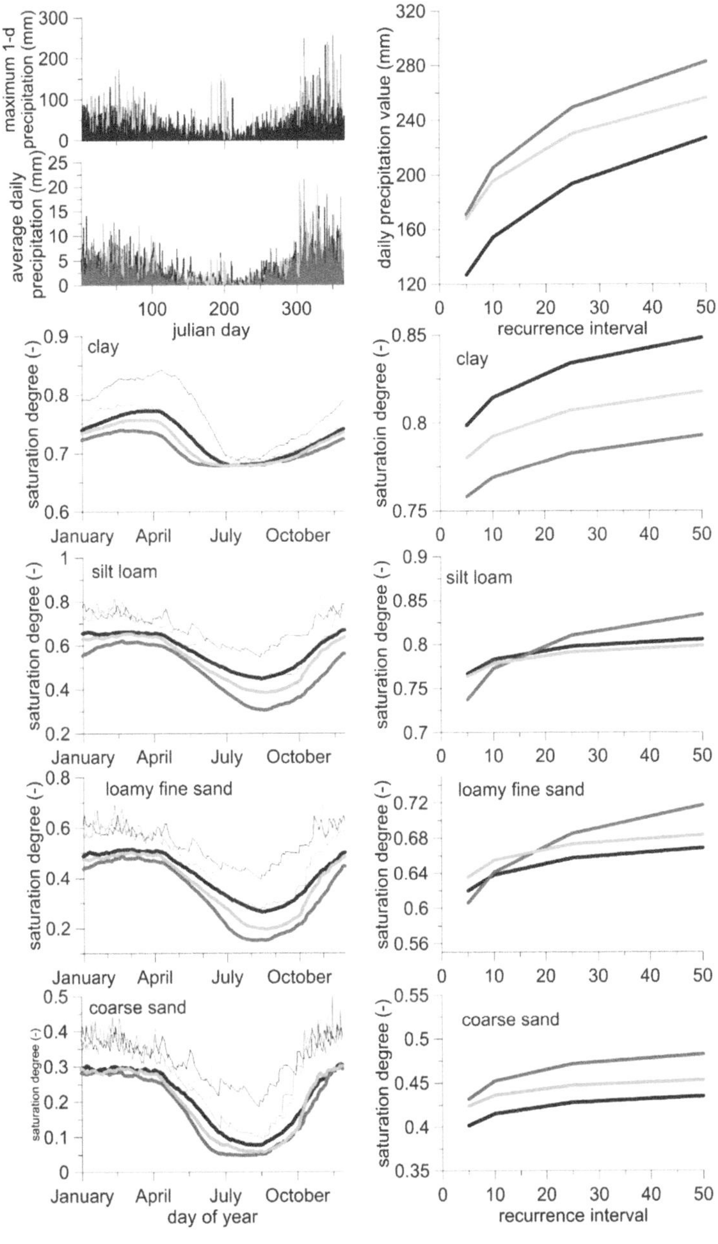

Figure 5.20 Left column: estimated mean and peak daily precipitation and mean (thick line) and peak (thin line) saturation degree for the investigated soils; right column: GEV distribution of yearly peak daily precipitation and GEV distribution of peak saturation degree. Blue: control period 1981–2010; green: RCP4.5 2071–2100; red: RCP8.5 2071–2100.

case, the calculated values strongly depend on the hypothesis of infinite thickness (for example, for coarse sand, the values of the saturation degree does not exceed 0.5). In fact, significant differences could be captured by modifying the lowermost boundary condition (Rianna *et al.*, 2016).

The yearly maximum values of daily precipitation (see also Figure 5.17) have been interpreted through a GEV distribution in order to estimate the corresponding recurrence intervals. The same procedure has been applied to peak yearly saturation degrees (right column in Figure 5.20). The differences between the most and less pervious soil (coarse sand and clay, respectively) are substantial. In fact, for the same rainfall distributions, and considering recurrence interval ranging from 5 to 50 years, the maximum expected values of the degree of saturation are considerably higher than the current ones for coarse sand, whereas the opposite occurs for clay. The other soils exhibit an intermediate behaviour.

5.5 LANDSLIDE RISK AND IMPLICATIONS FOR MITIGATION AND ADAPTATION STRATEGIES

In Italy, landslides represent a serious threat to the population. In the 66-year period 1950–2015, 6407 persons were killed (4074), went missing (32), or were injured by landslides (2301), including 1910 (estimated) fatalities caused by the 9 October 1963, Vajont landslide, the most destructive slope failure in Europe in historic time (Salvati *et al.*, 2010). Figure 5.21 portrays the geographical and temporal distributions of fatal landslides between 1950 and 2015. In this period, 681 (8.4%) of the 8103 municipalities in Italy have experienced landslides with casualties (deaths, missing persons, injured people). The number of municipalities increases to 1655 (20.4%) if evacuees and the homeless are considered. In Italy, fatal landslides are typically very rapid to extremely rapid movements, primarily soil slips, debris flows, rock falls and rock slides.

In Italy, landslides also cause considerable economic and societal damage. Single large landslides have repeatedly caused direct damage to structures or the infrastructure exceeding several million euros. As an example, the 1982 Ancona deep-seated slide in stiff marine clays caused damage estimated at 1,000 billion 1982/Lit (Alexander 1989), €516 million, and the 2010 Montaguto earthflow involving structurally complex clays produced economic losses in excess of €60 million. Single small landslides also proved capable of producing significant damage. On 17 December 2014, a small slide derailed a train along the railroad connecting Genova (Italy) to Nice (France). Although the direct human consequences were limited, the railroad remained closed for several months, with economic consequences estimated in excess of €5 million. On 24 October 2015, another small landslide destroyed a few meters of the aqueduct that delivered water to the city of Messina, Sicily, leaving 80,000 people with water for several days. Populations of landslides caused by a single trigger (i.e., an intense or prolonged rainfall event or a rapid snowmelt event) have repeatedly caused damage to private and public properties, heritage sites, roads, railroads and other infrastructural systems, to the agriculture, tourism and the environment. Despite the efforts, the economic costs of event-triggered landslides remain poorly known and largely underestimated (Guzzetti *et al.*, 2014).

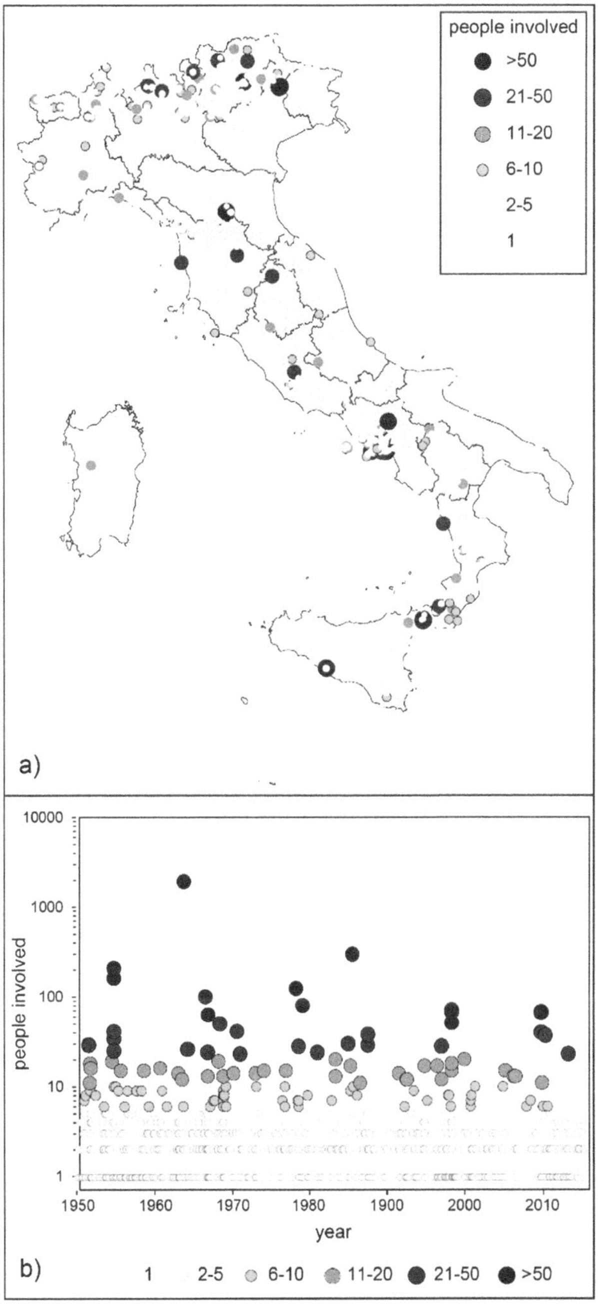

Figure 5.21 Geographical (a) and temporal (b) distributions of the fatal slope failures in the period 1950–2015 with indication of the number of people involved (killed, missing, injured).

An important question is whether, or to what extent, landslide risk to the population and the impact of landslides on built-up areas, the infrastructure, the agriculture, the environment and – more generally – the landscape, the society and the economy, will change in Italy in response to the predicted climatic and environmental changes. Answering the question is difficult, because of (i) the significant physiographical, geological, and meteorological variability that limits the reliability of the climate and landslide projections, (ii) the lack of proper models to describe (and forecast) the vulnerability to landslides of different elements at risk, including the population, and (iii) the uncertainty associated with the existing climate and landslide forecasting models. The physiographical complexity and the uncertainty in the climate and the landslide scenarios makes it difficult to predict accurately the possible changes in landslide risk, to design effective mitigation measures, and to implement adequate adaptation strategies. In the following, we base the discussion on the expected changes of landslide risk on existing information on landslides affecting primarily natural slopes, and on reasonable local and regional landslide hazard scenarios. We further consider the possible changes in the abundance and distribution of the vulnerable elements and the effects of the changing climate on artificial slopes and landslide retaining and stabilization works.

In Italy, risk posed by landslides to the population depends on (i) the type, abundance, distribution, and frequency of the landslides, and (ii) the abundance and distribution of the population. The population of Italy has increased from 28 million in 1861 to 59 million in 2011 and it is expected to grow to 61 million in 2065 (http://www.istat.it). Due to the increase in the size of the population and the larger proportion of people living in urban areas, mortality to landslides has decreased steadily from 1860 to 1980 and marginally after 1980 (Guzzetti, 2000; Guzzetti *et al.*, 2005b). Given the fact that the population is not expected to grow significantly in the next decades (http://www.istat.it), we do not expect a significant increase in the landslide mortality. Variations in landslide mortality will depend on the balance between the augmented landslide hazards and the adoption of adequate mitigation and adaptation strategies. Local changes in mortality are expected in response to geographical and temporal variations in the number of people exposed to landslides, e.g., in touristic areas in the mountains and along the coasts. In these areas landslide risk is expected to increase if the number of tourists will increase. Landslide risk may also change (increase or decrease) if the seasonal period of the harmful landslides will change in response to the changing climate.

Vegetation (or the lack of vegetation) influences the occurrence of rainfall-induced landslides. Modifications in vegetation cover due to land use changes are known to impact landslide occurrence (Glade, 2003). Disturbances of the extent and the status of the vegetation (by forest fires, agricultural or forest practices) reduce the ability of the vegetation to preserve slope stability during rainfall events due to the interception of water by the canopy and soil cohesion strengthening produced by the roots. Falcucci *et al.* (2007) found that from 1960 to 2000 land use change has accelerated in Italy and attributed the changes primarily to the expansion of agricultural and urban areas and of forests. Santini & Valentini (2011) modelled potential land use changes in the year 2100 and confirmed the heterogeneous response of the territory to climate and socio-economic modifications, with the largest changes predicted in Southern Italy, Sicily and Sardinia.

5.5.1 Expected variations in landslide risk due to climatic and environmental changes

Landslide risk depends on hazard and on the vulnerability to landslides, which is in turn affected by the type, size and velocity of the movement (Galli & Guzzetti, 2007). For the same landslide type, we can assume that the vulnerability of the different elements at risk will not change significantly in the near future. Should the landslide types change in an area in response to changing climate or environmental drivers, the vulnerability to landslides will change as well. The size of the landslide also affects the vulnerability of specific elements at risk (structures and infrastructures). On the other hand, the hazard depends on the landslide susceptibility, on the frequency of the events and on the size of the expected landslides, considered a proxy for landslide destructiveness (Guzzetti *et al.*, 2005a, b). It is safe to assume that morphological and geological factors conditioning landslide susceptibility (slope, relief, lithology, structure, seismicity) will not change significantly in the next 50 to 100 years. However, soil, vegetation cover and land use may change significantly in response to climatic, environmental and human-induced (i.e., agriculture, forestry) drivers (Falcucci *et al.*, 2007; Santini & Valentini, 2011). The changes may affect landslide susceptibility and chiefly the susceptibility to small and medium size landslides.

Variations in the frequency or intensity of the triggering weather events are expected to affect the rate of occurrence of the landslide events. For the coming decades, in South of the Alps climate models practically estimate the same total amount of annual rainfall concentrated in a fewer number of high intensity events. As shown above, this is expected to result in more abundant shallow landslides in the most pervious soils and in less frequent deep-seated failures (Buma & Dehn, 1998). The climate models also predict a higher air temperature. This will favour evapotranspiration and reduce the amount of water in the sub-surface. However, higher temperatures will also favour forest fires, which are known to increase the frequency of landslides in a period of variable length after the fire (Moody *et al.*, 2013).

Modifications in land use and land cover induced by changes in agricultural and forestry practices are also expected to influence the rate of occurrence of landslides, and hence landslide hazard. For the probability of landslide size (area, volume, Malamud *et al.*, 2004), the available information indicates that the frequency-area statistics of landslide areas does not change significantly in a multi-decadal period (Guzzetti *et al.*, 2006). However, Brunetti *et al.* (2009) have argued that the type of failures conditions the statistics of landslide volumes. Where the type of landslide changes in response to changing drivers, the frequency-size statistics will also change, the associated hazard will be modified and landslide risk will change as well. Cardinali *et al.* (2006) have shown that in SW Umbria, central Italy, intense rainfall has triggered soil slips/debris slides in volcanic soils that were previously undetected in the area, changing landslide risk significantly. In the same general area, intense rainfall events may increase the abundance of shallow slides in the weathered cover that mantles clay and silt sediments, also changing landslide risk.

The effects of the changing climate are first observed in mountain areas, which are "sentinels" of the changes. In the Italian Alps, the frequency of rock slope failures (rock falls, topples, rock slides) and ice avalanches is increasing and the increased frequency is correlated to a higher air temperature (Huggel *et al.*, 2012). The same may occur in high-mountain areas in the Apennines. In NW Italy, the seasonal period when shallow slope failures occur is becoming longer, with landslides most abundant at the beginning and towards the end of the warm, summer period (Stoffel *et al.*, 2014). Further, in mountain areas rock failures align at the elevation of the melting permafrost (Chiarle *et al.*, 2014). A warmer climate should then favour rock failures in high mountain areas, increasing landslide risk. This is expected to have an impact on the safety of tourists, inhabited areas, transportation routes and touristic resorts and facilities. In mountain areas, snowmelt is a known trigger of different types of landslides. Cardinali *et al.* (2000), working in Central Italy, have shown that rapid snowmelt is a primary trigger of diffused landslides also at lower elevations and in hilly terrain. In the recent years, snowmelt induced landslides proved to be abundant in the Emilia-Romagna region. A change in climate that will result in more frequent snowfall events at lower elevations will result in an increased frequency of snowmelt-induced landslide events. The current ability to forecast snowmelt and rain-on-snow induced landslides remains limited, complicating vulnerability assessments and risk management. In addition, snow-measuring stations are located chiefly at high elevations and are rare in hilly, low-elevation terrain where rapid snowmelt has shown to be able to trigger abundant landslides.

It is known that following wildfires the amount of rainfall required to initiate shallow landslides and associated surface erosion processes decreases (Moody *et al.*, 2013). A more arid and hot climate will favour wildfires. A higher frequency and broader geographical distribution of wildfires will increase the probability of shallow landslides. The modified rainfall characteristics that initiate shallow landslides in burnt areas will also jeopardize the use of existing rainfall thresholds for possible landslide occurrence in regional or national landslide early warning systems. Analysis of recent high-intensity rainfall events in the Marche Region, Central Italy, revealed the complication of chained effects, e.g., prolonged rainfall events combined with rain-on-snow events affecting burnt areas in steep terrain. For these multiple and complex events, which may increase due to the expected climate and environmental changes, we lack a sufficient understanding and adequate models to evaluate the consequences.

In the Central Apennines the expected increase in the rainfall intensity, coupled with a generalized lack of maintenance of debris flow control structures, may increase the frequency of large debris flow events with catastrophic consequences in areas that are currently considered at low to moderate risk. In the Alps and the Apennines, several campgrounds, touristic resorts and facilities, inhabited areas and transportation routes are built on debris fans generated by debris flows and hyper-concentrated floods. In these areas, the assessment of debris flow risk will have to be re-evaluated considering the expected higher frequency – and possibly larger volume – of the debris flow events.

Italy has more than 7,000 km of seacoasts, with a large proportion of the population, the built-up areas and the infrastructures along the coasts. A significant proportion of the coastal areas is affected by landslides, of all types, from small rock falls, chiefly from rocky cliffs, to very large, deep-seated slides most abundant in hilly and mountain coasts. Although the sea certainly affects landslide activity along the coasts, the expected sea-level changes in the upcoming decades are too small to modify the activity of coastal landslides significantly and particularly of the deep-seated landslides. An increased/decreased frequency of sea storms and of the wave heights may change erosion and sediment transport locally and consequently the frequency and intensity of landslides and erosion processes. This may have societal and economic consequences in coastal areas. Touristic resources may be affected, with an increased risk to the population. Transportation networks may also be affected with societal and economic consequences.

Due to climate and environmental changes, vulnerability to landslides is increasing in Italy. The forecasted climate and environmental changes are expected to favour very to extremely rapid failures and chiefly shallow slides, debris flows, rock falls and rock slides. Notably, these are the types of landslides that most frequently cause casualties in Italy (Guzzetti *et al.*, 2005b). Urbanization in hazardous areas and the abandonment of rural and mountain areas, has increased the impact of landslides. Systematic lack of maintenance of structures and infrastructures, of engineered slopes, natural and human made drainage networks (including aqueducts, sewage systems and drains along roads), terraced slopes and changed agricultural and forest practices, have also increased landslide risk. Hence, the changing climate drivers will act on a landscape that was long modified by human actions. This adds to the complexity of predicting future changes in landslide risk.

5.5.2 General remarks

Despite the known and unknown unknowns, and the considerable uncertainties, the following general remarks can be made on landslide risk in Italy.

- Landslide risk to the population, measured by the number of fatal events, is expected to increase due to an augmented frequency of very rapid to extremely rapid failures, chiefly rock falls, shallow slides, and debris flows. The geographical areas where very to extremely rapid failures will occur are expected to increase, further augmenting risk.
- Landslide mortality is expected to remain about the same in Italy, because the population should not increase significantly and improved information and landslide early warning systems should contribute to reduce the risk.
- The societal and economic impact of slow moving landslides in overconsolidated clays and in structurally complex formations is expected to slightly decrease, due to a decrease in pore pressure and, consequently, in the displacement rates. However, this will not reduce the need to monitor active slow moving landslides and to maintain defensive measures.
- Environmental and economic consequences of landslide events are expected to increase due to the augmented frequency of high intensity rainfall, of rapid snowmelt events, of rain-on-snow events, of rainfalls affecting burnt areas and of multiple and/or chained events.
- Populations of landslides caused by a single trigger are expected to cause damage chiefly to transportation networks, utilities and built-up areas. Where the failures are very or extremely rapid they may increase the threat to the population.
- The impact of slope failures affecting engineered slopes, including earth dams, road cuts, retaining walls, embankments, levees and terraced slopes, are also expected to change. The safety conditions of all engineered slopes and the efficacy of existing landslide mitigation/stabilization works should be re-evaluated considering the expected climate changes.

5.5.3 Recommendations

The following general recommendations can be proposed to reduce landslide risk in Italy, in the context of the expected climate, environmental, and societal changes. Reduction of the risk through mitigation efforts and adaptation strategies will have to be pursued at different temporal and geographical scales, adopting a mix of structural (hard) and non-structural (soft) measures.

Most commonly, structural (hard) mitigation measures are designed and constructed considering a reference "return period" for the design event. This is evaluated assuming a stationary time series (i.e., of precipitation or discharge) used to select the design event and to determine its return period. In the context of a changing climate, this hypothesis is not valid and other approaches would be useful for the design of the engineering structures (Pauling & Pauleth, 2007). Structural measures are expensive to implement and require constant maintenance. For these reasons, they should be limited to areas where landslide risk is large and where the adoption of the mitigation measures will reduce the risk significantly. Structural (hard) measured should be used primarily to reduce the risk posed by rapid to extremely rapid landslides (i.e., shallow slides, debris flows and avalanches and rock falls). To identify the sites or areas where landslide risk is large and can be mitigated effectively with hard measures, a systematic screening of the existing landslide risk conditions is necessary. The screening should consider the possible changes in the risk induced by the forecasted climate, environmental and societal changes. The structural measures should also be used to reduce the vulnerability of specific elements at risk, increasing their robustness i.e., the ability to sustain landslide events.

Soft (i.e., non structural) measures should include a comprehensive review of the existing landslide hazard and risk assessment and management plans, including the Geo-Hydrological Plans (PAI, an Italian acronym for *Piano di Assetto Idrogeologico*) available for all major catchments, and the civil protection plans prepared for individual municipalities. The existing plans were prepared ignoring or underestimating the effects of the potential climate and environmental changes. This limits their effectiveness in a changing climate. In some areas, the plans ignore or underestimate the risk posed by specific landslide types and particularly the types of failures that are expected to increase in response to the climate change. In these areas, mitigation actions and adaptation strategies based on the existing assessments might be inadequate.

Monitoring and early warning are probably the most effective forms of soft ("non-structural") measures. The ability of the existing monitoring networks of meteo-hydrological sensors to measure the climate variations is limited and should be improved. Also, the existing networks are insufficient for landslide hazard assessment and risk evaluation. The networks (i) are blind to snowfall at lower terrain elevations, (ii) have an insufficient density of rain gauges in elevated areas and in small mountain catchments, where debris flows form, (iii) do not measure accurately high intensity rainfall events, and (iv) typically, they do not record soil moisture and pore water pressure. In addition, at only a few (largely experimental) sites meteorological and hydrological measurements are coupled with surface and sub-surface topographic and deformation measurement records. These limitations need to be corrected to improve the current ability to forecast landslide risk.

A significant effort is required to expand the currently limited ability to forecast landslides, and to establish and operate successfully landslide early warning systems, at all scales, from the local (site specific) to the national (synoptic) scale. This requires new research on the landslide triggers, on the uncertainty associated with the landslide forecasts, on methods for the acquisition of accurate and systematic information on event landslides, mandatory information to validate the landslide forecasts and on common tools and protocols for the evaluation of the performances of the landslide early warning systems. Lastly, to mitigate landslide risk, an adequate behaviour of individuals and groups of individuals is mandatory. To foster the adoption of positive behaviours, a proper understanding of landslide phenomena and the associated risk is necessary. This can be achieved through new research, and targeted education and dissemination efforts.

5.6 CONCLUSIONS

The assessment of the effects of the expected climate changes on landslide occurrence and evolution is becoming a frequent discussion topic, especially by media. Unfortunately, a reliable prediction of the problems that could be encountered in the near future is a very complex exercise, because of the uncertainty that characterises any involved discipline and still more, the ensemble of disciplines as a whole. On the other hand, the natural growth of any branch of knowledge is providing new instruments for, at least, drawing rational scenarios capable to account for the major variables involved. Often, the analysis of the effects of climate change is performed following two complementary approaches (Crozier, 2010). One approach is based on the interpretation of paleo-evidences and the second approach consists in the comparative frequency analysis between features of the precipitation and landslides, including an application of downscaled climate predictions to both physically-based and statistical models. Today, the analysis of historic records covering significant time intervals (e.g., precipitation records, catalogues of landslide events), the use of modern sensors to measure atmospheric variables and the adoption of powerful numerical codes capable of simulating the complex slope-atmosphere interaction, facilitate the adoption of the second approach.

The goal of this report was to summarise the available knowledge on landslides and their expected variations due to climate change. In particular, the typical landslide types in Italy were considered, based on their interaction with atmospheric forcing, stressing their effects on landslide behaviour. Future rational scenarios of the effects of the expected climate changes were depicted in two typical geomorphological settings, characterised respectively by (i) the presence of slow active slides in fine-grained soils and (ii) the diffusion of essentially granular unsaturated sloping soils susceptible to rapid flowslides. The results show that the expected climate changes could determine a moderate deceleration of active slides in clay and a substantial worsening of the slope stability conditions in unsaturated granular soils. Moreover, such analyses, albeit preliminary, allow to stress the following additional issues:

- The current climate simulations cannot be directly used as input for impact tools without the previous adoption of bias correction approaches.

- After testing the reliability of climate data, the tools currently adopted to assess the hydraulic and mechanical behaviour of unstable slopes can be used conveniently.
- The adoption of too simplified methods, as those aimed at the assessment of future changes in cumulative precipitation over reference intervals uniquely, could lead to antithetical results with respect to those provided by more complex physically-based models.

It is worth to mention that the expected climate trends depend on the adopted emission scenarios. The scenario RCP 8.5 represents an extreme condition for the increase of the radiative forcing. Due to the considerable uncertainties on the adopted assumptions, it is appropriate to consider also simulations provided by different concentration scenarios. Should climate significantly change as some indicators suggest, society should be ready to face the new challenges accounting for associated changes in landslide risk, including demographic and socio-economic aspects. The global situation concerning Italy has been summarised, accounting for the tools that land managers already deal with. An update of such instruments is necessary. To this aim, some lines to pursue were suggested.

REFERENCES

Agliardi, F. & Crosta, G.B. (2014) Long and short-term controls on the Sprian a rockslide (Central Alps, Italy). In: Sassa, K., Canuti, P. & Yin, Y.-P. (eds.) *Landslide Science for a Safer Geoenvironment*. Cham, Springer International Publishing, Vol. 2. pp. 243–249.

Aleotti, P. (2004) A warning system for rainfall-induced shallow failures. *Engineering Geology*, 73, 247–265.

Alexander, D. (1989) Urban landslides. *Progress in Physical Geography*, 13 (2), 157–191.

Allen, R.G., Pereira, L.S., Raes, D. & Smith, M. (1998) *FAO Irrigation and Drainage Paper No. 56: Crop Evapotranspiration (Guidelines for Computing Crop Water Requirements)*.

Alonso, E.E., Gens, A. & Delahaye, C.H. (2003) Influence of rainfall on the deformation and stability of a slope in overconsolidated clays: A case study. *Hydrogeology Journal*, 11, 174–192.

Alpert, P., Ben-Gai, T., Baharad, A., Benjamin, Y., Yekutieli, D., Colacino, M., Diodato, L., Ramis, C., Homar, V., Romero, R., Michaelides, S. & Manes, A. (2002) The paradoxical increase of Mediterranean extreme daily rainfall in spite of decrease in total values. *Geophysical Research Letter*, 29 (11), 31/1–31/4.

Angeli, M.G. (1992) Observed and predicted critical hydraulic conditions in natural inhomogeneous slopes. In: Chowdury, R.N. (ed.) *Geomechanics and Water Engineering in Environmental Management*. Rotterdam, Balkema. 636 pp.

Angeli, M.G., Buma, J., Gasparetto, P. & Pasuto, A. (1998) A combined hillslope hydrology stability model for low-gradient clay slopes in the Italian Dolomites. *Engineering Geology*, 49 (1), 1–13.

Antronico, L., Gullà, G. & Borrelli, G. (2004) Shallow instabilities for sliding flow: Regional influence and area effects. In: *Proceedings of the 9th Int. Symp. on Landslides, Rio de Janeiro*, Vol. 2. pp. 1381–1388.

Bates, B.C., Kundzewicz, Z.W., Wu, S. & Palutikof, J.P. (2008) *Climate Change and Water*. Int. Panel on Climate Change, Technical Report 63.

Belloni, L. & Gandolfo, M. (1997) La frana di Spriana. *Geologia Tecnica e Ambientale*, 3, 7–36.

Berti, M. & Simoni A. (2010) Field evidence of pore pressure diffusion in clayey soils prone to landsliding. *Journal of Geophysical Research*, 115.

Bertini, T., Cugusi, F., D'Elia, B. & Rossi Doria, M. (1984) Climatic conditions and slow movements of colluvial covers in central Italy. In: *Proceedings of the 4th Int. Symp. on Landslides, Toronto*. pp. 81–83.

Bertini, T., Cugusi, T., D'Elia, B. & Rossi Doria, M. (1987) Groundwater regime and slow movements of natural clayey slopes. In: *Proceedings of the 9th Europ. Conf. Soil Mechanics and Foundation Engineering, Dublin*. pp. 385–388.

Bertini, T., Cugusi, F., D'Elia, B., Lanzo, G. & Rossi Doria, M. (1992) Slow movements investigations in clay slopes. In: *Proceedings of the 6th Int. Symp. on Landslides, Christchurch, New*. pp. 329–334.

Bertolini, G. & Pizziolo, M. (2008) Risk assessment strategies for the reactivation of earth flows in the Northern Apennines (Italy). *Engineering Geology*, 102, 178–192.

Blasi, C., Michetti, L., Del Moro, M.A., Testa, O. & Teodonio, L. (2007) Climate change and desertification vulnerability in Southern Italy. *Phytocoenologia*, 37, 495–521.

Boldini, D., Comegna, L., Rianna, G. & Tommasi, P. (2014) Evapotranspiration estimate in a clayey slope affected by landslide phenomena. *Rivista Italiana di Geotecnica*, 48, 21–33.

Borgatti, L. & Soldati, M. (2010) Landslides as a geomorphological proxy for climate change: A record from the Dolomites (northern Italy). *Geomorphology*, 120, 56–64.

Braganza, K., Karoly, D., Hirst, A., Mann, M., Stott, P., Stouffer, R. & Tett, S (2004) Simple indices of global climate variability and change: Part II – Attribution of climate change during the 20th century. *Climate Dynamic*, 22, 823–838.

Breugem, W.P., Hazeleger, W. & Haarsma, R.J. (2007) Mechanisms of northern tropical Atlantic variability and response to CO_2 doubling. *Journal of Climate*, 20, 2691–2705.

Brunetti, M., Buffoni, L., Maugeri, M. & Nanni, T. (2000) Trends of minimum and maximum daily temperatures in Italy from 1865 to 1996. *Theoretical and Applied Climatology*, 66, 49–60.

Brunetti, M., Maugeri, M., Navarra, A. & Nanni, T. (2002) Droughts and extreme events in regional daily Italian precipitation series. *International Journal of Climatology*, 22, 543–558.

Brunetti, M., Buffoni, L., Mangianti, F., Maugeri, M. & Nanni, T. (2004a) Temperature, precipitation and extreme events during the last century in Italy. *Global and Planetary Change*, 40, 141–149.

Brunetti, M., Maugeri, M., Monti, F. & Nanni, T. (2004b) Changes in daily precipitation frequency and distribution in Italy over the last 120 years. *Journal of Geophysical Research*, 109, D05102.

Brunetti, M., Maugeri, M., Monti, F. & Nanni, T. (2006) Temperature and precipitation variability in Italy in the last two centuries from homogenized instrumental time series. *International Journal of Climatology*, 26, 345–381.

Brunetti, M.., Guzzetti, F. & Rossi, M. (2009) Probability distributions of landslide volumes. *Nonlinear Processes in Geophysics*, 16, 179–188.

Bucchignani, E., Montesarchio, M., Zollo, A.L. & Mercogliano, P. (2016) High-resolution climate simulations with COSMO-CLM over Italy: Performance evaluation and climate projections for the XXI century. *International Journal of Climatology*, 36 (2), 735–756

Buma, J. & Dehn, M. (1998) A method for predicting the impact of climate change on slope stability. *Environmental Geology*, 35 (2–3), 190–196.

Cardinali, M., Ardizzone, F., Galli, M., Guzzetti, F. & Reichenbach, P. (2000) Landslides triggered by rapid snow melting: The December 1996-January 1997 event in Central Italy. In: *Proceedings of the EGS Plinius Conference, Maratea*. pp. 439–448.

Cardinali, M., Galli, M., Guzzetti, F., Ardizzone, F., Reichenbach, P. & Bartoccini, P. (2006) Rainfall induced landslides in December 2004 in south-western Umbria, central Italy: Types, extent, damage and risk assessment. *Natural Hazards and Earth System Sciences*, 6, 237–260.

Cascini, L., Sorbino, G. & Cuomo, S. (2003) Modelling of flowslide triggering in pyroclastic soils. In: Picarelli, L. (ed.) *Proceedingsof the Int. Conf. on Fast Slope Movements – Prediction and Prevention for Risk Mitigation, Napoli.* Patron, Bologna. pp. 93–100.

Chiarle, M., Iannotti, S., Mortara, G. & Deline, P. (2007) Recent debris flow occurrences associated with glaciers in the Alps. *Global and Planetary Change*, 56, 123–136.

Chiarle, M., Paranunzio, R., Laio, F., Nigrelli, G. & Guzzetti, F. (2014) Recent slope failures in the Dolomites (Northeastern Italian Alps) in a context of climate change. *Geophysical Research Abstracts*, 16, EGU2014-4017.

Cola, S., Gabrieli, F., Marcato, G., Pasuto, A. & Simonini, P. (2016) Evolutionary behaviour of the Tessina landslide. *Rivista Italiana di Geotecnica*, 1, 51.

Comegna, L., Picarelli, L. & Urciuoli, G. (2007) The mechanics of mudslides as a cyclic undrained-drained process. *Landslides*, 4 (3), 217–232.

Comegna, L., Picarelli, L., Bucchignani, E. & Mercogliano, P. (2013) Potential effects of incoming climate changes on the behaviour of slow active landslides in clay. *Landslides*, 10, 373–391.

Cotecchia, V. (2006) Experience drawn from the great Ancona landslide of 1982. *Bulletin of Engineering Geology and the Environment*, 65, 1–41.

Crosta, G.B. & Agliardi, F. (2003) Failure forecast for large rock slides by surface displacement measurements. *Canadian Geotechnical Journal*, 40 (1), 176–191.

Crosta, G.B. & Frattini, P. (2009) Probabilistic rainfall threshold for shallow landslide triggering. In: *Proceedings of the 1st Italian Workshop on Landslides, Napoli*, Vol. 2. pp. 45–54.

Crosta, G.B., Chen, H. & Lee, C.F. (2004) Replay of the 1987 Val Pola Landslide, Italian Alps. *Geomorphology*, 60 (1–2), 127–146.

Crozier, M.J. (2010) Deciphering the effect of climate change on landslide activity: A review. *Geomorphology*, 124, 260–267.

Cubasch, U., Wuebbles, D., Chen, D., Facchini, M.C., Frame, D., Mahowald, N. & Winther, J.G. (2013) Introduction. In: Stocker, T.F., Qin, D., Plattner, G.-K., Tignor, M., Allen, S.K., Boschung, J., Nauels, A., Xia, Y., Bex, V. & Midgley, P.M. (eds.) *Climate Change 2013: The Physical Science Basis. Contribution of Working Group I to the Fifth Assessment Report of the Intergovernmental Panel on Climate Change.* Cambridge and New York, Cambridge University Press.

Damiano, E. & Mercogliano, P. (2013) Potential effects of climate change on slope stability in unsaturated pyroclastic soils. In: Margottini, C., Canuti, P. & Sassa, K. (eds.) *Proceedings of the Landslide Science and Practice, Rome*, Vol. 4. Springer-Verlag Berlin-Heidelberg.

Damiano, E., Olivares, L. & Picarelli, L. (2012) Steep-slope monitoring in unsaturated pyroclastic soils. *Engineering Geology*, 137–138, 1–12.

Dehn, M., Bürger, G., Buma, J. & Gasparetto, P. (2000) Impact of climate change on slope stability using expanded downscaling. *Engineering Geology*, 55 (3), 193–204.

Del Ventisette, C., Casagli, N., Fortuny-Guasch, J. & Tarchi, D. (2011) Ruinon landslide (Valfurva, Italy) activity in relation to rainfall by means of GBInSAR monitoring. *Landslides*, 9, 497–509.

Di Maio, C., Vassallo, R., Vallario, M., Pascale, S. & Sdao, F. (2010) Structure and kinematics of a landslide in a complex clayey formation of the Italian Southern Apennines. *Engineering Geology*, 116, 311–322.

Di Maio, C., Vassallo, R. & Vallario, M. (2013) Plastic and viscous shear displacements of a deep and very slow landslide in stiff clay formation. *Engineering Geology*, 162, 53–66.

Dixon, N. & Brook, E. (2007) Impact of predicted climate change on landslide reactivation: Case study of Mam Tor, UK. *Landslides*, 4, 137–147.

Duca, S. (2013) *Design of an Experimental Procedure and Set Up for the Detection of Ice Segregation Phenomena in Rock by Acoustic Emissions*. PhD Thesis. Politecnico di Torino, Italy. 227 pp.

Ehret, U., Zehe, E., Wulfmeyer, V., Warrach-Sagi, K. & Liebert, J. (2012) HESS opinions "should we apply bias correction to global and regional climate model data?" *Hydrology and Earth System Sciences*, 16, 3391–3404.

Esu, F. (1977) Behaviour of slopes in structurally complex formations. In: *Proceedings of the Int. Symp. on the Geotechnics of Structurally Complex Formations, Capri*, Vol. 2. pp. 292–304.

Falcucci, A., Maiorano, L. & Boitani, L. (2007) Changes in land-use/land cover patterns in Italy and their implications for biodiversity conservation. *Landscape Ecology*, 22, 617–631.

Feser, F., Rockel, B., von Storch, H., Winterfeldt, J. & Zahn, M. (2011) Regional climate models add value to global model data: A review and selected examples. *Bulletin of the American Meteorological Society*, 92, 1181–1192.

Fischer, L., Purves, R.S., Huggel, C., Noetzli, J. & Haebe W. (2012) On the influence of topographic, geological and cryospheric factors on rock avalanches and rockfalls in high-mountain areas. *Natural Hazards and Earth System Sciences*, 12, 241–254.

Galli, M. & Guzzetti, F. (2007) Landslide vulnerability criteria: A case study from Umbria, Central Italy. *Environmental Management*, 40 (4), 649–665.

Gariano, S.L., Petrucci, O. & Guzzetti, F. (2015) Changes in the occurrence of rainfall-induced landslides in Calabria, southern Italy, in the 20th century. *Natural Hazards and Earth System Sciences*, 15, 2313–2330.

Giannecchini, R., Galanti, Y. & D'Amato Avanzi, G. (2012) Critical rainfall thresholds for triggering shallow landslides in the Serchio River Valley (Tuscany, Italy). *Natural Hazards and Earth System Science*, 12, 829–842.

Giordan, D., Allasia, P., Manconi, A., Baldo, M., Santangelo, M., Cardinali, M., Corazza, A., Albanese, V., Lollino, G. & Guzzetti F. (2013) Morphological and kinematic evolution of a large earthflow: The Montaguto landslide, Southern Italy. *Geomorphology*, 187, 61–79.

Giorgi, F. (2006) Climate change hot spot. *Geophysical Research Letters*, 33, L08707.

Giorgi, F., Jones, C. & Asrar, G. (2009) Addressing climate information needs at the regional level: The CORDEX framework. *WMO Bullettin*, 58, 175–183.

Giusti, G., Iaccarino, G., Pellegrino, A., Russo, C., Urciuoli, G. & Picarelli, L. (1996) Kinematic features of earthflows in Southern Apennines, Italy. In: *Proceedings of the 7th Int. Symp. on Landslides, Trondheim*, Vol. 1. pp. 457–462.

Glade, T. (2003) Landslide occurrence as a response to land use change: A review of evidence from New Zealand. *Catena*, 51 (3–4), 297–314.

Govi, M., Mortara, G. & Sorzana, P.F. (1985) Eventi idrologici e frane. *Geologia Applicata e Idrogeologia*, 20-parte 2, 359–375.

Grana, V. & Tommasi, P. (2014) A deep-seated slow movement controlled by structural setting in marly formations of Central Italy. *Landslides*, 11, 195–212.

Greco, R., Comegna, L., Damiano, E., Guida, A., Olivares, L. & Picarelli, L. (2013) Hydrological modelling of a slope covered with shallow pyroclastic deposits from field monitoring data. *Hydrology and Earth System Sciences*, 17, 4001–4013.

Gudmundsson, L., Bremnes, J.B., Haugen, J.E. & Engen Skaugen, T. (2012) Downscaling RCM precipitation to the station scale using statistical transformations – A comparison of methods. *Hydrology and Earth System Sciences*, 9, 6185–6201.

Guida, D. & Iaccarino, G. (1991) Fasi evolutive delle frane tipo colata nell'alta valle del F. Basento (Potenza). *Studi Trentini di Scienze Naturali, Acta Geologica*, 68, 127–152.

Guzzetti, F. (2000) Landslide fatalities and the evaluation of landslide risk in Italy. *Engineering Geology*, 58 (2), 89–107.

Guzzetti, F. & Cardinali, M. (1991) Debris-flow phenomena in the Central Apennines of Italy. *Terra Nova*, 3, 619–627.

Guzzetti, F., Reichenbach, P., Cardinali, M., Galli, M. & Ardizzone, F. (2005a) Probabilistic landslide hazard assessment at the basin scale. *Geomorphology*, 72 (1–4), 272–299.
Guzzetti, F., Stark, C.P. & Salvati, P. (2005b) Evaluation of flood and landslide risk to the population of Italy. *Environmental Management*, 36 (1), 15–36.
Guzzetti, F., Galli, M., Reichenbach, P., Ardizzone, F. & Cardinali, M. (2006) Landslide hazard assessment in the Collazzone area, Umbria, central Italy. *Natural Hazards and Earth System Sciences*, 6, 115–131.
Guzzetti, F., Peruccacci, S., Rossi, M. & Stark, C.P. (2007) Rainfall thresholds for the initiation of landslides in central and southern Europe. *Meteorological Atmospheric Physics*, 98, 239–267.
Guzzetti, F., Zoboli, R., Salvati, P., Bianchi, C. & Mazzanti, M. (2014) Quanto sono costate e quanto costano le calamità idrogeologiche in Italia? In: *Proceedings of the XIII Giornata Mondiale dell'Acqua, Calamità Idrogeologiche: Aspetti Economici, Accademia Nazionale dei Lincei, Roma*. pp. 13–31.
Huggel, C., Clague, J.J. & Korup, O. (2012) Is climate change responsible for changing landslide activity in high mountains? *Earth Surface Processes and Landforms*, 37, 77–91.
Hungr, O., Leroueil, S. & Picarelli, L. (2014) The Varnes classification of landslide types, an update. *Landslide*, 11, 167–194.
IPCC (2007) Climate change 2007: The physical science basis. In: Solomon, S., Qin, D., Manning, M., Chen, Z., Marquis, M., Averyt, K.B., Tignor, M. & Miller, H.L. (eds.) *Contribution of Working Group I to the Fourth Assessment Report of the Intergovernmental Panel on Climate Change*. Cambridge and New York, NY
Jacob, D., Petersen, J., Eggert, B., Alias, A., Christensen, O.B. & Bouwer, L.M. (2014) EURO-CORDEX: New high-resolution climate change projections for European impact research. *Regional Environmental Change*, 14, 563–578.
Kostopoulou, E. & Jones, P.D. (2005) Assessment of climate extremes in the eastern Mediterranean. *Meteorology and Atmospheric Physics*, 89, 69–85.
Kottek, M., Grieser, J., Beck, C., Rudolf, B. & Rubel, F. (2006) World map of the Köppen-Geiger climate classification updated. *Meteorologische Zeitschrift*, 15, 259–263.
Lafon, T., Dadson, S., Buys, G. & Prudhomme, C. (2013) Bias correction of daily precipitation simulated by a regional climate model: A comparison of methods. *International Journal of Climatology*, 33 1367–1381.
Laternser, M. & Schneebeli, M. (2003) Long-term snow climate trends of the Swiss Alps (1931–99). *International Journal of Climatology*, 23, 733–750.
Lembo Fazio, A., Manfredini, M., Ribacchi, R. & Sciotti, M. (1984) Slope failure and cliff instability in the Orvieto hill. In: *Proceedings of the 4th Int. Symp. on Landslides, Toronto*, Vol. 2. pp. 115–112.
Lenderink, G., Mok, H.Y., Lee, T.C. & Van Oldenborgh, G.J. (2011) Scaling and trends of hourly precipitation extremes in two different climate zones – Hong Kong and the Netherlands. *Hydrology and Earth System Sciences*, 8, 4701–4719.
Malamud, B.D., Turcotte, D.L., Guzzetti, F. & Reichenbach, P. (2004) Landslide inventories and their statistical properties. *Earth Surface Processes and Landforms*, 29 (6), 687–711.
Maraun, D., Wetterhall, F., Ireson, A.M., Chandler, R.E., Kendon, E.J., Widmann, M., Brienen, S., Rust, H.W., Sauter, T., Themessl, M., Venema, V.K.C., Chun, K.P., Goodess, C.M., Jones, R.G., Onof, C., Vrac, M. & Thiele-Eich, I. (2010) Precipitation downscaling under climate change. Recent developments to bridge the gap between dynamical models and the end user. *Reviews of Geophysics*, 48, RG3003.
Mariotti, A., Zeng, N., Yoon, J., Artale, V., Navarra, A. Alpert, P. & Li, L.Z.X. (2008) Mediterranean water cycle changes: Transition to drier 21st century conditions in observations and CMIP3 simulations. *Environmental Research Letters*, 3, 044001.

Montesarchio, M., Zollo, A.L., Bucchignani, E., Mercogliano, P. & Castellari, S. (2014) Performance evaluation of high-resolution regional climate simulations in the Alpine space and analysis of extreme events. *Journal of Geophysical Research: Atmospheres*, 119, 3222–3237.

Moody, J.A., Shakesby, R.A., Robichaud, P.R., Cannon, S.H. & Martin, D.A. (2013) Current research issues related to post-wildfire runoff and erosion processes. *Earth-Science Reviews*, 122 (C), 10–37.

Olivares, L. & Picarelli, L. (2003) Shallow flowslides triggered by intense rainfalls on natural slopes covered by loose unsaturated pyroclastic soils. *Géotechnique*, 53 (2), 283–288.

Olivares, L. & Damiano, E. (2007) Post-failure mechanics of landslides: Laboratory investigation of flowslides in pyroclastic soils. *Journal of Geotechnical and Geoenvironmental Engineering ASCE*, 133 (1), 51–62.

Pauling, A. & Pauleth, H. (2007) On the variability of return periods of European winter precipitation extremes over the last three centuries. *Climate of the Past*, 3, 65–76.

Picarelli, L. & Olivares, L. (2001) Innesco e formazione di colate di fango in terreni sciolti di origine piroclastica. In: *Forum on Fenomeni di colata rapida di fango nel maggio '98, Napoli.* Commissariato di Governo per l'Emergenza Idrogeologica in Campania. pp. 26–38

Picarelli, L. & Napoli, V. (2003) Some features of two large earthflows in intensely fissured tectonized clay shales and criteria for risk mitigation. In: *Proceedings of the Int. Conf. on Fast Slope Movements-Prediction and Prevention for Risk Mitigation, Sorrento*, Vol. 1. pp. 431–438.

Picarelli, L. & Di Maio, C. (2010) Deterioration processes of hard clays and clay shales. In: Calcaterra, D. & Parise, M. (eds.) *Weathering as Predisposing Factor to Slope Movements.* London, Geological Society Publishing House. pp. 15–32.

Picarelli, L., Olivares, L., Andreozzi, L., Damiano, E. & Lampitiello, S. (2004) A research on rainfall-induced flowslides in unsaturated soils of pyroclastic origin. In: *Proceedings of the 9th Int. Symp. on Landslides, Rio de Janeiro.* Leiden, CRC Press, Vol. 2. pp. 1497–1503.

Picarelli, L., Urciuoli, G., Ramondini, M. & Comegna, L. (2005) Main features of mudslides in tectonised highly fissured clay shales. *Landslides*, 2, 15–30.

Picarelli, L., Leroueil, S., Olivares, L., Pagano, L., Tommasi, P. & Urciuoli, G. (2012) Groundwater in slopes. In: Clague, J.J. & Stead, D. (eds.) *Landslides: Types, Mechanisms and Modeling.* Cambridge, Cambridge University Press. pp. 235–251.

Piccarreta, M., Capolongo, D., Boenzi, F. & Bentivenga, M. (2006) Implications of decadal changes in precipitation and land use policy to soil erosion in Basilicata, Italy. *Catena*, 65, 138–151.

Piccarreta, M., Capolongo, D., Miccoli, M.N. & Bentivenga M. (2012) Global change and long-term gully sediment production dynamics in Basilicata, southern Italy. *Environmental Earth Sciences*, 67, 1619–1630.

Piervitali, E., Colacino, M. & Conte, M. (1998) Rainfall over the central-western Mediterranean basin in the period 1951–1995. Part I: Precipitation trends. *Nuovo Cimento della Società Italiana di Fisica*, 21, 331–344.

Pirone, M. (2010) *Analysis of Slope Failure Mechanism in Unsaturated Pyroclastic Soils, Based on Testing Site Monitoring.* PhD Thesis University of Naples Federico II. 309 pp.

Pirone, M., Damiano, E., Urciuoli, G, Olivares L. & Picarelli, L. (2012) Groundwater-atmosphere interaction in unsaturated pyroclastic slopes at two sites in Italy. *Rivista Italiana di Geotecnica*, 66, 21–29.

Pisani, G., Castelli, M. & Scavia, C. (2010) Hydrogeological model and hydraulic behaviour of a large landslide in the Italian Western Alps. *Natural Hazards Earth System Science*, 10, 2391–2406.

Polemio, M. & Petrucci, O. (2010) Occurrence of landslide events and the role of climate in the twentieth century in Calabria, southern Italy. *Quarterly Journal of Engineering Geology and Hydrogeology*, 43, 403–415.

Rianna, G., Comegna, L., Mercogliano, P. & Picarelli, L. Potential effects of climate changes on soil-atmosphere interaction and landslide hazard. *Natural Hazards*, in press. http://doi.org/10.1007/s11069-016-2481-z

Salvati, P., Bianchi, C., Rossi, M. & Guzzetti, F. (2010) Societal landslide and flood risk in Italy. *Natural Hazards and Earth System Sciences*, 10, 465–483.

Santini, M. & Valentini, R. (2011) Predicting hot-spots of land use changes in Italy by ensemble forecasting. *Regional Environmental Change*, 11, 483–502.

Schroeder, P.R., Dozier, T.S., Zappi, P.A., McEnroe, B.M., Sjostrom, J.W. & Peton, R.L. (1994) *The Hydrologic Evaluation of Landfill Performance (HELP) Model: Engineering Documentation for Version 3, EPA/600/R-94/168b, US.* Cincinnati, OH, Environmental Protection Agency, Risk Reduction Engineering Laboratory.

Skempton, A.W. & Hutchinson, J.N. (1969) Stability of natural slopes and embankment foundations. In: *Proceedings of the 7th Int. Conf. on Soil Mechanics and Foundation Engineering, Mexico City*, Vol. 4. pp. 291–340.

Soldati, M., Corsini, A. & Pasuto, A. (2004) Landslides and climate change in the Italian Dolomites since the Late glacial. *Catena*, 55, 141–161.

Stoffel, M., Tiranti, D. & Huggel, C. (2014) Climate change impacts on mass movements – Case studies from the European Alps. *Science of the Total Environment*, 493, 1255–1266.

Teutschbein, C. & Seibert, J. (2012) Bias correction of regional climate model simulations for hydrological climate change impact studies: Review and evaluation of different methods. *Journal of Hydrology*, 456–457, 12–29.

Tiranti, D., Rabuffetti, D., Salandin, A. & Tararbra, M. (2013) Development of a new translational and rotational slides prediction model in Langhe hills (north-western Italy) and its application to the 2011 March landslide event. *Landslides*, 10, 21–138.

Tommasi, P., Ribacchi, R. & Sciotti, M. (1997) Geotechnical aspects in the preservation of the historical town of Orvieto. In: *Proceedings of the Int. Symp. on Geotechnical Engineering for the Preservation of Monuments and Historic Sites, Napoli.* pp. 849–858.

Tommasi, P., Pellegrini, P., Boldini, D. & Ribacchi, R. (2006) Influence of rainfall regime on hydraulic conditions and movement rates in the overconsolidated clayey slope of the Orvieto hill (central Italy). *Canadian Geotechnical Journal*, 43 (1), 70–86.

Tommasi, P., Boldini, D., Caldarini, G. & Coli, N. (2013) Influence of infiltration on the periodic re-activation of slow movements in an overconsolidated clay slope. *Canadian Geotechnical Journal*, 50 (1), 54–67.

Trenberth, K. (2011) Changes in precipitation with climate change. *Climate Research*, 47, 123–138.

USDA Soil Conservation Service (1985) *National Engineering Handbook, Section 4, Hydrology*. Washington, DC, US Government Printing Office.

Valt, M., Cagnati, A., Crepaz, A. & Marigo, G. (2005) Snow precipitation in the last years on Italian Alps. *Hrvatski Meteoroloski Casopis*, 40, 654–657.

Van Vuuren, D.P., Edmonds, J., Kainuma, M., Riahi, K., Thomson, A., Hibbard, K., Hurtt, G.C., Kram, T., Krey, V., Lamarque, J.F., Masui, T., Meinshausen, M., Nakicenovic, N., Smith, S.J. & Rose, S.K. (2011) The representative concentration pathways: An preview. *Climatic Change*, 109 (1–2), 5–31.

Varnes, D.J. (1978) Slope movement types and processes. In: Schuster, R.L. & Krizek, R.J. (eds.) *Landslides, Analysis and Control, Special Report 176*. Washington DC Transportation Research Board, National Academy of Sciences. pp. 11–33.

Viero, A., Furlanis, A., Squarzoni, C., Teza, G., Galgaro, A. & Gianolla, P. (2013) Dynamics and mass balance of the 2007 Cima Una rockfall (Eastern Alps, Italy). *Landslides*, 10, 393–408.

von Storch, H., Zorita, E. & Cubasch, U. (1993) Downscaling of global climate change estimates to regional scales: An application to Iberian rainfall in wintertime. *Journal of Climate*, 6, 1161–1171.

Weidner, S., Moser, M. & Lang, E. (1998) Influence of hydrology on sagging of mountain slopes ("Talzuschube") – New results of time series analysis. In: *Proceedings of the 8th IAEG International Congress, Vancouver*. pp. 1259–1266.

Westra, S., Alexander, L.V. & Zwiers, F.W. (2013) Global increasing trends in annual maximum daily precipitation. *Journal of Climate*, 26, 3904–3918.

Wilby, R.L. & Dessai, S. (2010) Robust adaptation to climate change. *Weather*, 65, 180–185.

Zollo, A.L., Rillo, V., Bucchignani, E., Montesarchio, M. & Mercogliano, P. (2016) Extreme temperature and precipitation events over Italy: Assessment of high resolution simulations with COSMO-CLM and future scenarios. *International Journal of Climatology*, 36 (2), 987–1004.

Chapter 6

Global warming and landslide disaster: evidence from Japan

G. Wang
Kyoto University, Kyoto, Japan

I. Towhata
University of Tokyo, Tokyo, Japan

ABSTRACT

In this report, we present the precipitation data that had been obtained in Japan since 1890. The long-term tendency of precipitation indicates that the annual precipitation in Japan is decreasing. However, extreme rainfall events (such as ≥50 and 100 mm/h) showed an increasing tendency, probably due to climate change. The statistical data of sediment-related disaster due to precipitation also showed a tendency for slight yearly increase. Some catastrophic landslides related to extreme weather condition in recent years in Japan were also introduced as examples. Finally, we analyzed the flooding occurrence of Kamo River in Kyoto City, where long-term records are available. The results indicate that the frequency of flooding of the Kamo River is not related with the climate change, and the effect of human activity on nature is more serious than the natural processes.

6.1 INTRODUCTION

At present, climate change is largely accepted as a global problem. To provide the world with a clear scientific view on the current state of knowledge in climate change and its potential environmental and socio-economic impacts, the Intergovernmental Panel for Climate Change (IPCC) was established in 1988 by the United Nations Environment Programme (UNEP) and the World Meteorological Organization (WMO). The IPCC reviews and assesses the most recent scientific, technical and socio-economic information produced worldwide relevant to the understanding of climate change with the involvement of both Administration official and expertise including academic scientists. A series of reports presented by IPCC gives a consensus that the global is warming at rates unprecedented in the historical record, and also the latest summary report addressed that the majority of the global warming in the past century originates from greenhouse gas emissions, and global warming will continue even if drastic emission controls are implemented (IPCC, 2013).

Global warming will inevitably result in the change of hydrologic cycle (IPCC, 2007), because a warming atmosphere and warmer ocean surface will enable higher rates of evaporation and then more precipitation. This change is likely to have a number of different potentially impacts on different types of landslides, because the groundwater and pore-pressure fluctuations in the slope could be changed and then affect the slope stability.

In this report, we present the statistical data of annual rainfall as well as the features of these rainfall in recent some decades in Japan, and also present the landslides triggered mainly by rainfall. We also introduce some catastrophic landslides triggered by heavy rainfall on areas with differing geological background in recent years in Japan. Finally, we presented a case study on the long-term variation of frequency of flooding in Kamo River, Kyoto, Japan, to examine possible evidence for climate change in a span more than 1000 years. Through this case study, some problems in sorting out the evidences for climate change are highlighted.

6.2 PRECIPITATION OVER JAPAN

The Japan Meteorological Agency (JMA) operationally observes precipitation at about 1,300 unmanned regional meteorological observation stations all over Japan (collectively known as the Automated Meteorological Data Acquisition System, or AMeDAS). Since 1996, JMA has published annual assessment reports titled *Climate Change Monitoring Report*. These reports present the atmospheric, oceanic and environmental monitoring and analysis, and provide up-to-date information on climatic conditions in Japan and also around the world. In these reports, the long term variation of annual precipitation had been analyzed in detail. Figure 6.1 shows the results for the period 1898-2013, where a 30-year average annual precipitation was calculated based on the precipitation from 1971 to 2000, and the precipitation anomaly refers to the departure of the annual precipitation from this average (the base line). The bars indicate the precipitation anomaly for each year, and the solid line indicates the five-year running mean. The tendency of this solid line suggests that the annual precipitation is likely to become less in recent years. During the period 1898–1923 and the period 1949–1959, there was much more precipitation than in recent decades. However, over the past three decades, the precipitation anomaly became more distinguished, and the difference between the precipitation anomalies in the wet and dry years became greater than the early three decades of the 20th century. This may be the evidence of climate change in Japan.

Based on the maximum daily rainfall data for the period 1901–2006, Okada (2009) reported that the maximum daily rainfall with a return period of 100 years was 100–200 mm and 200–400 mm in Northern Japan and Pacific side of Western Japan, respectively, the maximum daily rainfall with a return period of 200 years was 150–250 mm and 400–500 mm in the Sea of Japan side (ranging from the Northern to Eastern Japan) and the Pacific side of Western Japan, respectively. Normally the southern part has much more precipitation than the northern part, and the western part has more precipitation than the eastern part.

Figures 6.2a and 6.2b present the annual number of rainfall events with their intensities ≧50 mm/h and 100 mm/h, respectively, over the period from 1976 to 2010. It is noted that these data are based on the records of 1000 AMeDAS points operated by JMA. From Figure 6.2a it is noticed that there were 168 rainfall events with intensities ≧50 mm/h over the period 1976–1986, this number increased to 195 over the period 1987–1998, and rose to 226 over the period 1999–2010, namely the rainfall event

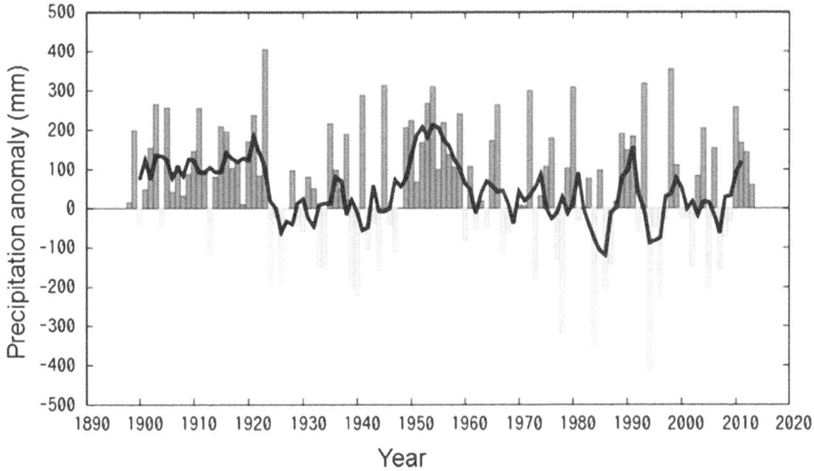

Figure 6.1 Annual anomaly in precipitation from 1898 to 2013 in Japan (derived from analysis of precipitation records from 51 observation stations) (JMA, 2014).

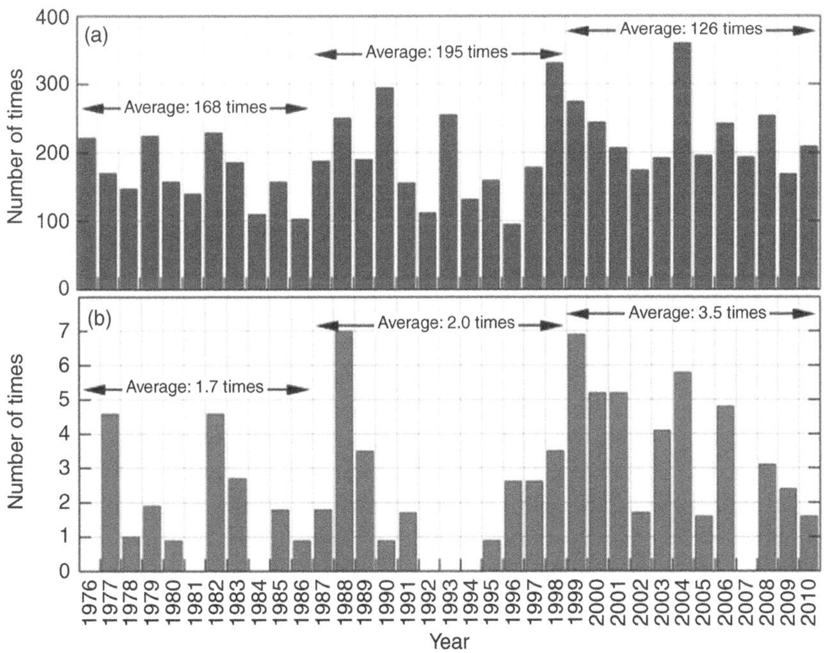

Figure 6.2 Annual number of events with precipitation exceeding 50 mm/h (a) and 100 mm/h (b), respectively, from 1976 to 2010 (per 1,000 AMeDAS stations).

with intensity ≧50 mm/h increased approximately 1.3 times in the past three decades. Further examining of Figure 6.2b results in a conclusion that the rainfall events with intensities ≧100 mm/h increased approximately twofold over the period 1976–2010. From Figures 6.1 and 6.2, we may conclude that although the total annual precipitation over Japan showed a decreasing trend, the number of abnormal rainfall events is increasing yearly.

6.3 LANDSLIDES IN JAPAN

As mentioned above, Japan is experiencing a higher frequency of extreme rainfall events. This could increase the frequency of occurrence of high pore pressure, and thus the activity of rainfall-triggered landslides. In Japan, sediment-related disasters have been divided into several types, such as debris flow, landslide, steep-slope failure, and others. Differing laws have been issued for dealing with different types of sedimentary disaster. In this report we use the term landslide to refer to all these types of sedimentary disasters.

Ministry of Land, Infrastructure, Transport and Tourism (MLIT), Japan has compiled statistics on landslide damage from 1982 to 2013 (MLIT, 2013), and the results are presented in Figure 6.3. It is noted that 225 landslides triggered by the Mid Niigata Prefecture Earthquake in 2004 were excluded from the statistical data for those landslides in 2004, considering that coseismic landslides had less relationship with the extreme weather, although they might have been affected by the precedent rainfall. According to the statistics, about 897 landslides occurred annually in an average during the period 1982–1991. This number rose to 1008 over the next ten years (from 1992 to 2001), and further reached 1088 in the most recent decade (from 2002 to 2013).

It is well known that the initiation of landslides depends on many factors, such as the geological and topographical features of the potential landslide area, and different types of landslide may respond either to meteorological changes in the long term (monthly or yearly rainfall) or short term (daily or weekly rainfall) (Asch, 1996; Buma and Dehn, 1998). Also, the difference in the scales of landslides may be in several orders. Therefore, it will be difficult to make a conclusive remark on the impact of climate change on landslide disaster. Nevertheless, many studies have reported the increasing tendency of landslide with time (Dixon and Brook, 2007; Jakob and Lambert, 2009; Chiang and Chang, 2011) with the conclusion that different areas will have different increasing rates. On the basis of these statistics shown in Figure 6.3, we may conclude that the number of landslides in Japan is likely to rise over the course of time. This trend shows consistency with the occurrence of extreme precipitation. Therefore, we may conclude that the increased number of landslide likely resulted from extreme precipitation due to the climate change.

In the following sections, we will introduce some catastrophic landslides that had been triggered by the heavy rainfalls associated with powerful typhoons in the past three years. One is the deep-seated landslides that occurred in Kii Peninsula in 2011, the second one is the large scale shallow landslide that occurred in Izu Oshima Island in 2013, and the third one is the 2014 Hiroshima landslide disaster.

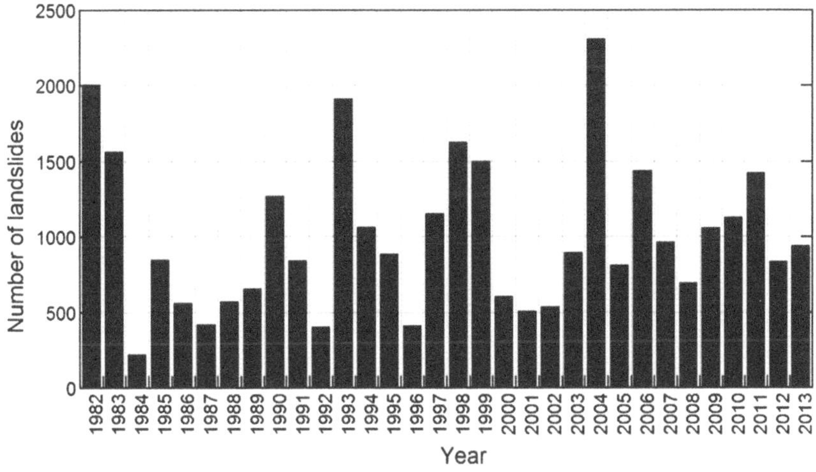

Figure 6.3 Number of landslides in Japan since 1982.

6.3.1 Deep-seated landslides in Kii Peninsula triggered by Typhoon Talas in 2011

Typhoon Talas formed as a tropical depression originating over the sea west of the Mariana Islands on 24 August 2011, and moved slowly northward and developed into a typhoon on 30 August. On 3 September, Talas made landfall on Shikoku Island and reached the Sea of Japan on the next day after crossing the Shikoku and Chugoku regions. Because Talas had a large scale strong wind area and moved very slowly, record-breaking heavy rain fell over a wide area from western to northern Japan, especially over a wide area of the Kii Peninsula (Figure 6.4) (Chigira *et al.*, 2013). From 31 August to 4 September, the total precipitation exceeded 1000 mm over two-thirds of the Kii Peninsula, and a maximum value of 2439 mm was recorded at Kami-Kitayama. This record-breaking heavy rainfall caused inundation and river flooding, and also triggered more than 3000 landslides. The total amount of sediment generated was estimated to be some 100 million m^3. In Nara and Wakayama Prefectures, 72 deep-seated catastrophic landslides were triggered (Hayashi *et al.*, 2013). Almost all these landslides are in large scale with their volume $\geq 100{,}000\ m^3$ (Ishizuka *et al.*, 2012; Chigira *et al.*, 2013). The displaced material of some landslides blocked the river, forming landslide dams, as exampled by Figure 6.5. Some landslide dams breached naturally immediately after the formation, but some big ones became a great threat to the lives and properties of residents living downstream, especially in the soon coming typhoons.

Concerning the rainfall features and landslide occurrence, Chigira *et al.* (2013) made a detailed study and stated that "the rainfall amounts were quite large in comparison with previous precipitation in the Kii Mountains, located in the center of the Kii Peninsula, where the average annual precipitation is 1300 mm in the northwest of the region, and over 3000 mm in the southeast, based on the 30-year record of the Automated Meteorological Data Acquisition System of the Japan Meteorological

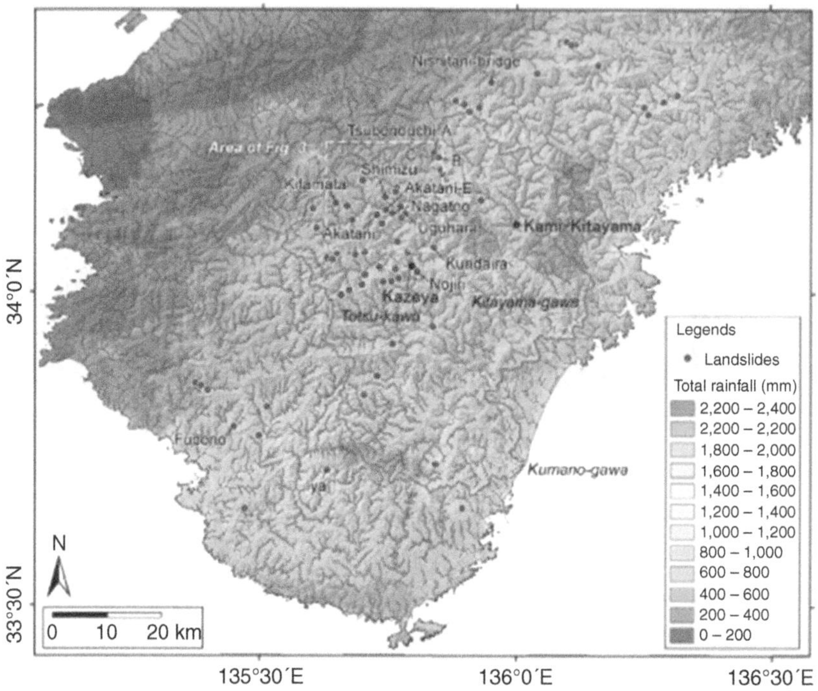

Figure 6.4 Distribution of landslides and cumulative rainfall between 31 August and 5 September 2011 (from Chigira *et al.*, 2013).

Agency. Despite such heavy rainfall over a wide area caused by Typhoon Talas, large landslides only occurred in a rather sporadic manner, and shallow landslides did not occur widely, which is a similar pattern to the landslides associated with Typhoon NABI in 2005, in the same Shimanto Belt in Kyushu. The rainfall-induced landslides were concentrated in the center of the Kii Peninsula, which was not the area of the highest rainfall. This could be attributed to the variations in the pattern of rainfall and geological conditions".

Seventeen landslide dams had been formed by the displaced landslide materials during the period of Typhoon Talas. MLIT conducted emergency investigation on five large landslide dams, which had been identified to be very likely to overflow and collapse. Temporary drainage and other emergency measures had been conducted to prevent the dams from overflowing or collapsing, and many of the drainage and other structures had been completed before the rainy season of 2012.

6.3.2 Landslides in Oshima triggered by Typhoon Wipha in 2013

Typhoon Wipha started as a tropical storm on 9 October 2013 over the sea near the island of Guam, and then it strengthened on October 11, and later on October 12, 2013, Wipha intensified into a typhoon. Typhoon Wipha weakened as it skirted Japan's

Figure 6.5 Some large landslides and landslide dams caused by Typhoon Talas in the Totsugawa Village, Nara Prefecture.

eastern coastline on October 15 as the ocean waters surrounding the typhoon began to cool. Typhoon Wipha reached extratropical status late on October 16, and finally disappeared on the north-east of the Pacific Ocean.

Typhoon Wipha dumped heavy rain and whipped up strong winds in many areas of East Japan. Though the storm caused only limited damage on the mainland, it triggered a massive landslide on the island of Izu Oshima (as shown in Figure 6.6), killing 37 and leaving 3 others missing. About 350 homes were destroyed throughout Japan, mostly on Izu Oshima.

The 24-hour amount of rainfall through 8:20 a.m. in Oshima on Oct. 16 totaled 824 millimeters, setting a new record for this island (Figure 6.7). This daily precipitation is equivalent to more than twice the amount Oshima had in all of October in an average year. The maximum hourly rainfall reached 125 mm. Also a maximum wind velocity was recorded as 35 m/s. As a result of the rainfall, several landslides were recorded in the thin volcanic ash layer of hillside surfaces. Some displaced materials traveled approximately 2.2 km, causing ruinous damage to the downslope area.

Field investigation revealed that the thickness of those displaced landslide materials on the source area was about 70–120 cm, and the outcropped sliding surfaces are loess (aeolian sediments), whose permeability is lower than the overlain volcanic ash layer. As a possible mechanism on the initiation of the shallow landslide, it is inferred that this permeability contrast might have enabled the formation of perched water layer, and increased water table resulted in the instability of the volcanic ash layer. The strong wind might be another trigger of the landslides, because wind velocity reached about 35 m/s (Figure 6.8). Those trees (about 20 m high on average with roots less than 70 cm beneath the soil layers) may suffer strong windthrow, which can increase the instability of the shallow slope.

6.3.3 Landslides in Hiroshima triggered by heavy rainfall in 2014

Unseasonable weather conditions occurred in Japan in August 2014, such that Western Japan experienced record-high precipitation and record-low sunshine durations. The

Figure 6.6 Large-scale shallow landslides in Oshima during Typhoon Wipha, 2013.

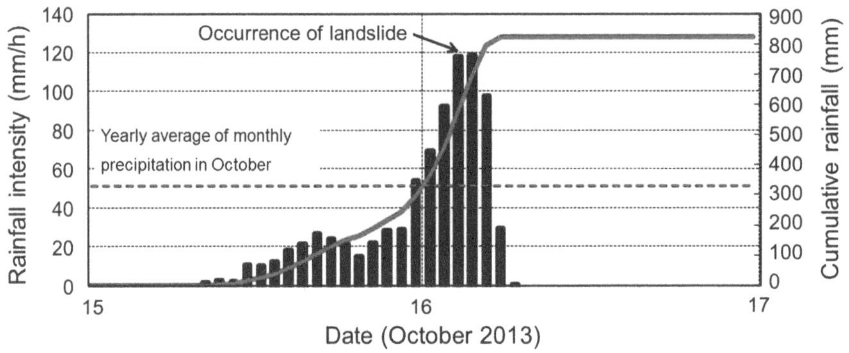

Figure 6.7 Rainfall in Oshima, 14–16 October 2013.

monthly precipitation ratio to the normal averaged over the Pacific side of Western Japan for August 2014 reached as high as 301%. Different types of natural disasters such as flooding and landslides had been triggered due to the occurrence of the record-high precipitation, among them the most catastrophic ones were those landslides that occurred in Hiroshima City on August 20, 2014, which resulted in 74 dead, 69 injured and more than 500 buildings destroyed (Fire and Disaster Management Agency, 2015). Figure 6.9 presents a view of the area (Asaminami Ward of Hiroshima city) most damaged by the landslides on 20 August 2014.

Early on the morning of August 20, as the Akisame-front passed through the northern Hiroshima city, it brought torrential downpours continuously and intensively to a narrow band as the result of a phenomenon called back-building. The AMeDAS data from the Miiri station showed that the maximum 3 hour precipitation (1:30–4:30) and the maximum 1 hour precipitation (3:00–4:00) recorded 217.5 mm and

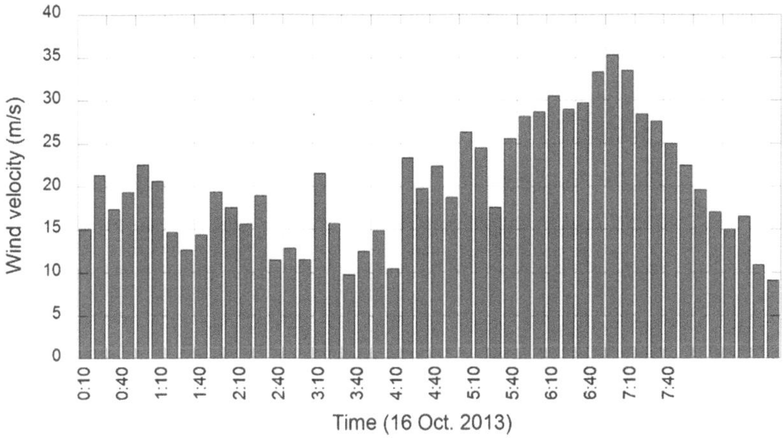

Figure 6.8 Wind velocities in Oshima (on 16 October 2013).

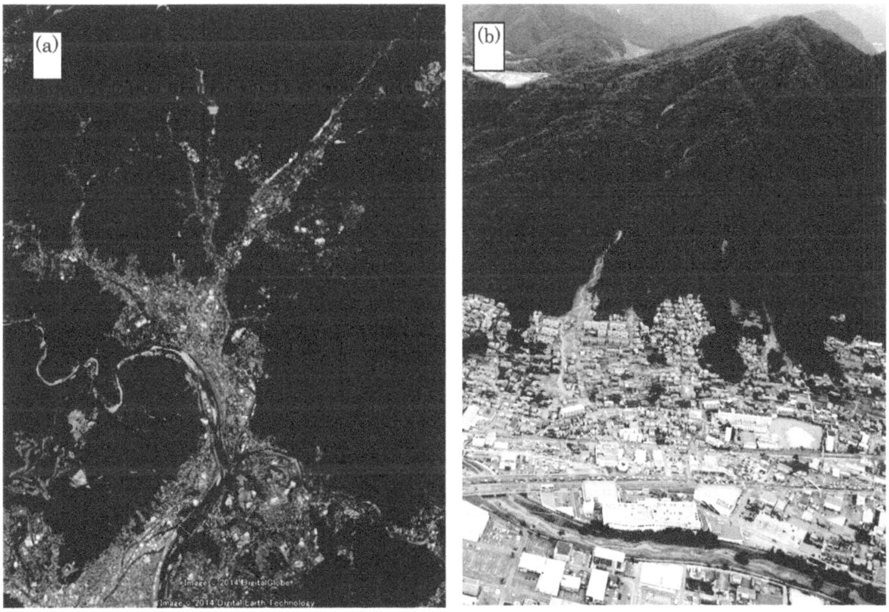

Figure 6.9 landslides in Hiroshima on 14–16 October 2014. (a) Locations (after GSI, 2014), (b) close-up view of the most catastrophic one in Yagi district, Asaminami Ward (taken by GSI on August 20, 2014).

101 mm, respectively. Based on the XRAIN (X-band polarimetric (multi parameter) Radar Information Network) data of MLIT, and the rainfall data of C-band Radar of JMA, the rainfall distribution had been analyzed (Yamamoto and Kobayashi, 2014), and the results showed that the rainfall with its 12-hour rainfall intensity greater than 200 mm was localized in a band of 23 km × 5 km (Figure 6.10).

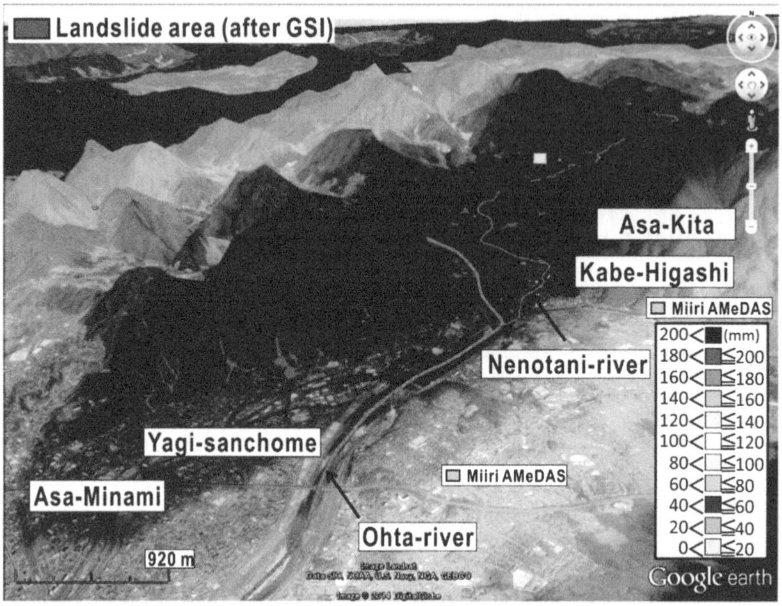

Figure 6.10 Accumulative rainfall (0:00–6:00 of 20 October, 2014) and locations of landslides (after Yamamoto and Kobayashi, 2014). AMeDAS: Automated Meteorological Data Acquisition System; GSI: Geospatial Information Authority of Japan.

Due to the torrential downpour, more than 160 landslides were triggered near the mountains beside the city of Hiroshima in a narrow area of 10 km × 2 km along the rain band. The distribution of cumulative rainfall for the period of 0:00~6:00 of August 20, 2014 had been detailed by Yamamoto and Kobayashi (2014) and is presented in Figure 6.10 together with the locations of the resultant landslides. The landslides were mainly localized in Asaminami Ward and Asakita Ward. The worst hit district was Yagi district of Asaminami Ward, where 52 people were killed, and more than 200 houses were completely or partially destroyed.

The landslide areas mainly consist of granites (late cretaceous felsic plutonic (new Ryoke granitic) rocks, and late cretaceous non-alkaline felsic volcanic rocks), hornfels (middle/late Jurassic Melange matrix (accretionary complex)), and Takada Rhyolites. Figure 6.11 presents the geological settings together with the locations of landslides. Although some landslides originated on the slopes of hornfels and a few landslides on the slopes of Takada Rhyolites, most of the devastating landslides were triggered on the areas composed of granite.

Most of the landslides are channelized debris flows with small and shallow landslides (with depths of a few meters) originating on the source areas. The initial landslides were very small, but they changed to debris flows before reaching the area of human settlement. The displaced materials from the landslide source areas were not great in volume, but they entrained the sediments (in some cases, with big boulders) along the torrent to become highly destructive flows.

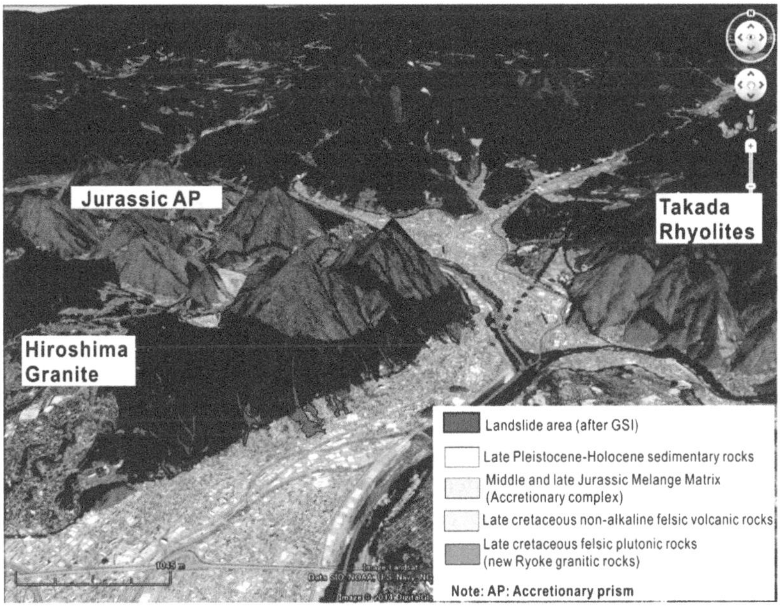

Figure 6.11 Geological settings and distribution of landslides (after Yamamoto and Kobayashi, 2014).

The August 20 landslide disaster in Hiroshima City caused great impact on the society of Japan, and alerted the nation again to the dangers associated with urbanization near mountains. It is still fresh in our minds that in 1999, many landslides were triggered by heavy rainfall in Hiroshima City, resulting in 31 dead, 1 missing, and 169 houses being destroyed (Chigira, 2001; Wang *et al.*, 2003). Against this background, in 2000 the Japanese government issued a law, "Act on Sediment Disaster Countermeasures for Sediment Disaster Prone Areas". However, the designation of warning zones under the framework of this law has not been performed speedily due to the financial and manpower shortages. With the increasing tendency of abnormal weather conditions, prompt and effective countermeasures become essential for the mitigation of such kinds of landslide hazard.

6.4 ON LONG-TERM VARIATION IN FREQUENCY OF FLOODING

One of the significant impacts of climate change on slope disaster is the intensity of precipitation. To shed light on this issue, we attempted to study the frequency of flooding that has possibly changed in the past 1000 years. Although the rain-gauge records or river water records are more suitable for a scientific study of precipitation change, the flooding was chosen by the authors because pre-modern people preferred to record the onset of flooding rather than of rainfall or river water level. It is important as well to study records over a substantially long period of time because the climate change occurs rather slowly.

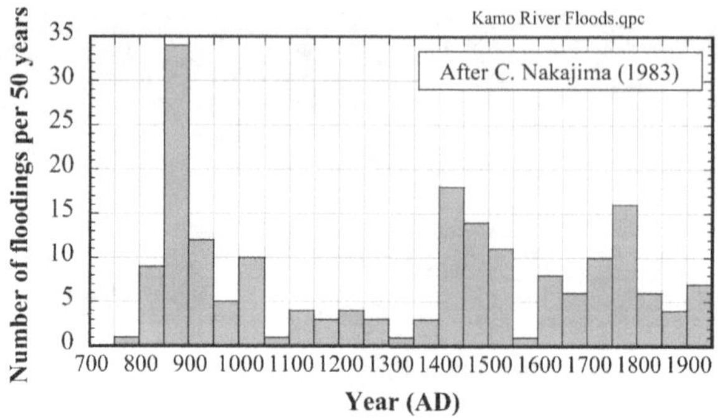

Figure 6.12 Frequency of flooding in the Kamo River over a historic time (C. Nakajima, 1983).

The present study was made of the flooding of the Kamo River in Kyoto, Japan, of which the flooding has been recorded since Kyoto became the capital of Japan at the end of the 8th century AD. Moreover, no study was made of events after the 1870s, when modern civil engineering was introduced to Japan and the idea of river levee construction started to change. Thus, the present study addresses the breaching or overtopping of the Japanese traditional levee which was a soil embankment with only light or no compaction. Noteworthy is that the amount of written records is relatively low in such a period as early 1100s and care must be taken not to pursue detailed quantitative information.

Figure 6.12 illustrates the histograms of Kamo River flooding in 50-year periods. It is noticeable that floods occurred more frequently in AD 800–900, 1400–1500 and 1700–1800. It is interesting that the first era of high flood vulnerability matches with what is called the medieval warm period. It may hence be inferred that the warm climate in those days induced a high rate of rainfall, and that this might be the case again in the near future in which global warming is anticipated. To further study this issue, it was attempted to compare this diagram with the historical knowledge of climate change or the change of temperature over the course of the historical age. The intended range of time is not as short as that of the modern meteorological records but not as long as typical geological studies, either. In this perspective, studies on tree-ring growth are suitable. Fig. 6.13 illustrates one such study in which Esper *et al.* (2002) indicated that the rate of tree growth was very high during the medieval time because of the warm climate. Although this simultaneous occurrence of high rate of flood and warmer climate is interesting, the tree growth was slow in 1400–1500 and 1700–1800 suggesting that the temperature was low in spite of the high rate of flooding.

It was supposed that the frequency of flooding in the Kamo River was related more closely with human activities. This is because Kyoto became the capital city of Japan at the end of the 8th century until it lost its political power in the 1200s. After a war period in 1300s, Kyoto again became the center of both politics and culture in 1400s but was damaged by many serious wars and loss of order in the 1500s. After the 1600s, Kyoto was one of the centers of both trading and culture, although the political power moved to present-day Tokyo. It is interesting in Fig. 6.14 that the population of Kyoto was big three times in pre-modern history such as 800–1000 when Kyoto

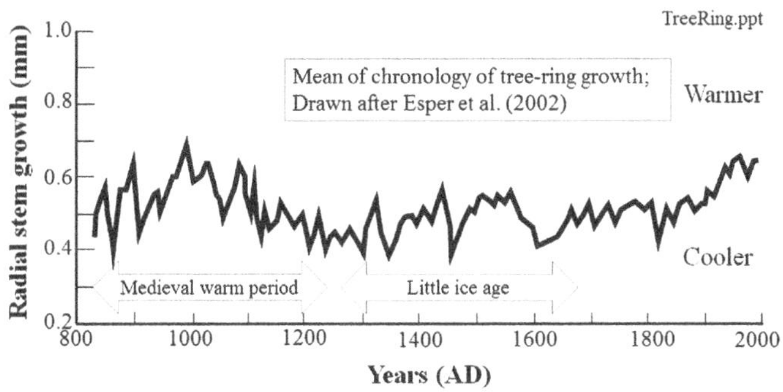

Figure 6.13 Mean change of tree ring over a historic time (drawn after Esper *et al.* (2002)).

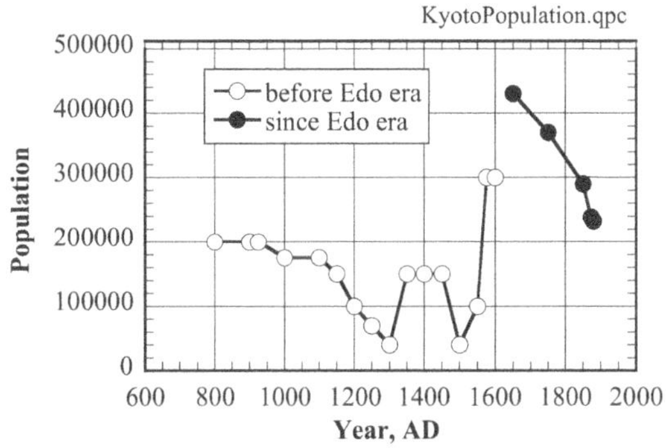

Figure 6.14 Historical change of population in Kyoto (after Wikipedia).

was the national capital, around 1400 when Kyoto was the economic center, followed by the 1600s when the population exceeded 400,000. These times of large population numbers are coincident with the times when Kamo River caused many floods.

From this brief study it may be concluded that the frequency of flooding of the Kamo River is not related with the climate change. The good correlation with the population of Kyoto suggests that many trees were cut down in the mountains around Kyoto in order to provide the city with fuel and construction materials. More materials were needed when the population was larger. Thus, the effect of human activity on nature is more serious than natural processes.

6.5 CONCLUSIONS

From the above-mentioned data and case studies, some conclusions can be drawn as follows.

1 The annual precipitation in recent years is likely to decrease. However, it is extremely likely that the numbers of days with precipitation of $\geq$50 mm and $\geq$100 mm will increase, showing an increasing trend of extreme precipitation.

2 The number of landslides in Japan is likely to rise as the years go by. This trend shows consistency with the occurrence of extreme precipitation, showing that the increased number of landslides likely results from extreme precipitation.
3 Extreme precipitation associated with typhoon and other climatic anomalies are continuing to trigger devastating landslides, bringing new challenges to the buildup of safe society.
4 The frequency of flooding of the Kamo River in Kyoto City is not related with the climate change, and the effect of human activity on nature is more serious than the natural processes.

REFERENCES

Buma, J. & Dehn, M. (1998) A method for predicting the impact of climate change on slope stability. *Environmental Geology*, 35, 190–196.

Chiang, S.H. & Chang, K.T. (2011) The potential impact of climate change on typhoon-triggered landslides in Taiwan, 2010–2099. *Geomorphology*, 133, 143–151.

Chigira, M. (2001) Micro-sheeting of granite and its relationship with landsliding specifically after the heavy rainstorm in June 1999, Hiroshima Prefecture, Japan. *Engineering Geology*, 59, 219–231.

Chigira, M., Tsou, Ch.Y., Matsushi, Y., Hiraishi, N. & Matsuzawa, M. (2013) Topographic precursors and geological structures of deep-seated catastrophic landslides caused by Typhoon Talas. *Geomorphology*, 201, 479–493.

Emergency Operation Center of Fire and Disaster Management Agency. (2015) *Damage by the Heavy Rainfall Starting from August 19 in Hiroshima Prefecture and the Activities of Fire and Disaster Management Agency.*

Esper, J., Cook, E.R. & Schweingruber F.H. (2002) Low-frequency signals in long tree-ring chronologies for reconstructing past temperature variability. *Science*, 295 (5563), 2250–2253

GSI (Geospatial Information Authority of Japan) (2014) *Information on the Damage Induced by the 2014 August Heavy Rainfall in Hiroshima.* Available from: http://www.gsi.go.jp/BOUSAI/h26-0816heavyrain-index.html [Latest access on 27th June 2015].

IPCC (Intergovernmental Panel on Climate Change) (2013) *IPCC Fifth Assessment Report: Climate Change 2013.* Geneva, IPCC.

Ishizuka, T., Okamoto, A. & Nakagomi, J. (2012) Emergency structural and non-structural measures against secondary disasters by landslide dams in Kii Peninsula, Japan caused by typhoon Talas in 2011. In: Chigira, M. (ed.) *Characterization, Prediction, and Management of Deep-Seated Catastrophic Landslides.* Kyoto Disaster Prevention Research Institute, Kyoto University. pp. 38–44 (in Japanese with English abstract).

Jakob, M. & Lambert, S. (2009) Climate change effects on landslides along the southwest coast of British Columbia. *Geomorphology*, 107, 275–284.

JMA (Japan Meteorological Agency) (2014a) *Climate Change Monitoring Report 2013.*

JMA (2014b) *Unseasonable Weather Conditions in Japan in August 2014.* Available from: http://ds.data.jma.go.jp/tcc/tcc/news/press_20140903.pdf [Latest access on 27th June 2015].

MLIT (2013) *Occurrence of Sediment-Related Disasters in 2013.*

Nakajima, C. (1983) History of flooding of Kamo River (1). *DPRI Annuals, Kyoto University*, 26B (2), 75–92 (in Japanese).

Okada, K. (2009) Global warming and landslide disaster. *Japanese Geotechnical Journal*, 57 (4), 6–9 (in Japanese).

Wang, G., Sassa, K. & Fukuoka, H. (2003) Downslope volume enlargement of a debris slide–debris flow in the 1999 Hiroshima, Japan, Rainstorm. *Engineering Geology*, 69 (3–4), 309–330.

Yamamoto, H. & Kobayashi, H. (2014) Characteristics of heavy rainfall and debris flow disaster in Hiroshima City by Akisame-front, 20 August 2014. *Journal of Japan Society for Natural Disaster Science*, 33 (3), 293–312 (in Japanese with English abstract).

Chapter 7

Landslide risk management in Norway

B. Kalsnes and F. Nadim
Norwegian Geotechnical Institute, NGI, Oslo, Norway

R.L. Hermanns
Geological Survey of Norway, NGU, Trondheim, Norway

H.O. Hygen
Norwegian Meteorological Institute, met.no, Oslo, Norway

G. Petkovic and B.K. Dolva
Norwegian Public Roads Administration, NPRA, Oslo, Norway

H. Berg
Norwegian Water Resources and Energy Directorate, NVE, Oslo, Norway

D.O. Høgvold
Norwegian Directorate for Civil Protection, DSB, Tønsberg, Norway

ABSTRACT

Landslides and snow avalanches have caused more than 2000 deaths in Norway in the past 150 years, with snow avalanches being responsible for about 1500 fatalities. The three landslide-related disasters causing the largest number of deaths in Norway in the 20th century, however, all involved large rock slides into bodies of water generating a displacement wave. Other important landslide types in Norway include quick clay slides, soil slides and debris flows. The weather-triggered landslides today are mainly of economical and environmental concern and the costs for landslides seem to steadily increase, both due to higher frequency and to higher exposure of infrastructure and constructed facilities. Climate research indicates that there will be a noticeable increase in average temperature, more precipitation and more instances with intense precipitation in Norway in the future. These are factors that probably contribute to increasing landslide activity. In order to assess and manage this increasing risk, there has been an intense research activity both on national and international fronts; including projects funded by the EU, the Norwegian Research Council and other national agencies.

A national strategy for dealing with floods, landslides and avalanches, was issued by the Norwegian Government in 2012 in the White Paper "How to live with the hazards". The Norwegian Water Resources and Energy Directorate (NVE) is the operational authority responsible for landslide risk management at the national level. The Directorate for Civil Protection (DSB) has a coordinating role in ensuring a well-functioning emergency preparedness and response. The municipalities are responsible for local civil protection and making sure natural hazards are considered and taken properly account of in land use planning and in building applications. The municipal level has also a key role in implementing disaster risk reduction policies.

7.1 INTRODUCTION

Landslides are a frequent natural hazard and a major threat to man and the environment in many parts of the world. According to the CRED statistics (EM-DAT, 2014), since 1900 about 130,000 persons have lost their lives because of landslides and flash floods and the material losses amounted to over US$ 50,000 M. In the period 2000–2014, corresponding figures are about 26,000 deaths and US$ 40,000 M. The actual figures are, however, much higher. In the CRED-EM database the losses due to earthquake-triggered landslides are attributed to earthquakes, and only a fraction of the incidents with material losses are reported. Recent catastrophic landslides in Afghanistan, United States, the Philippines and India illustrate that the landslides are still a major threat in developed as well as developing countries.

The CRED-EM database defines disaster as an event with more than 10 people reported killed or more than 100 people affected (displaced or evacuated). The reported numbers of occurrence of landslide disasters per country in the period 1974–2003 are shown on Figure 7.1. This figure suggests that the most exposed countries are located

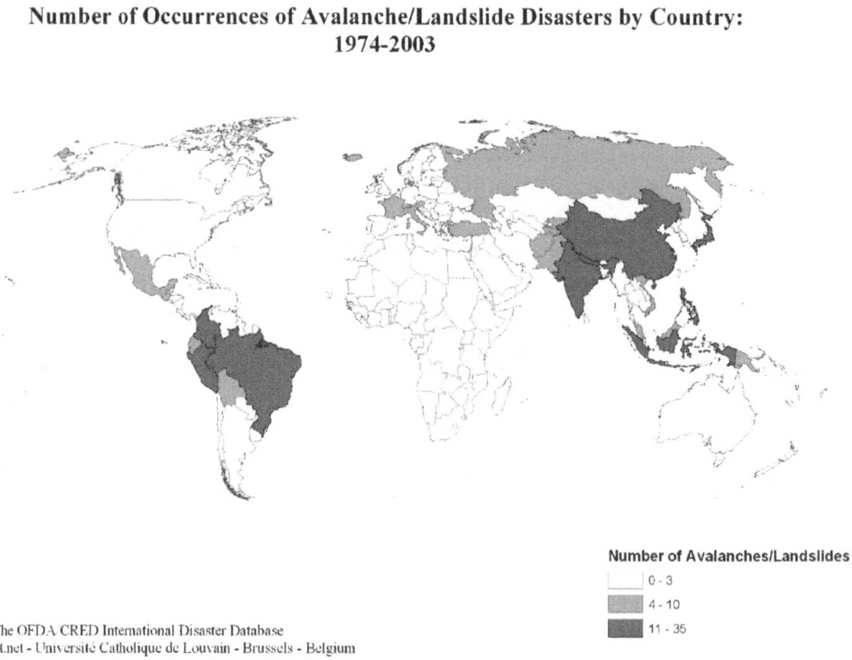

Figure 7.1 Number of occurrence of landslides (EM-DAT, 2014).

in eastern Asia, the Mediterranean region, and Central and South America. Even though Norway is not among the most landslide-prone countries, landslide studies on a European scale showed that the landslide hazard in Norway is significant, especially along the western and northern coast lines, as shown in Figure 7.2 (Jaedicke *et al.*, 2013). The landslide events in Norway rarely end up as disasters, mainly because of low exposure in the most landslide-prone areas, but also because of our increasing ability to forecast and mitigate for landslide events.

As a consequence of environmental change and social developments, such as an increase in exposure in many parts of the world, the risk associated with landslides is changing. Climate change, increased susceptibility of surface soil to instability, anthropogenic activities, growing urbanization, changes in land use, and uncontrolled urbanisation with increased vulnerability of population and infrastructure as a result, all contribute to the change – and in most instances growth – of the landslide risk. There is therefore a worldwide need for developing effective strategies for landslide risk management. These are related both to improved capabilities for forecasting the occurrence of landslides and the risk associated with them, and to risk mitigation activities. Mitigation and prevention of the risk posed by natural hazards have not attracted widespread and effective public support in the past. However, the situation has changed dramatically during the past decade, and it is now generally accepted that a proactive approach to risk management is required to significantly reduce loss of lives

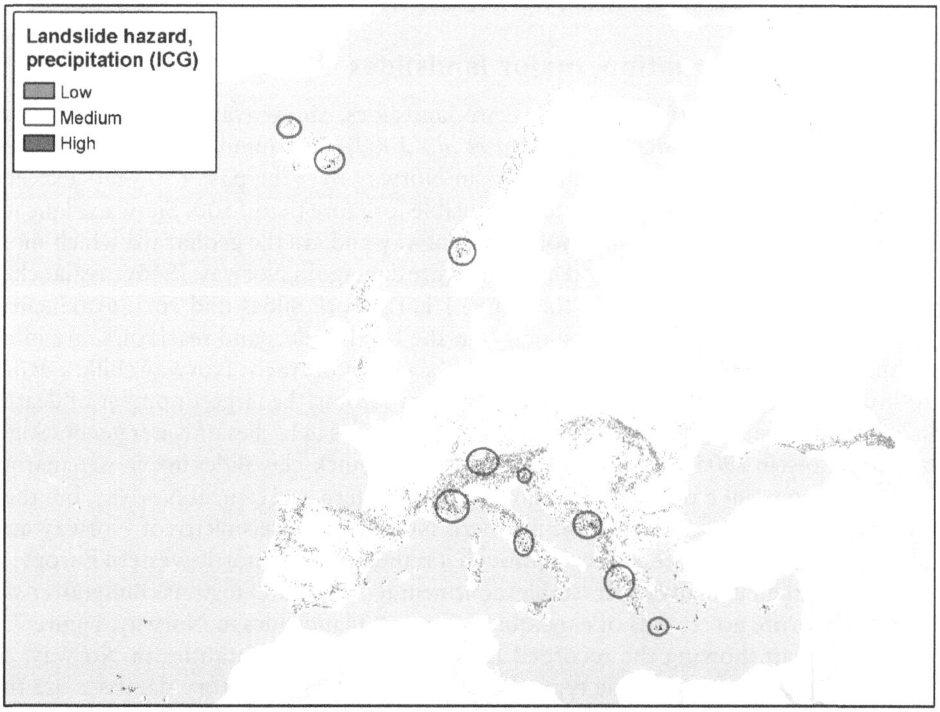

Figure 7.2 Landslide hazard in Europe (Jaedicke *et al.*, 2013).

and material damage associated with natural hazards. The wide media attention on major disasters caused by natural hazards during the last decade has clearly changed people's mind in terms of acknowledging risk management as an alternative to crisis management. A milestone in recognition of the need for natural disaster risk reduction was the approval of the "Hyogo Framework for Action 2005–2015: Building the Resilience of Nations and Communities to Disasters" (ISDR, 2005). This document, approved by 164 UN countries during the World Conference on Disaster Reduction in Kobe in January 2005, clarifies international working modes, responsibilities and priority actions for the 10 year period 2005–2015.

The JTC1 initiative of presenting country/regional reports of landslide risk management with a perspective on establishing international good practice with a climate change adaption strategy, is a positive contribution in this context. There is a need for sharing experience, ideas and general knowledge from all landslide-prone areas of the world, in order to choose the most appropriate strategies for a given geographical, economical and cultural setting. This paper presents the Norwegian contribution in this respect. It includes an overview of the present landslide risk situation, the Norwegian procedures for handling the landslide risk, as well as recent research done, both in Norwegian and in international settings, for how to handle landslide risk in changing climate.

7.2 LANDSLIDE HAZARD AND RISK IN NORWAY

7.2.1 General situation, major landslides

The main natural hazards in Norway are landslides, snow avalanches, floods and, to a lesser extent, earthquakes (Nadim *et al.*, 2009). The number of deaths caused by all types of landslides and avalanches in Norway over the past 150 years exceeds 2000, of which snow avalanches are responsible for more than 1500. Snow avalanches affect large parts of western and northern Norway and are the geohazard which most frequently leads to loss of lives and infrastructure damage in Norway. (Snow avalanches will not be discussed further in this paper.) Large rock slides and rock avalanches, which can generate devastating tsunamis in the fjords, lakes and reservoirs, are most common in western Norway and in the Troms area in northern Norway (Blikra *et al.*, 2006). The three natural hazard related disasters causing the largest number of deaths in Norway in the 20th century involved large rock slides in bodies of water generating a tsunami (Loen in 1905 and 1936; Tafjord in 1934). Quick-clay slides in exposed marine sediments represent a particularly high hazard in eastern and central Norway, but they also occur in parts of western and northern Norway. The seismicity of Norway and adjacent areas is moderate and, even though it is the highest of north-western Europe, it is still lower than in many other 'stable' continental (intraplate) regions (Bungum *et al.*, 2005). There are no records of earthquake-triggered landslides in Norway. Figure 7.3 presents a map showing the recorded landslide events with fatalities in Norway, as well as a classification by slide type. The partners for the inventory database are the National Road Authority, the National Railway Authority, NGU and the Norwegian Geotechnical Institute (NGI). The road and railway authorities collect information about events that have an impact on their infrastructure. NGU has contributed to

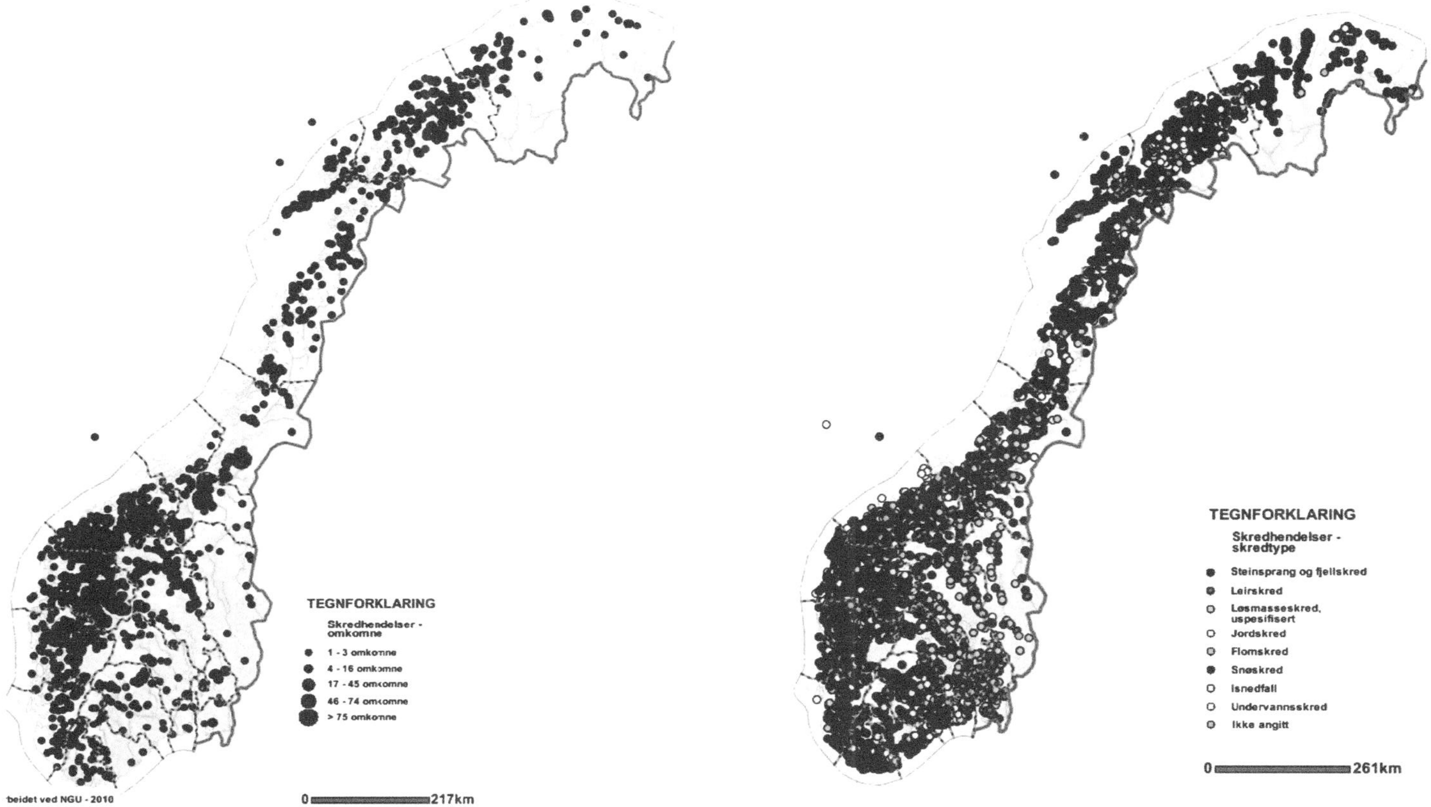

Figure 7.3 Maps showing all reported sliding events in Norway. The first map illustrates the number of fatalities from each slide, while the second shows the events classified by slide type (www.skrednett.no).

Figure 7.4 The Rissa quick clay slide of 1978 (Nadim *et al.*, 2009).

acquire historic information on events where the consequences have involved persons or major damages on buildings/infrastructure/farming (see also Furseth, 2006). NGI has contributed information mainly on snow avalanches and quick clay landslides. The database is hosted at NVE since January 2014.

About 5,000 km^2 of Norway is covered by soft marine clay deposits. Nearly 20% of this area consists of highly sensitive or quick clay. Landslides in quick clay represent a common and serious geohazard in Norway and Sweden. They frequently initiate without warning and turn into a flowing liquid in minutes. Statistically, large quick clay slides involving several million m^3 occur in Norway at intervals of about 4 years (Aas, 1981). The largest quick clay slide in Norway in the 20th century occurred at Rissa near Trondheimsfjord in 1978 (Figure 7.4). It covered an area of 330,000 m^2 and 5–6 million m^3 of clay poured out of the slide area. The largest historical quick clay in Norway occurred in Verdal, north of Trondheim, in 1893. It involved 55 million m^3 of clay and caused 116 fatalities, making it the most destructive natural catastrophe in Norway in modern times. A typical quick clay slide is a combination of a minor initial slide and a progressive failure developing very rapidly in all directions from the first slide. For example, the Rissa quick clay slide started with the failure of a small fill at the lakeside. The initial slide involved only 200 m^3 of sediments. It grew to about 6 million m^3 in a few hours through a retrogressive sliding process.

There is evidence of large submarine landslides along parts of the long coastline of Norway. The Storegga slide, which took place around 8,200 years ago, is one of

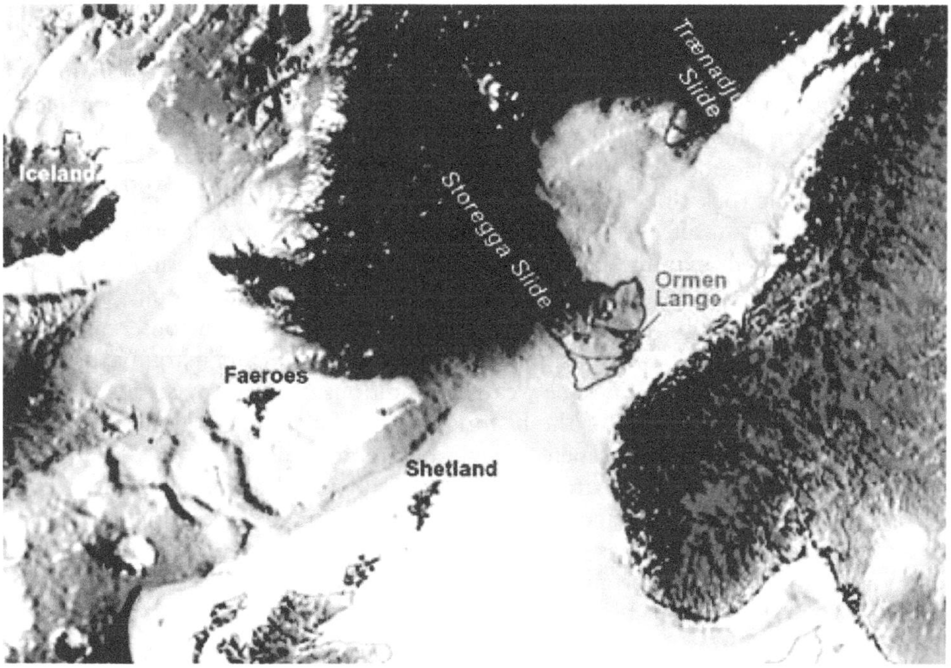

Figure 7.5 Extent of the Storegga slide (Solheim *et al.*, 2005).

the world's largest known submarine slides with an estimated slide volume in excess of 3,000 km^3 (e.g., Solheim *et al.*, 2005). Evidence of a major tsunami generated by the Storegga slide has been found along the coasts of Norway, Scotland and the Faeroe Islands. Figure 7.5 shows the extent of the Storegga slide.

7.2.2 Catastrophic rock slope failures and displacement waves

Catastrophic rockslope failures are destructive due to the high release of energy in a short time, and normally they cannot be mitigated by any physical measures. As long as they are of smaller volume (30–100.000 m^3), the impact area could be indicated by rock fall hazard maps, and building restrictions in those areas would ensure that nobody settles in the hazard zone. However if the volume becomes larger, those failures have excessive run-out length that can increase the hazard zone significantly. Worse is the situation when a rockslope failure hits a water body and offsets a displacement wave. These waves could be tens to more than 100 m high in the near-field (<10 km) and several metres to more than 10 m high in the far field (>10 km) (Hermanns *et al.*, 2014).

Deep erosion during multiple glacial cycles together with Holocene isostatic rebound produced an extensive fjord system in Western Norway (Vorren and Mangerud, 2008) comprising more than hundred thousands of kilometres of either high, steep coastline or sensitive marine sediments. Most valleys and fjords are settled, with populations concentrated at low elevation close to lakes, marine shorelines and rivers. Consequently, Norway's population is more exposed to displacement waves generated by impact of catastrophic rock slope failures than any other nation. Indeed, five out of the ten largest landslide disasters in the Norwegian history have been related to either catastrophic rock slope failures or the combination of catastrophic rock slope failures and displacement waves (Hermanns *et al.*, 2012).

Catastrophic rock slope failures can be triggered by earthquakes, intense precipitation or result from progressive failure due to the fatigue of rock bridges through time. In general, an extreme failure event is preceded by a long period of progressive failure (NGU, 2012). In Norway there is no account of a catastrophic rock slope failure being triggered by a seismic event. This is consistent with the observation that the seismic activity of Norway positioned at a passive continental margin is relatively low (Bungum *et al.*, 2000). Extreme weather conditions prior to catastrophic failures have been reported for some of the historic events, but most of the historic events have been linked to no special meteorological conditions (Furseth, 2006). However, for all recorded events, there are accounts of visible rock slope deformation with opening of cracks for several months, years or decades in advance of the event. This makes mapping for rock slopes with gravitational slope deformation and monitoring of deforming rock slopes a promising mitigation measure to prevent loss of life during future events.

7.2.2.1 Case: the 1934 Tafjord disaster

Catastrophic rock slope failures have occurred repeatedly in Tafjord and seismic data indicate that at least deposits of two prehistoric events lie under the mountain of Langhammaren in Tafjord (Hermanns *et al.*, 2006). The Tafjord, Storfjord, Geirangafjord region is one of the areas of Norway with most displacement wave disasters triggered by rock slope failure in Norway (Hermanns *et al.*, 2014). The largest previous event sourced from Skafjellet and caused a displacement wave running into Stranda on February 8th, 1731, resulting in multiple casualties (Furseth, 2006). This disaster caused caution and for more than 150 years the coastal areas were avoided for settlement. In the beginning of the 20th century, people moved back in these areas. On Langhammeren mountain in Tafjord people living in summer mountain settlements knew of a crack opening at least since 1870. This crack had opened by at least 1.5 m in 1930 and rock fall activity also increased at that time. In March 1934 rock fall activity became very intense (Furseth, 2006), but these signs were not taken as a warning in the villages around Tafjord. The only precaution was that the boats were drawn higher on land. At 3:10 in the night of 7 April 1934, about 3 million cubic metres of rock fell from an elevation of 730 m on the Langhammaren mountain and dropped into Tafjord, causing a displacement wave. The wave had the highest run-up with 62.3 m (Figure 7.6) close to the impact of the rockslide itself and the run-up elevation at a distance of about 5 km, where the villages of Tafjord and Fjora are located, was still greater than 10 m. Forty people lost their lives in this event.

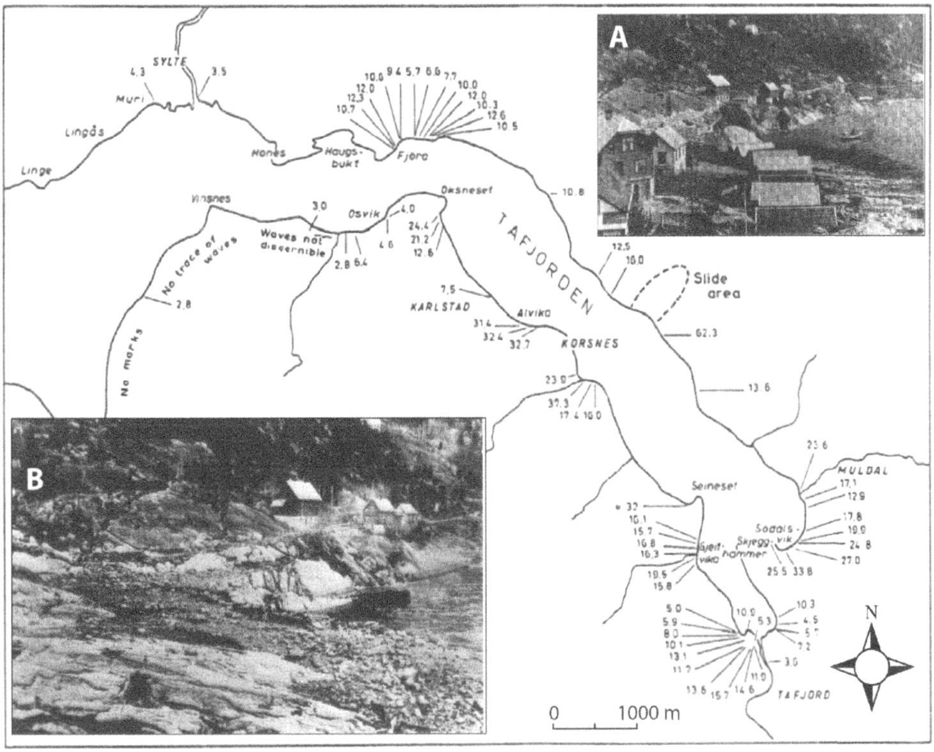

Figure 7.6 Wave height in meters after rockslide, April 7th, 1934. Map after Jørstad (1968). Photo inset A shows the village of Fjora (see map) before the 1934 event the photo inset B after the event (Photos from NGU Photo archive).

7.2.3 Water-triggered landslides

Landslides can be triggered by different factors (geological, morphological, physical and human among others). The term "trigger" commonly refers to an external stimulus that causes an immediate response in terms of landslide activity. Main weather-related landslide triggering factors in Norway are:

1 *Rainfalls.* The main trigger for most landslides is heavy or prolonged rainfall. Generally, a rainfall-triggered landslide is usually related to an extreme short duration event, such as the intense rainfall associated with a thunderstorm, or to a long duration event with lower intensity, or to a combination of both. Reduction of effective stresses by percolating water, and the corresponding reduction in the soil shear strength, is generally considered as the primary cause of rainfall-induced landslides.

2 *Erosion.* Landslides can be triggered by undercutting of the slope by a river, especially during a flood, or by bank and lateral erosion in coastal settings, especially

within clay slopes and fissured material. Undercutting and excavation reduce stability by increasing the gradient of the slope and by removing toe weighting. Some types of landslides such as debris flows may initiate by mobilization of a channel bed due to surface erosion caused by water flow. In Norway, erosion of riverbanks in quick clay areas is a serious threat.

3 *Snowmelt*. Particularly in mountain areas, a sudden increase of temperature in late winter or early spring, leading to rapid melting of the snow pack, can be a key mechanism in the landslide initiation. The water from the snowmelt infiltrates into the ground and, in the presence of underlying impervious layers of frozen soil or rock, leads to a rather rapid increase of soil pore pressure. Such an effect can be enhanced by precipitation, adding groundwater to the system and accelerating at the same time the rate of thawing.

4 *Weathering*. Prolonged weathering of bedrock causes the reduction of material strength, leading to the creation of a regolith layer weaker than the parent rock. This weak layer may become a preferential slippage plane during a slide event. An increase in freeze-thaw cycles due to climate change may increase the effect of weathering on landslide frequency.

The ways weather affects the slope behaviour are related to a number of factors such as slope angle, thickness of the deposit, stratigraphy, mechanical and hydraulic properties of the soil, initial conditions, boundary conditions, etc. The same event can therefore produce totally different consequences for different combinations of these factors. For example, some landslides are activated by short-duration, intense rainfalls, while others are induced by long lasting precipitation. The landslide can correspondingly involve the entire deposit or only a superficial part of it.

In the case of precipitation-induced landslides occurrence and features depend on the rainfall characteristics (frequency, magnitude and recurrence intervals) in combination with other initiatory natural (slope morphology, stratigraphy and lithology, soil moisture etc.) or anthropogenic (built up and forested surfaces evolution, modification of slope geometry etc.) factors. The number of landslides, triggered in a recurrence cycle, is directly proportional to the recorded amount of rainfall. A similar relationship has been observed between the number of triggered landslides in a given area and the rainfall duration. In some cases, it has been demonstrated that landslides are more likely to occur (i) after a drought period followed by heavy rainfall; and (ii) after snowmelting or long periods of rain followed by low amount of precipitations.

In Norway, the coupling between meteorological factors and landslides was investigated in the GeoExtreme project. A database of more than 20,000 recorded historical events were coupled with a meteorological database to assess the predictability of landslides and avalanches caused by meteorological conditions (Jaedicke *et al.*, 2008a). The statistical analyses of the historical events showed strong regional differences in the country. While precipitation is the most important trigger for snow avalanches in the coastal western part of the country, wind plays an important role in northern Norway. Debris flows can be explained both by short-term intensive precipitation events and rain accumulating over a longer period up to 15 days. Most likely, this can be explained with two different geomechanical release processes; surface erosion during short and intense storm events, and slow build up of pore pressure over long longer time periods with lower rainfall intensity. In addition, snowmelt in spring is an important triggering factor for debris flows in Norway. Consequently, there may be a geographical variation of the most intense landslide periods; spring in the south-eastern part of Norway (rain and snow-melt), fall in western part of Norway (intense storm related rainfall). Some examples of landslide events of various types are presented below.

7.2.3.1 Case 1: Kvam 2011/2013

Heavy rain combined with intense snow melt led to several landslides as well as a serious flood event in South East Norway during the period May 16th until June 2nd 2013. The settlement Kvam was severely hit by a debris flood causing damage to a lot of buildings and infrastructure. About 200 people were evacuated. A very similar event happened the same place in June 2011. The extreme bed load was caused by the many landslides in the upstream valley Veikledalen. The many landslides were caused by heavy rain (up to 70 mm in a few hours), high ground water level and intense snowmelt. Figure 7.7 shows water supply, ground frost depth and temperature during May 2013 in Kvam (NVE, 2014). Approximately 150000 m^3 of deposited sediment were removed from the river bed and the alluvial fan at Kvam after each of the 2011 and the 2013 events.

7.2.3.2 Case 2: Byneset 2012

In the morning of January 1st 2012 a quick clay slide was triggered in thick clay deposits at Esp in an agricultural area outside the city of Trondheim. The slide had a volume of 350000 m^3, an area of approximately 3.5 ha and a thickness of 10 m. The clay mass was deposited in the ravine downstream in a length of 900 m with a thickness of 6–8 m in the upper parts and 3 m at the lower end. The main triggering factor was considered to be river erosion in the ravines (NVE, 2012). Nine farms were

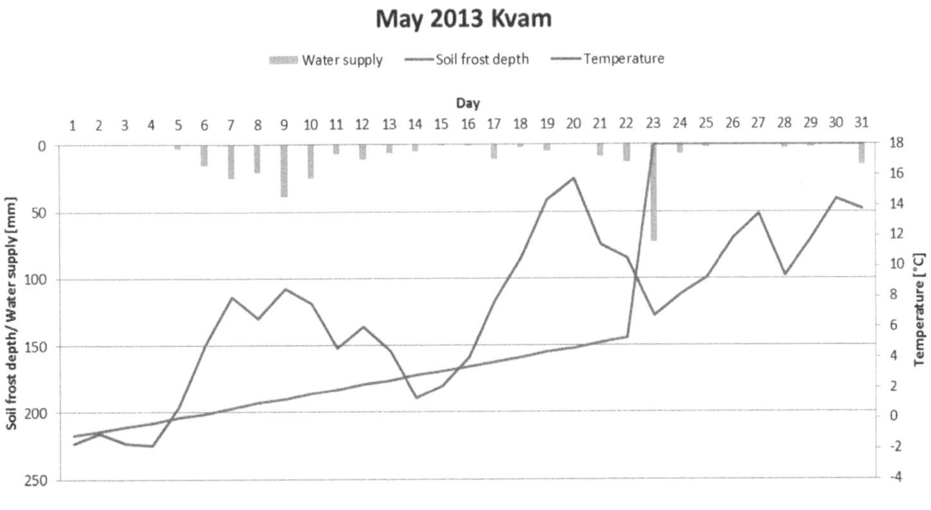

Figure 7.7 Water supply, ground frost depth and temperature during May 2013 in Kvam (NVE, 2014).

Figure 7.8 Aerial photo of Veikledalen and the downstream alluvial fan with settlement Kvam (Photo: NVE).

Figure 7.9 The quick clay slide at Esp, Trondheim. Photo: NVE.

considered threatened and consequently evacuated. To prevent new slides and the slide from developing further, protection measures against erosion has been installed.

7.2.3.3 *Case 3: Southeast Norway 2000 (Jaedicke et al., 2008b)*

During October and November 2000, southeast Norway experienced heavy rainfall with exceptionally long duration. The records from meteorological stations in the area show precipitation records up to 400% of normal values for the month of November. However, no single event with exceptionally high precipitation intensity was observed. Nevertheless, the accumulated precipitation resulted in severe slide activity in the region due to long time infiltration. Most of the slides were situated in the low lands of southeastern Norway. In this region, the terrain is characterized by flat open areas that are cut by steep river ravines. The terrain was formed from river and marine sediments after the last ice age. The ground consists mainly of a thin layer of soils of grain sizes varying from gravel to clay. The area is ideal for farming and comprises more than 60% of the farmland in Norway. The first slides occurred on 19 and 20 October 2000. The slide activity started again at the beginning of November and increased gradually until 20 November 2000 when 13 slides occurred within 1 day. After that peak, the slide activity continued until 21 December. Additional slides were released during the spring 2001, when slopes already weakened in autumn failed during the spring thaw. The plan area of the slides ranged from 5 to 80 m width and up to 150 m length. The volumes of the slides ranged from 5 to 20.000 m^3. The majority of the slides occurred in silt and clay deposits. Most of the slides were slumps located at man-made fills and cuts. Only a few occurred in natural slopes and terrain. The release mechanism for the majority of the slides was evaluated to be infiltration from precipitation. In some cases the slides were directly caused by river bank or surface erosion, or by human activities such as excavation.

7.2.4 Landslides triggered by human activities

In addition to precipitation, human activity is a major trigger for clay slides. This is particularly apparent for quick clay slides which do not seem to depend on topography and climate. Quick clay is extremely sensitive to changes in stress. It completely loses its strength when it is remoulded and starts behaving like a fluid. Owing to these material properties, quick clay slides typically display a retrogressive failure mode, and are therefore often large in extent. This can result in very destructive slides causing much damage, like the large quick clay slide in Verdalen in 1893 that killed 116 people.

Human impact on ground stability plays an increasingly large role as development expands, and more roads and houses are built. Especially alarming is the increase in construction of cabins and roads in areas where no building development previously existed. Common anthropogenic factors that could lead to destabilization of slopes are cut and fill construction, blasting and other construction activity, poor drainage and increased supply of water, changes in slope geometry, dam failure, traffic and mining, deforestation, removal of vegetation cover and concentration of water runoff (Skrednett.no). Human-induced landslides are the types of events that we have the best knowledge to prevent. It is therefore important to place focus on this topic to investigate what can be done to most effectively reduce the frequency of this type of landslides.

The causes of documented historical landslides in Norway were studied in the EC funded research project SafeLand (Nadim *et al.*, 2011). In that study only clay slides and earth slides were included, since these landslide types were regarded as most sensitive to human activity. Clay slides also included quick clay slides, as these two types are not separated in the database. Earth slides are debris slides consisting of various mixtures of soil, including the grain sizes cobble, gravel and sand; and typically involve glacial till deposits in Norway. Earth flows (flood slides) mobilized by excessive amounts of water were also included in the earth slides reported in the database. Such slides most often occur during flooding of rivers and streams, causing increased erosion and change in the course of the river. The database describes 380 historical clay slide disasters and 950 earth slide disasters (NGI 2009, Skrednett.no). The results of the study are given in Table 7.1.

The historical data, as collected by Astor Furseth (Furseth, 2006) and dating back to the 12th century, do not give a complete description of the triggering events of the reported landslides. In the referred study the researchers made an attempt to identify the triggering factor for the 380 reported clay slides and the 950 reported earth slides. The results of the analysis are shown in Table 7.1 below.

Of the clay slides triggered by natural factors most are due to intense and/or prolonged rainfall (22), flood (13), river changing its course (11) and water erosion from sea or rivers (7). Almost all of the earth slides are triggered by rainfall or flooding, often in connection with spring thaw. As many as 104 earth slides are reported to be caused by Ofsen; the great flood in 1789. When it comes to the registered human-induced landslides, the majority (24) are categorized as due to construction work in connection to road or railway. 19 slides have been explained as being related to landfills or embankments either being placed at an unfavourable location or having slid out during heavy rain etc. The other major categories for triggering events include excavation (10), dam failure (9), change in natural runoff/drainage (8), and work in gravel/sand pits, mines etc. (8).

An evaluation of the flood in Eastern Norway May 2013, carried out by the Directorate for Civil Protection DSB, revealed that landslide risk in the region is strongly interconnected with man-made changes in the topography. Such changes include construction of forest roads or local roads which cut off natural water runoff, deforestation which may trigger erosion, drainage of forests and moors, or expanding urbanization. What these elements have in common is that they affect the way water runs through the hillsides, water piles up where it is not supposed to, and floods and landslides occur in areas not previously exposed to such hazards. Forest roads were pointed out as a particular problem; they are planned and built as temporary roads for forestry, but they are left remaining in the hillsides for decades, often without maintenance (DSB, 2013a).

The evaluation concluded that in order to prevent, or limit the consequences of, future floods and landslides, a mapping of the scope of such man-made changes should be done in the affected areas. This requires cooperation across sectors and government levels, since the impact of changing water runoff may take place far away from the original cause. An important project in this respect is the project *Natural Hazards,*

Table 7.1 Results from the analysis done on triggering events of historical landslide events reported in the NGU database by Astor Furseth.

	Human	*Partly human*	*Natural*	*Unknown*
Clay slide	40	3	79	253
Earth slide	28	21	546	350

Infrastructure, Floods and Landslides (NIFS), a cooperation project involving the Norwegian Water Resources and Energy Directorate NVE, the Norwegian Road Authority, and the Norwegian Railway Authority. The project aims to provide an overview of how floods and landslides affect critical infrastructure for transportation, to dcfinc preventive measures for floods and landslides, and to improve disaster management.

7.3 CLIMATE AND DEMOGRAPHICAL CHANGES IN THE COMING YEARS

7.3.1 General research on climate change in Norway

The projections of expected climate change in IPCC (2013) paints a clear picture of a region in the arctic and sub-arctic, including Norway, which will be strongly affected by global warming. The reason for this stronger effect at the higher latitude is the melting of the polar ice. Selected global models have been downscaled to give more detailed information on Norway and an overview of these results, together with other results are presented in the report “Klima i Norge 2100” (Hanssen-Bauer *et al.*, 2015). The main results in this report are, see Figure 7.10:

- There will be a clear temperature increase in Norway, about 2.5°C to 5°C dependent on global emission scenario
- There is going to be more precipitation in Norway, about 10 to 20% dependent on global emission scenario, with distinct seasonal and regional variations
- There is going to be more instances with strong precipitation, again with seasonal and regional variations
- The results regarding wind is still very uncertain
- The amount of snow is going to decline during the century. Timing of this decrease is very dependent on region, and especially height above sea-level.

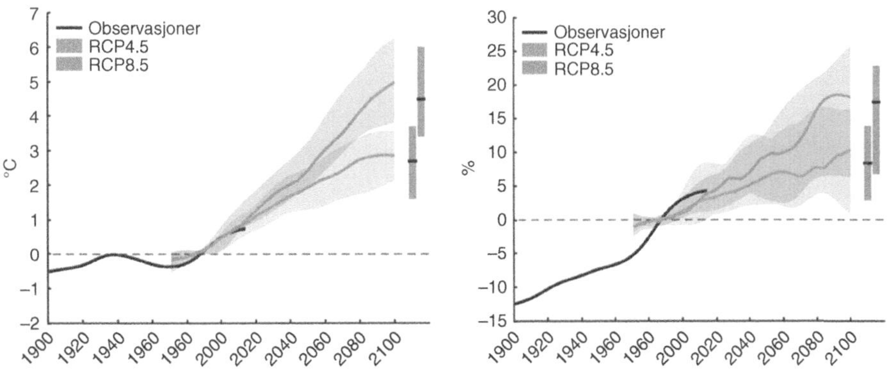

Yearly average temperature in Norway based on ensembles of future climate for the scenarios of RCP 8.5 and RCP 4.5.

Yearly amount of precipitation in Norway based on ensembles of future climate for the scenarios of RCP 8.5 and RCP 4.5.

Figure 7.10 Past and future temperature and amount of precipitation in Norway (Hanssen-Bauer *et al.*, 2015).

- The flood pattern is going to change from strong spring floods due to snowmelt towards a domination of flashfloods in small rivers
- Some regions, especially the south-eastern part of Norway might experience summer drought.
- The sea level is going to increase, with large regional variations in the local effect of this increase.

7.3.2 Changes in weather patterns in the context of slope stability

Extreme rainfall is identified as a major trigger for landslides. The historic observations show a clear pattern of increasing precipitation, and especially an increase in the intensity of the extreme events. In the InfraRisk project (NGI, 2013) this tendency has been documented for the historic data on daily extreme rainfalls. Figure 7.11 is from InfraRisk and documents the geographic distribution of extreme daily precipitation in Norway. A pattern with the stronger precipitation on the mountainous areas exposed to circulation patterns with advection from the west is here shown.

In InfraRisk there were 6 projections for future extreme daily rainfalls explored. The projections were:

- HAD-A2: Hadley centres model with emission scenario A2
- SMHI-A1B: Rossby Centre at SMHI's model with emission scenario A1B

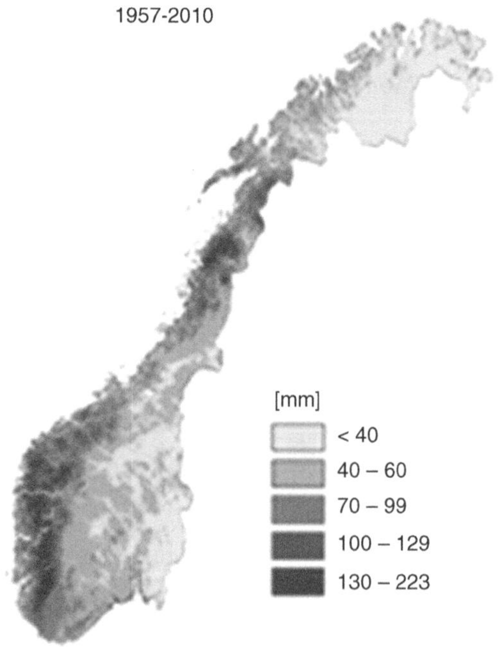

Figure 7.11 Geographic distribution of extreme daily precipitation in Norway (NGI, 2013).

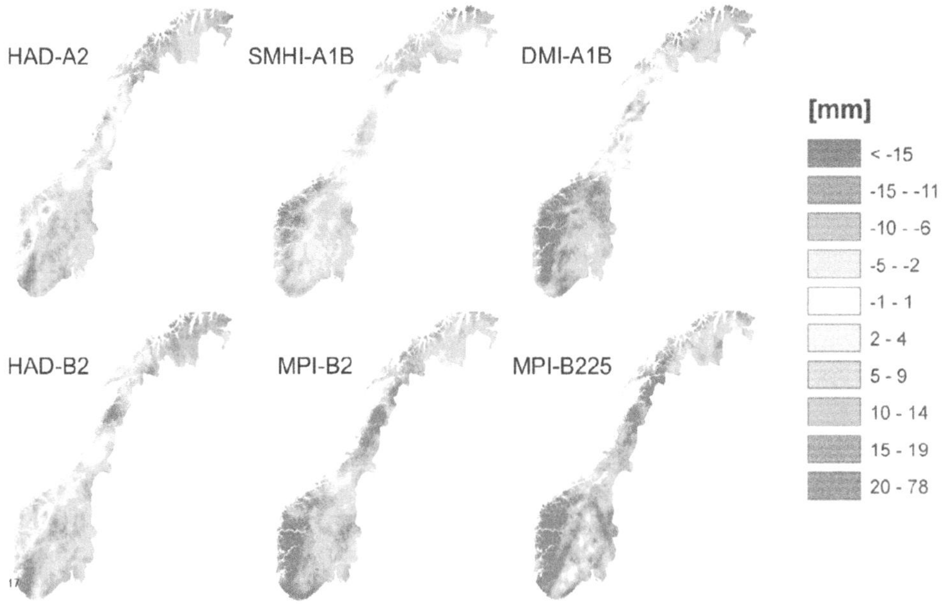

Figure 7.12 Projections for changes in max 24 hour rainfall during this century (NGI, 2013).

- DMI-A1B: DMI's model with emission scenario A1B
- HAD-B2: Hadley centres model with emission scenario A2
- MPI-B2: ECHAM 5 model with emission scenario B2
- MPI-B225: ECHAM 5 model with emission scenario B2

All the global results have been downscaled by MET Norway to Norwegian conditions and further adjusted with a bias correction down to $1 * 1$ km resolution. The collective pattern for these possible futures is a distinct increase in extreme daily precipitation for Norway, see Figures 7.12 and 7.13. Another striking feature in the comparison of these futures is the great uncertainty of the pattern in changed precipitation. The variations between the models is due to different expectations in changes in circulation patterns over Norway.

7.3.3 Changes in demography and infrastructure

Any major change in demography and infrastructure is expected to have an impact on the landslide risk pattern in Norway. Estimates of demographical changes in the coming century are being made regularly by Statistics Norway (www.ssb.no). In their last study in 2011 (Brunborg *et al.*, 2011), the estimated population in Norway in 2100 using alternative assumptions varied between 5 million and 12.5 million. This represents an increase of 0 to 150% compared to today's population of 5.0 million people. For their main alternative, the estimated population in 2100 is 7.5 million,

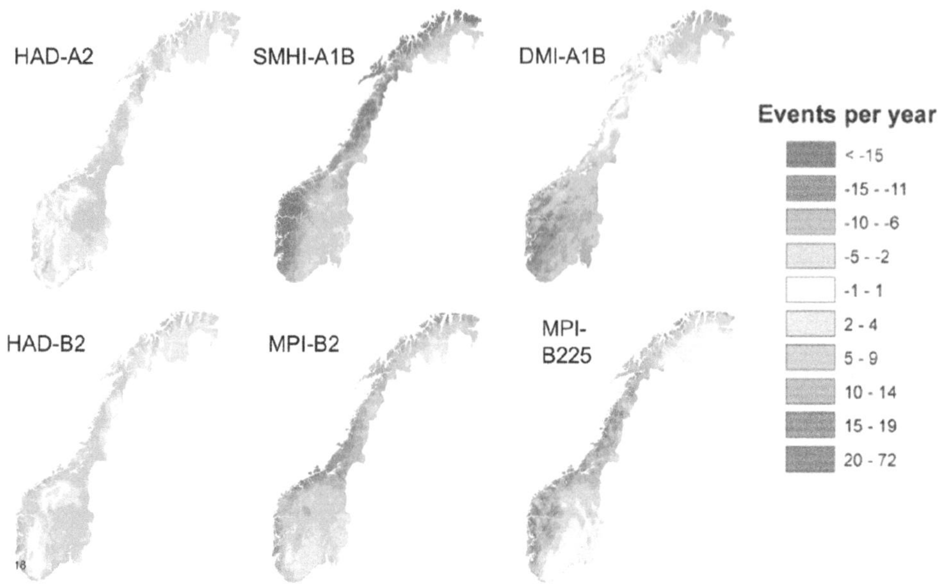

Figure 7.13 Projections for changes in number of days with minimum 10 mm rainfall during this century (NGI, 2013).

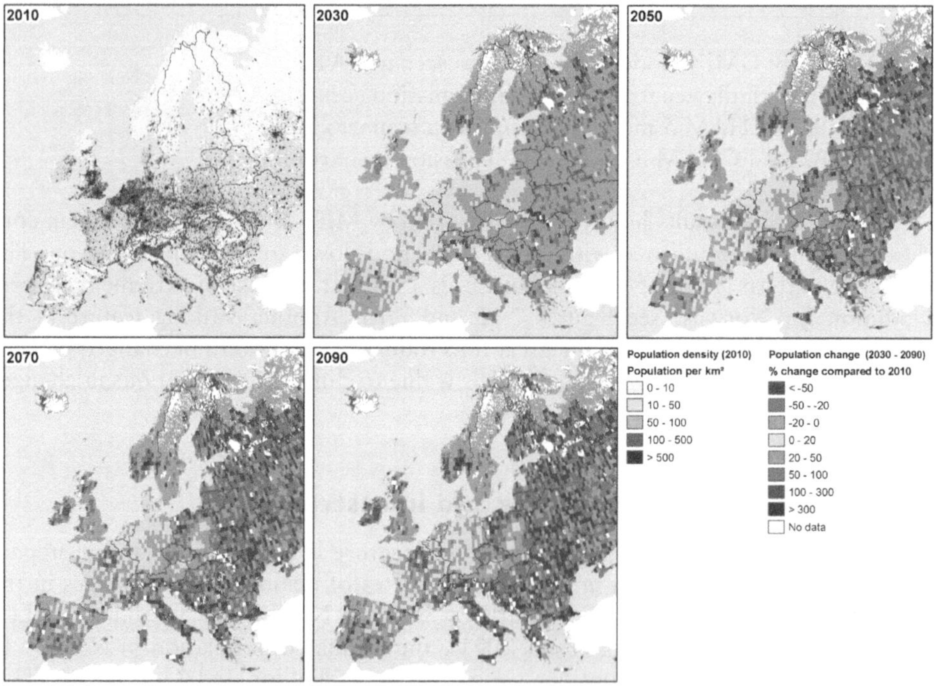

Figure 7.14 Demographical changes in Europe in the coming century (Jaedicke *et al.*, 2011).

representing an increase of 50% compared to today's situation. Statistics Norway's study does not include any regional estimates, nor does it provide any indications of rural versus urban settlements. However, a national increase of population of 50% may most probably lead also to increased exposure of people and infrastructure in landslide-prone areas.

Demographical changes in a more regional setting were provided in the SafeLand project. Figure 7.14 presents the estimated demographical changes for Europe. For Norway, there is a clear tendency to increased urban population in regions close to major cities, such as Bergen and Stavanger along the western coast, the capital Oslo in the south-eastern region and Trondheim in the central region. The need for new settlements in urban regions may therefore cause an increased exposure of man, buildings and infrastructure in landslide prone areas in these regions.

7.4 LANDSLIDE RISK ASSESSMENT AND MANAGEMENT IN A CHANGING CLIMATE

7.4.1 State-of-the-art landslide risk assessment methodologies in use

There are various definitions of the term risk, depending on which scientific community one belongs. For landslide risk assessment, risk is normally seen as a measure of the probability and severity of an adverse effect to health, property or the environment, and is estimated as the product of probability and consequences. A risk analysis is thereby performed by identifying hazards (temporal probability of adverse events, magnitude and intensity), and analysing and evaluating the causes and consequences of these. Risk analysis is further used in decision making regarding strategies that are intended to prevent accidents and to manage the adverse events. The purpose of risk assessment is to provide a basis for decisions, such as:

- Is there a need for remediation or can the risk be accepted?
- In what areas is the need for risk-reducing measures largest?
- How does the estimated landslide risk compare to risks associated with other hazards in the community?

The ranking of risk can be carried out by categorizing risk and consequence classes, as done by the Directorate for Civil Protection, DSB (DSB, 2010). Furthermore, the risk may be assessed and illustrated in a risk matrix, which may also contain acceptance criteria. Once the risk ranking is performed, the resulting risk is classified as acceptable or unacceptable depending on the colour of the cell:

- Red: unacceptable risk
- Green: acceptable risk
- Yellow: further investigations are needed; and an "as low as reasonably practicable" (ALARP) principle applies

The risk matrixes in Figures 7.15 and 7.16 illustrate risk matrixes in use in Norway for various situations (infrastructure, rockslide).

Yearly nominal probability for landslide per unit length of road							
	I ≤ 1/2						
	II ≤ 1/5						
	III ≤ 1/10						
	IV ≤ 1/20						
	V ≤ 1/50						
	VI ≤ 1/100						
	Daily traffic amount	**A** < 200	**B** 200 - <500	**C** 500 - <1500	**D** 1500- <4000	**E** 4000- <8000	**F** ≥ 8000

Figure 7.15 Risk matrix for landslides along roads (for 1 km length) used by the National Road Authorities (NIFS, 2013).

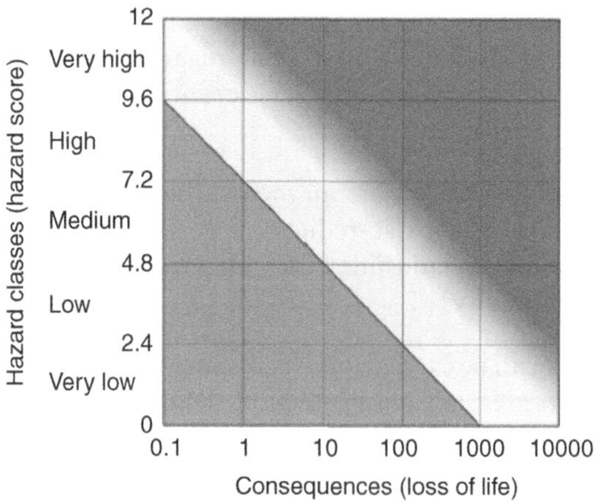

Figure 7.16 Risk classification matrix for follow up with monitoring and further investigations of unstable rock slopes in Norway (NGU, 2012).

7.4.2 Research

There has been an intense research activity, both nationally and internationally, related to landslide hazard and risk in a changing climate during the last decade. A key promoter of landslide risk research was the establishment of International Centre for Geohazards (ICG) in Norway in 2003. ICG was one of the first 13 Centres of Excellence established and funded by the Research Council of Norway (NRC). ICG was hosted by the Norwegian Geotechnical Institute (NGI), with partners from the University of Oslo (UiO), the Norwegian University of Science and Technology (NTNU), the Geological Survey of Norway (NGU) and NORSAR. The goal of ICG was to be an international arena for conducting scientific and technological research on identification, assessment and mitigation of geohazards. The ICG research focused on landslides, offshore geohazards, earthquakes and tsunamis. In the funding period 2003–2013 ICG fulfilled its main goals, one of which was being represented in a large number of international research projects. One of the main outcomes of ICG was a better cooperation between key Norwegian organisations in this field, causing good scientific interaction and synergy effects.

In 2008 the Research Council in Norway established a 5-year programme, NORKLIMA, where the primary objective was to generate vital new knowledge about the climate system, about climate trends in the past, present and future, and about the direct and indirect impacts of climate change on the natural environment and society, as a basis for adaptive responses by society. The effects of climate changes on the physical environment was key part in this study, resulting in a varied type of projects also involving landslide studies. A new program, KLIMAFORSK, was initiated in 2013, with the aim of generating essential future-oriented knowledge about the climate to the benefit of society, of national and international significance.

Table 7.2 summarises some of the key Norwegian, or international with significant Norwegian content, research projects related to landslide risk assessment and management in the last decade.

7.4.2.1 Climate and transport

"Climate and Transport" was a research and development programme carried out by the Norwegian Public Roads Administration (NPRA) from 2007 to 2013. The main objectives of the programme were to investigate the effect of climate change on the road network and recommend actions concerning planning, design, construction and maintenance to sustain both the safety and the accessibility of the Norwegian transportation network in a more demanding climate.

The programme resulted in a revision of design rules, i.e. NPRA's own manuals, the publication of several new manuals for relevant topics, 38 reports (in Norwegian) and a number of conference papers.

'Climate and Transport' looked into all types of landslides and the possible effect of climate change on their frequency and characteristics. These included rock falls and rock avalanches, earth slides, debris flows, snow and slush avalanches and quick clay slides. Each of the landslide types has its "most important" climate factors. With increasing rain intensity in some places and increasing frequency of freeze-thaw processes in the cracks, a higher frequency of rock falls and rock slides can be expected. Increased rainfall and storm frequency and increased temperature may affect the frequency, volume and range of snow avalanches (a frequent cause of road closures). Water saturated avalanches, slush avalanches, can be triggered by sudden temperature increases, for example in spring or during heavy rain on snow-covered ground

Table 7.2 Key landslide risk research projects with Norwegian participation.

Project	*Duration*	*Coordinator*	*Partners*	*Funding/ Budget*	*Research topic*
Climate and Transport	2007–2013	The Norwegian Public Roads Administration (NPRA)	The Norwegian Public Roads Administration (NPRA) and more than 10 Norwegian institutions.	NPRA	Effect of climate change on the road and rail network.
GeoExtreme	2005–2008	ICG/NGU/ NGI	Bjerknes Centre, Norwegian Meteorological Institute (met.no), Centre for International and Environmental Research (CICERO)	NRC/1 mill EUR	Relationships between climate scenarios and landslides in Norway, assessment of social-economic consequences of future landslide.
InfraRisk	2010–2013	ICG/NGI	Met.no, CICERO, The Institute of Transport Economics (TØI), National Emergency Planning College (NUSB), The Norwegian Rail Administration (JBV), NPRA	NRC/1 mill. EUR	Impact of extreme weather events on Norwegian infrastructure.
SafeLand	2009–2012	ICG/NGI	27 European organisations (universities, national geological surveys, private institutions)	EC FP7/ 8.8 mill EUR	Landslide risk assessment and management in Europe a changing climate and demographic perspective.
Matrix	2010–2013	GFZ (Germany)	NGI, and 11 other European organisations	EC FP7/4 mill EUR	Multihazard risk assessment.
Natural hazards – infrastructure for floods and slides (NIFS)	2012–2015	NPRA	NVE, JBV	NIFS/5.5 mill EUR	Reduce vulnerability and prevent accidents and injuries caused by landslides, avalanches and floods.

The various projects outlined in Table 7.2 are presented in some more detail below.

in winter. Such conditions are likely to occur more frequently in the future, given the expected climate scenarios. Climate change may also lead to increased ice formation and ice fall.

Intense rainfall events, especially after prolonged rain, will increase the debris flow hazard. We will see more major debris flows occurring along streams and rivers.

There will be more sediment transport in streams and rivers, which will fill up the drainage pipes and culverts. An increase in water flow in rivers may increase the risk of naturally triggered quick clay slides. Finally, a warmer climate leading to thawing of permafrost in the mountains, could also increase the occurrence probability of major rock avalanches and impact waves (tsunamis) where they hit lakes or fjords.

In the changing climate, the landslide types most likely to increase in frequency are "wet" landslide types, such as debris flow and slush avalanches. These landslide types are usually not taken into account for when planning new roads. "Wet" avalanches are often triggered by human activity in the release area. Prevention methods are drainage, open or closed control dams and/or bridges in avalanche course, deflectors, sedimentation basins, alternative drainage paths in the runout area, and moving the road. Alternative drainage paths are an important remedial measure for sediment transport along streams and rivers, for ensuring operative drainage systems. A risk assessment of existing roads with respect to increased erosion and sediment transport should be carried out. One of the conclusions from the programme was that more effort should be put into operation of drainage systems than is done today.

7.4.2.2 GeoExtreme (Jaedicke et al., 2008a)

The aim of the GeoExtreme project was to study the changing geohazard pattern in Norway due to a changing climate. A major focus in the project was to consider national datasets and use the study areas as verification rather than the other way round. Contrary to conventional research approach, to go into detail, the national picture was in focus. This objective was formulated out of the request from decision makers and politicians to present results that can serve as a basis for national priorities in climate change adaptation. This approach prohibited detailed process studies in selected areas, selected slopes or on single rocks. The results from GeoExtreme gave a general overview over the whole of Norway. The study showed that the limitations of inhomogeneous data collection of landslide and avalanche events can be bypassed by combining the event database with a homogeneous meteorological dataset. This allows then statistical analysis to find the most important meteorological trigger elements for the various types of landslides and avalanches. The identified most important triggering element for different types of landslides and avalanches show a wide spatial variation on national scale in Norway. This implies that adaptation strategies need to consider regional differences already on the national planning level.

The GeoExtreme project showed that for future scenarios these regional differences can only be addressed with more detailed climate modelling. The new climate modelling approaches on a stretched grid successfully model extreme value frequencies on scales that correspond to the results from the statistical analysis of the slide events.

More detailed investigations of study areas and slope stability modelling give an impression of the consequences of a changing climate geohazard level for communities and settlements in the future. Again, regional differences will ask for adjustments of national climate adaptation strategies to the local conditions. The socio economic analysis shows that future geohazard risks may be limited by an unambiguous connection of the local authorities' right to decide and approve new development and their responsibility for damages by geohazards.

As a consequence, the GeoExtreme project recommended to consider possible future changes to the geohazard level already in the planning phase of any new development and to adjust these considerations to the local and regional differences in the expected changes.

7.4.2.3 InfraRisk (NGI, 2013)

InfraRisk focused on a range of questions concerning the impact of extreme weather events (EWEs) on the Norwegian transportation infrastructure. The background was limited systematic knowledge about which EWEs affect the Norwegian infrastructure adversely, and the climatic parameters that control these events and their spatial and temporal distribution.Furthermore there were large gaps in the mapping of natural hazards that can be related to EWEs. An assessment of the risk to the infrastructure from EWEs also requires identification of the different types of values infrastructure objects have to society and how these values may degrade in case of a given event. Lastly there was a need to investigate whether the relevant decision making systems are able to convert an improved knowledge about the risk from EWEs into effective adaptation.

In order to address these needs the InfraRisk project had the following objectives:

- To gain a more solidified understanding of how changing climate affects the frequency, intensity and distribution patterns of EWEs in Norway.
- To identify current knowledge about the impact of EWEs on infrastructure in Norway.
- To quantify the vulnerability and societal value of Norwegian infrastructure.
- To investigate which mitigation measures that could be implemented in order to more effectively adapt to these types of events.
- To develop a methodological framework for assessment of the total risk to infrastructure from the combination of EWEs.
- To propose guidelines for pro-active behaviour with regard to mitigation actions against climate induced natural hazards

7.4.2.4 SafeLand (Nadim and Kalsnes, 2014)

As a consequence of climate change and increase in exposure in many parts of the world, the risk associated with landslides is growing. In areas with high population density, protection works often cannot be built because of economic or environmental constraints, and is it not always possible to evacuate people because of societal reasons. The growing hazard and risk, the need to protect people and property, the expected climate change and the reality for society in Europe to live with hazard and risk and the need to manage risk were the reasons for the SafeLand research. The effects of global change on the evolution of landslide risk are shown schematically on Figure 7.17.

Main results achieved in SafeLand include (see www.safeland-fp7.eu):

- Various guidelines related to landslide triggering processes and run-out modelling.

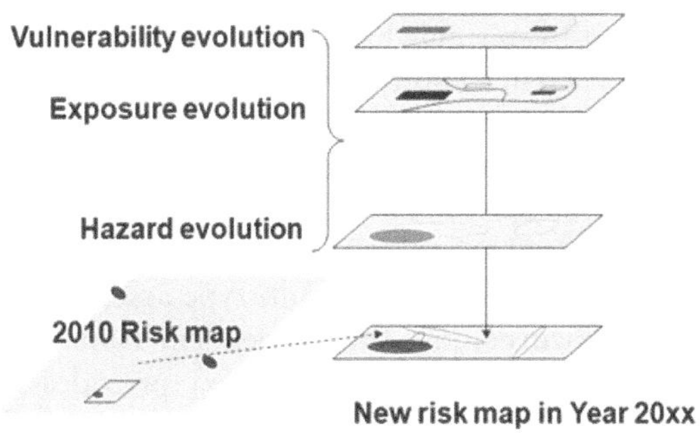

Figure 7.17 Schematic diagram: Evolution of landslide risk caused by changes in vulnerability, frequency of landslides (hazard) and exposure of elements at risk.

- Development and testing of several empirical methods for predicting the characteristics of threshold rainfall events for triggering of precipitation-induced landslides, and development of an empirical model for assessing the changes in landslide frequency (hazard) as a function of changes in the demography and population density.
- Guidelines for landslide susceptibility, hazard and risk assessment and zoning.
- New methodologies for physical and societal vulnerability assessment.
- Identification of landslide hazard and risk hotspots for Europe. The results show clearly where areas with the largest landslide risk are located in Europe and the objective approach allows a ranking of the countries by exposed area and population.
- Different regional and local climate model simulations over selected regions of Europe at spatial resolutions of 10×10 km and 2.8×2.8 km. These simulations were used to perform an extreme value analysis for trends in heavy precipitation events, and subsequent effects on landslide hazard and risk trends.
- Guidelines for use of remote sensing techniques, monitoring and early warning systems.
- Development of a prototype web-based "toolbox" of innovative and technically appropriate prevention and mitigation measures. The toolbox does a preliminary assessment and ranking of up to 60 structural and non-structural landslide risk mitigation options.
- Case histories and "hotspots" of European Landslides have been collected and documented. Data for close to fifty potential case study sites have been compiled and summarized. Most of the case study sites are located in Europe (Italy, France, Norway, Switzerland, Austria, Andorra, and Romania); but they also include one site in Canada and one in India. Almost every type of landslide and every type of movement is represented in these sites.
- Research on stakeholder workshops and participatory processes to involve the population exposed to landslide risk in the decision-making process for choosing the most appropriate risk mitigation measure(s).

7.4.2.5 *MATRIX*

People suffer losses not just from single hazards, but also from multiple events in combination. In both their occurrence and their consequences, different hazards are often causally related. Classes of interactions include triggered events, cascade effects, and rapid increases of vulnerability during successive hazards. Effective and efficient risk reduction, therefore, often needs to rest on a place-based synoptic view. The aim of MATRIX was to tackle multiple natural hazards and risks in a common theoretical framework. It integrated new methods for multi-type assessment, accounting for risk comparability, cascading hazards, and time-dependent vulnerability. MATRIX identified the conditions under which the synoptic view provides significantly different and better results – or potentially worse results – than established methods for single-type hazard and risk analysis. Three test cases (Naples, Cologne and the French West Indies), and a "virtual city" provided MATRIX with all characteristic multi-hazard and multi-risk scenarios, including landslide risk. The MATRIX IT-architecture for performing, analysing and visualising relevant scenarios generated tools to support cost-effective mitigation and adaptation in multi-risk environments.

7.4.2.6 *NIFS*

Natural hazards – infrastructure for floods and slides (NIFS) is a multi-disciplinary project based on the White Paper with the title "How to live with the hazards" (Norwegian Ministry of Petroleum and Energy, 2012)., and the subsequent parliamentary debate. The Norwegian National Rail Administration (JBV), the Norwegian Water Resources and Energy Directorate (NVE) and the Norwegian Public Roads Administration (NPRA) have been working well in partnership over a number of years and within a number of areas, which has resulted in this targeted R&D initiative. The project has a wide remit and will be throwing light on challenges ranging from the strategic to the operative sphere in connection with natural hazards. The project period has been set to 2012–2015, and NOK 42 mill. as well as considerable internal resources have been allocated to the task.

Significant importance is being attached to joint solutions and collaboration across agencies, as well as liaison with other resources in society within this area. This includes the research and teaching organisations, trade and industry and other public institutions. The project outcomes are being published on an ongoing basis on the joint website www.naturfare.no. Relevant research outcomes will be implemented in the three organisations. Because there is a close connection between the weather and natural hazards, each sub-project and each activity will take account of the climate and climate change. By strengthening the competence of each agency in providing protection against natural hazards more robust infrastructure will be developed. This allows us to reduce the risk of damage caused by natural hazards, to improve the traffic flow and to protect the infrastructure.

The overall objective is a safer society with more robust infrastructure, safe buildings, safe transport and good forecasting systems for landslides, avalanches and floods.

The programme's main objective is to build and propose good partnership platforms by coordinating the activities of the agencies in order to reduce vulnerability and prevent accidents and injuries caused by landslides, avalanches and floods.

Impact targets are used to describe the future situation the project should seek to attain. We wish to reach them by:

- Improving the level of safety for the population by providing infrastructure which is better designed to withstand the impacts of natural hazards, at their present level and in the future with higher intensities and frequencies as a consequence of climate change.
- Step-by-step contingencies to ensure better preparedness in natural disaster situations.
- Better cooperation between agencies in disaster situations.
- Optimal use of financial and professional resources to solve shared challenges associated with natural hazards.
- Raised levels of safety and robustness, and improved regularity.

The organisational structure of the Government Agency Programme and its sub-projects:

The programme plan is available in English at our website as report no 57 in 2013.

7.5 REGULATION AND POLICY ISSUES

7.5.1 Institutional framework

A White Paper with the title "How to live with the hazards" was issued by the Norwegian Government in 2012 (Norwegian Ministry of Petroleum and Energy, 2012). This White Paper outlines how responsibilities for dealing with floods, landslides and avalanches are distributed among the main actors in Norway. The importance of cooperation between the relevant actors is highlighted, and a national strategy for cooperation and coordination will be developed on these bases.

Each individual, land owner or infrastructure owner has a responsibility to protect herself and her property. In order to limit the losses to private stakeholders, different compensation systems are established. One of these is the insurance against natural hazards. According to the Natural Hazards Insurance Act, all buildings that are insured against fire in Norway are automatically insured against natural hazards such as floods, landslides and storms. For objects that could not otherwise be insured, such as farm land, private roads, etc., the Norwegian National Fund for Natural Damage Assistance can provide compensation for damage.

The municipalities are responsible for local civil protection. This includes an overall risk and vulnerability analysis as basis for an emergency management plan. According to the Planning and Building Act, the municipalities are responsible for making sure natural hazards, including landslides, are considered and taken properly account of in land use planning and in building applications. This includes risk and vulnerability analyses related to new developments. For existing constructions and infrastructure, there is no clear legal duty on the municipalities to protect against landslides. Still the necessary legal instruments to implement protection measures are in place in the Natural Hazards Act. The municipalities also have responsibilities as infrastructure owners.

There are many small municipalities in Norway and, in many of them, the capacity for planning and implementation of the above Acts is limited. It is therefore essential to provide planners and decision-makers with tools and examples of best practices, which can simplify the integration of disaster risk reduction into local planning. The Norwegian Directorate for Civil Protection (DSB) has, on behalf of the Ministry of Justice and Public Security, an overall coordinating role for disaster risk reduction in Norway. DSB provides regulations, guidance and competence building for planners at the municipal level. It carries out audits of all ministries and agencies, as well as the county governors, on disaster management. Norway's 18 county governors play an important role in linking the national knowledge, strategies and priorities with local planning processes carried out by the municipalities.

The Ministry of Petroleum and Energy is responsible for landslide risk management at the national level, with the Norwegian Water Resources and Energy Directorate (NVE) as the operational authority. NVE's work in preventing damage from floods and landslides is structured according to the following tasks:

- Hazard and risk mapping
- Assistance and control of land use planning in the municipalities
- Planning and construction of structural protection measures
- Monitoring and early warning services
- Assistance to the police and municipalities in emergency situations
- Research & Development, Communications

Every governmental body has an independent responsibility within its sector to prevent damage and deal with the landslide risk. The Ministry of Local Government and Modernisation together with the Directorate for Building Quality are national planning and building authorities. The Norwegian Meteorological Institute is responsible for weather forecasts, including extreme weather warnings. The Geological Survey of Norway is the governmental agency for geo-scientific matters, and performs parts of the mapping related to landslide hazard, compile and stores landslide relevant geological data and makes it accessible to the public. The Norwegian Mapping Authority has nationwide responsibility for geographical information. The Directorate for Civil Protection (DSB) has a coordinating role in ensuring a well functioning emergency preparedness and response.

The County Governor represents the Government at regional (county) level. This includes supervision of the municipalities in emergency preparedness according to the Civil Protection Act, as well as ensuring that safety against hazards are taken into account in land use planning.

Governmental infrastructure owners such as the Norwegian Road Authority and the Norwegian Railroad Authority have the same responsibility for safety against landslides on their infrastructure as all other infrastructure owners. This includes all aspects of risk management, i.e. mapping hazards, decisions on location, technical solutions and protection as well as emergency management.

The research and education institutions play an important role in increasing the knowledge on landslide risk and prevention measures. The private consultants and construction companies deliver analyses and structural measures for the management of landslide risk.

7.5.2 Role of municipalities

The 428 municipalities are the foundation of disaster risk reduction in Norway. The municipalities are responsible for key public services and the coordination of these before, during and after disasters. Such services include, health services, schools and kindergartens, care for the elderly and other vulnerable populations, information to the public, etc. In accordance with the principles of responsibility and similarity, the main responsibility for preventive planning and disaster management lies with the municipalities within their territorial borders. Holistic risk and vulnerability assessments, physical planning, emergency plans and exercises are the cornerstones of disaster risk reduction at the local level. The Civil Protection Act of 2010 requires Norwegian municipalities to establish systems for crisis prevention, preparedness and response. Furthermore, according to the Planning and Building Act (2008), they are required to carry out risk and vulnerability analyses prior to any new physical developments such as urban expansion.

The municipalities have access to some of the most important tools for dealing with the impacts of weather extremes and climate change. These tools assist them to implement risk reduction measures through land-use planning, and development and maintenance of important infrastructure and societal functions.

7.5.3 Legal framework

The Civil Protection Act (2010) and the Planning and Building Act (2008) are closely connected. Together, they provide the essential legal framework for disaster risk reduction at local level. The aim of the Civil Protection Act is the development of resilient local communities, and that the municipalities address disaster risk reduction in a systematic and holistic way. One of the requirements in the act is for municipalities to carry out holistic risk and vulnerability assessments. Such assessments should be updated on a regular basis – at least every four years when the municipal plans are revised, or in case of changes in risk and vulnerability. Access to new knowledge, such as updated climate change projections, may make it necessary to update the assessment. A holistic risk and vulnerability assessment should provide an overview of possible undesirable events that would exceed the municipality's day-to-day capacity for preparedness, and cover the most serious existing and future risks that affect the local community. Furthermore, it should describe vulnerabilities that may affect the municipality's capacity to manage risks, and provide a framework for planning of preventive measures as well as preparedness and response. The assessments will provide input to regional or national risk assessments in various sectors.

Under the Planning and Building Act, the Technical Regulations of 2010 (TEK10) provide safety standards for buildings and infrastructure, corresponding to different return periods for floods and landslides. An important element of the Planning and Building Act is the possibility for municipalities to define 'zones that require special consideration'. Building restrictions can be applied to such areas, local authorities might find it necessary to carry out more thorough risk assessments, and special building codes may apply in order to maintain safety. Flood and landslide risk, or the uncertainties about such risks, can be used to define these zones.

The county governors have the authority to stop development plans if disaster risk has not been sufficiently assessed or taken into consideration in the planning process.

For landslide risk the Norwegian Water Resources and Energy Directorate has the same authority.

7.5.4 Mapping of landslide hazards

To be able to work systematically with the prevention of damage caused by landslides, it is essential to know the hazards and identify the areas prone to such hazards. Municipal planners require information on disaster risk in order to fulfill their requirements for risk and vulnerability assessments. For the planning of protection measures and for emergency management hazard and risk maps are important for the identification of areas at risk and the prioritization of measures.

The Norwegian Water and Energy Resources Directorate (NVE) is responsible for the national mapping programs. These include susceptibility maps, hazard maps and risk maps. Susceptibility maps for rock fall were issued in 2009 and susceptibility maps for debris slides/shallow landslides were issued in May 2014. These maps identify areas potentially prone to such landslides and cover the whole country. Detailed mapping of flood and landslide risk is time consuming and requires expertise and resources. In 2011, NVE started a programme of detailed hazard mapping for rock fall, debris flow and snow avalanches. A plan including areas in 119 municipalities has been made.

The mapping of areas prone to quick clay slides was initiated after the large quick clay slide at Rissa in 1978. By 2012, some 1750 zones with the potential for quick clay slides were identified. These zones have been classified both with regard to hazard and with regard to risk. The risk maps are important as basis for the planning of protection measures.

Systematic mapping of unstable rock slopes has been performed by the Geological Survey of Norway (NGU) since 1990. Mapping has focused on three counties, but is currently being expanded according to priorities made by NVE. The mapping results in a hazard and risk classification (NGU, 2012).

7.5.5 Land use planning

Land use planning which takes account of the risk is considered the best and most efficient measure to prevent damage from landslides to new development (Norwegian Ministry of Petroleum and Energy, 2012). In lack of other suitable areas for development or due to conflict with other interests, development in hazard areas may be decided. In such cases, protection measures must be installed in order to bring the safety to an acceptable level according to the safety standards in the Planning and Building Act. Technical Regulations of 2010 (TEK10) provide safety standards for buildings and infrastructure, corresponding to different return periods for floods and landslides. Three different safety classes are defined by maximum annual probability of damage as shown in Table 7.3 below.

For quick clay slides the standards are applied using the calculated stability represented by material coefficient γ_M. This must be larger than 1.4 or alternatively the stability must be improved by a minimum percentage, dependent on the initial situation and type of development. Emergency institutions cannot be located in flood, avalanche or landslide risk areas.

Every municipality is obligated according to the Civil Protection Act to perform an overall Risk and Vulnerability (RAV) analysis for its territory as a basis for preparedness

Table 7.3 Safety classes (TEK10).

Safety class for landslides	*Consequences*	*Largest accepted annual probability*
S1	Small (garages, sheds, small storage buildings)	1/100
S2	Medium (residential houses <10 units, other buildings with <25 persons, medium economic or societal consequences)	1/1000
S3	Large (residential houses >10 units, other buildings with >25 persons, large economic or societal consequences)	1/5000

to deal with harmful events and for land use planning. The municipality is responsible for making sure that natural hazards are being evaluated and taken properly into account in every new development scheme, according to the Planning and Building Act.

7.5.6 Protection measures

The national mapping programs has revealed that more than 150000 people in Norway live in areas prone to floods or landslides. In addition there are schools, kinder gardens, hospitals etc. in the hazard areas. Growth in population and the economy leads to an increased damage potential. Climate change will strengthen this effect.

To improve the safety in these developed areas and reduce the costs of the society for the future, structural protection measures could be installed. There is no legal obligation to build protection measures in these areas, as long as the development was established according to the standard of its time. NVE provides assistance in structural protection in already developed areas where the challenges are larger than what could be expected to be dealt with by private owners or the municipalities. The prioritisation is based on cost-benefit to the society. Funding is shared between the government and the municipalities, normally with a 80/20 split. For new development structural protection must be financed by the developer.

Proper land use planning as well as structural protection reduces the probability of damage. It is still not economically or practically feasible to protect everywhere against landslide hazard. In addition there will always be a residual risk. Through monitoring, early warning and emergency preparedness it is possible to reduce the consequences of the events when they occur. This could be through evacuation of people or other risk reducing efforts.

7.6 CRISIS PREPAREDNESS

7.6.1 Emergency preparedness

Crisis preparedness in Norway is organized in accordance with four main principles (Norwegian Ministry of Justice and Public Security, 2012):

Subsidiarity: Risks should be managed at the lowest possible organizational level. In practice, this means that the municipal level has the responsibility for disasters that exceed their capacity for day-to-day preparedness.

Similarity: During a crisis, the organization which manages it should be as similar as possible to the organization under normal circumstances.
Responsibility: The organization responsible for a service under normal circumstances also has the responsibility for disaster prevention, preparedness and response.
Cooperation: Authorities, agencies and other stakeholders have a responsibility to seek cooperation with each other in disaster prevention, preparedness and response.

According to these principles, disaster risk reduction is integrated into all levels of planning and in all sectors. The municipal level has a key role in implementing disaster risk reduction policies, as they have the main responsibility for crisis preparedness as well as prevention through land use planning. Municipalities are also imposed to carry out exercises every year, based on their own holistic risk and vulnerability analyses. The county governors provide guidance and advice to municipalities, and also carry out audits of their crisis preparedness. They also follow up local planning processes to ensure that they take relevant risks into account. In cases of flood and landslide risk, the Norwegian Water Resources and Energy Directorate (NVE) is the national authority and their regional offices work in close collaboration with the county governors and the municipalities.

There is a huge variety of stakeholders for emergency preparedness in Norway. In general, the municipal level is obliged to establish and maintain crisis preparedness, and make use of all relevant local resources available, including NGOs and the private sector. Due to variations in climate, topography and populations, there are local variations in risk, vulnerability and capacity.

Landslide risk also varies across the country, and thus local preparedness has to be fit to the type of landslides that might occur in each particular area. Two types of landslides are of particular interest and are described in Norway's national risk assessment: Huge rock slides and quick clay slides.

7.6.2 Regional early warning service for landslides

NVE in cooperation with NPRA, National Rail Administration and Meteorological Institute runs a national landslide early warning system (EWS) to assist road and railways authorities as well as local authorities in taking measures before potentially dangerous events occur. The service was made operational in the autumn of 2013. The EWS has been developed to warn about debris flows, debris slides, debris avalanches and slush flows at regional scale. The service is organised in synergy with the flood warning service. It is national, operational 24/7 and is supervised by 10 hydrologists/geologists. Four warning levels are used: green, yellow, orange and red.

The EWS uses hydrological models and web tools that were developed in the flood and snow avalanche warning systems (Devoli *et al.*, 2014). Some of the computer tools are developed in collaboration with the meteorological institute, road and railway authorities as well as private consultants. The first web tool *xgeo.no* is an expert tool used for natural hazards forecasts. The portal assists the forecasters in the prediction and the monitoring phase with data on the hydro-meteorological situation and quantitative prognosis. The portal also includes information from the real-time database *regObs.no*. This is a web-tool developed for registration of observations related to natural hazards. It was designed as a public tool supporting crowd sourcing, and is currently available to the public as a website and app.

The EWS is complemented by a web portal *varsom.no*. The tool is used to issue and distribute alert messages to both decision makers and the public. The portal presents daily warning messages (bulletins) for snow avalanches, floods, landslides and ice conditions in rivers 3 days ahead. To make the bulletin page as user friendly and educational as possible, it includes relevant information such as landslide types, real-time weather radar images etc. The page always features a list of the regions and a map showing their respective warning levels, see Figure 7.18.

7.6.3 Early warning service – large rockslides

In some occasions, physical mitigation is impossible, difficult, or economically too costly due to the size and complexity of the landslide. This is relevant for the large rockslides in Norway. Large rockslides are rare but have been those landslides that caused highest number of fatalities in the 20th century and loss of property due to large run-out, the possibilities for river damming and the generation of disastrous displacement waves in fjords. DSB has carried out a comparison of 14 different potential disasters that may affect the Norwegian society (DSB 2013a) showing that a rock slope failure hitting a fjord, although it would be predicted, represents one of the largest risks. The fastest moving known unstable mountain, Åkneset in Western Norway, represents a great threat to the nearby communities of Hellesylt, Stranda and Geiranger as wave heights might there be several tens of meters high following laboratory experiments and numerical modelling (Harbitz *et al.*, 2014). The volume of the unstable masses is approximately 30–40 million m^3 or more, and it moves 3–10 centimeters per year (Blikra, 2012).

As a result of the national mapping program for large rock slides, four objects have so far been identified with high risk. These are Åknes, Mannen and Heggurak-sla in Western Norway and Jettan in Northern Norway. At these sites monitoring and early warning systems have been installed. The systems are operated by the inter-municipal organizations Åknes/Tafjord Beredskap IKS and Nordnorsk fjellovervåking IKS. A real-time operational handling is established on both the technical and geological part. All the monitoring data achieved from the unstable areas are processed and analysed in order to find movement behaviour and velocity trends. This is also the basis for establishing threshold values for definition of warning levels. Velocity and acceleration is the criteria used for the different levels. The implementation of these thresholds has been based on the historical data from Åknes and the experience from other international studies.

The early-warning system also includes the implementation of warning methods to the affected population. For the Åknes rockslide this include two lines. Firstly a system of phone messages based on a continuous updated database in addition to automatic SMS messages based on regional coverage from the mobile network companies. In addition for the Åknes case, electronic typhoons are established in all the villages situated in the displacement wave hazard zones.

The investigations, monitoring and early-warning that have been designed and implemented for the Norwegian rockslides follow requirements guided in national

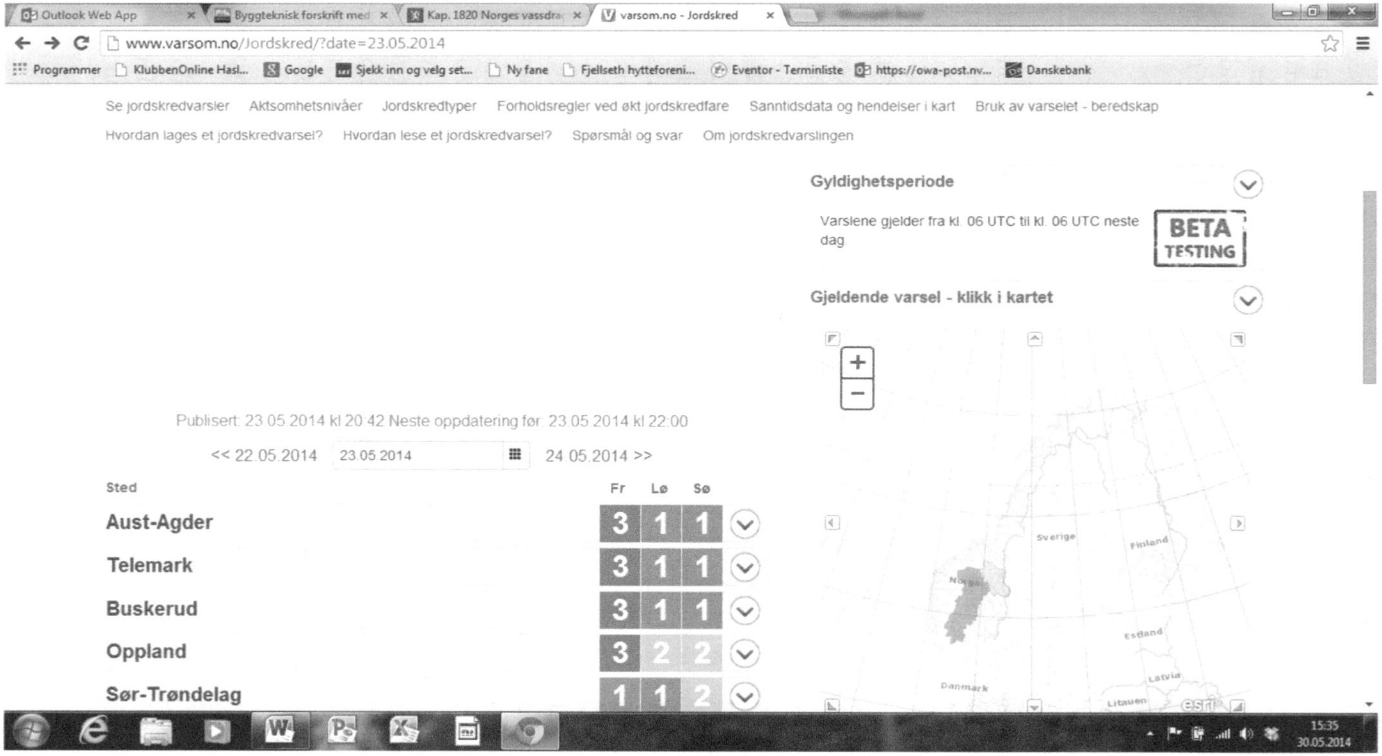

Figure 7.18 Landslide warning message presented at county scale (www.varsom.no).

codes and international standards for geotechnical design. An exception has been made from the normal safety standards according to the Technical Regulations of 2010. Further development could be performed in the hazards zones even if the probability of damage is higher than the standard levels. This is provided sufficient monitoring and warning systems are in place to make sure that people will be evacuated in time before the rock slide. Without this exception large areas along the fjords in question would be banned for further development.

7.6.4 Quick clay slides in a city

One of the scenarios in the National Risk Assessment (DSB, 2013b) is the occurrence of a quick clay landslide in a city. The scenario takes place in a known quick clay zone in the highest risk class, in a densely populated urban area in the city of Trondheim. According to the scenario, an evacuation is implemented after an initial landslide of 10×100 meters into the river Nidelva, before a landslide of approximately 3 million m^3 of clay covering an area of $0.5\,km^2$ causes a flood wave upstream and downstream the river Nidelva, which runs through the city. The water level upstream rises quickly to 12 metres above sea level, and an area of $1.5\,km^2$ with around 1,100 citizens is flooded, including the central business district. The assessment of consequences estimates around 200 fatalities as direct or indirect consequence of the landslide, and 2,500 injured or ill people. Cultural artifacts of great national importance will be lost and there will be economic and material losses, including damage of critical infrastructure, exceeding NOK 30 billion.

It has been assessed that a quick clay slide in this zone would have an annual probability of 1/2,000–1/3,000. However, human activity could often trigger quick clay landslides, but they could also be prevented by erosion protection measures and construction regulations.

7.7 FINAL REMARKS

Norway will be warmer and wetter in the coming years, and there will be more extreme weather. Consequently, the weather-triggered landslide hazard is likely to increase. There are signals that this tendency has already started. In order to meet the challenges posed by the increasing landslide hazard, there has been a strong focus both on research and on policy issues related to landslide risk management in recent years. Norway is a long country with spread population and a large number of small communities. The key state and private institutions have a strong focus on developing procedures and strategies for dealing with the increasing landslide hazard. However, there is a lack of technical competence and capacity in the municipalities, who, by Norwegian law, are responsible for implementing preventive measures and for disaster management. Furthermore, considering that extreme weather events have a local character and are very difficult to predict precisely, there is still a long way to go before the weather-triggered landslide risk can be handled in an optimal way.

REFERENCES

Aas, G. (1981) Stability of natural slopes in quick clays. In: *Proceedings of the 10th International Conference on Soil Mechanics and Foundation Engineering, Stockholm, Sweden.*

Blikra, L.H. (2012) The Åknes rockslide, Norway. In: Clague, J.J. & Stead, D. (eds.) *Landslides: Types, Mechanisms and Modeling.* Cambridge, Cambridge University Press. pp. 323–335.

Blikra, L.H., Longva, O., Braathan, A., Anda, E., Dehls, J.F. & Stalsberg, K. (2006) Rock slope failures in Norwegian fjord areas; examples, spatial distribution and temporal pattern. In: Evans, S.G., Scarascia Mugnozza, G., Strom, A. & Hermanns R.L. (eds.) *Landslides from Massive Rock Slope Failures, NATO Science Series IV (49), Earth and Environmental Sciences.* Dordrecht, Springer. pp. 475–496.

Brunborg, H. & Texmoen, I. (2011) *Befolkningsframskrivning 2011-2100: Nasjonale resultater.* SSB, Økonomiske analyser, 4/2011

Bungum, H., Lindholm, C., Dahle, A., Woo, G., Nadim, F., Holme, J., Gudmestad, O., Hagberg, T. & Karthigeyan, K. (2000) New seismic zoning maps for Norway, the North Sea and the UK. *Seismological Research Letters.* 71, 687–697.

Bungum, H., Lindholm, C. & Faleide, J.I. (2005) Postglacial seismicity offshore mid-Norway with emphasis on spatio-temporal-magnitudal variations. *Marine and Petroleum Geology,* 22, 37–148.

Devoli, G., Kleivane, I., Sund, M., Orthe, N.-K., Ekker, R., Johnsen, E. & Colleuille, H. (2014) Landslide early warning system and web tools for real-time scenarios and for distribution of early warning messages in Norway. In: *Proceedings IAEG XII CONGRESS, September 15-19, 2014, Torino, Italy.*

DSB (2010) *Samfunnssikkerhet i arealplanlegging.* Kartlegging av risiko og sårbarhet, Temahefte. Available from: www.dsb.no.

DSB (2013a) *National Risk Analysis. Disasters that May Affect the Norwegian Society.* Norwegian Directorate for Civil Protection. ISSN: 978-82-7768-318-8.

DSB (2013b) *Evaluering av myndighetenes forebyggingsarbeid og håndtering av flommen i mai 2013.* Norwegian Directorate for Civil Protection. ISBN: 978-82-7768-325-6.

EM-DAT (2014) Available from: http://www.emdat.be/database.

Furseth, A. (2006) *Skredulykker i Norge.* Oslo, Tun Forlag.

Hanssen-Bauer, I. (ed.) (2015) *Klima i Norge 2100.* Norwegian Centre of Climate Services (Met.no, Bjerknessenteret, Nansensenteret, Havforskningsinstituttet, NVE).

Harbitz, C.B., Glimsdal, S., Løvolt, F., Kveldsvik, V., Pedersen, G.K. & Jensen, A. (2014) Rockslide tsunamis in complex fjords: From an unstable rock slope at Åkerneset to tsunami risk in western Norway. *Coastal Engineering,* 88, 101–122.

Hermanns, R.L., Blikra, L.H., Naumann, M., Nilsen, B., Panthi, K.K., Stromeyer, D. & Longva, O. (2006) Examples of multiple rock-slope collapses from Köfels (Ötz valley, Austria) and western Norway. *Engineering Geology,* 83, 94–108.

Hermanns, R.L., Hansen, L., Sletten, K., Böhme, M., Bunkholt, H., Dehls, J.F., Eilertsen, R., Fischer, L., L'Heureux, J.-S., Høgaas, F., Nordahl, B., Oppikofer, T., Rubensdotter, L., Solberg, I.-L., Stalsberg K. & Yugsi Molina, F.X. (2012) Systematic geological mapping for landslide understanding in the Norwegian context. In: Eberhardt, E., Froese, C., Turner, A.K. & Leroueil, S. (eds.) *Landslide and Engineered Slopes: Protecting Society Through Improved Understanding.* London, Taylor & Francis Group. pp. 265–271.

Hermanns, R.L., Oppikofer, T., Roberts, N.J., Sandøy, G. (2014) Catalogue of historical displacement waves and landslide-triggered tsunamis in Norway. In: Lollino, G. (ed.) *Engineering Geology for Society and Territory,* Vol. 4. Berlin, Springer.

IPCC (2013) *Climate Change 2013. The Physical Science Basis. 5th Assessment Report, Working Group 1.* Available from: http://www.climatechange2013.org/images/report/WG1AR5_ALL_FINAL.pdf

ISDR (2005) *Hyogo Framework for Action 2005-2015: Building the Resilience of Nations and Communities to Disasters*. Available from: http://www.unisdr.org/2005/wcdr/intergover/official-doc/L-docs/Hyogo-framework-for-action-english.pdf.

ISO (2000) *Risk Management Terminology*. Working draft for ISO (the International Organization for Standardization) guide. 34, Rev. 2000-11-01.

Jaedicke, C. & Kleven, A. (2008) Long-term precipitation and slide activity in south-eastern Norway, autumn 2000. *Hydrological Processes*, 22, 495–505.

Jaedicke, C., Solheim, A., Blikra, L.H., Stalsberg, K., Sorteberg, A., Aaheim, A., Kronholm, K., Vikhamar-Schuler, D., Isaksen, K., Sletten, K., Kristensen, K., Barstad, I., Melchiorre, C., Høydal, Ø.A. & Mestl, H. (2008) Spatial and temporal variations of Norwegian geohazards in a changing climate, the GeoExtreme project. *Natural Hazards and Earth System Sciences*, 8, 893–904.

Jaedicke, C., Sverdrup-Thygeson, K., Syre, E., Nadim, F., Kalsnes, B. & Vangelsten, B.V. (2011) *SafeLand Deliverable D3.7. Expected Changes in Climate-Driven Landslide Activity (Magnitude, Frequency) in Europe in the Next 100 Years*. Available from: http://www.safeland-fp7.eu/results/Documents/D3.7.pdf.

Jaedicke, C., van den Eeckhaut, M., Nadim, F., Hervas, J., Kalsnes, B., Vangelsteb, B.V., Smith, J.T., Tofani, V., Ciurean, R., Winter, M.G., Sverdrup-Thygeson, K., Syre, E. & Smebye, H. (2013) Identification of landslide hazard and risk "hotspots" in Europe. *Bulletin of Geological Environment*, 73, 893–904.

Leroi, E., Bonnard, Ch., Fell, R. & og McInnes, R. (2005) Risk assessment and management. In: Hungr, O., Fell, R., Couture, R. & Eberhardt, E. (eds.) *Landslide Risk Management*. London, Taylor & Francis Group. pp. 159–199. ISBN: 04 1538 043 X.

Nadim, F. & Kalsnes, B. (2014) Progress of living with landslide risk in Europe. In: *Proceedings World Landslide Forum 3, 2-6 June 2014, Beijing, China*.

Nadim, F., Schack Pedersen, S.A., Schmidt-Thomé, P., Sigmundsson, F. & Engdahl, F. (2009) Natural hazards in Nordic countries. *Episodes*, 31 (1), 176–184.

Nadim, F., Høydal Ø.A., Hefre, H. & McClean, A. (2011) *SafeLand deliverable D1.6. Analysis of Landslide Triggered by Anthropogenic Factors in Europe*. Available from: http://www.safeland-fp7.eu/results/Documents/D1.6.pdf.

NGI (2013) *Impacts of Extreme Weather Events on Infrastructure in Norway (InfraRisk)*. NGI report 20091808-01-R, 2013. Available from: http://www.ngi.no/upload/Prosjektweb/InfraRisk/Sluttrapport/20091808-01-R_InfraRisk-prosjekt-sluttrapport.pdf.

NGU (2012) *Recommended Hazard and Risk Classification System for Large Unstable Rock Slopes in Norway*. NGU report 2012.029.

NIFS (2013) *Naturfareprosjektet. Delprosjekt 1 Naturskadestrategi. Sammenligning av risikoakseptkriterier for flom og skred*. Available from: http://www.naturfare.no/_attachment/619730/binary/958523.

Norwegian Ministry of Justice and Public Security (2012) White paper no. 29. *Meld St 39 (2011-2012) Melding til Stortinget. Samfunnssikkerhet*.

Norwegian Ministry of Petroleum and Energy (2012) White paper no. 15. *Meld St 15 (2011–2012) Melding til Stortinget. Hvordan leve med farene – om flom og skred*.

NVE (2012) Norwegian Water Resources and Energy, January 2012. *Kvikkleireskredet ved Esp på Byneset i Trondheim*. NVE Rapport nr 1 – 2012.

NVE Norwegian Water Resources and Energy Directorate. *Landslides in Southern Norway May 2013. Jordskred i Sør-Norge mai 2013*. Rapport nr 52 – 2014, in preparation.

Solheim, A., Berg, K., Forsberg, C.F. & Bryn, P. (2005) The Storegga slide complex: Repetitive large scale sliding with similar cause and development. *Marine Petroleum and Geology*, 22, 97–107.

Vorren, T.O. & Mangerud, J. (2008) Glaciations come and go. In: Ramberg, I.B., Bryhni, I., Nøttvedt, A. & Rangnes, K. (eds.) *The Making of Land: Geology of Norway*. Trondheim, Geological Society of Norway. pp. 481–533.

Chapter 8

Overview of landslides distribution in Russian Federation and variations of their activity due to climate change

O.V. Zerkal
Moscow State University, Geological Department, Laboratory of Engineering Geodynamics, Moscow, Russia

A.L. Strom
Geodynamics Research Center – branch of JSC "Hydroproject Institute", Moscow, Russia

8.1 INTRODUCTION

Landslides are spread quite nonuniformly over the territory of the Russian Federation. From 1500 to 4000 active landslides (ignoring small-scale events) are reported annually, increasing to 7000 some years. Also up to 300 new landslides are generated annually (Fig. 8.1). In some regions slope deformations affect up to 60% of the total area and their study and monitoring are of high importance, while in other parts of the country landslides can be considered as rare, "exotic" phenomena (Fig. 8.2).

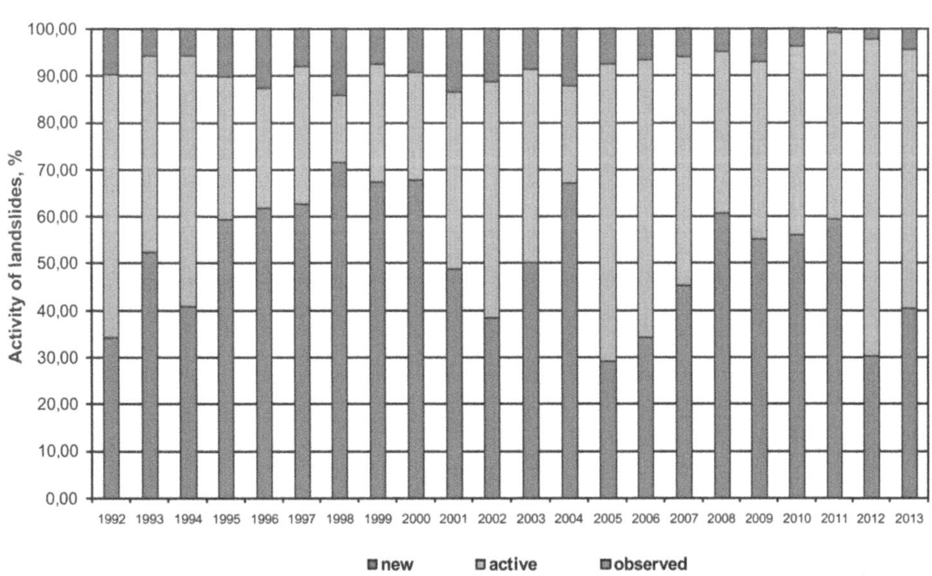

Figure 8.1 Overall landslide activity in Russian Federation.

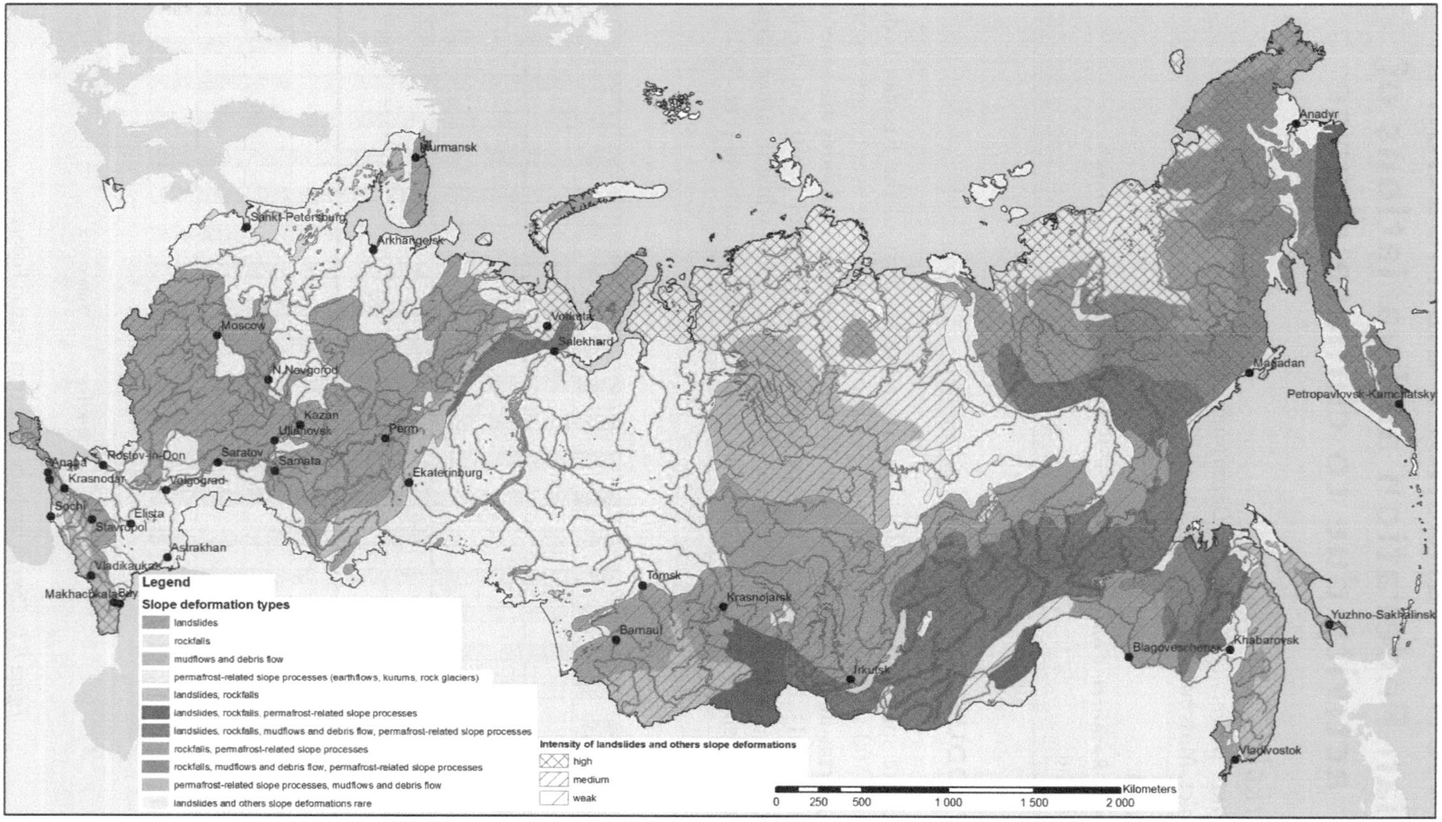

Figure 8.2 Distribution and intensity of landslides and other slope processes in Russian Federation.

Landslides distribution and activity, prevalence of the particular types of slope processes depend on the environmental conditions, which are quite variable over the entire territory of Russia:

- orographic conditions, varying from lowlands, sometimes below mean sea level (the Peri-Caspian Depression with elevation −28 m) to mountain systems with alpine-type relief and elevations exceeding 5000 m (Caucasus);
- climatic conditions, varying from arctic deserts to subtropics;
- geological conditions, varying from stable almost aseismic Archean-Proterozoic shields and ancient platforms to modern collision belts, subduction and rift zones with extremely high seismic activity and, locally, with active volcanoes (Kamchatka, Kuril Islands, Central Caucasus);

Three main groups of environmental factors (combination of the geological medium components, after Bondarik (1981)) influencing the landslide formation, evolution and modern activation have been selected (Zolotariov, 1995; Zerkal, 2003): slowly changing factors, rapidly changing factors and triggering factors.

Environmental factors evolving in the geological time scale, which are practically constant considering human life time scale, such as geological, tectonic and geomorphic conditions, predetermine location of the landslide-prone regions and areas and general regularities of modern slope processes spatial distribution. It is obvious that such regional factors are considered as constant when analyzing variability of landslide activity influenced by climate change.

Slowly changing factors, such as modern tectonic movements, secular climate variations recorded by century- and decade-long observations determine modern general tendencies in landslides evolution. Such slowly changing factors of landslides activation can be divided into regional and zonal factors.

Landslide size and activity at any particular area are determined by natural factors characterized by annual cycle, which, in turn, depend on long-term variations and rhythms caused by solar cycles in particular. One of such rapidly changing factors is the annual weather variability affecting rock massifs watering and soils temperature regime. Usually rapidly changing natural factors have zonal distribution (zonal-geological factors according to Trofimov (2002).

It is reasonable to analyze effect of climate change on landslide activity in Russia for each province, considering their characteristic features.

We do not analyze hereafter triggering factors (phenomena) leading to activation of any particular slope deformation.

8.2 REGIONAL FACTORS INFLUENCING LANDSLIDES DISTRIBUTION WITHIN THE TERRITORY OF RUSSIAN FEDERATION

General distribution of landslides is controlled by the spatially distributed lithological and tectonic peculiarities of the territory, predetermining both composition of the superficial rocks and soils and relief. Significant influence on modern landslide activity in some parts of Russia has their Neogene-Quaternary geological evolution.

Several large mountain systems are located in Russia completely or partially (Fig. 8.3) (Khain, 2001; Gatinskiy, 2004; Zerkal, 2013):

I. High-altitude mountains:

- The Caucasus (the Elbrus peak – 5642 m a.s.l.), that have been formed due to collision of the East-European sector of the Eurasian Plate and Arabian Plate;
- The Altai (the Belukha peak – 4509 m a.s.l.), which is a part of the Central-Asian Mountain belt;
- The volcanic chain of the Kamchatka and Kuril Islands (the Kliuchevskoi volcano – 4835 m a.s.l. after the last eruption on August 15, 2013), which is formed by the subduction of the Pacific Plate under the Okhotsk Plate.

II. Middle-altitude mountains:

- The Sayan Mountain System that includes two mountain ranges – the 650 km long Western Saysan with elevations up to 3121 m a.s.l. and the 1000 km long Eastern Sayan up to 3491 m. a.s.l. It is located east from the Altai Mountains and also belongs to the Central Asia Mountain belt;
- The Verhoyansk-Chukotka Mountainous area that includes the Verhoyanksk Range (about 1200 km long and up to 250 km wide, highest peak – 2389 m a.s.l.) and the Cherskiy Range (about 1500 km long with the highest peak – 3003 m a.s.l.) formed due to interaction between the Eurasian and North American Plates;
- The Stanovoi Range more than 700 km long and up to 180 km wide with maximal altitude up to 3000 m a.s.l. It has been formed by the interaction between the Eurasian and Amur Plates;
- The Baikal Mountain System – ranges up to 2840 m a.s.l. bounding the continental Baikal Rift zone;
- The Amur Mountain System represented by series of ranges up to 400 km long with highest peak of 2167 m a.s.l.);
- The Sikhote-Alin Mountain System up to 1200 km long with highest peak of 2090 m a.s.l.;
- The Urals mountain System more than 2000 km long and up to 150 km wide with the highest peak of 1895 m a.s.l.). It divides the East European ancient platform and West-Siberian epi-Paleozoic platform.

Most of these mountains, except the Sikhote-Alin and Urals are characterized by high and medium seismic activity.

Other large orographic units of Russia are:

- The East-European plain. Most of its territory have altitudes up to 200–300 m a.s.l. though some parts can be classified as low-hill terrains such as the Kola Peninsula up to 800 m a.s.l. high and the Cis-Ural region with elevations up to 480 m a.s.l. East-European plane is dissected by large valleys of the Volga, Oka, Kama, Don, Northern Dvina, Dniepr Rivers and their tributaries, locally with high steep slopes;
- West-Siberian plain with elevations up to 150 m a.s.l. occupying epi-Paleozoic platform;

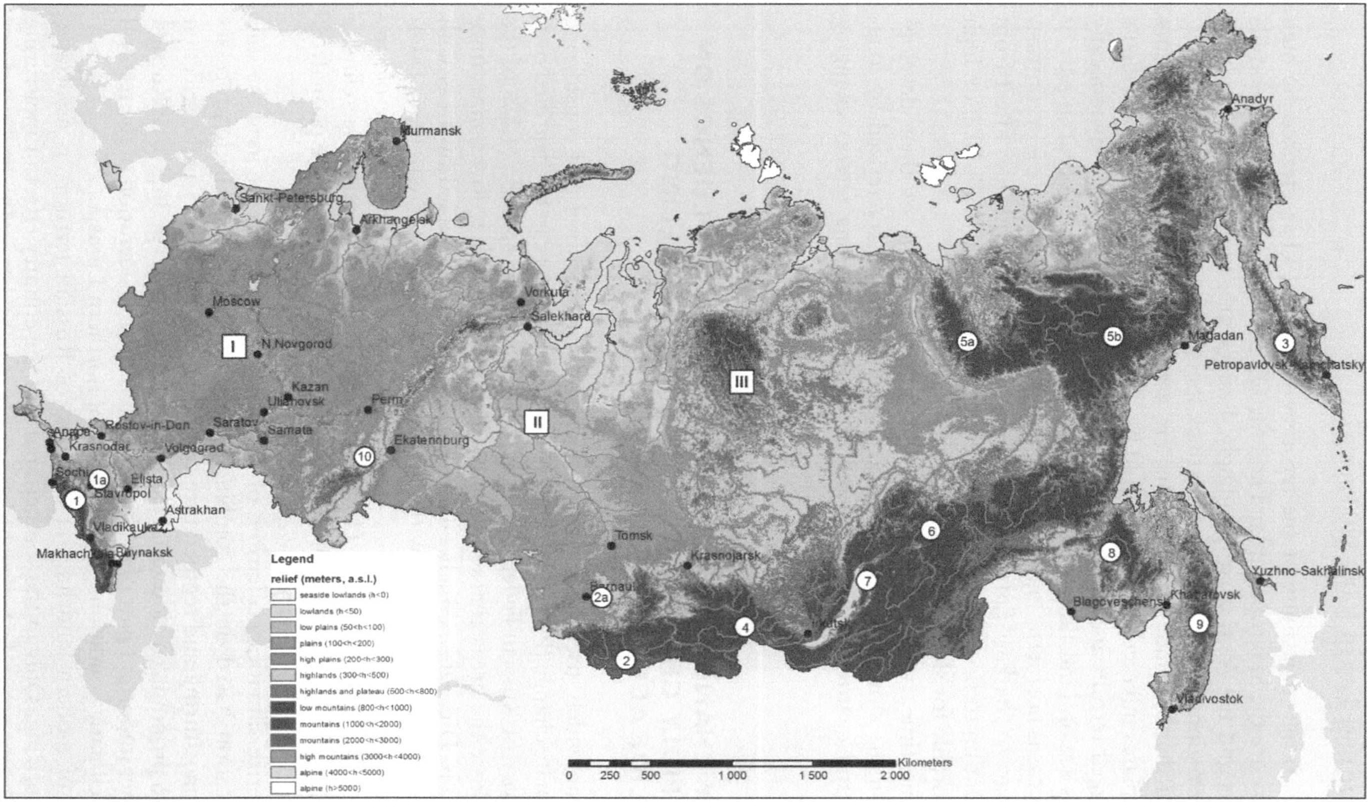

Figure 8.3 Relief of Russian Federation. 1 – the Great Caucasus, 1a – the Stavropol upland; 2 – the Altay mountain system, 2a – the Ob Plateau; 3 – The volcanic chain of Kamchatka and Kuril Islands; 4 – the Sayan Mountain system; 5 – the Verkhoyansk-Chukotka Mountainous system: 5a – the Verkhoyansk mountains, 5b – the Cherskiy Range system; 6 – the Stanovoy Range; 7 – the Baikal Mountain system; 8 – the Amur Mountain system; 9 – the Sikhote-Alin Mountain system; 10 – the Urals Mountain system. I – the East-European plain; II – the West-Syberian plain; III – the Middle-Siberian upland.

- Central Siberian Plateau with elevations of about 500–600 m mainly, corresponding to the Siberian ancient platform. Some parts of it, the Anabar and Putorana highland, are raised up to 1000–1200 m a.s.l.

The smaller orographic units of the territory of Russia, characterized by high landslide activity are the Stavropol upland and the Ob plateau located at the foothills of the Caucasus and the Altai correspondingly. Besides other factors high landslide activity of these regions is predetermined by thick loess deposits.

It is evident that landslide prone regions are associated, at first, with the zones of the tectonic plates interaction characterized by contrasting neotectonic movements resulting in rugged terrain and high seismicity. For example, in the geodynamically active Caucasus region landslides affect from 25% of the area (in the central part of the Greater Caucasus northern slope) up to 60% (in the western part of it). In contrast, in the central part of the East-European or Siberian platforms landslides affect from tenth of a percent up to first percent of the area increasing locally to 9–10% on high river banks favorable for slope instabilities.

At the same time, types of slope deformations and scale of their development depend on the slow changing zonal-geological factors characterized by long-term (decades and centuries) variations and on rapidly changing factors changing annually.

8.3 SLOWLY CHANGING FACTORS AND THEIR INFLUENCE ON THE RAPIDLY CHANGING FACTORS CONTROLLING LANDSLIDE ACTIVITY IN RUSSIA

According to the widely used Köppen-Geiger climate classification and considering influence of climate change (Peel *et al.*, 2007) 16 climatic classes can be identified over the entire territory of Russian Federation (Fig. 8.4). Most wide spread (74,8% of the entire territory) are the areas characterized by the humid relatively low temperature climate (Dfa, Dfb, Dfc, Dfd). Much smaller part (14,9%) is occupied by areas with relatively low temperate climate but with dry winter season (Dwa, Dwb, Dwc, Dwd) and by the polar tundra (ET – 7,4%). Areas with other classes of climate occupy less than 3% of the entire territory. However, based on the Köppen-Geigerclimate climate classification only, it is difficult to explain wide distribution of the permafrost in Russia.

To characterize climatic peculiarities of Russian territory better we used the soil-climate classification based on the combined analysis of the atmospheric processes and of the soil moisturing and temperature conditions. Its peculiarity is the selection of the areas with pronounced altitudinal climatic zonality typical of 36,8% of Russian territory. Another important part of such approach is the consideration of the sectoral distribution of climatic conditions for vast areas of northern Eurasia.

Formation of modern landslides at the most of Russian territory is associated with overmoisturizing. Only in the southern parts of the country effect of moisturing becomes insufficient or weak. It should be pointed out that the sublatitudinal (from

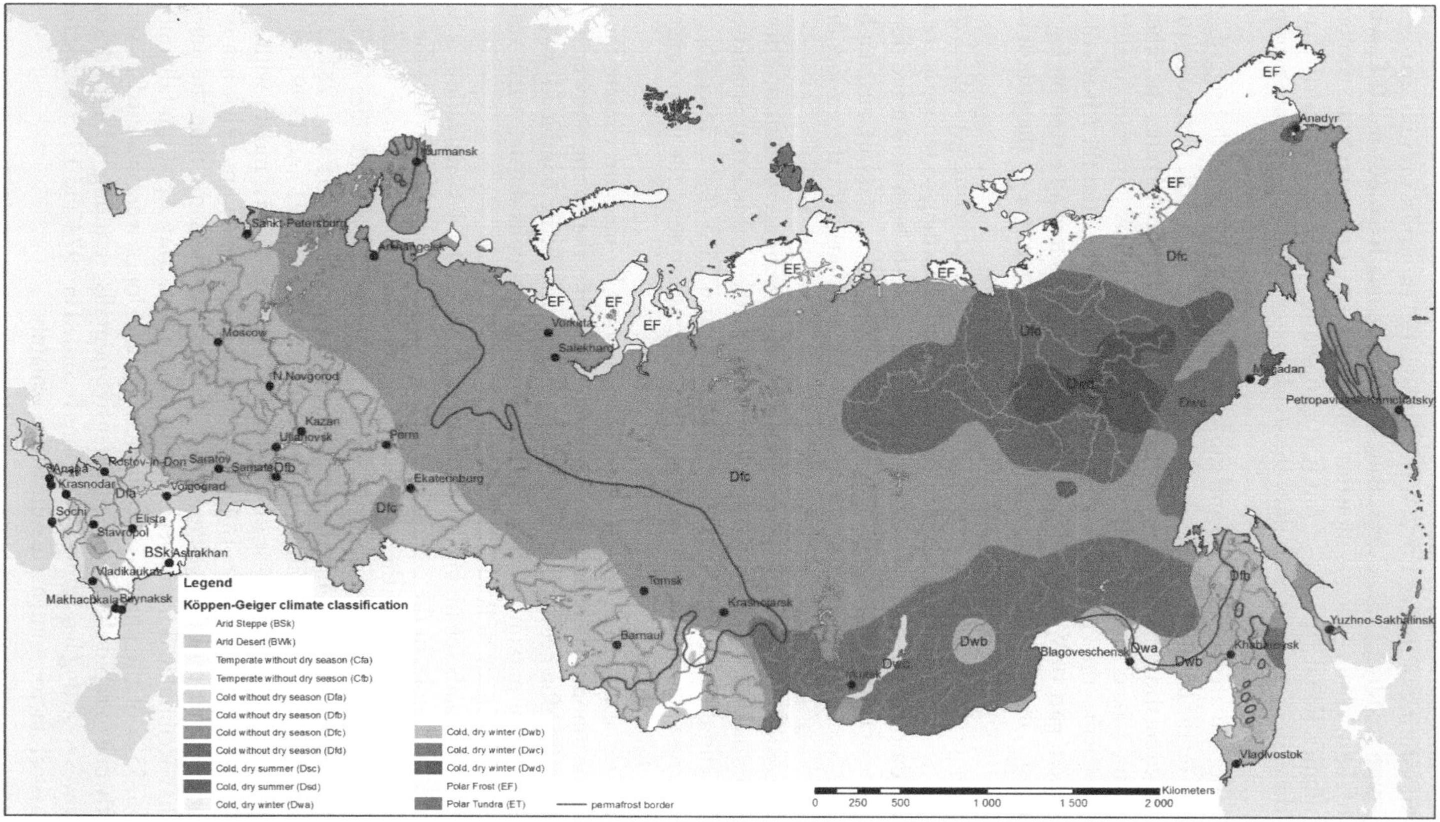

Figure 8.4 Climate zoning of Russian Federation (according to Köppen-Geiger climate classification) (after Peel *et al.*, 2007).

west to east) moisture transport prevails during the whole year, predetermining conditions of moisturing of the territory. Western (Atlantic) transport prevails up to 60°E (to north-south trending Urals), decreasing and weakening from 60° to 90°E, where influence of the opposite – Pacific moisture transport starts increasing being combined with the submeridional (Arctic – from north to south) moisture transport (Zerkal, 2013). In the eastern part of the country the south-eastern (Pacific) moisture transport prevails.

The arctic, sub-arctic, moderate and sub-tropical climatic belts are distinguished in Russia based on the latitudinal zonality of solar insolation (Fig. 8.5). According to the atmospheric circulation the following sectors are identified: the Atlantic, the Siberian (East-Siberian), mixed (affected by both the Atlantic and East-Siberian circulation regimes), the Pacific (the Pacific monsoon in the southern part) and the Middle-Asian for the southern part of Western Siberia (Zerkal, 2013). As it can be seen on Fig. 8.6 the permafrost is distributed within the zone of East-Siberian type of the atmospheric circulation strongly affected by air coming from the Arctic.

Three engineering-geological zones (climate-related engineering-geological zone, according to Trofimov (2002)) are identified in Russia according to temperature and moisture conditions – the continuous permafrost zone, zone of both frozen and thawed soils, and zone of the thawed and unfrozen soils (Zerkal, 2013). It is obvious that both conditions of landslides formation and effects of climate change for these zones differ significantly.

For the continuous permafrost zone the temperature change is critical. Effect of climate change on slopes stability can be exemplified by the landslide that originated in 2008 in the central part of the Vorkuta town located at 67.5°N, 64°E. Intensification of the permafrost degradation due to global warming was one of the main causes of these slope deformations.

For the frozen and thawed soils zone characterized by medium climate the key factors are the amount of winter snow accumulation and the spring snowmelt rate when meltwater influences soil moisturizing significantly being combined with the spring precipitation maximum. Besides, the abnormal rainfalls during warm season act as a 'classical' factor of landslides activation in this zone.

Effect of annually changing climatic factors on landslides activation is well expressed, for example, in the central part of the East-European plain where the Oka River falls into Volga River (the so-called *Nyzhniy Novgorod-Vasilsursk landslide-prone province*). Landslides activations when number of active landslides increased 2–2.3 times over the mean annual values were recorded here in 1966, 1981, 1994, 1995, and 2001. Intensive landslides were reported in 1845, 1892 and 1899 too. Just these years the significant increase of temperature between the last month with mean negative temperature (March, in 1845 – April) and the following month when mean temperature became positive reached 9–11° Celsius. At the same time the amount of snow that accumulated from December to March plus May-April rainfall amount reached 46–55% of the annual precipitation. Valley slopes were watered in spring significantly due to rapid snowmelt and intensive rainfalls, resulting in landslides activation. However, strong landslides activations recorded in 1974 and 1979 were caused by the 'classical' factor – intensive rainstorms – mainly.

Figure 8.5 Zonal and sectoral climate distribution of Russian Federation.

Figure 8.6 Climate-related engineering geological zoning of Russian Federation (after Trofimov, 2002).

8.4 VARIATION OF LANDSLIDES ACTIVITY IN DIFFERENT REGIONS OF RUSSIAN FEDERATION DUE TO CLIMATE CHANGE

8.4.1 Zone of the thawed and unfrozen soils

8.4.1.1 Southern region of the European part of Russia

This region is located between the Black and Azov Seas in the west and the Caspian Sea in the east and includes the downstream part of the Don River basin, the Ciscaucasia and the Caspian Lowland, Crimea and the Greater Caucasus mountain system. Landslides are generally widespread over this territory and slope deformations of almost all types are known here – blockslides and flowslides, rockfalls, debris flows and mud flows, rock avalanches, ice and rock-ice avalanches. In some districts part of the area affected by landslides reaches 44–60% and even more. First reports about slope processes in this region came from the last quarter of the 18th century and describe formation of ice and rock-ice avalanches from the Devdorak Glacier (42.7°N, 44.52°E) that had blocked the main road between Russia and Trans-Caucasia. The special Landslide Research Centre (the so called "Landslide Station") that performed regular landslide monitoring in this area was organized in the 30s of the 20th century on the Black Sea shore of the Caucasus. At the beginning of the 80th regular monitoring of landslides and other exogenous processes was arranged over the entire region in question. More than 9800 landslides, over 1400 rockfall-prone sites and over 2000 debris flow prone gullies have been identified and monitored here.

The following landslide-prone provinces have been selected in this region: the Azov, the Scythian, the Taman, the Black Sea Coastal, the Northwestern-, Central- and Eastern-Caucasian (Zerkal & Korolev, 2009) (Fig. 8.7).

The *Azov landslide-prone province* is situated in the north of the region. It covers northern part of the Don River valley and northern coast of the Azov Sea. The largest landslides in the province have been observed along the 90–100 m high right bank of Don river valley. Towards the west these geomorphic step forms a high cliff of the Azov Sea northern coast. According to (Zerkal *et al.*, 1999) the following types of landslides exist in the Azov landslide-prone province:

- rotational landslides and surface earthflows in loess which volumes varies from 15–25 thousand m^3 to 100–120 thousand m^3,
- large rotational landslides, affecting the Neogenic (clay, chalkstone) and Quaternary (sand, loess) deposits. Their volume varies from 1–2 million to 30 million m^3.

The *Taman landslide-prone province* covers northern coast (a high cliff) of the Taman peninsula. The Taman landslide-prone province includes rotational landslides and surface earthflows in loess (similar to loess landslides of the Azov landslide-prone province).

The *Scythian landslide province* is situated south-east from the Azov one. It occupies southern coast of Azov Sea, the Kuban piedmont plain and the Don-Sal and the Stavropol highlands. The largest modern landslides in this province are observed in the Kuban River valley. Their size is smaller within the Stavropol highlands. Landslide affected area varies from 10% to 60% (Fig. 8.8).

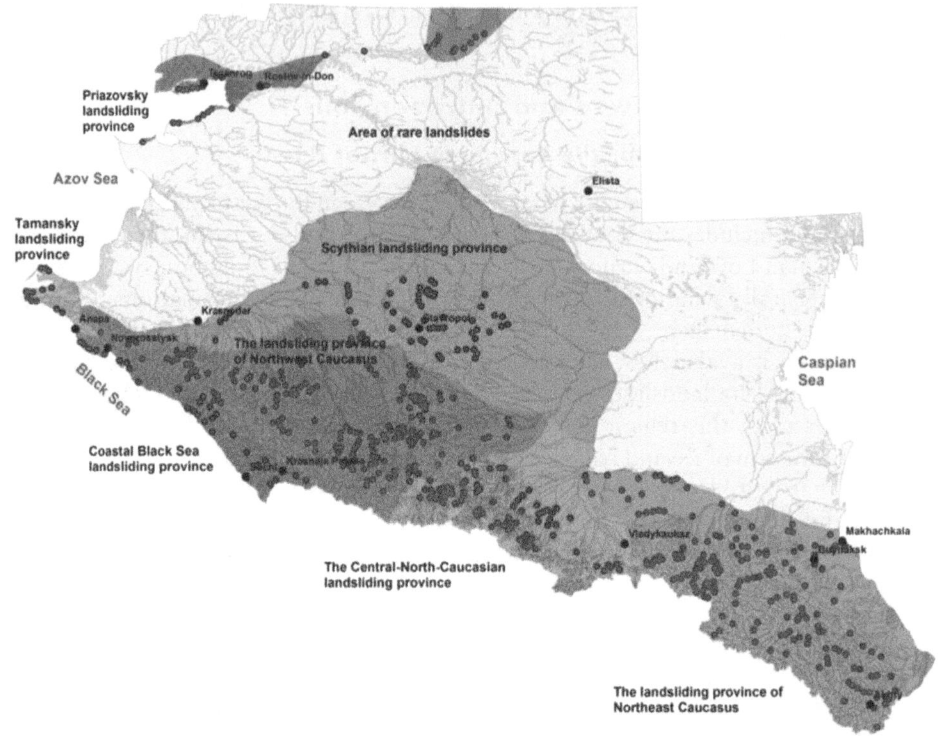

Figure 8.7 The landslide-prone provinces of the Greater Caucasus and Ciscaucasia region (after (Zerkal & Korolev, 2009)).

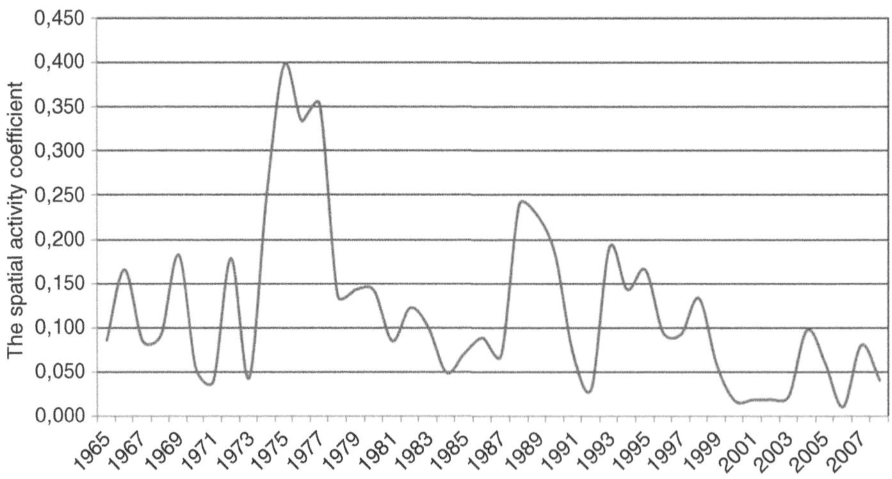

Figure 8.8 Time variation of landslide activity in the Stavropol highlands of the Scythian landslide-prone province (according to the State geological monitoring survey).

East boundary of this landslide-prone province corresponds to the Caspian Sea Early Pleistocene paleocliff.

Large rotational landslides in loess, ranging from 2–3 million to 18 million m^3 in volume, are representative of the Kuban River valley. Their activity is caused by the river erosion. Slope deformations of the Stavropol highlands include:

- rotational landslides and surface earthflows in loess with volumes from 15–25 thousand m^3 to 100–120 thousand m^3,
- surface earthflows in Neogen clay with volumes up to 100–120 thousand m^3 and more.

We want to mention that some of the past (Pleistocene) events had been several orders of magnitude larger (Pánek *et al.*, 2012).

The *Black Sea Coastal landslide-prone province* is situated in a subtropical climatic zone. The maximal annual precipitation for this territory is 2500 mm at the coastal areas, and increases up to 4000 mm in its mountainous part. The *Black Sea Coastal landslide-prone province* is one of the most investigated. Several landslide zones can be allocated within this province: the Sochi, the Tuapse, and the Jankhotsk zones. Landslides affect from 57–61% up to 73–74% of the area of this zones. Variation of the landslide activity in time for the Sochi landslide zone is shown on Fig. 8.9.

Two main types of landslides are typical of this zone: surface earthflows, and compound landslides up to 30 m thick and even more with volumes up to 10 million m^3.

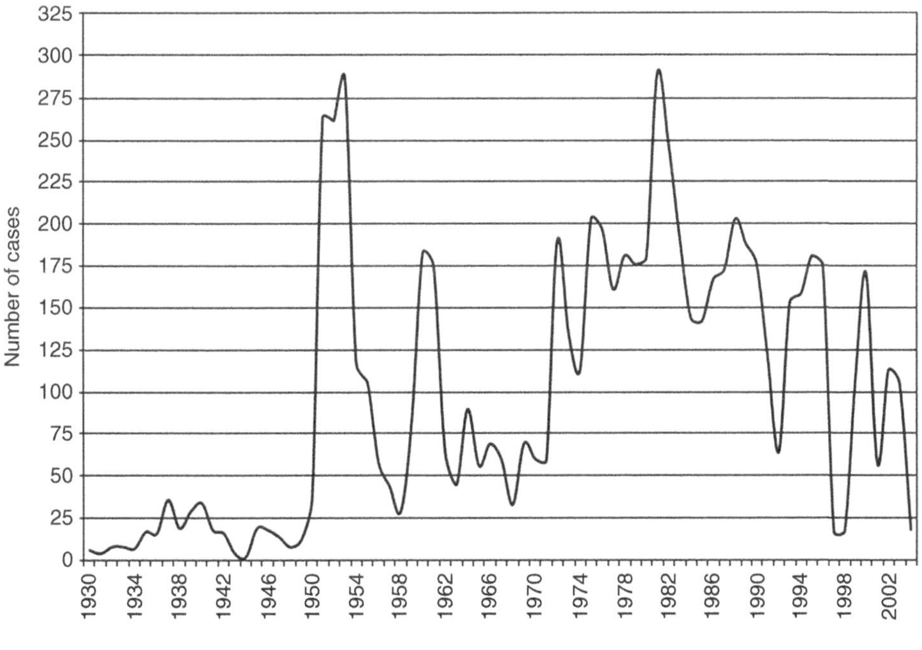

Figure 8.9 Time variation of landslide activity in the Sochi landslide-prone zone (according to the State geological monitoring survey).

They deform slopes up to 15 degrees composed of the Upper Cretaceous-Paleogen terrigeneous flysch.

The *Northwest Caucasus landslide-prone province* is situated within the western termination of the Greater Caucasus orogen. The prevailing type of landslides in this zone are earthflows, while the rotational landslides are less frequent. Landslides affect from 1–3% to 10–12% of the area and, locally, up to 39% of the area in different parts of this province.

The Central-Caucasian landslide-prone province occupying the central part of the Greater Caucasus northern slope is characterized by variable conditions favourable for landslides. Practically all known types of landslides could be identified here. Their origination and evolution vary significantly depending on elevation and geological background. This province includes the Low-mountain, Medium-altitude and High-mountain zones. Within the Low-mountain zone landslides affect from 7–13% to 54–75% of the area. The largest landslides in the Low-mountain zone are in the Baksan River valley, such as the "Zayukovo" rotational landslide up to 160 million m^3 in volume (43.6°N, 43.3°E), 19 million m^3 of which was activated in 2005.

In the Medium-altitude zone landslides affect from 10–13% (in the Northern Jurassic depression) to 67–78% (at the southern cliff of the Rocky ridge) of the area. At some places it reaches 89–100% of the area (at the Lower-Chegem volcanic upland where Q_{3-4} lava flows overlay Albian clay). Large rotational landslides (7.5 million m^3 and more) and compound landslides affect the sub-escarp (sub-cuesta) part of the Rocky ridge. The largest modern compound landslide with total volume more than 400 million m^3 have been identified at the Jissa mountain (43.92° N 41.59°). Some old (likely Pleistocene) landslides were even larges, which can be exemplified by the Karivhoh rock avalanche in the Ardon River valley (42.875°N, 44.2°E) about 2 billion cubic meters in volume that had blocked the river and climbed up to 500 m up on its opposite bank (Strom, 2004).

On the contrary, territory affected by active landslides in the High-mountain landslide-prone zone is up to 4–11% of its area only. Large rotational landslides (ranging from millions to tens of millions m^3) and compound landslides could be found here mainly. The rotational Urup landslide (44°N, 41.24°E) for example, was 80.7 million m^3 in volume.

The landslide-prone province of Northeastern Caucasus embraces the territory of the Chechen and Dagestan Republics. In the foothills earthflows and compound landslides in Jurassic, Cretaceous and Paleogene clay deposits prevail. Volumes of earthflows are up to 2,7–3 million m^3. Active landslides affect up to 40–54% of the entire area. Sometimes it increases up to 72%. Their occurrence is quite variable (Fig. 8.10).

Slope deformations in the so called "Calcareous Dagestan" composed of folded Upper Jurassic – Late Cretaceous sequence seems to be the most hazardous. 4 active landslides are more than 10 million m^3 in volume (the largest one is the "Khunzakh" landslide 50,4 million m^3 in volume – 42.52°N, 46.7°E) and large number of them exceed 1 million m^3.

Regional activation of landslides took place in the Southern part of European Russia in 1936–1937, 1952–1953, 1967, 1988–1990, 2002–2003, and also, apparently, in 1914–1915 (few data are available for the period of the World War 1).

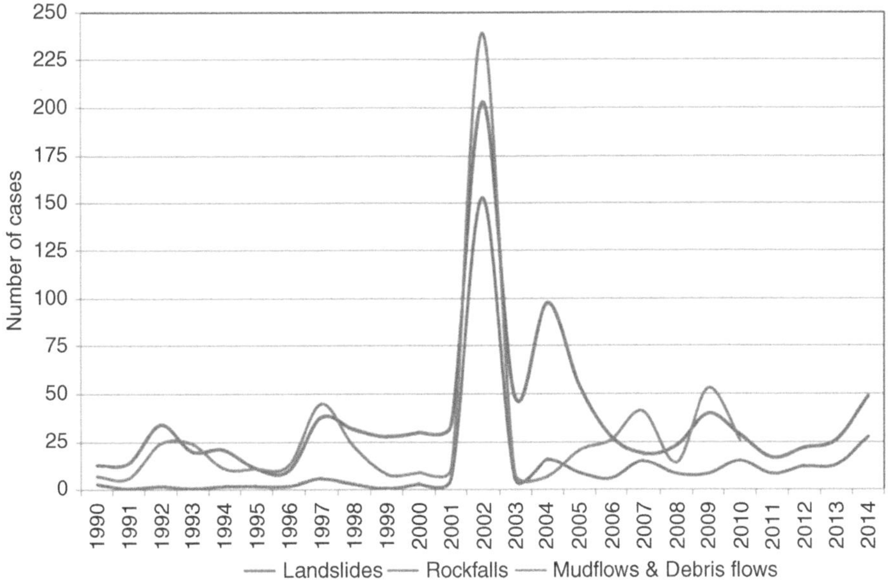

Figure 8.10 Variation of landslide activity within the Mountainous Dagestan (according to the State geological monitoring survey).

In July–August, 2002 the abnormally intensive precipitation caused activation of 457 landslides of the total number of the 615 that have been surveyed. 89 active landslides exceeded 1 million m^3 and 32 of them exceeded 10 million m^3. Besides, 95 new landslides were formed during this period.

In 1944–1945, 1960–1962, 1973–1974, 1980–1982, 1995–1996, 2001, 2004 and 2009 landslide activation occurred within some parts of the region in question (subregional activation). As a rule, landslides have activated annually within some local zones during hazardous period of year.

Activation of landslides of various types depends on meteorological and climatic conditions. The leading factor of regional and subregional activation is the abnormally high precipitation, while for mountain areas it depends on annual temperature change too.

The Effect of Climate Variation on Landslide Activity. The analysis of landslides evolution in the South of the European Part of Russia has allowed to suspect, that, due to global climate change, since the middle of 90th of XX century there is a tendency of essential changing of meteorological and climatic conditions in the region. It should be noted that according to IGECC definition climate change is the statistically representative variability of climate parameters during the long period of time, such as some decades or more.

A great number of temperature and precipitation records from more than 100 meteorological stations have been assembled and analysed to determine if these parameters really influence landslide evolution. After that the comparative statistical data

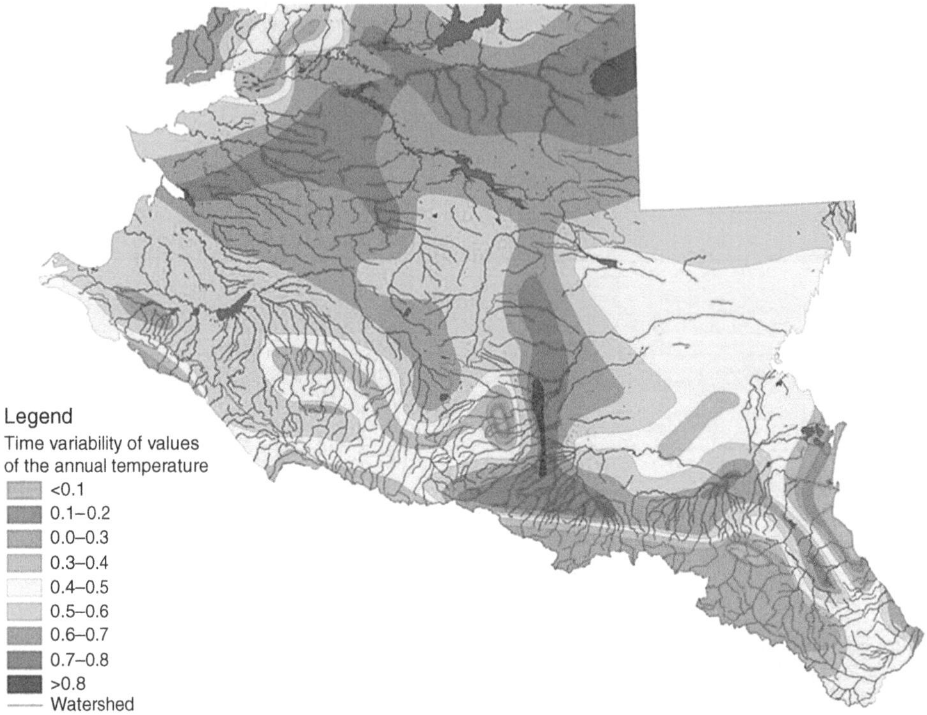

Figure 8.11 Time variability of values of the annual temperature for the15-year period in the Greater Caucasus and Ciscaucasia region.

processing for last two 15-years periods have been done (this is the minimal time interval that was recommended by IGECC for the statistical analysis of meteorological and climatic data). To obtain regional estimates the parameters from the particular meteorological stations were interpolated.

It has been revealed, that mean temperature changed from +0.5°C to +0.8°C for vast flat areas north from the Great Caucasus Range (Fig. 8.11). Time variability of the amount of precipitation – the mid-annual sums and the sums of precipitations during the landslide-prone season – are very different for various parts of the region. That is why the time variability of amount of precipitation analysis was conducted in relative values to obtain a unified representation for the whole region. It was revealed that for the 15-year time period variability of values of the mid-annual sums of precipitation varies from 20% decrease to 30% increase (Fig. 8.12). For the sums of precipitations in landslide-prone season results were very similar (Fig. 8.13).

The changes of the temperature regime recorded over the most of the southern part of European Russia (Fig. NN8), likely will not affect the landslide activity significantly. However, considering that some part of this region belongs to up to 5-km high Greater Caucasus Mountain range, the increase of the mean annual temperature forms the prerequisites to glaciers degradation, activation of the thermokarst phenomena, and

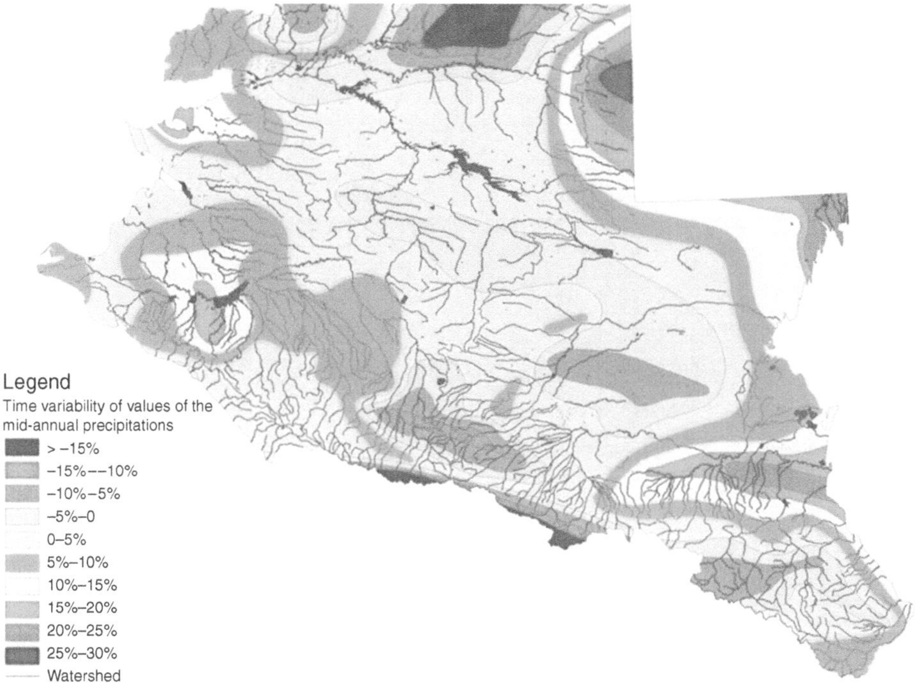

Figure 8.12 Time variability of values of the mean annual sums of precipitation for the15-year period in the Greater Caucasus and Ciscaucasia region.

formation of the new and growth of the existing glacial lakes – the potential sources of GLOFs. This hazard increases significantly with the increase of the precipitation. It should be pointed out that mean annual temperatures became above 0°C at the elevations of about 3000 m a.s.l. (see Fig. 8.14 D,H).

Joint effect of the above factors can be exemplified by the significant activation of debris flows in the Gerhojan-su valley (43.32°N, 42.93°E) associated with the glaciers in its upper reaches. While only 4 debris flows were recoded there during the prior 30-year period (1965–1994), 5 such events occurred in the decade from 1995 to 2004: on 15.07 and 15.08.1995, on 20.08.1999, the multi-stage one on 18–26.07.2000, and on 05.07.2002. The last one occurred during the regional landslides activation. The next debris flow in the upper reaches of this valley took place on 15.07.2008. Thus there is real increase of the geological hazard (and risk) caused by the mean annual temperature growth in the high mountainous part of the region.

Change of the soil moistening regime has been recorded for the most of the European Russia Southern part. At the same time amount of precipitation decreases during the landslide-prone season, which results in the decrease of the overall landslide activity. But for the Northern Caucasus landslide-prone province the significant increase of the area affected by landslides was revealed. The remote sensing analysis showed that in the mid-elevation zone of the Dagestan (North-Eastern Caucasus) where the amount

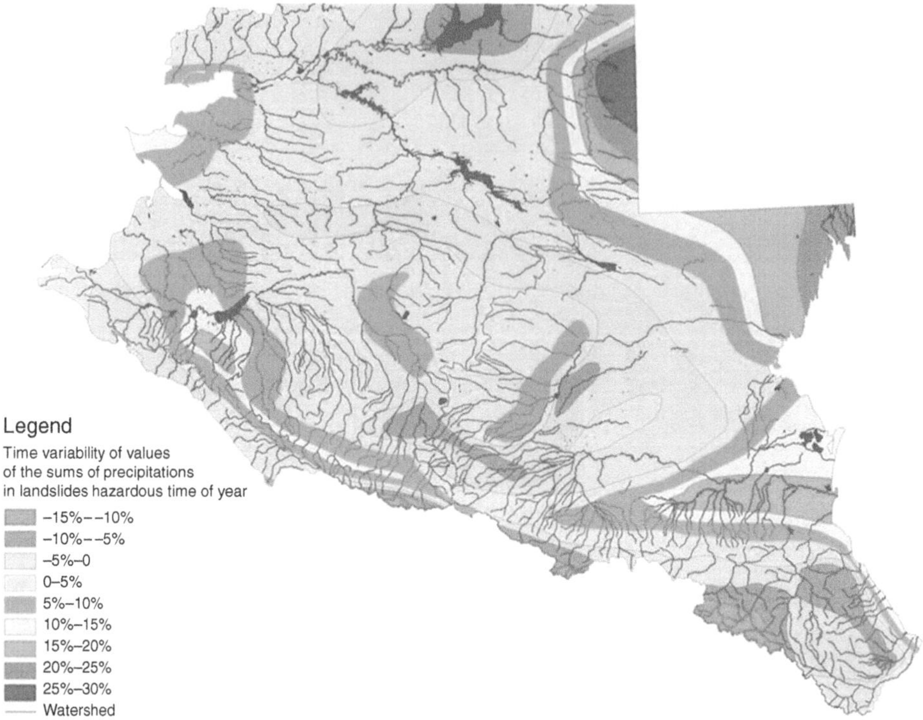

Figure 8.13 Time variability of values of the sums of precipitation in landslide-prone season for the 15-years period in the Greater Caucasus and Ciscaucasia region.

of precipitation during the landslide-prone season increases significantly (Fig. 8.13) area affected by landslides increased up to 1.3 times since its engineering-geological reconnaissance performed in the mid-1980s (Tagirov *et al.*, 1984).

Anomalous rainstorms when up to 2–3 monthly precipitation norms has been released within one day became much more frequent in the subtropical zone (close to the Black Sea). Due to short period of such rainstorms most of water flows down along the erosional gullies, which cause activation of numerous small shallow landslides, as well as debris and mud flows in the foothill zone.

The Crimea landslide-prone province. Data on the Crimean landslides are available since the end of the 18th century when the Kuchuk-Koi landslide more than 50 Mm^3 in volume (44.417°N, 33.91°E) was described and mapped. In February 1786 it activated and moved up to 170 m. After that several abrupt activations of this landslide, which permanent slip rate is up to 20 cm/year, occurred in 1817, 1915, 1925, and 1938.

At present about 1590 landslides were described at the peninsula, more than 570 of which have been formed due to technogenic impacts (slopes overloading, undercutting, artificial watering). Up to 230–260 landslides are active annually on an average. During regional activations more than 30% of landslides move.

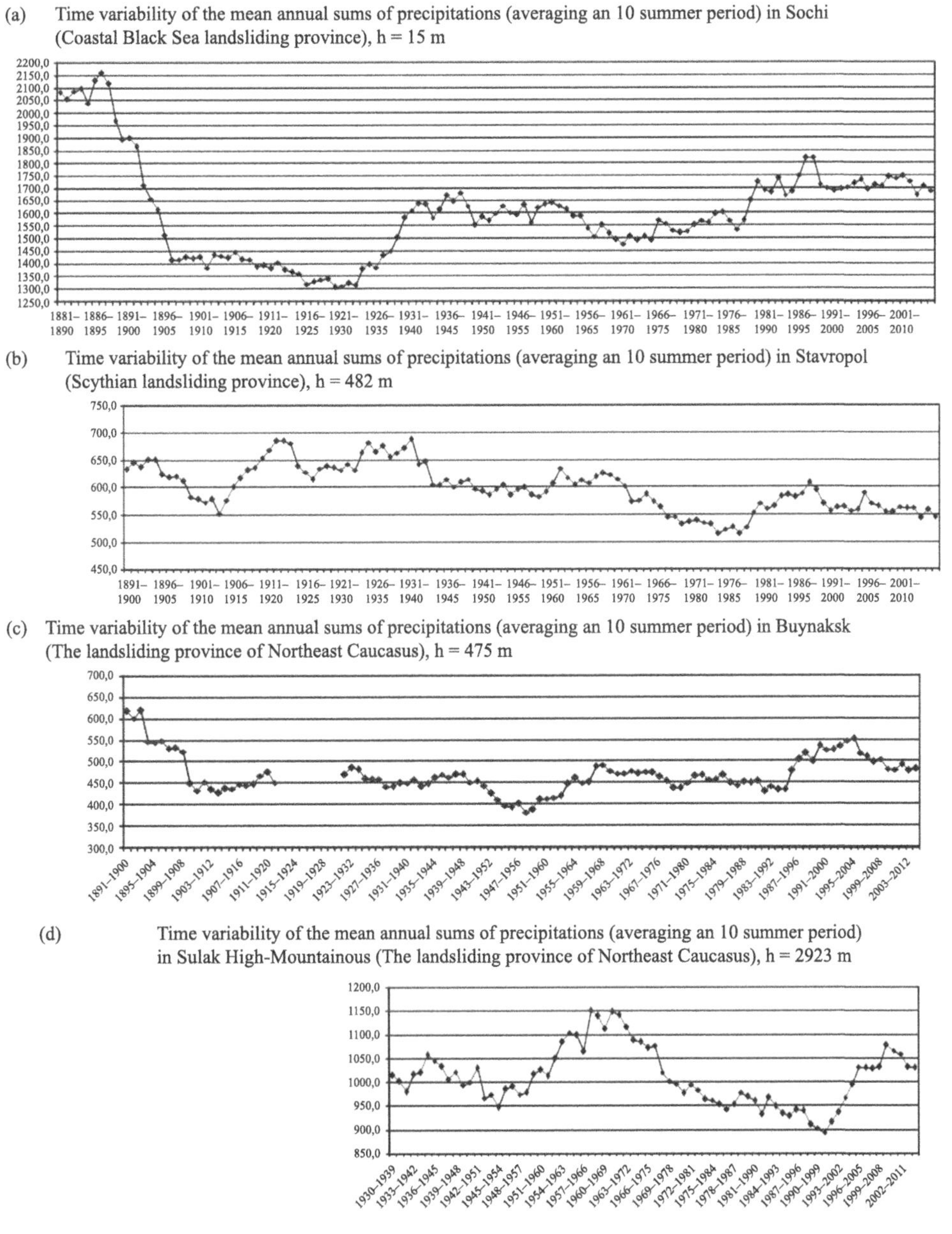

Figure 8.14 Time variability of the climatic factors in the south of the European part of the Russian Federation (After Zerkal *et al* (2006) with additions (the climatic data for the period of 2005–2013) from the European Climate Assessment & Dataset project – http://eca.knmi.nl/и www.ww24.ru were used).

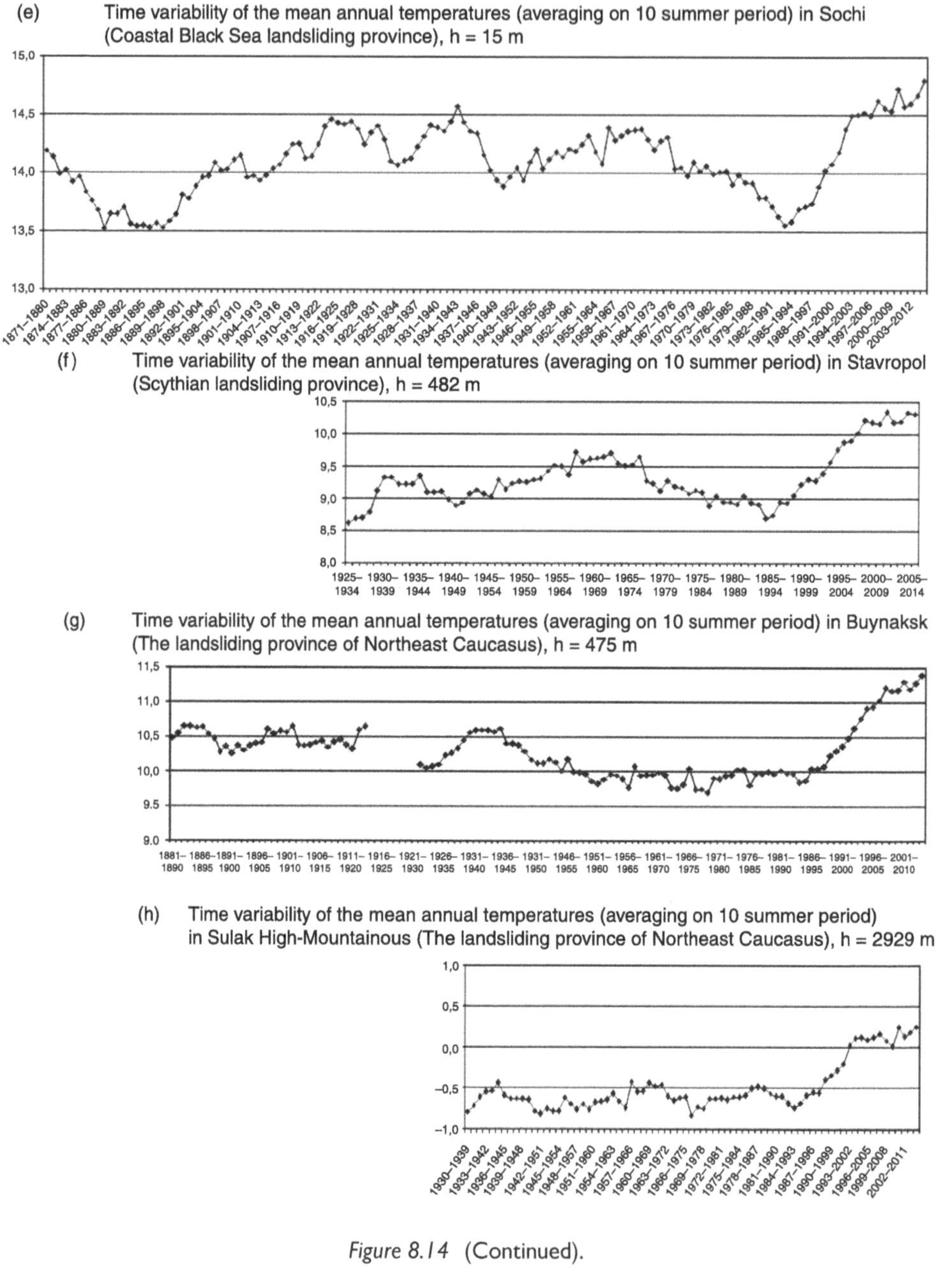

Figure 8.14 (Continued).

The Crimea landslide-prone province includes Western, South-Western, Southern, Central and Eastern (Kerch) landslide-prone subprovinces.

Western subprovince embraces coastal cliffs up to 25–30 m high composed of non-lithified soils and bounding the Northern-Crimea plain with step landscape.

Rotational, translational and compound landslides and earthflows are widely developed here. Most of them do not exceed few thousand cubic meters and their activity is controlled by coastal abrasion.

It is evident that the assumed change of the sea level due to global warming would cause activation of sea abrasion and, consequently, activation of slope processes.

The South-Western landslide-prone subzone is located within the northern lowest chains of the Crimea Mountains. Several ancient large-scale landslides up to 50 Mm^3 in volume have been identified here (Pánek *et al.*, 2008a, b). It was hypothesized that they could be triggered by large prehistoric earthquakes. Rockfalls 17–30 thousand m^3 in volume were triggered by the 1927 M6.6 Crimea earthquake. Modern rotational, translational and compound rockslides and rockfalls up to several ten thousand cubic meters are widely developed on the beach cliffs.

The Southern landslide-prone subzone embraced south-western and southern slopes of the main chain of the Crimea mountains up to 1500 m high. Its crest is at a distance of about 10 km from the sea shore. This area lies within the subtropical (Mediterranean) climatic zone with mean annual sums of precipitation varying from 400–600 mm at the coastal part to 800–1000 mm in the mountainous part, and with peak values reaching 700–1100 mm and 1100–1800 mm respectively. Maximal daily precipitation can reach 90–210 mm.

About 1100 landslides of practically all known types have been identified within this landslide-prone region and nearly 10% of them are active annually (mean estimate). During activation periods this number increases up to 30–35%. According to Erysh *et al.* (1999) 47% of landslides activated in 1998. 10 of the permanently active landslides are more than 10 Mm^3 in volume. Intensive formation of large landslides took place in the beginning of 20th century being caused by intensive watering of the slopes by precipitation. During the 1905–1915 decade annual sums of precipitation exceeded mean 4-year long summer period values 1,3–1,6 times twice.

At present, according to the instrumental monitoring data mean annual displacement rates of large landslides are of the order of few cm/year. However, during activation periods associated with abnormally high precipitation mainly it increases up to meters per hour and landslides move up to tens meters per event (in 1938 the above mentioned Kuchuk-Koi landslide moved 120 m). Besides, debris flows up to 70–100 thousands m^3 occur in high-water years in some gullies.

The largest Holocene rockslide up to 30 Mm^3 in volume had occurred on the southern slope of the Demerdji Mountain (44.75°N, 34.41°E). Largest (up to 300 thousand m^3) historical rockfall occurred there in April of the high-water 1894. Large (up to 25 thousand m^3) rockfalls were recoded in 1966 and 1988. During 1977–2004 62 rockfalls exceeding 1000 m^3 occurred on the southern side of the Crimea Mountains Main chain (Erysh *et al.*, 1999). Most of them – up to 4 in one year – took place in the intensive moistening years (1981–1982, 1987, 1997) when the prolonged spring periods were characterized by frequent temperature variations from negative to positive and vice versa, which support freezing and melting of water in fractures and their widening.

Several rock avalanches have been identified in the Southern landslide-prone region too, but none of them are historical. The preliminary data allow assumption of their pre-Holocene age.

Analysis of possible influence of climate on landslide activity, of the precipitation amount and intensity at first, shows that the recorded maximums of landslide activity in 1969, 1982, 1988, 1998, and 2008 occurred with one year lag after the periods when precipitation exceeded mean annual values for 20% and more. However, the debris flow and rockfall activations coincide with high moistening periods. At present, along with mean annual temperature variations, the annual precipitation sums decrease, which likely, should lead to decrease of landslide activity. At the same time, considering increase of the abnormal climatic events, some local landslide activations could occur.

The Central landslide-prone subzone is located on the northern slopes of the Main (Southern) Crimea Mountain chain and on the lower chains northward from it. This area is characterized by step landscape with some deficit of moistening. Rather small rotational, translational and compound rockslides and rockfalls occur in the river canyons cross-cutting these chains and on the southern cliffs of the cuesta mountains. Slides in soil (rotation, translation, compound slides, and earthflows) are distributed locally on the slopes composed of the Neogene-Paleogene clay and Quaternary loess. Their peak activity is associated with intensive rainfalls like in spring, 2015. It can be assumed that climate change could have major effect on landslide hazard just within the Central landslide-prone region.

The Kerch (Eastern) landslide-prone subzone is located within the Kerch Peninsula that belongs to the moistening-deficit part of the medium climate zone. About 250 landslides have been mapped along the sea-shore, 110 of them – at the Kerch Strait bank. Rotational, translation and compound slides and earthflows up to hundreds thousands cubic meters occur here in soils forming beach cliffs. Their activity is controlled, at first, by beach abrasion, and, thus, possible change of sea level that could be caused by climate change would have major effect on landslide hazard.

8.4.1.2 Central region of the European part of Russia

This region is characterized by the uneven landslides distribution. The largest and most active landslides are concentrated in the valleys of large rivers – the Volga, the Oka, the Sura (the latter two are Volga left tributaries), the Kama (Volga right tributary) and others (including the Moskva River valley) with up to 150–200 m high western and southern banks. Most of these valleys have been developed for a geologically long time and in some periods, when the Caspian Sea level was much lower than at present, valleys formed rather deep canyons. According to Goretskiy (1964), at some localities east from the Samara city the buried Upper Pliocene valley bottom was found out by drilling at 300 m bellow sea level. Western banks of these paleovalleys had been affected by numerous large-scale landslides. Construction of the hydraulic power plants and infilling of their reservoirs (Gorkiy and Kuibyshev in 1955–1957, Volgograd in 1958–1961, Saratov in 1967–1968, Lower Kama in 1979 and Cheboksary in 1981) affected local climatic conditions and caused reservoir banks reworking that led to activation of some old landslides. The Nyzhniy Novgorod – Vasilsursk, the Middle-Volga, the Saratov and the Volgograd landslide-prone provinces can be identified in this region (Fig. 8.15).

The *Nyzhniy Novgorod – Vasilsursk landslide-prone province* embraces the downstream part of the Oka River valley and the right bank of the Volga Rover between

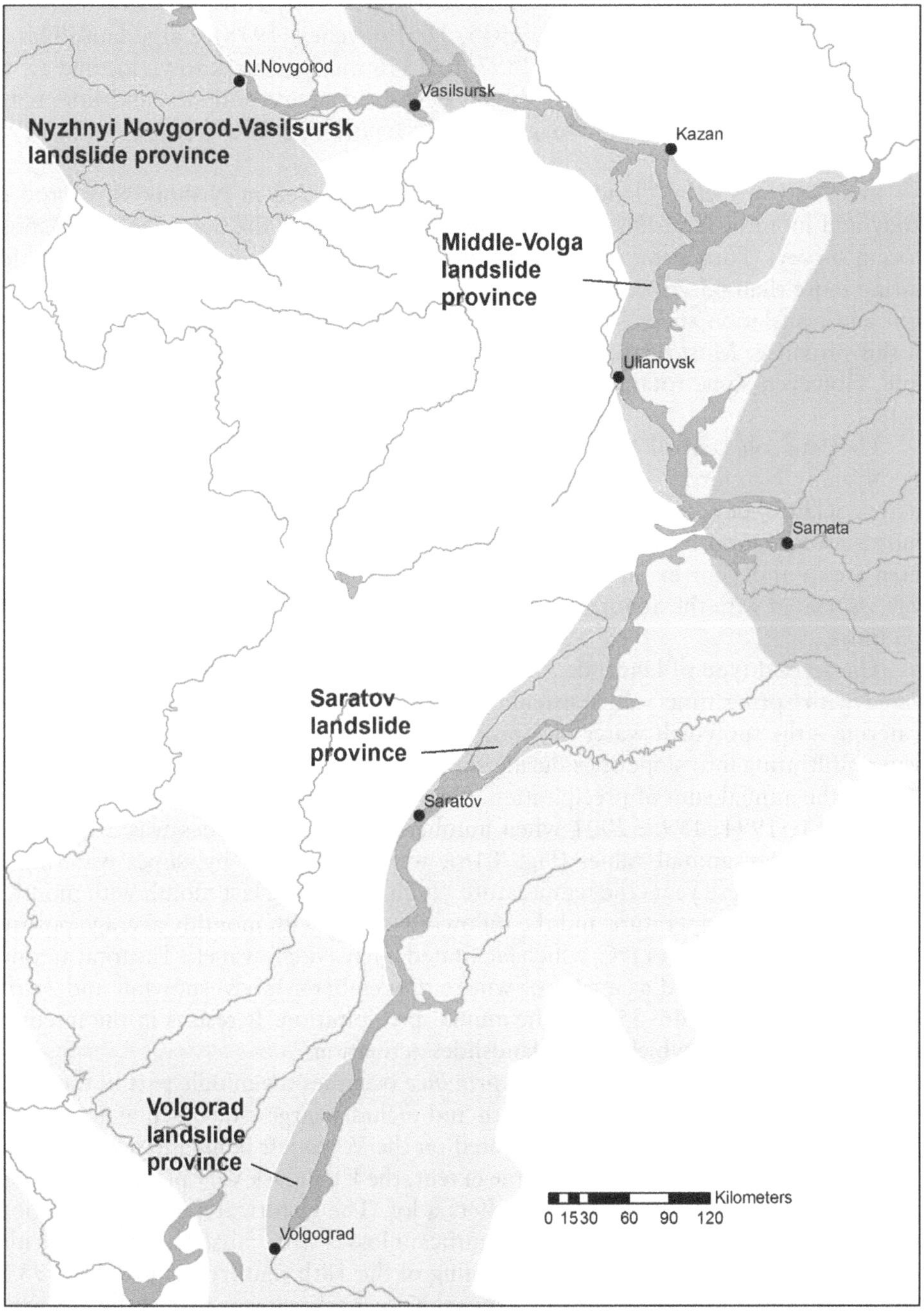

Figure 8.15 Landslide-prone provinces in the Central Region of European Russia.

the Oka and Sura River mouths. The first historical data about landslides in this area are from the 1329 A.D. when large landslide in the Nishniy Novgorod City destroyed about 150 homesteads (Landslides, 1935, Haritonychev, 1978). Large landslides in the City occurred also in 1447 and 1597. In 1556 the Vasilsursk town located at the junction of the Sura and Volga Rivers was destroyed significantly by landslide. Later on large landslides in Nishniy Novgorod were reported in 1807, 1833, 1845, 1892, and in 1899.

In 1946 the special "landslide station" was organized in Nishniy Novgorod to study and monitor landslide processes (now it is a part of the State geological monitoring survey). Formation of 529 landslides within this province has been recorded during more than 65 years long observation period (Information Bulletin). The rotation and translation slides in soil, compound slides and earthflows are most typical of this province. Most of them are up to few tens of thousands cubic meters in volume. However, some rotation and compound slides are up to several millions cubic meters.

The data collected allow revealing effect of climatic factors on landslide activity in the Nyzhniy Novgorod – Vasilsursk landslide-prone province. The rapidly changing factors of landslide activation follow the annual cycle. Two types of landslide activation due to precipitation have been identified. The first – the "classical" one – is due to intensive rainfall. For example, significant activation of landslides recorded in 1974 was associated with the abnormal rainfalls in May that exceeded mean monthly value 2.3 times.

The second type of landslide activation, most common in the study area is associated with spring time soil moistening. Two components affect increase of slopes watering – the snowmelt water infiltration and spring rainfalls. The total amount of water infiltrating into slope soils during short period of time (0,5–1 month) can reach 50% of the annual sum of precipitation. Significant landslide activations recorded in 1966, 1981, 1994, 1995, 2001 when number of active landslides was 2–2,3 times more than mean annual values (Fig. 8.16), were caused just by slopes watering in spring period. These years the temperature "jump" from the last month with monthly average negative temperature and the following month with monthly average positive temperature was up to 9–11°C, which facilitated intensive snow melt. The total amount of precipitation counted as a sum of winter (December–March) snowfall and April–May rainfall reached 46–55% of the annual precipitation. It results in the intensive moistening of slopes, which causes landslides activation.

The *Middle-Volga landslide-prone province* occupies the middle part of the Volga River valley near the Kama River mouth and includes large cities such as Kazan and Ulianovsk. While the Kazan City is located on the Volga left bank and have not been affected by landslide processes at a large extent, the Ulianovsk City placed on the high right bank of the Volga River valley suffers a lot. The historical data available, both published and unpublished, show that significant losses were caused here by landslides starting from the end of the 17th – beginning of the 18th century (Landslides, 1935). Largest landslides were recorded in Ulianovsk City (earlier it was named the Simbirsk) in 1724, 1743, 1785, 1866 and 1868. Ten large destructive landslides occurred here from 1877 to 1902. Later on large landslides occurred in 1913 and 1914, and the catastrophic one with volume of about 3 million m^3 in May, 1915. Significant losses were caused by landslides in 1927, 1945, 1955, 1957 and 1979.

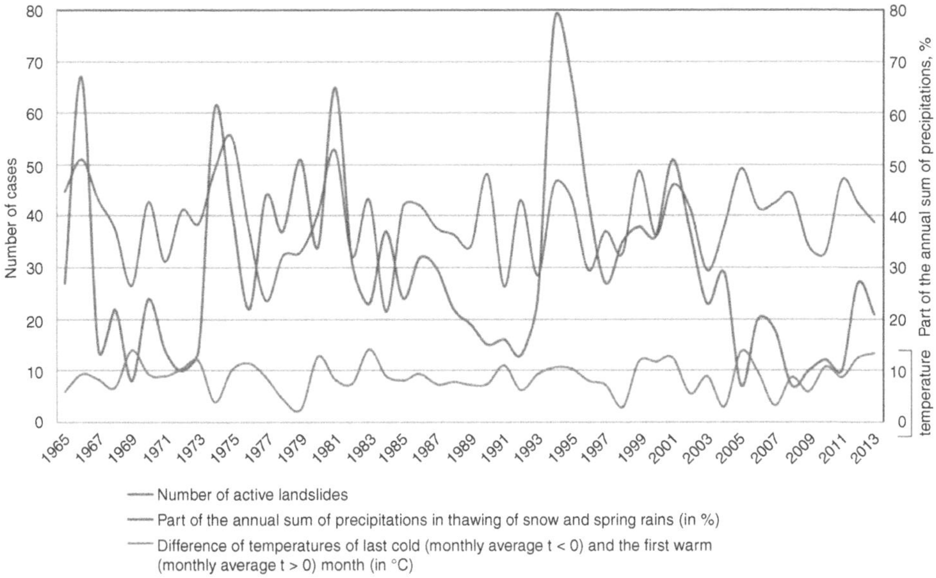

Figure 8.16 Variation of activity of landslides and time variability of values of the climatic factors in the Nyzhi Novgorod area.

At present more than 960 rotation and translation slides in soils, the compound slides and earthflows of different volumes have been identified on the high right bank of the Volga River in the Ulianovsk and the adjacent areas. Every year from 40–42% to 64–66% of the landslides monitored along the banks of the Kuibyshev (Samara) Reservoir are active with the mean partial factor of activity of 58%. Here the anthropogenic factor – i.e. fluctuation of the reservoir level plays the critical role. Influence of the climatic factors just increase or decrease the anthropogenic effect.

The *Saratov landslide-prone province* is located downstream from the Middle-Volga one. The first descriptions of landslides in the Saratov City are from 1783. Large-scale movements were recorded in 1811, 1818, 1846, 1869, 1884 (the catastrophic one that damaged about 300 houses), 1913–1915, and 1927. Study of landslides in the Saratov City started in 1895 in the course of the engineering-geological investigations for the railroad bridge across the Volga river. Large displacements were reported in 1905, 1935–36 and 1963–68. About 260 landslides of various size have been identified within this province. Among them 36 landslides ranging from 50000 m^3 to more than 10 Mm^3 are in the Saratov City (Information Bulletin).

The climatic factors affecting landslide activity in the Saratov province are similar for those, identified in the Middle-Volga province – the abnormal rainfalls and the spring snow melt combined by rainfalls. There are, however, some differences. The threshold temperature jump between last month with mean negative and the first month with mean positive temperature reaches here 12–14°C and even more.

The *Volgograd landslide-prone province* is located within the lower reaches of the Volga River valley. At present about 110 rotation and compound soil slides and

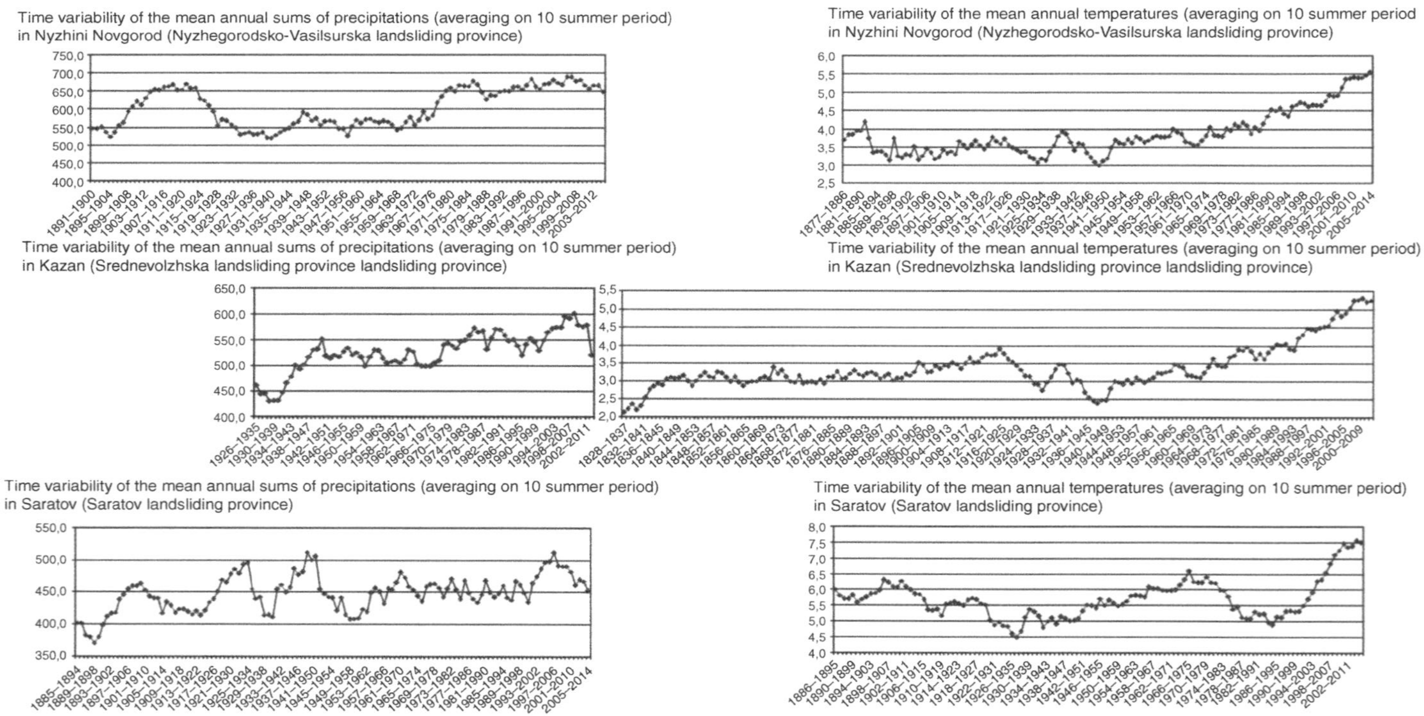

Figure 8.17 Time variability of values of the climatic factors in the center of the European part of the Russian Federation (according to the European Climate Assessment & Dataset project – http://eca.knmi.nl/и www.ww24.ru).

earthflows have been identified in this province, in the Volgograd City mainly. The main triggering factor is the variation of the Volgograd reservoir level, which is just supplemented by climatic factors.

8.4.1.3 The Southern Siberian region

This region is located within the engineering-geological zone of thaw and unfrozen soils, in the medium continental climatic zone and is characterized by the uneven landslides distribution. This region is composed of smooth uplands cut by large river valleys such as the Ob and the Enesei Rivers, and their tributaries. Data on landslides distribution here are incomplete. Slope processes have been studied with reasonable level of detail just in few landslide-prone regions – in the Barnaul, Tomsk and Krasnoyarsk Cities.

Landslides in the Barnaul landslide-prone area are developed on the high left bank of the Ob River predominantly within the Ob loess plateau where more than 360 rotational and compound landslides in soils and earthflows have been identified. Most of them do not exceed several tens of thousands cubic meters in volume.

Data available on landslide activity in this area allow assumption on the prevailing role of the climatic factors in the slope processes activation (Fig. 8.18). It can be seen that landslide activation is in direct correlation with rainfall intensity. All recorded periods of higher landslide activity in 1979, 1990, 1992–1994, 2000–2002 have been associated with intensive moistening caused by atmospheric precipitation.

Analysis of climate change for the Barnaul landslide-prone area showed that gradual rise of the annual mean temperatures has been recoded here since 30th of the XX Century, i.e. during the entire observational period. In parallel the decrease of the annual mean precipitation has been recorded, which results in the decrease of the overall landslide activity. Thus, the tendencies of landslides evolution in this region are in line with the long-term climate change.

About 70 modern rotational and compound landslides ranging in volume from several tens thousands up to few million cubic meters have been identified in the Tomsk landslide-prone region within the Tom River valley. They affect the 12-km long section of the high IV terrace of the Tom River just in the Tomsk city and 6–12 of them activate annually.

Study of the climate variations typical of the Tomsk region showed that significant change of the climatic conditions occurred at the end of the 1970s–beginning of the 1980s. During this period mean annual temperatures have changed from the negative to stable positive values. And just this period was characterized by sharp landslides activation. For example, the mean annual retreat rate of the "Camp garden" landslide headscarp (56.45°N, 84.95°E) was 0,64 m/year in 1968–1980, 5,5 m/year in 1980–1983, and 10 m/year in 1983–1987 (Zolotariov *et al.*, 1994). However, the present-day landslide activity in this region have decreased and headscarp retreat rate ranges from few dozen centimeters to 2 meters per year. Effect of the climatic factors (precipitation amount), which remains more or less stable, just intensifies or damps the technogenic influence.

About 300 active landslides were identified in the Krasnoyarsk landslide-prone region. Most of them are in the valleys of the Enesei River and its tributaries.

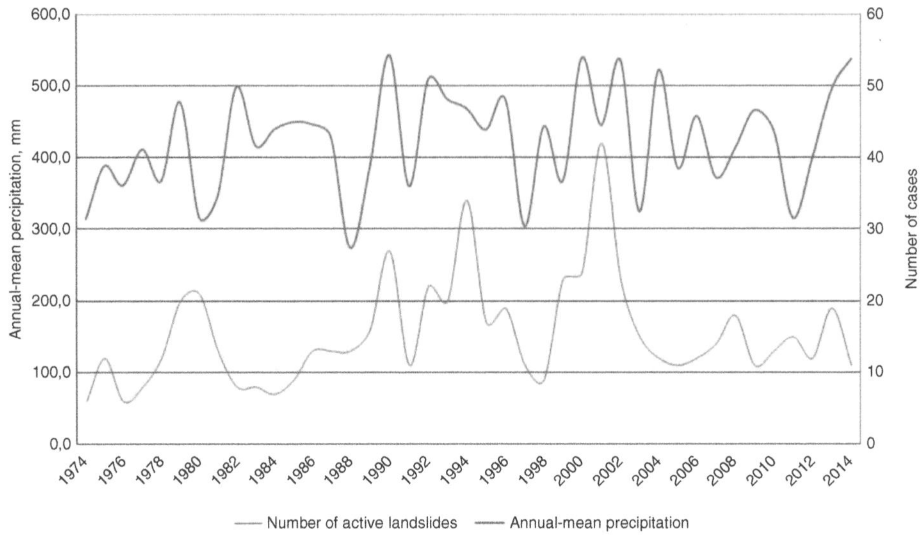

Figure 8.18 Variation of activity of landslides and time variability of values of the climatic factors in Barnaul area.

The *Sakhalin Island* – one more landslide-prone region of Russia – is located east from the continental part of Russian Far East. It is rather narrow (up to 135 km wide), but stretches from South to North for 948 km. The southern part of the Island is occupied by low and medium high (up to 1609 m a.s.l.) mountains mainly; the northern one is more flat and low. The climate of its southern part is medium monsoon and belongs to the Pacific sector of atmospheric circulation. Most of the Island, except for the Shmidt Peninsula at the extreme north, is within the engineering-geological zone of the thawed and unfrozen rocks.

The rotational, translational and compound landslides in soils, earthflows and mudflows are widely developed in the central and southern parts of the island mainly. Most of them are up to few thousand cubic meters in volume. However, rotational rockslides up to 5 Mm^3 in volume have been identified (Lobkina, 2006). In some parts of the island up to 30% of the entire area is affected by landslides.

Activation of the existing landslides and formation of new slides in the southern part of the Sakhalin Island is caused by typhoons passing over the island and accompanied by the extreme rainfalls – up to 2–3 monthly rainfall in few days. The recorded maximal rainfall during one of typhoons was 1277 mm (while mean annual sum of rainfall for the Yuzno-Sakhalinsk area is about 900 mm/y). Massive formation of Sakhalin landslides took place in 1875, 1928, 1954, 1964, 1970, 1978, 1981, 1993, 1996, 2002, 2009 (Lobkina, 2006). The most catastrophic activation (more than 500 landslides and debris flows) was associated with the typhoon Fillies in August, 1981. We must mention that landslide activation associated by the 2002 typhoon Chataan and with the series of cyclones in 2009 was much lower than in 1981.

The assumed climate changes that would be accompanied, in particular, by the increase of the anomalous meteorological phenomena, including increase of Pacific typhoons frequency and intensity will, obviously, increase landslide hazard in Russian Far East located within the medium monsoon climatic zone with the Pacific type of atmospheric circulation.

8.4.2 The permafrost zone

The widely developed permafrost is typical of Arctic desert, polar, wooded and mountain tundras. The sectoral climatic distribution and vertical differentiation play the important role in the location of the continuous permafrost southern margin, which is about 52°–54°N within the Siberian atmospheric circulation sector and shifts up to 65°–66°N (for more than 1000 km) within the Atlantic sector (see Fig. 8.5).

No uniform classification of the cryogenic slope processes that would consider all aspects of these variable and complicated phenomena exist up to now. One of classifications considering effect of ice was proposed by Sharpe (1938) who distinguished "rock glacier cap", "solifluction" and "debris avalanche". Most widely known classifications of soil landslides in permafrost zone in Russia were proposed by Kaplina (1965), Garagulia (1983), Bolikhovskiy & Kiuntsel (1990), and by Leibman (2007). Classifications of slope processes in rocks in the permafrost zone were elaborated for rock streams and rock seas by Tiurin (1979) and for rock glaciers by Galanin (2009). General classification of landslides and related phenomena typical of the permafrost zone is presented in Table 8.1.

Within the continuous permafrost zone the subzone with seasonal rock and soil thawing can be identified (Trofimov, 2002). Intensity and variability of slope processes within this subzone, which southern boundary is also determined by the atmosphere circulation, is depressed significantly in comparison with more southern regions. No data on the present activity of rockslides are available and it is likely that the assumed climate change will not affect the frozen state of rock slopes significantly.

However, in the areas within this subzone composed of frozen soils the temperature variation intensifying soil moistening caused by ice melting would play a leading role in the increase of landslides activity. Main processes affecting slopes stability are the thermo-erosion and thermo-abrasion resulting in krio-earth slides and retrogressive thaw slumps. Ice soils widely distributed in this subzone precondition intensive degradation of its ice component in the coastal zone, being intensified by global warming.

Mean annual rates of the shore retreat due to slope deformations caused by thermal erosion are 2–3 m/year for the Barents and Karsk sea shores, from 1,9–2,0 to 4–6 m/year for the Laptev sea shore (up to 6,8 m/year on average with maximum more than 20 m/year on the islands) and from 1,6 to 5–7 m/year for the Eastern-Siberian sea shore (Zerkal, 2013; Zhelezniak, 2014). At the same time on sea banks where thermal-gravitational processes are not developed the shore retreat rate does not exceed 0,3 m/year (Zhelezniak, 2014). The total portion of the Arctic seas shoreline (excluding the Chukotka Sea with different geological conditions) affected by the permafrost-related landslides and related phenomena reaches 38% (Kalinina, 1992).

Table 8.1 The classification scheme of landslides and related phenomena in the permafrost zone.

		Material Type									
		Frozen							*Seasonal thawing*		
				Frozen Debris & Soil							
		Frozen Rock	*Soil-ice mixture*	*Quasi-solid soil*	*Frozen soil*	*Plastic frozen soil*	*Loosely frozen soil*	*Cooled soil*	*Rock*	*Debris*	*Earth*
Movement Type	Creep						Kurum		Do not differ from the "classical" types slope processes		
	Fall	Rockfall and Ice-Rockfall									
	Topple	Not									
	Slide	Rockslide and Ice-Rockslide*								Fast Kurum**	Krio-Earth slide***
	Spread						Rock sea	Cooled earth spread			Active-layer detachments spread
	Flow	Rock glacier	Ice-Rock glacier		Ice-Earth glacier	Ice-Earth flow	Rock stream				Slow and Fast solifluction earth flow****
	Complex	Ice-Rock and Ice-frozen soil avalanche								Ice-Debris avalanche	

*Including translational and rotational slides
**motion of debris (including pieces of ice) over the surface of frozen rock/soil and (or) ice (boundary of the frozen and thaw material)
***predominantly block (translational) motion of the seasonal thawed soil over the surface of ice and/or ice soils (boundary of the frozen and thaw material) including retrogressive thaw slump. All krio-earth slides are shallow since their thickness is limited by the seasonal thaw thickness.
****flow of the water-saturated thawed soil (including daily thawed) over the boundary between the thawed and frozen soils, including solifluction lobes, formation of thermal circuses and thermo-erosional cauldrons with "laminae" earthflows.

Rate of the destruction of the 6–40 to 100–120 m high sea shore cliffs of the Laptev and Eastern Siberian seas caused by thermal-erosional and thermal-gravitational phenomena varies in different periods due to climate change (Grigoriev *et al.*, 2006):

- in the 40s–50s of the 20th century (period of relative warming with 1,5–1,6°C increase of mean annual temperature at Diksom meteo station) shore retreat rate was from 1–1,5 to 5–6 m/year and up to 11–15 m/year at some sections at the Kolyma River mouth and up to 30–55 m/year at the shore line of the Novosibirsk islands.
- in the 60s–70s of the 20th century (period of relative cooling – 2,5–2,8°C decrease of mean annual temperature at Diksom meteo station) shore retreat rate was 2–4 m/year for north-facing cliffs, 0.5–6 m/year for south and east-facing cliffs and 3–7 m/year for west-facing cliffs. At the Kolyma River mouth it was 1.4 m/year on an average;
- starting from the 1990s (next period of relative warming – up to 3°C increase of mean annual temperature at Diksom meteo station) shore retreat rate increased up to 4–5 and 2–7,5 m/year for north-facing and east-facing cliffs correspondingly and decreased up to 1,8–4,8 m/year for west-facing cliffs. At the Kolyma River mouth this rate increased up to 5.0–9,8 m/year for 0,7–15 m high cliffs and up to 2,4–4,8 m/year for 20–35 m high cliffs. Some years maximal rate was up to 15–23 m/year.

Further dynamics of Arctic sea shores retreat will depend on the character of climate change and should increase with relative warming and decrease with relative cooling.

In the large river valleys of the Eastern Siberia (Lena, Indigirka, Yana, Kolyma Rivers and their tributaries) mean annual rate of their banks retreat due to slope processes caused by thermal erosion can reach several tens of meters. In the intra-continental part of the region besides the krio-earth slides the active-layer detachment spreads at bank edges and also on the slopes with angles less than 5° often take place. At slopes inclined more than 5–7°, where thickness of the seasonal thaw layer reaches 0,4–1,0 m slow solifluction earthflows (plastic and visco-plastic flows moving from 8–10 to 30 cm/year) and fast solifluction earthflows (plastic and visco-plasic flows moving several m/day and even faster) are typical (Garagulia, 1983; Zerkal, 2013; Zhelezniak, 2014). According to Garagulia (1983) slopes in the continuous permafrost zone are divided into low hazardous (5–7°), hazardous (7–15°) and very hazardous (>15°), depending on the extent of solifluction development.

Slow solifluction earthflows and fast solifluction earthflows form the specific microforms – tongue-shape local "terraces" several tens of meters wide and up to 2 m high named by Leibman and Kiziakov (2007) the solifluction lobes.

At tundra highlands and low mountains the solifluction earthflows can affect up to 16% of the entire area (Leibman & Kiziakov, 2007). Southward, in the southern tundra zone with more developed vegetation such slope processes are less typical and related geomorphic features are, likely, the relicts.

Evolution of the solifluction earthflows is the seasonal process that takes place in the warm period of the year and depends on maximal mean monthly temperature on the soil surface, mean annual temperatures and on annual daylight surface temperature variations. It leads to the solifluction earthflows thickening as well as

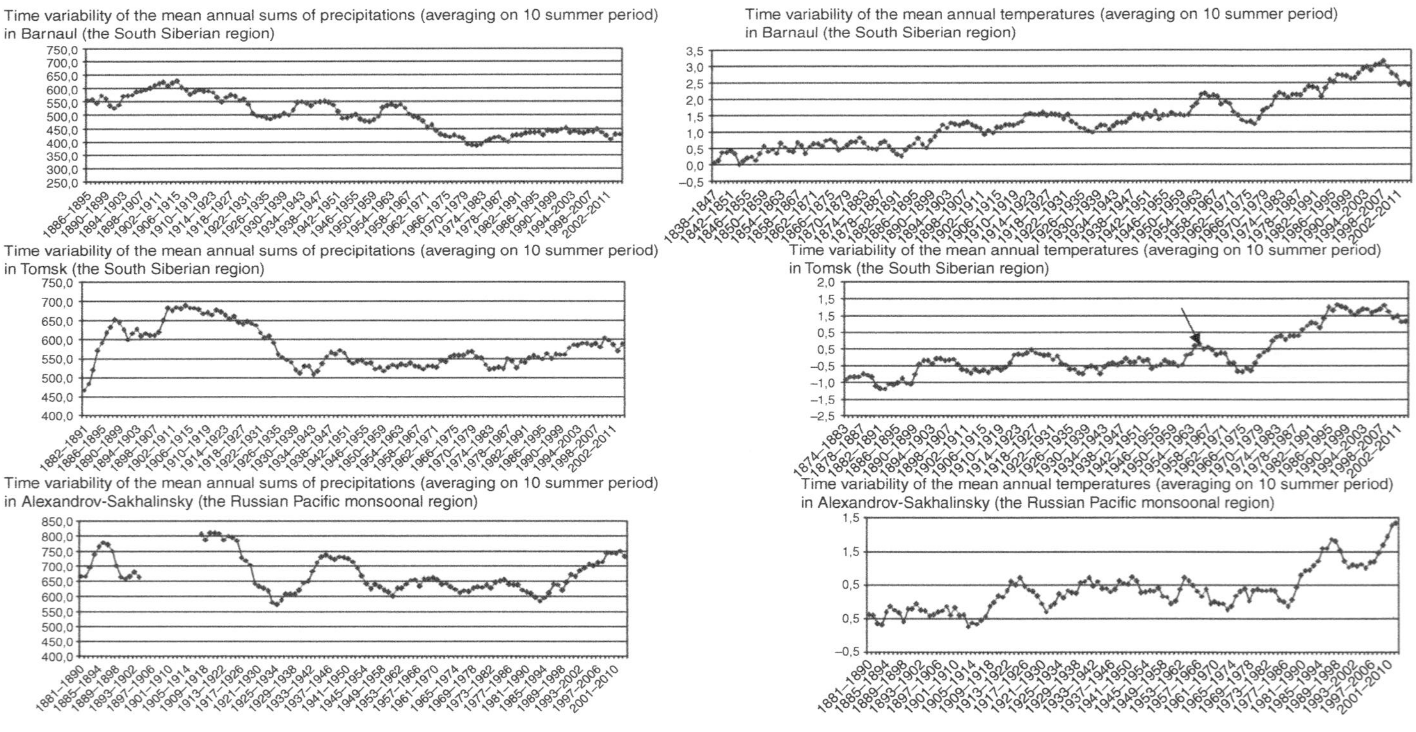

Figure 8.19 Time variability of values of the climatic factors (averaging on 10 summer period for precipitation and temperature) in South Siberian and Russian Pacific monsoonal regions (according to the European Climate Assessment & Dataset project – http://eca.knmi.nl/и www.ww24.ru).

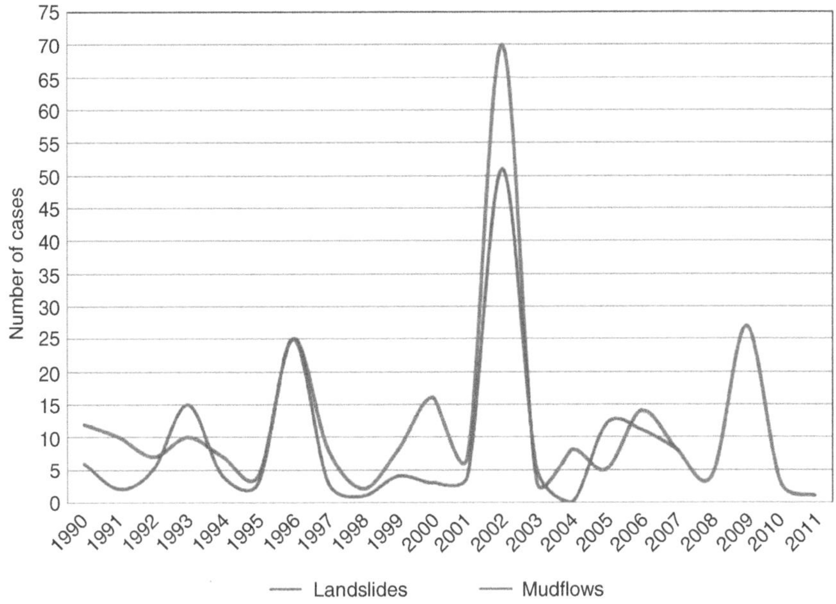

Figure 8.20 Variation of activity of landslides in time within the Southern Sakhalin region (the Makarov district, Sakhalin Island) (by data of State system of geological monitoring [Information bulletins].

to their formation at the new sites. As for the effect of climate change we can note (following Garagulia, 1983) that possible increase of maximal mean monthly and mean annual temperatures for 1°C will increase the thickness of the seasonal thaw layer for 0,2–0,25 m if annual variation of soil surface temperature is up to 7,5°C (maritime type of climate) and for 0,15–0,2 m if temperature variation is more than 21°C (sharp continental climate). It can be assumed that the most intense variation of the cryogenic slope flows (slow and fast solifluction) would occur in the areas with continental and sharp continental climate (the Siberian atmospheric circulation type – see Fig. 8.5).

For the rugged terrain within the seasonal thawing subzone the kurum formation is typical. The "laminae" kurums – thin (up to 30–40 cm) blanket of loose frozen soil at the elevations of 50–160 m a.s.l. moving downslope are typical of the Vrangel, Novaya Zemlia and Severnaya Zemlia islands (Ershov, 2001).

Ice and rock-ice avalanches and rock glaciers are typical of the Arctic and sub-Arctic mountains and for the highest parts of the Central Caucasus and Altai mountains where the climatic conditions are similar to the seasonal thawing permafrost subzone.

About 6500 rock glaciers of different types (slope blade-like, tongue-like, complex, according to Galanin (2009)) have been identified in the North-Eastern Asia. Areas of their present day formation are characterized by mean annual precipitation exceeding 500 mm/year (in the Eastern Siberia within the Pacific atmospheric circulation type). No rock glaciers are known in the areas with mean annual precipitation less than 300 mm/year.

Intensive physical weathering in high mountains nival zone predetermines wide distribution of rockfall and scree formation. Besides, ice and rock-ice avalanches occur in high mountains characterized by high potential energy of the relief.

Kurums up to 1 m thick are formed also in the subzone of sporadic permafrost in mountainous regions above 400–450 m a.s.l. in the north and above 700 m a.s.l. in the south of the subzone. Their thickness increases southward (Dynamic, 2001). Intense weathering typical of the mountainous parts of this subzone stipulates rockfall and scree formation. This zone is also the source of the glacial debris flows and mudflows.

In the areas characterized by very sporadic permafrost occurrence intensity of landslide processes increases due to damping of the cryogenic influence, while the surficial rocks are still heavily disintegrated by the seasonal freezing-thaw processes and increasing role of chemical weathering. In the plains within this subzone shallow blockslides became more frequent, which can be exemplified by the Ob River valley near the Hanty-Mansyisk City.

Kurum formation in this subzone was reported at the elevations from 1100–1200 m a.s.l. in the west to 1200–1300 m a.s.l. in the east (the bald mountain level of the Stanovoi Range). Southward this level raises up to 1600–2300 m a.s.l. (the Kusnetsk Alatau Mountains, South-western Transbaikalia) and even up to 3500 m a.s.l. (Tyva region). Kurums affect up to 70% of the area like as in the Hamar-Daban Range (Dynamic, 2001). Piping of fines in the areas with rugged terrain facilitates formation of this coarse debris carapace, Thickness of kurums in Southern Yakutia and northern Transbaikalia increases up to 2.0–2.5 m and the solifluction is replaced gradually be typical flowslides. High mountains with the very sporadic permafrost are also the source zone of the debris flows and mud flows.

8.5 CONCLUSIONS

Significant variability of the geomorphic, geological and climatic conditions typical of the vast territory of Russian Federation predetermines variability of landslides processes associated with climate change. The present day global warming results in quite variable changes of landslide activity and, thus, landslide-related hazard. Depending on the specific set of conditions, climate change results either in increase or in decrease of landslide activity. It should be mentioned, however, that in the permafrost zone embracing about 60% of Russian territory raise of mean temperatures most often intensifies slope processes.

REFERENCES

Alisov, B.P. (1956) *Climate of the USSR*. Moscow, MSU Publishing House 126 pp. (in Russian)

Bondarik, G.K. (1981) *General Theory of the Engineering (Physical) Geology*. Moscow, Nedra Publishing House 256 pp

Ershov, E.D. (ed.) (2001) *Dynamic geocryology, Part 4. Basics of the Geocryology*. Moscow, Moscow State University 688 pp. (in Russian).Erysh, I.F. & Salomatin, V.N. (1999) *Crimean Landslides* Simferopol, Apostrof Publishers. Part 1, 247 pp., Part 2, 175 pp. (in Russian).

Galanin A.A. (2009) Rock glaciers of north-eastern Asia: Mapping and geographical analysis. *Earth Cryosphere*, XIII (4), 49–61 (in Russian).

Garagulia, L.S. (1983) Classification of solifluction types for the estimation mapping of the territory. *Engineering Geology*, (4), 10–17 (in Russian).

Gatinskiy, Y.G. & Runkvist, D.V. (2004) Geodynamics of the Eurasia – Plate tectonics and block tectonics. *Geotectonics*, (1), 3–20 (in Russian).

Goretskiy, G.I. (1964) Alluvium of the great Anthropogenic rivers of Russian plane. Moscow, Nauka 415 pp. (in Russian).

Grigoriev, M.N., Razumov, S.O., Kunitzkiy, V.V. & Spektor, V.B. (2006) Dynamics of the Russian East Arctic sea coast: Major factors, regularities and tendencies. *Earth Cryosphere* (4), 74–94 (in Russian).

Haritonychev, A.T. (1978) Nature of the Nizhniy Novgorod Volga region: History, usage, protection. Gorkiy, Volga-Viatka Publishers. 175 pp. (in Russian).

Hromov, S.P. & Petrosiants, M.A. (2006) *Meteorology and Climatology* 7th edition. Moscow, Moscow State University Publishers. 582 pp. (in Russian).

Information Bulletin "Exogenous geological processes on the territory of Russian Federation", issues 1–8, VSEGINGEO, 1991–1998 (Sheko, A.I. *et al.*), issues 9–14, FC Geomonitoring, 1999–2004 (Zerkal, O.V. *et al.*) (in Russian).

Information Bulletin on the state of the interiors on the territory of Russian Federation. Part II. Exogenous Geological Processes, issues 28–37, FSUGE Hydrospecgeology, 2005–2013 (Markarian V.V. *et al.*) (in Russian)

Interstate Standard GOST 25100-2011 Soils (2011) *Classification*. Euro-Asian Council for Standardization, Metrology and Certification (EASC) (in Russian).

IPCC (2013) Climate change 2013: The physical science basis. In: Stocker, T.F., Qin, D., Plattner, G.-K., Tignor, M., Allen, S.K., Boschung, J., Nauels, A., Xia, Y., Bex, V. & Midgley, P.M. (eds.) *Contribution of Working Group I to the Fifth Assessment Report of the Intergovernmental Panel on Climate Change* Cambridge and New York, NY Cambridge University Press, 1535 pp.

Kliukin, A.A. (2005) Extreme manifestations of the adverse and hazardous exogeneous processes in the XX Century in Crimea. *Geopolitics and Ecogeodynamics of the Regions*, (1), 27–38 (in Russian).

Köppen, W. (1936) Das geographisca System der Klimate. In: Köppen, W. & Geiger, G. (eds.) *Handbuch der Klimatologie* Berlin, Verlag von Gebrüder Borntraeger pp. 1–44.

Koroliov, I.B. & Zerkal, O.V. (2006) Activity of hazardous exogenous geological processes in the Northern Caucasus region at the end of XXth – Beginning of the XXIst centuries. In: *Proceedings of the International Conference on Hydrogeology in the Beginning of the XXIst Century 14–16 September 2006, Novocherkassk*. pp. 127–130 (in Russian).

Koroliov, I.B. & Zerkal, O.V. (2007) Main results of the landslide and debris flow monitoring in the Southern Federal District during last decade. In: *Proceedings of the International Conference on Monitoring of the Geological Lithotechnical and Ecological-Geological Systems, 24–25 May 2007* Moscow. Moscow State University Publishers pp. 83–84 (in Russian).

Kuznetsova, L.P. (1978) Кузнецова Л.П. *Moisture Transfer in the Atmosphere Over the Territory of the USSR*. Moscow, Nauka 91 pp (in Russian).

Leibman, M.O & Kiziakov, A.I. (2007) *Cryogenic Landslides of Yamal and Yugorsky Peninsula*. Moscow-Tiumen 205 pp.

Lobkina, V.A., Kazakova, E.N., Zhiruev, S.P. & Kazakov, N.A. (2013) Methods of landslide risk assessment for territory of settlements of Sakhalin Region (Makarov city, Sakhalin). *Russian Journal of Pacific Geology*, 32 (5), 100–109 (in Russian).

Milanovskiy, E.V. & Semionov, M.P. (eds.) (1935) *Landslides of the Middle and Lower Volga Region and Methods of Landslides Protection*. Leningrad, ONTI. 252 pp. (in Russian).

Pánek, T., Hradecký, J., Smolková, V. & Šilhán, K. (2008a) Gigantic low-gradient landslides in the northern periphery of the Crimean Mountains (Ukraine). *Geomorphology*, 95, 449–473.

Pánek, T., Hradecký, J., Smolková, V. & Šilhán, K. (2008b) Giant ancient landslide in the Alma water gap (Crimean Mountains, Ukraine): Notes to the predisposition, structure and chronology. *Landslides* 5, 367–378.

Pánek, T., Šilhán, K., Hradecký, J., Strom, A., Smolková, V. & Zerkal, O. (2012) A megalandslide in the Northern Caucasus foredeep (Uspenskoye, Russia): Geomorphology, possible mechanism and age constraints. *Geomorphology*, 177–178, 144–157.

Peel, M.C., Finlayson, B.L. & McMahon, T.A. (2007) Updated world map of the Köppen-Geiger climate classification. *Hydrology and Earth System Sciences*, 11, 1633–1644.

Romanovsliy, N.N., Turin, A.I. & Sergeev, D.O. (1989) *Kurums of the Buld Mountain Belt.* Novosibirsk, Nauka Publishers 152 pp. (in Russian).

Strom, A.L. (2004) Rock avalanches of the Ardon River valley at the southern foot of the Rocky Range, Northern Caucasus, North Osetia. *Landslides* 1, 237–241.

Trofimov, V.T. (2002) Zonality of the engineering-geological conditions of the continents of Earth. Moscow, Moscow State University Publishers 348 pp. (in Russian).

Zerkal, O.V. (2013) Regularities of modern natural geological processes distribution. In: Trofimov V.T. & Kalinin E.V. (eds.) *Engineering Geology of Russia* Vol. 2. Moscow, Engineering Geodynamics of Russian Territory pp. 674–696 (in Russian).

Zerkal, O.V. & Markarian, V.V. (2003) Short-term forecast of the activity of hazardous geological processes in Russian Federation and its realization in 2002. In: *Problems of the Emergency Prediction. II Scientific-Practical Conference Proceedings, Moscow, October 23, 2002*. Moscow, URSS Publishers pp. 140–151 (in Russian).

Zerkal, O.V. & Koroliov, I.B. (2008) Influence of the global climate change on the activity of hazardous exogenic geological processes in the Northern Caucasus region territory. *Georisk* 3, 16–20 (in Russian).

Zerkal, O.V. & Koroliov, I.B. (2009) Typization of the Northern Caucasus region territory by the character of landslide processes. In: *Diversity of Modern Geological Processes and Their Engineering-Geological Assessment. Proceedings of the International Conference* Moscow, Moscow State University. pp. 125–126 (in Russian).

Zerkal, O.V., Koroliov, I.B. & Antipina, I.C. (2006a) *Assessment of Regional Changes of EngineeringGeological of the Northern Caucasus Region as the Basics for Prediction of Hazardous Geological Processes Activation (1:500 000 Scale)* Moscow. Unpublished Technical Report (in Russian).

Zerkal, O.V., Markarian, V.V. & Goncharov, A.A. (2006b) Some statistical parameters and estimates of the evolution of the exogeneous geological processes of the territory of Russian Federation. In: *Proceedings of the International conference "Problems of engineering Geodynamics and Ecological Geodynamics", Moscow, February 02–03, 2006*. pp. 39–40 (in Russian).

Zhelezniak, M.N., Grigoriev, M.N., Shepeliev, V.V. & Skachkov, Y.B. (2014) Modern state and evolution of Arctic Criolitozone: Ecological aspects of development. In: *Proceedings of the Conference, Yakutsk* (in Russian).

Zolotariov, G.S. & Zerkal, O.V. (1995) Factors influencing change of loess slopes stability and prediction of loess landslides. In: *Problems of Seismology and Engineering Geology. Proceedings of the International Conference Dedicated to 85 Years of the Academician G.A. Mavlianov, Tashkent, May 23–26, 1995*. Tashkent, Uzbekhydrogeology. pp. 126–131 (in Russian).

Zolotariov, G.S., Osiyuk, V.A., Root, P.E., Burlina, T.E., Zerkal, O.V. & Emelianov, S.N. (1994) Engineering-geological protection and objects from hazardous natural processes (problems of engineering-geological substantiation). *Hydrogeology, Engineering Geology. Overview*. Geoinformark, Ltd. Issue 5, 69 pp. (in Russian).

Chapter 9

Slope safety preparedness in Southeast Asia for effects of climate change

H. Rahardjo and A. Satyanaga
Nanyang Technological University, Singapore

K. Hoon
Singapore Land Authority, Singapore

W.L. Sham, Aaron and C.L., Ong
Building & Construction Authority, Singapore

B.B.K. Huat, M.H. Fasihnikoutalab and A. Asadi
Universiti Putra Malaysia, Selangor Darul Ehsan, Malaysia

P.P. Rahardjo
Parahyangan Catholic University, Indonesia

A. Jotisankasa
Kasetsart University, Bangkok, Thailand

T.M. Thu and T.T. Viet
Thuyloi University, Vietnam

9.1 INTRODUCTION

9.1.1 Global climate change and its impact on slope stability

Climate change is a significant change in the statistical distribution of weather patterns over periods of time ranging from decades to millions of years. The issue on climate change has been discussed and reported comprehensively in the first Intergovernmental Panel on Climate Change report (IPCC, 1990), fifth IPCC report (2013), Kyoto Protocol (Oberthur and Ott, 1999) and United Nation Framework Convention on Climate Change (UNFCCC) report (United Nations, 1993). The reports indicated that 90% of human activities caused the increase in the concentrations of greenhouse gases (carbon dioxide, CO2; methane, CH4; nitrous oxide, NO2; ozone, O3 and halocarbons), resulting in the destabilizing effect on the global climate, known as global warming. The study by Stern (2007) showed that the impacts of climate change were across nations and sectors, such as: food, water (i.e. increase in groundwater table, ice melting), ecosystems and extreme weather events (i.e. intense rainfall, high temperature, strong wind and floods). Among all sectors, water and extreme weather events were the main factors that affect slope stability around the world.

Previous studies indicated that rainfall events have become more intense in number of countries, i.e. United States (Kunkel, 2003), Canada (Francis and Hengeveld, 1998), Australia (Suppiah and Hennessy, 1998), South Africa (Mason *et al.*, 1999), United

Kingdom (Osborn *et al.*, 2000) and Europe (Forland *et al.*, 1998). Zhai *et al.* (2005) reported increasing trends in daily extreme rainfalls in western China, and also parts of southwestern and southern China using the data from the entire period of 1951–2000. Fujibe (2008) found significant upward trends in heavy daily rainfalls in Japan using the data over the period of 1901–2006. Groisman *et al.* (2005) found an upward trend in number of days with heavy rainfalls (greater than 99.7 percentile) in south eastern Australia. Gallant *et al.* (2007) investigated rainfall trends including heavy rainfall events (95th and 99th percentiles) over two periods (i.e. 1910–2005 and 1951–2005) in southwest and east of Australia.

Since rainfall characteristics are often used to design water structures, reviewing and updating rainfall characteristics (i.e., Intensity–Duration–Frequency (IDF) curves) for future climate scenarios is necessary (Mirhosseini *et al.* 2013). A lot of studies have been carried out to analyze factors for assessment, adaptation and mitigation of climate change, and to enhance and sharpen the disaster management policies for the many and various stakeholders. El-Hadji and Singh (2002), using long-term data (from 1951 to 1990) of rainfall and annual runoff, have developed model of spatial and temporal variability of rainfall and runoff in the Casamance River basin, located in southern Senegal, West Africa. Chowdhury and Beecham (2010) have analyzed the effects of changing rainfall patterns in Australia, in order to define the design parameters of Water Sensitive Urban Design (WSUD) technologies, like bio retention basins and permeable pavements, describing how these systems behave under varying rainfall conditions. Kuhn *et al.* (2011) have analyzed the effect of climate change and land use change in the Gallocanta Basin, Spain. Sherif *et al.* (2011) have analyzed spatial and temporal characteristics of rainfall in the United Arab Emirates (UAE). The rainfall patterns, rainfall probability of occurrence, rainfall intensity-duration-frequency (IDF) relationship, probable maximum precipitation (PMP) and drought scenarios were investigated, using standard statistical techniques, such as Gumbel, log Pearson, GEV, log normal, Wakeby and Weibull probability distributions.

The effects of climate change are also of major concern in the tropical regions, like South East Asia (SEA). SEA is mainly covered by residual soil with a significant thickness of unsaturated zone above the groundwater table. An unsaturated soil can be defined as a soil which has four phases (i.e., solid, water, air, and contractile skin) and the pore-water pressure is negative relative to the pore-air pressure (Fredlund and Rahardjo, 1993). The unsaturated zone is a dynamic interface of the slope with the environment (Figure 9.1) and as a result, factor of safety of residual soil slope is affected dynamically by climate change. In practice, geotechnical engineers commonly ignore the effect of climate change in their design to avoid the complexity in determining the flux boundary conditions at the soil-atmosphere interface. In most geotechnical designs such as slope stability, the actual flux boundary conditions are very important since the negative pore-water pressure in unsaturated soil is highly influenced by the flux boundary condition changes (i.e., infiltration, evaporation and transpiration) resulting from the changes in climatic conditions (Figure 9.2). Therefore, the effect of climate change, which is translated into flux boundary conditions, should be considered not only in geotechnical design, but also during the evaluation of natural structures like slopes.

In recent years, many research works have been carried out to study the effects of climate change on slope stability. In United Kingdom, the BIONICS project (biological

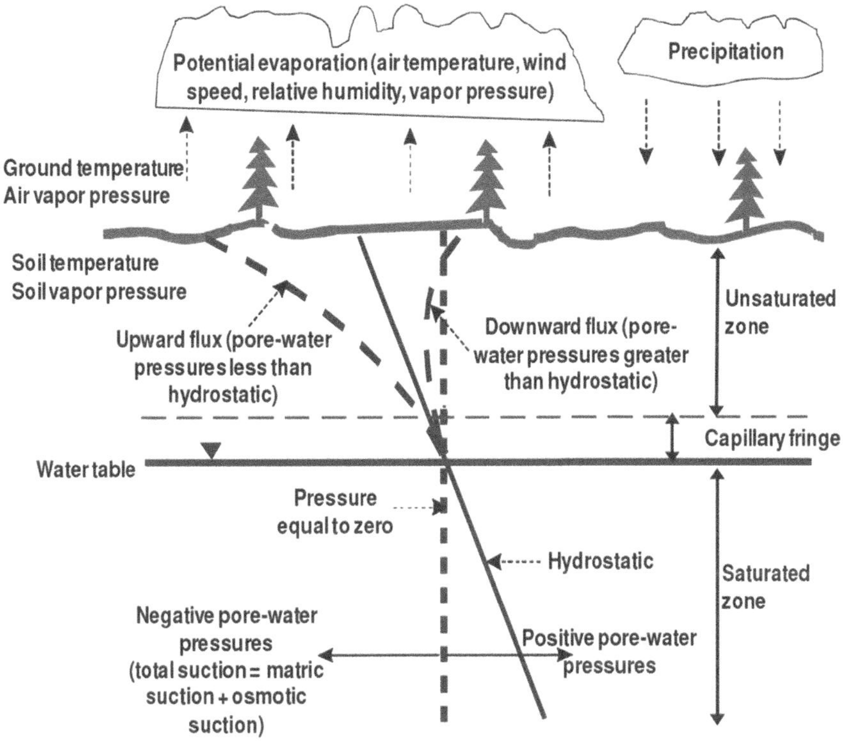

Figure 9.1 Effect of climate change on the behaviour of unsaturated soil (after Fredlund and Rahardjo, 1993).

and engineering impacts of climate changes on slopes) was established to investigate the characteristics of climate changes (Mendes, 2011) and their effects on pore-water pressure distribution and deformation within soil layer (Rouainia *et al.*, 2009; Kilsby *et al.*, 2009) and sustainability of slope infrastructure (Glendinning *et al.*, 2009). Springman *et al.* (2003) observed that the infiltration of the snowmelt was the main cause behind the rising groundwater table which triggered the majority of deep seated failures in Switzerland. Alonso *et al.* (2003) observed that hydraulic properties of the unsaturated soil played an important role in addition to the rate of rainwater infiltration in affecting slope stability. Research works in Asia indicated that slope failures due to frequent and heavy rainfalls were commonly shallow slides as observed in Malaysia (Liew *et al.*, 2004), Hong Kong (Brand, 1992) and Singapore (Pitts, 1985; Toll *et al.*, 1999; Rahardjo *et al.*, 2010).

The effects of climate change (i.e. changes in rainfall intensity, amount and pattern, in groundwater level and temperature) can be incorporated in geotechnical analyses by quantifying the flux boundary conditions near the ground surface and understanding groundwater level variations through field measurement. Rahardjo *et al.* (2014a) installed comprehensive instruments to study the effect of climate change on

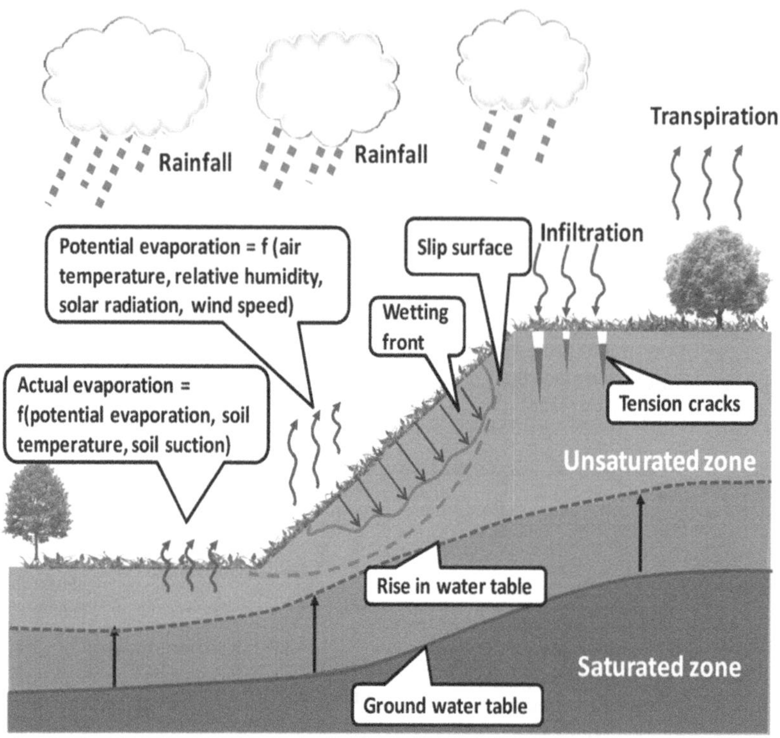

Figure 9.2 Mechanism of rainfall-induced slope failures.

the stability of the residual soil slopes in Singapore during periods of 2006 to 2008. Residual soils in Singapore are weathered soil that can be classified into three major formations: sedimentary rocks of Jurong Formation in the western part of Singapore, Bukit Timah granite or igneous rocks in the centre of Singapore and semi-hardened Old Alluvium in the eastern part of Singapore (Rahardjo *et al.*, 2012a) as shown in Figure 9.3. Common instruments for measurement of climatic data are weather station, tensiometer, piezometer, soil moisture sensor and soil temperature sensor. Weather station (Figure 9.4) consists of apparatuses to measure rainfall intensity (rainfall gauge), wind speed, air temperature, relative humidity and solar radiation. Tensiometer is required for monitoring the changes of negative pore-water pressure in the unsaturated zone. Piezometer is needed to observe the variation of groundwater table within soil layer during dry and rainy periods. Time-domain reflectometry is installed to measure the soil water content changes within soil layer whereas soil temperature sensor is required for measuring the variation of soil temperature for quantifying actual evaporation from the ground surface. All data are collected and presented in real time so that changes in the flux boundary conditions can be calculated at all times (Figure 9.3).

The effect of rainfall on slope stability can be quantified through the characterization of water infiltration into soil layers during rainfall. The rainwater infiltration characteristics can be determined indirectly via measurement of runoff and pore-water

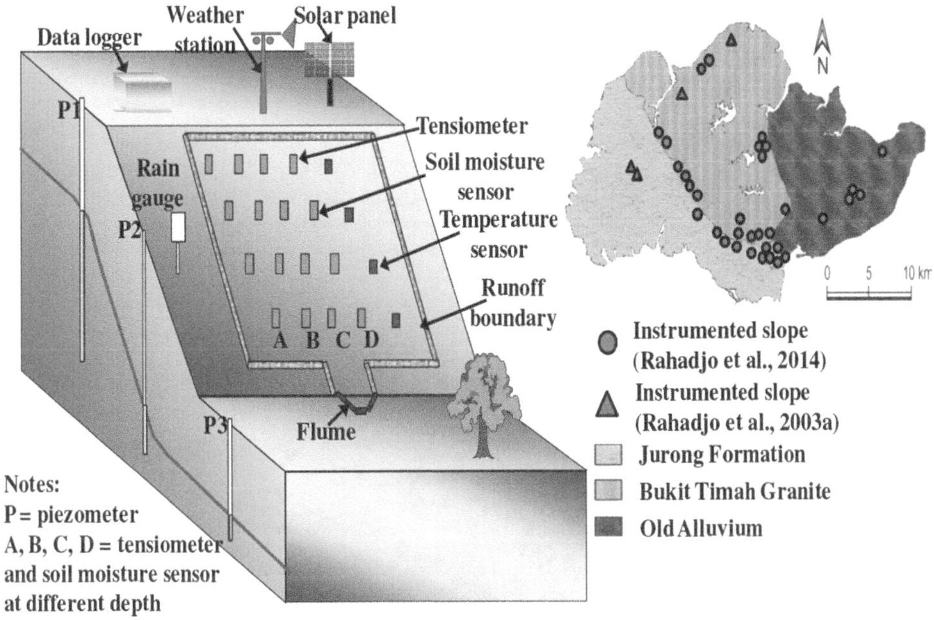

Figure 9.3 Field instrumentation for the study of the effect of climate change on slope stability in Singapore.

pressure changes during rainfall. The measurement of runoff during rainfall was carried out by Rahardjo *et al.* (2003a) using a Perspex flume with a water-depth probe for monitoring of water level inside the flume during rainfall. The set-up of field instrumentation for monitoring of runoff during rainfall is illustrated in Figure 9.3. The monitoring results of natural rainfall and runoff on 12 December 1998 for a residual soil slope from the sedimentary Jurong Formation are shown in Figure 9.5. The results indicated that an increase in total amount of rainfall caused an increase in total runoff. The results from various field measurements also showed that a large proportion of rainfall (40% to 100% of total rainfall) contributed to infiltration in the residual soil slope (Rahardjo *et al.*, 2005, 2013a) (Figure 9.6). The measurement of pore-water pressures using tensiometers and piezometers to study the effect of rainfall on slope stability has been carried out by many researchers (Lim *et al.*, 1996; Rahardjo *et al.*, 1998; Li *et al.*, 2005; Ng *et al.*, 2008; Fredlund *et al.*, 2012; Rahardjo *et al.*, 2013a). The changes of pore-water pressures during dry and rainy periods can be incorporated into slope stability analyses to obtain the variation of factor of safety with time. Similar studies can be carried out to investigate the effects of changes in rainfall intensity, pattern and duration due to climate change on slope stability.

The changes in climatic condition resulted in the variation of groundwater table during dry and rainy periods. Rahardjo *et al.* (2010) analyzed the monitoring results from Cassagrande Piezometers installed in several residual soil slopes in Singapore for over a two-year period (June 2006 until September 2008) (Figure 9.3). The

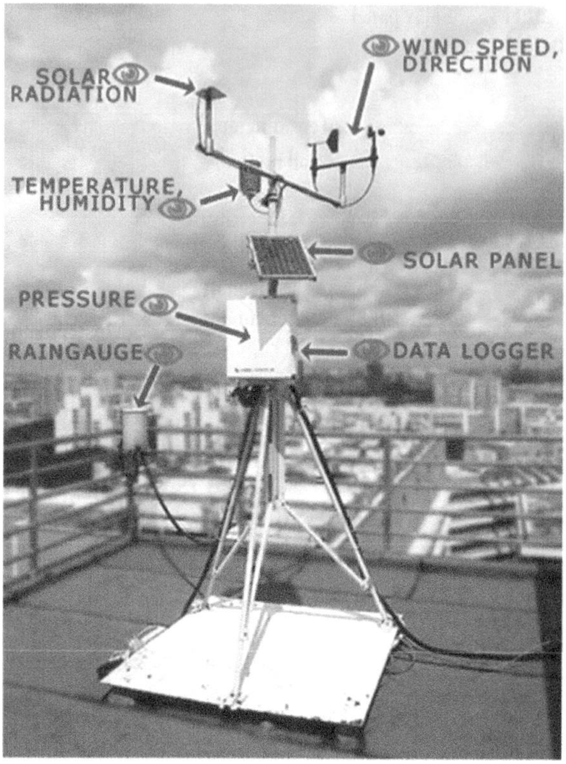

Figure 9.4 Weather station for measurement of climatic data (after Rahardjo *et al.*, 2013).

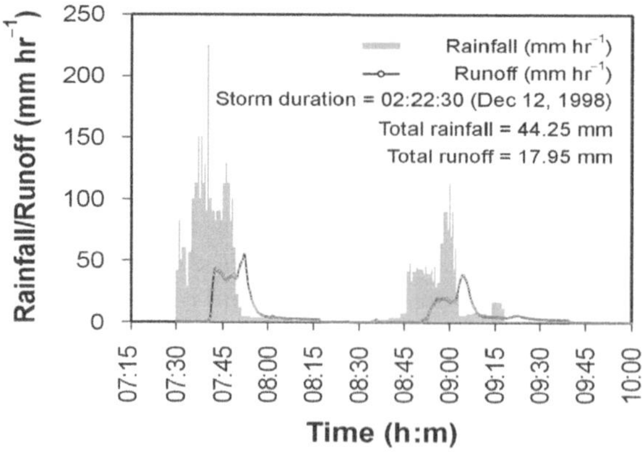

Figure 9.5 Rainfall and runoff hydrograph of a storm event on 12 December 1998 recorded at residual soil slope from the sedimentary Jurong Formation.

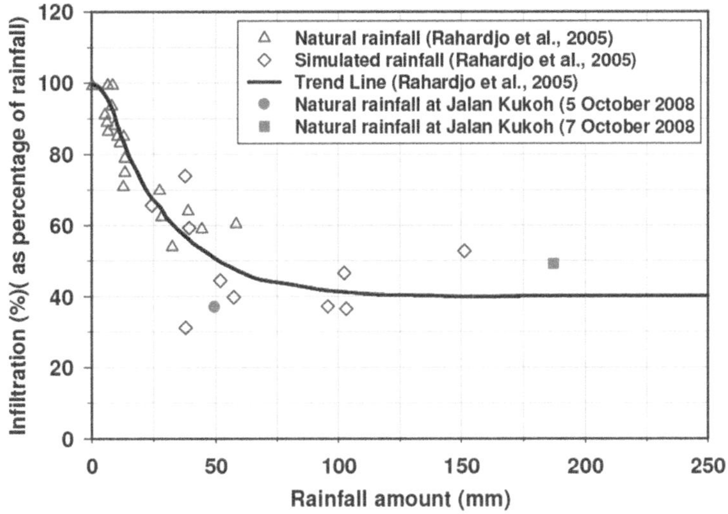

Figure 9.6 Relationship between infiltration and rainfall amount (Rahardjo *et al.*, 2013a).

relationships between variations of groundwater table depths and the heights of the residual soil slopes during dry and rainy periods are illustrated in Figure 9.7 and Figure 9.8. It was observed that the groundwater table of the residual soil slope from the sedimentary Jurong Formation (Figure 9.7) was generally deeper than that of the residual soil slope from the Bukit Timah Granite (Figure 9.8). However, the groundwater table of the residual soil slope from the sedimentary Jurong Formation indicated a larger difference between dry and wet periods as compared to that of the residual soil slope from the Bukit Timah Granite. This condition may be attributed to the high variability of soil types in the residual soil slope from the sedimentary Jurong Formation that was derived from the sedimentary rocks of different types, from conglomerate to shale, siltstone and sandstone. On the other hand, the residual soil slope from the Bukit Timah Granite was derived mainly from the same granitic rock. The variations of groundwater table can be incorporated into the slope stability analyses to obtain the variation of factor of safety during dry and rainy periods for the respective soil types. Similar studies can be carried out to investigate the changes in groundwater table variations due to climate change on slope stability.

Climate change results in an increase in the global temperature including the tropical region such as Southeast Asia. Previous study by Rahardjo *et al.* (2013a) indicated that the difference in soil and air temperatures was the main factor affecting the evaporation rate in Singapore. Prolonged evaporation from long periods of dry weather may result in the desiccation cracks to occur in the ground surface (Heath and Lehr, 1987; Morris *et al.*, 1992; Hewitt and Philip, 1999). The development of crack networks increases the rate of water infiltration into slope during rainfall (Krisnanto *et al.*, 2014). As a result, slopes with desiccation cracks are more susceptible to failure since the rainwater easily infiltrates into the soil layers through the tension cracks of the crack

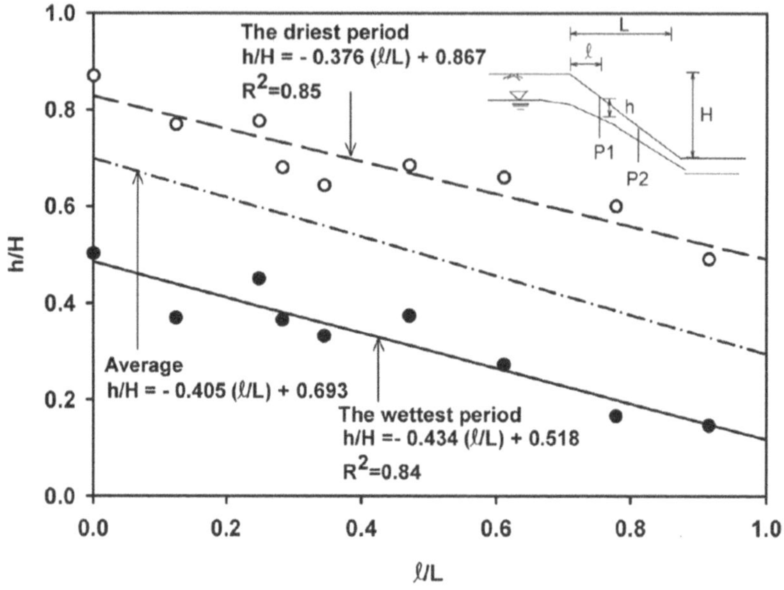

Figure 9.7 Variation of groundwater table position for the residual soil slopes from sedimentary Jurong Formation (after Rahardjo *et al.*, 2010).

network on the slope surface and rapidly increases pore-water pressures within soil layer. Evaporation can be divided into two categories, which are: potential evaporation and actual evaporation. Potential evaporation is defined as the maximum evaporation that can occur and can be quantified using Penman (Penman, 1948) method via measurements of air temperature, relative humidity, wind speed and solar radiation. Actual evaporation is usually less than the potential evaporation (Wilson *et al.*, 1997) as it also takes into account the soil temperature and negative pore-water pressure of soil near the ground surface in addition to the climatic factors for quantification of potential evaporation. Actual evaporation can be quantified using Wilson *et al.* (1993) equation. Rahardjo *et al.* (2012) compared the air and soil temperatures from the field measurements during the period of October 1998 to December 1998 in the residual soil slope from the Bukit Timah Granite, Singapore (Figure 9.9). The typical variation of the differences in air and soil temperatures during day and night is shown in Figure 9.10. It was observed that the maximum and minimum differences between air and soil temperatures occurred at noon (12.00 pm) and midnight (12.00 am), respectively. Similar observations were reported in Rahardjo *et al.* (2014b). The results showed similar trends with the variations of the air and soil temperatures reported in Rahardjo *et al.* (2013a). The variations of climatic factors and negative pore-water pressures, as parts of the actual evaporation, during dry and rainy periods are incorporated into the seepage analyses as flux boundary conditions on the slope surface. The pore-water pressure changes from the seepage analyses are then used in the slope stability analyses to obtain the variations of factor of safety with time. This method can be used to quantify the effect of climate change with respect to temperature variations on slope stability.

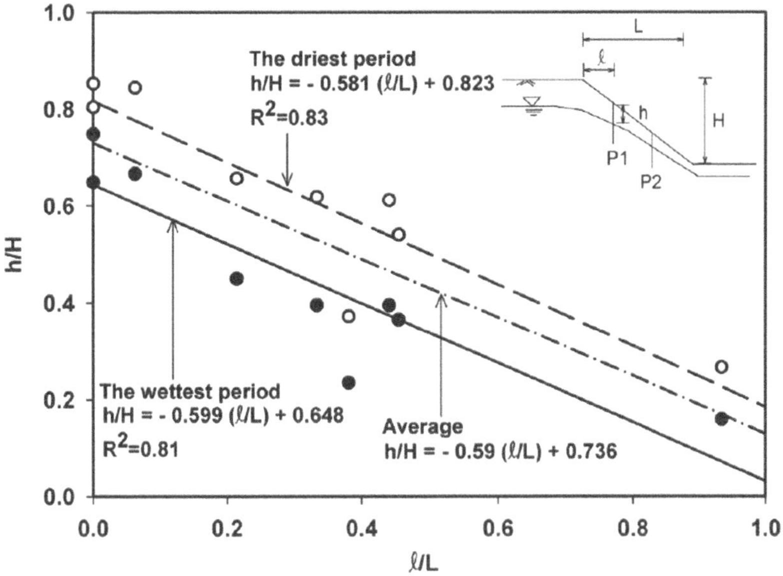

Figure 9.8 Variation of groundwater table position for the residual soil slopes from Bukit Timah Granite (after Rahardjo *et al.*, 2010).

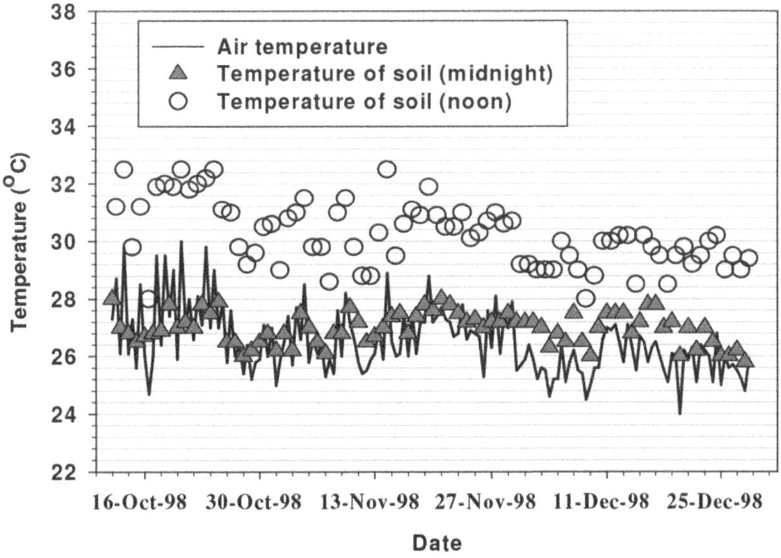

Figure 9.9 Variation of air and soil temperatures for a residual soil slope from the Bukit Timah Granite (after Rahardjo *et al.*, 2013a).

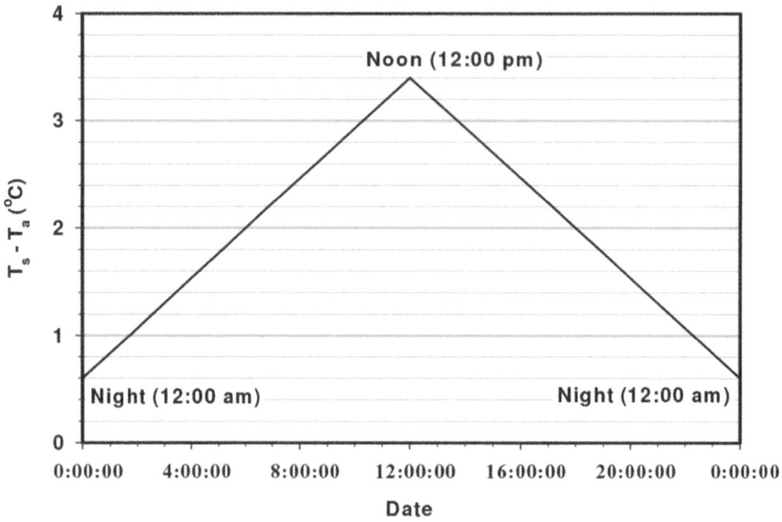

Figure 9.10 Typical variation of differences between air and soil temperatures for a residual soil slope from the Bukit Timah Granite (after Rahardjo et *al.*, 2013a).

Many researchers carried out numerical analyses to study the effect of flux boundary conditions on slope stability (Potts *et al.*, 1997; Alonso *et al.*, 2003; Rouainia *et al.*, 2009; Rahardjo *et al.*, 2013a). Flux boundary conditions are applied to the slope model for the transient seepage analyses. The hydraulic properties of unsaturated soil (soil-water characteristic curve and permeability function) need to be incorporated in the transient seepage analyses. The unsteady seepage flow equation (Figure 9.11) is used in the analyses to calculate the change in pore-water pressure during dry and rainy periods. Typical numerical model for seepage analyses and the example of rainfall intensity from field instrumentation results are shown in Figure 9.11 from Rahardjo *et al.* (2011). The variation of pore-water pressures with time as the output from the seepage analyses are incorporated into slope stability analyses to obtain the variation of factor of safety during dry and rainy periods. A typical numerical model for slope stability analyses and an example of critical slip surface during rainfall are shown in Figure 9.12. The saturated and unsaturated shear strengths of the investigated soils must be incorporated in the stability analyses. One common method used to calculate the variation of factor of safety of slope is Bishop's simplified method of slices as shown in Figure 9.12. The minimum factor of safety might occur at the end of rainfall or several hours after the rain stopped depending on soil properties within the slope (Rahardjo *et al.*, 2011).

Slope failures have high socio-economic impacts on many countries (Brabb & Harrod, 1989). The total global cost associated with slope movement has exceeded 20 billion US$ per year (Leroueil, 2000). Human casualties directly resulting from slope failures have been enormous, with an estimated 110,000 casualties in China alone in the 20th century (Li, 1989). Researchers compiled the data of the effects of landslips in Hong Kong and found that more than 470 fatalities and hundreds of public facilities

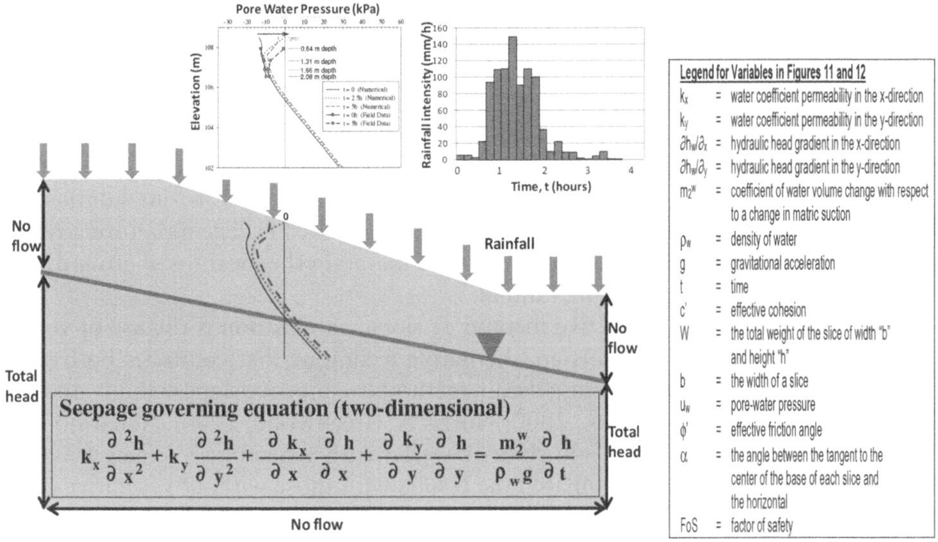

$$k_x \frac{\partial^2 h}{\partial x^2} + k_y \frac{\partial^2 h}{\partial y^2} + \frac{\partial k_x}{\partial x}\frac{\partial h}{\partial x} + \frac{\partial k_y}{\partial y}\frac{\partial h}{\partial y} = \frac{m_2^w}{\rho_w g}\frac{\partial h}{\partial t}$$

Figure 9.11 Typical numerical model for seepage analyses.

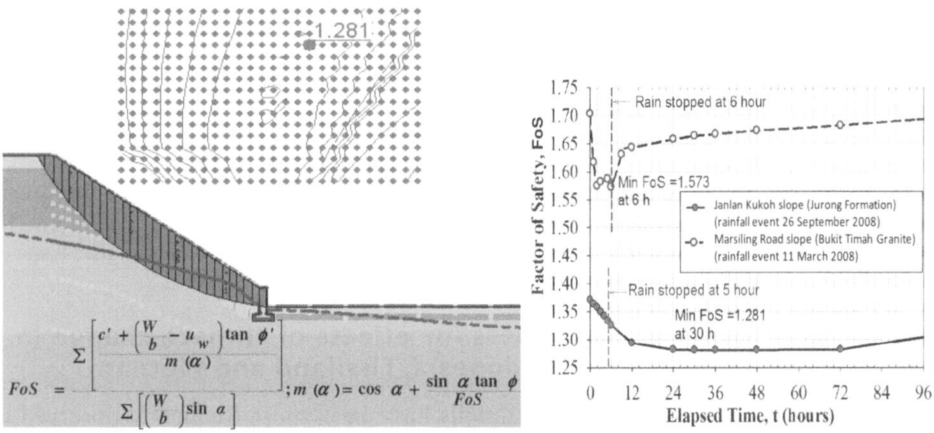

$$FoS = \frac{\Sigma\left[\frac{c' + \left(\frac{W}{b} - u_w\right)\tan\phi'}{m(\alpha)}\right]}{\Sigma\left[\left(\frac{W}{b}\right)\sin\alpha\right]};\ m(\alpha) = \cos\alpha + \frac{\sin\alpha\tan\phi}{FoS}$$

Figure 9.12 Typical numerical model for slope stability analyses.

were destroyed by landslides since 1948 (Malone, 1997). Slope failures may pose a risk to environmental sustainability. The cost of repairing slopes is more expensive than the cost of preventive measures (Bennett and Doyle, 1997). To protect slopes against the possibility of rainfall-induced slope failures, preventive measures are necessary to ensure the safety of nearby buildings or public facilities. Horizontal drains for lowering groundwater levels within the slope have been recognized as one of the most economical and effective method available. Horizontal drains for the stabilization of slopes have been used extensively around the worlds, such as in Australia (Snowy Mountains Hydro-Electric Authority, 1983), Austria (Veder and Lackner, 1985), Brazil (Costa Nunes, 1985), France (Cartier and Virollet, 1980), United Kingdom (Hutchinson, 1977), United Sates (Smith, 1980) and Hong Kong (McNicholl *et al.*, 1986). Many researchers studied the effectiveness of horizontal drains in improving the stability of slope using finite element analyses. Two-dimensional finite element modelling of horizontal drains (Rahardjo *et al.*, 2003b) has been conducted and a residual soil slope from the sedimentary Jurong Formation in Singapore, installed with horizontal drains, has been instrumented comprehensively by Rahardjo *et al.* (2011). The monitoring results showed that horizontal drains were effective to maintain the location of groundwater table and stability of slope during rainfall.

Another efficient and effective method of slope stabilization is the use of vegetation or green technology. This is an alternative technique that combines both living biological elements and engineering design methods to improve slope stability and erosion control (Montgomery *et al.*, 2000; Norris & Greenwood, 2000). Rahardjo *et al.* (2014c) instrumented residual soil slopes covered with vegetation (Vetiver grass and Orange Jasmine shrub) and compared the results with those without vegetative cover. Figure 9.13 shows that the overall factor of safety of slopes covered with deep rooting Vetiver grass and Orange Jasmine shrub are higher than those without vegetative cover. This indicates that vegetative covers performed well in maintaining slope stability by reducing the rainwater infiltration, decreasing pore-water pressures (transpiration) and reinforcing the soil (deep rooting). Another method that can be used to reduce the rainwater infiltration into slope during rainfall is capillary barrier system (CBS). CBS, a slope cover designed using unsaturated soil mechanics principles, consists of a fine-grained layer of soil overlying a coarse-grained layer of soil. Rahardjo *et al.* (2013b) instrumented a residual soil slope covered with CBS and compared the results with those obtained from the slope without CBS. Figure 9.14 shows that the overall factor of safety of slopes covered with CBS are higher than those without CBS indicating that CBS performed well in reducing the rainwater infiltration into the slope. Further investigations are required to study the performance of CBS and vegetative covers for the anticipated changes in rainfall intensity, amount and duration due to climate change.

9.1.2 Slope safety preparedness for effects of climate change in Singapore, Malaysia, Indonesia, Thailand and Vietnam

In Singapore, prolonged and heavy rainfalls have been more frequently observed in recent years. Singapore prepares to adapt to the impacts of climate change by adopting four approaches: reducing carbon emissions, infrastructural improvement, harnessing the green growth opportunities and forging international partnerships (NCCS, 2012). Singapore Land Authority (SLA) is a statutory board under the Ministry of Law who manages some 14.000 hectares of State land and about 5,000 State buildings in Singapore. SLA is committed in addressing the effects of climate change on slope safety in Singapore by combining two software tools (Geographic Information System or GIS) and Analytical Hierarchy Process (AHP), to create a Slope Hazard Analysis and Repository System (SHARES). SHARES uses geo-analytics to identify locations prone to slope failure and prioritise slope protection works to be carried out, via a real-time forum which promotes consultation and collaboration within the government. The

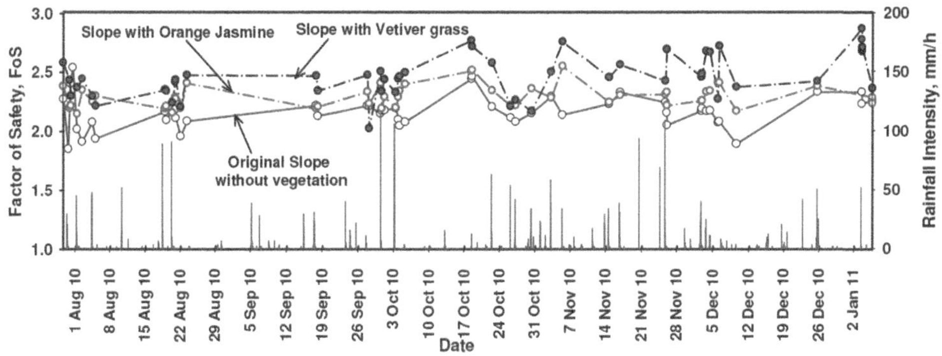

Figure 9.13 Factor of safety variations for slope with and without vegetative cover (after Rahardjo *et al.*, 2014c).

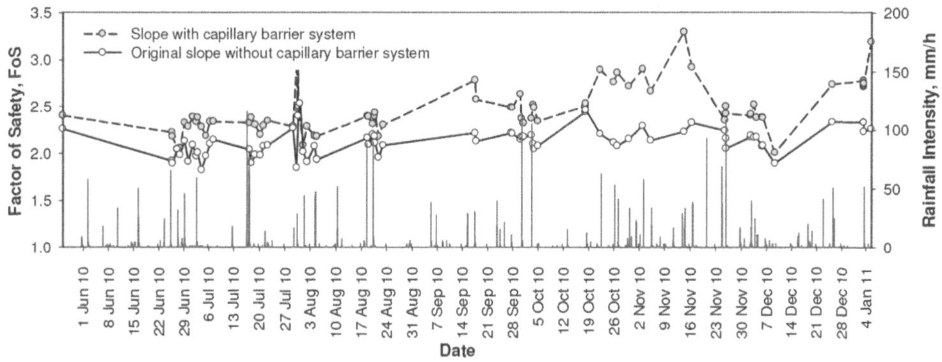

Figure 9.14 Factor of safety variations for slope with and without capillary barrier system (after Rahardjo *et al.*, 2013b).

use of SHARES provides a good foundation for current efforts in addressing impacts of climate change. Moving forward, new factors will be incorporated in SHARES to further enhance the analytical capabilities and robustness of this system. Those new factors predicted future rainfall generated through statistical downscaling and temporal disaggregation from high-resolution climate-change scenarios (Simonovic and Peck 2009), or use of soil data properties reflecting micro-climatic conditions in Singapore.

Building and Construction Authority (BCA), an agency under the Ministry of National Development, Singapore, regulates all buildings, engineered earth retaining works and structures empowered by the Building Control Act of Singapore. BCA addresses the effect of climate change on slope safety in Singapore by adopting four approaches: reviewing the building control regulation, initiating the 3 stages slope risk management framework, conducting periodic structural inspection on slope and its protection structures and designing a systematic incident management work flow during slope failure (Ong, 2008).

Climate change in Malaysia has resulted in the increase of rainfall intensities, durations and frequencies. As a result, the overall national groundwater level increases and many landslides occur in Malaysia. The most common type of landslide in Malaysia is the shallow slide where the slip surface is usually less than 4 m deep and occurs during or immediately after an intense rainfall (Singh *et al.*, 2008). One method to prepare for the impacts of climate change in Malaysia is the use of Live Pole technique to maintain the slope stability against shallow-seated failures. Live Pole technique utilize the ability of certain woody plants (short trees or shrubs) to reinforce the soil slope by root binding action, to improve the shear strength of the soil slope by increasing the apparent cohesion of soil and to reduce the soil moistures via evapotranspiration. In order to achieve the immediate and long term slope stabilization, these hardy species must be planted in close spacing and have roots that penetrate deep into the slope (Huat and Mafian, 2011).

The effects of climate change on groundwater table, temperature and rainfall intensity in Indonesia have caused the changes in the frequency, distribution and mode of landslides. Recent study by Tohari (2012) indicated that during rainy period, slopes are more prone to failure in West Java region whereby the soils are classified as water sensitive soil (i.e. clay shale) and expansive soil. Many landslides have occurred in West Java region in the past before climate change. With climate change, severe rainfalls resulted in the reactivation of old landslides in this region. Preparations to address the impact of climate change in Indonesia have been carried out through numerical analyses and instrumentation of the water sensitive and expansive soils. Numerical analyses are required for predicting the location, direction and magnitude of movement of the area prone to landslide based on characteristics of the old landslides. Then, preventive measures can be designed based on the analyses results. Several geotechnical instruments such as inclinometer and piezometer are installed to monitor the soil movements and groundwater table within the critical slope. Warning and evacuation order can be issued once the stability of the slope approaches a dangerous level. In addition, protection of foundations and structures installed within the critical slope are carried out to ensure the safety of those structures.

Climate change in Thailand results in rainfall-induced landslides which affect people's lives and infrastructure. Two critical rainfall envelopes (accumulated rainfalls of 150 mm and 300 mm) are empirically drawn in order to classify the localized landslide events and more destructive wide spread debris flow events. Field instrumentations (i.e. tensiometers, piezometers) on critical slopes and laboratory measurements (i.e. soil-water characteristic curve, permeability function, unsaturated shear strength) have been carried out to understand the relationship between the climate change, hydro mechanical properties of unsaturated residual soils and the slope stability. Several attempts have been made to develop technologies that will enable Thailand to adapt to climate change, such as: dynamic landslide hazard map, low-cost landslide warning box (Mairaing & Thaiyuenwong, 2010) and telemetry system for landslide warning (Jotisankasa *et al.*, 2010). In addition, soil bio-engineering (i.e. vetiver grass) have been widely used for slope stabilization and erosion control in Thailand (Truong *et al.*, 2008).

Global climate change has significantly affected the temperature, rainfall and sea level in Vietnam. The temperature in Vietnam rises by 0.5 to 0.7°C over 50 years period (1957–2007) as reported by Tai *et al.* (2009). The annual rainfall in Vietnam has increased to the range of 1400–2400 mm (Quang, 2012). The sea level around Vietnam has risen by 5 cm between 1960s and 1990s (UNEP, 1993) and it is estimated to keep rising at an average of 2 mm per year. The increases in rainfall and sea level have caused an increase in landslide frequencies in Vietnam. Hanh (2009) reported that landslides and flash floods have destroyed 13280 houses, killed 988 people and resulted in significant damages on farming production and infrastructures over the period of 16 years, from 1990 to 2005. Several methods have been used to prepare Vietnam in addressing the effect of climate change on landslides and the rising of sea level. Landslide hazard maps are developed to categorize the hazard zone that are prone to landslides. Then, preventive measures can be installed on the critical slopes to prevent landslide during rainfall. Many sea dykes have been upgraded or built to accommodate the rising of the sea level near Vietnam.

The details of climate change and its impact on Singapore, Malaysia, Indonesia, Thailand and Vietnam are presented in Sections 9.2 to 9.6. The preparations to adapt to the climate change from each country are also described briefly in those sections. The report from each country is presented as it is without any amendments in its content.

9.2 SLOPE SAFETY PREPAREDNESS FOR EFFECTS OF CLIMATE CHANGE IN SINGAPORE

9.2.1 Singapore's slope preparedness

Dealing with climate change is complex, as it is extremely difficult to predict what the future "state of the climate" look like and how it may impact us. The real world, as Emery and Trist (1965) viewed it to be, is a "turbulent environment", where the level of interactions between individual constituents, as well as between the system and its environment as a whole cannot be fully understood simply by analyzing its constituents and it will take more than the use of traditional tools, thinking and leadership to deal with such complexity (Uhl *et al.*, 2007). However, if one could find comfort is that the world now is at least not as ignorant and clueless about climate change, as compared to its recent past (10 to 15 years ago). Today, there is at least global scientific consensus via the Intergovernmental Panel on Climate Change (IPCC) on the following:

(a) There are growing signs to indicate that the climate is indeed changing;
(b) If carbon emissions from human activities continue to grow, climate change will accelerate;

The world that we know now will be very different from the future that is to come. A recent Price water house Coopers (PWC) study predicts the evolution of supply chains, changing patterns of mobility, emerging technology and opportunities, and the creation of new business models in response to the impacts of climate change by 2030 (PWC, 2009).

9.2.2 Singapore's resilience strategy

Singapore, being a responsible global citizen within the international community, is committed to do its part in the global effort to address climate change. This was

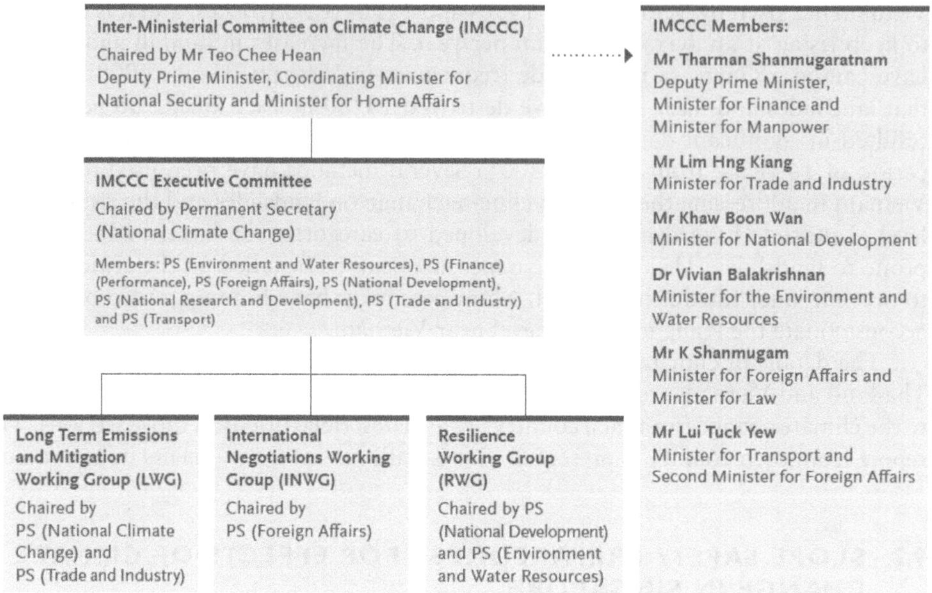

Figure 9.15 Formation of committees and workgroups.

underlined during the Singapore Parliament Committee of Supply Debate in 2012, where Deputy Prime Minister Teo Chee Hean spoke on Singapore's vision "to be a climate resilient global city that is well positioned for green growth (needing) a whole-of-nation effort – involving the people, the private and the public sectors to realize (this) vision together, (ensuring) Singapore remains a vibrant and liveable nation for our future generations" (NCCS, 2012).

The National Climate Change Secretariat (NCCS), a dedicated unit established under the Prime Minister's Office (PMO) ensures effective coordination on Singapore's Domestic and international policies, plans and actions on climate change. A national strategy, supported by committees and workgroups to address climate change-related issues, is shown in Figure 9.15 (NCCS, 2012).

In 2007, a national study examining the long-term effects of climate change on Singapore was carried out. It was found that the mean sea level around Singapore could rise by up to 0.65 m, and temperatures could increase by up to 4.2°C in 2100. These findings are comparable to those reported in the Fourth Assessment Report (AR4) and Fifth Assessment Report (AR5) by the IPCC. Key findings are summarized in Table 9.1. Over time, findings will be updated as climate science and models improve. Adaptation planning would benefit significantly from robust scientific analyses of climate change effects and impacts. Therefore, in March 2013, the Meteorological Service Singapore (MSS) established the Centre for Climate Research Singapore (CCRS) to build up in-house capability in climate science and climate modelling. This will augment Singapore's national preparedness for climate change and help policy makers make informed decisions. Facing a relatively uncertain future, yet not having the luxury

Table 9.1 Key findings of the Singapore Climate Change Study (NEA, 2013).

Climate Change Projections (in 2100 relative to present)[1]	*Singapore 1st National Climate Change Study Local Findings*	*IPCC AR4 Projections*	*IPCC AR5 Projections*
Change in Average Temperature (°C)	+2.7 to +4.2 (A1B Scenario)	+1.7 to +4.4 (A1B Scenario, Global)	+0.3 to +4.8 (All Scenarios, Global)
Change in Mean Sea Level (m)	+0.24 to +0.65 (B1, A1B, A2 Scenarios)	+0.18 to +0.59 (All Scenarios, Global)	+0.26 to +0.82[2] (All Scenarios, Global)

Notes: [1] Studies to assess future change in rainfall for Singapore are ongoing.
[2] Ice sheet melt processes have been better quantified in AR5.

Figure 9.16 Singapore's adaptation approach (NCCS, 2012).

of time to debate on the causes of climate change to be completely settled before taking any action, Singapore takes on early preparations to adapt to climate change impacts – via a four-fold approach: reduce carbon emissions in all sectors; be ready to adapt to climate change; harness green growth opportunities; and, forge partnerships (NCCS 2012). Working alongside mitigation efforts, adaptation measures are strategically implemented to reduce the damaging impact of climate change and increase resilience to potential future effects, via the adoption of a resilience framework, as shown in Figure 9.16.

In recent years, intense rainfall has been observed in Singapore. During the period of December 2006 to January 2007, Singapore has recorded one of the highest average prolonged rainfall over the last 80 years of 345 mm for over 20 hours, of which the 20 hours of continuous rainfall of 765.9 mm from 19 to 20 December 2006 was the highest recorded for the past 137 years. Increased rainfall can increase likelihood of failure of slopes. It is common that a slope might fail during or soon after being subjected to prolonged and intense rainfall, independent to the presence of slope stabilizing features (i.e.: cut-off drains, vegetation or retaining wall). The occurrence of landslides arising from exceptional torrential rainfall occurred in different parts of Singapore served to building and/or land owners as a solemn reminder and has prompted the need to both government agencies and private owners to further work on, be prepared and vigilant in maintaining their slopes and slope protection structures. Depending on proximity to surrounding properties of the slopes and slope protection structures, including those originally designed adequately to engineering standards but deteriorate to the point where they may fail if any structural defects are left unattended. Their failure would inevitably cause damages to properties, loss of lives and financial losses as can be seen from the past landslide incidents in Singapore and overseas. In Singapore, land or property owners, be it private or government agencies are responsible to properly maintain their slope and slope protection structures in a good and safe condition. Therefore, it is important that every owner plays a proactive role in fulfilling their duty by committing to regular inspection, maintenance and explore various measures to safeguard the stability of slopes and earth retaining structure in a good and safe condition.

9.2.3 Singapore Land Authority (SLA)

The Singapore Land Authority (SLA) is a statutory board under the Ministry of Law, focusing on land resource optimization. SLA manages some 14.000 hectares of State land and about 5,000 State buildings. Playing a critical role in optimizing land resources for the economic and social development of Singapore, SLA ensures the best use of State land and buildings, provides an effective and reliable land management system, including issuance and guarantee of land titles and geo-spatial demarcation of land; and enabling full use of land information for better land management and creation of new business opportunities (SLA, 2013). SLA is one of many governmental agencies involved, whom collectively ensure Singapore remains resilient to the adverse effects of climate change.

9.2.4 SHARES – Slope Hazard Analysis & Repository System

Following a prolonged intense rainfall in December 2006, which resulted in a record number of slope failures island-wide, SLA explored into the use of GIS based coincidence modelling to prepare a "Slope Susceptibility Map" for spatial analysis. A geo-statistical model to map slope hazard zones, using contributory factors such as topography, climate and other factors, was looked into. A comprehensive repository of past slope failures was also developed (Ng *et al.*, 2009). This solution was a first for Singapore in the usage of such a methodology to identify locations prone to slope failure and facilitate slope protection works to be carried out.

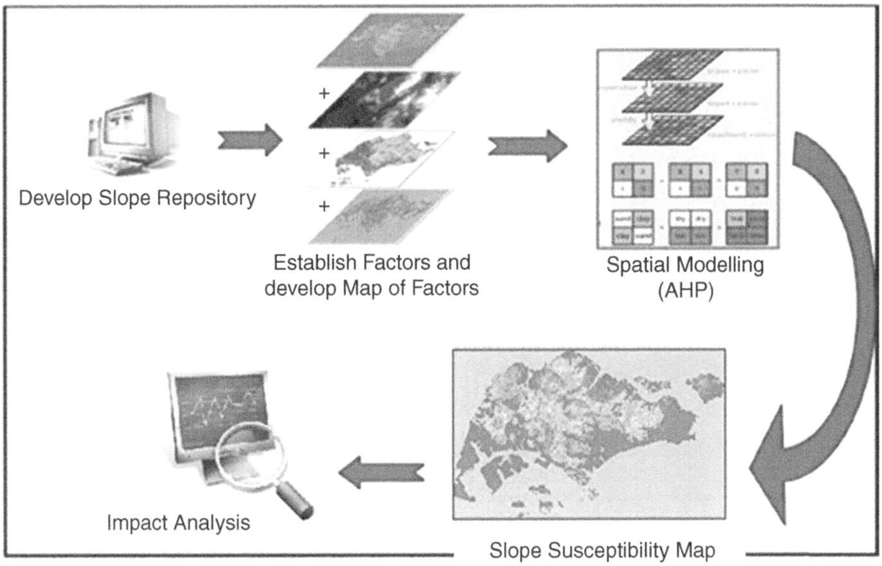

Figure 9.17 Conceptual idea of SHARES.

Only recently, SLA has taken a step further by incorporating this model (now known as SHARES – Slope Hazard Analysis & Repository System) onto GeoSpace, a common portal and platform for government agencies to share and access geospatial data collectively, via a Geocommunity where collaborative efforts between land management public agencies can easily take place. Led by SLA as part of geospatial thrust for whole-of-government, GeoSpace serves as a data clearinghouse and forms the backbone of Singapore Geospatial Collaborative Environment (SG-SPACE), Singapore's national spatial data infrastructure (Loh & Khoo, 2010; SLA, 2011; SLA, 2012). The launch of the SHARES, as an inter-agency sharing model, saw the creation a real-time, map-based forum, which promotes consultation and collaboration within government. Figure 9.17 illustrates the entire conceptual idea of SHARES. Data from past slope failures such as incident date, location, affected area (which can be drawn digitally), slope dimensions, description of failed slope/vicinity and point of contact are first collected, forming a repository, which can thereafter be spatially displayed in Figure 9.18 and Figure 9.19.

The development team analyzed the collected data and 8 factors (cross-referenced with leading academia, experts and Professional Engineers) influencing the occurrence of slope failure, which were: elevation levels (captured on a Digital Elevation Model – DEM), slope steepness (slope angle/gradient), slope aspect (direction as to where the slope is facing towards), rainfall, geology, presence of geological faults and fractures, presence of road network and presence of natural drainages (hydrology). Referencing upon the collected data on past slope failures, development of the model required extensive data preparation and packaged into raster-based GIS layers, which thereafter are overlaid and combined. Categories are formed with each layer, with numerical

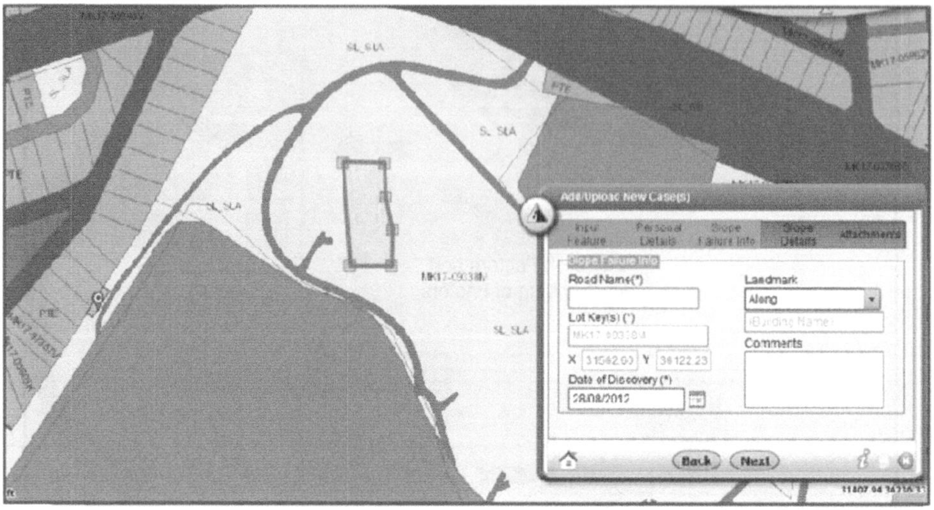

Figure 9.18 Capturing of data from past slope failures.

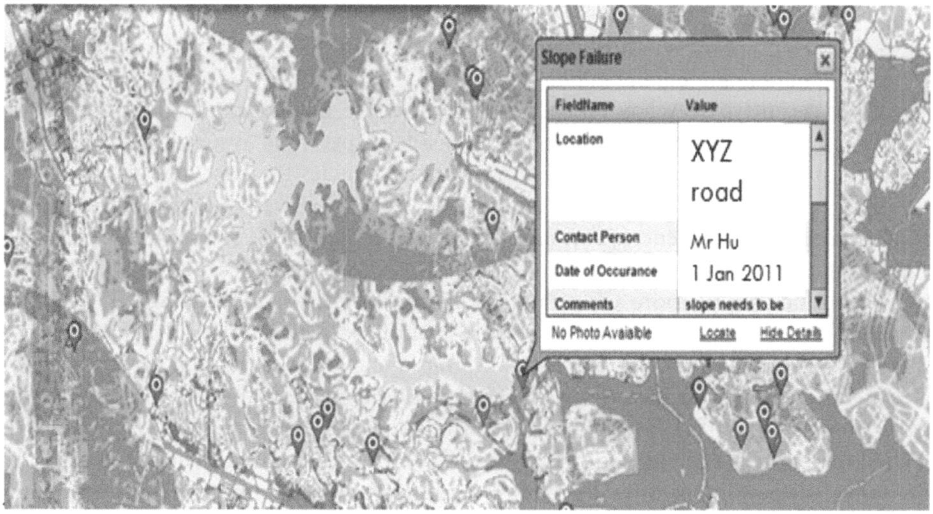

Figure 9.19 Displaying captured past slope failure on a map.

ranking assigned based on a statistical tool – Analytical Hierarchy Process (AHP), so as to determine the relative influence of the various factors. A geospatial coincidence model was then used to combine the different GIS layers together. The resultant map is classified into categories depicting various levels of susceptibility to slope failure, island-wide – as shown in Figure 9.20. A proximity-impact analysis (a 10 m buffer zone around slopes) is then performed to sieve out where the high susceptible locations are to areas with properties, by using another GIS layers such as land ownership, building

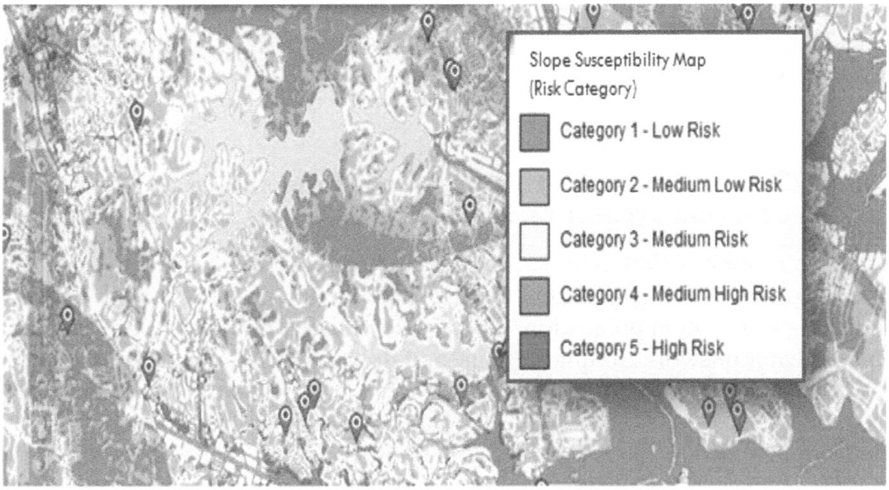

Figure 9.20 Resultant map classified into different categories.

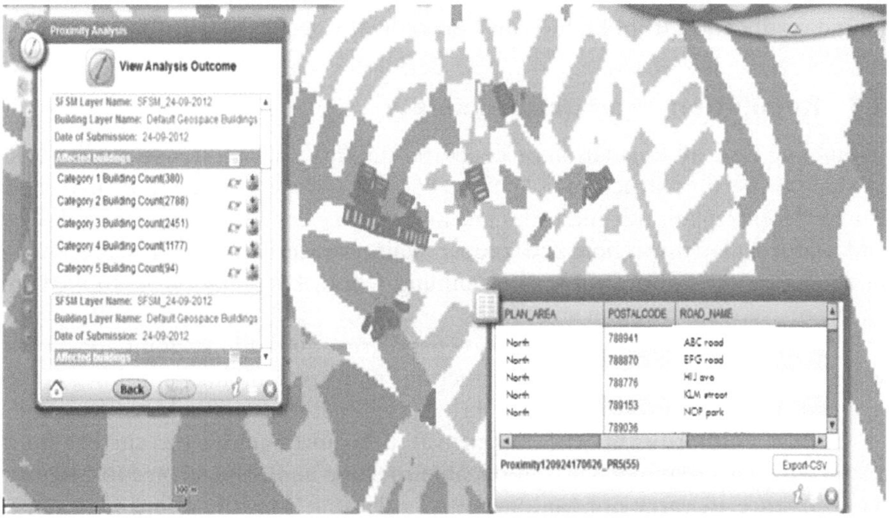

Figure 9.21 Proximity-impact analysis.

locations/footprints together with the generated "Slope Susceptibility Map", as shown in Figure 9.21.

9.2.5 Building and Construction Authority (BCA)

The Building and Construction Authority (BCA) is an agency under the Ministry of National Development, championing the development of an excellent built environment for Singapore. "Built environment" refers to buildings, structures and infrastructure in our surroundings that provide the setting for the community's activities (BCA, 2014). Besides, BCA regulates all buildings, engineered earth retaining works and structures empowered by the Building Control Act of Singapore to shape a safe, high quality, sustainable, friendly and future-ready built environment for Singapore.

9.2.6 BCA preparedness to safeguard slope safety and slope protection structure resilience

As safety is mission critical to BCA, continuous efforts like review and step up the Building Control Regulations, initiating the 3 s-stage slope risk management framework, conducting public communication programme and setting up the efficient incident management framework to enhance slope and slope protecting structure safety regardless of the impacts of climate change. In preparing Singapore build environment for the changing climate with exceptional heavy rainfall, under the Singapore building regulation slopes and slope protection structures, slopes are to be designed to withstand the most onerous ground water conditions. In addition, the Professional Engineering Board (PEB) also set up enhanced requirements in qualifying PEs to register as Professional Engineer (Geotechnical) (PE Geo). On the regulatory framework, BCA also revised the submission guideline for engineered slope and slope protection which required certification from PE Geo and Accredited Checker (AC) submission.

9.2.7 Regulation of structural safety

The Building Control Regulation of structural stability and safety also applies to slope and slope protection structures. The lifecycle of a slope protection structure starts from conceptual plan, adequate design, and quality construction. It is followed by mandatory regular inspection and proper maintenance before it is demolished and rebuilt. Figure 9.22 shows how the Building Control Regulation is applied to the lifecycle of a slope protection structure.

In the design stage, developers (or owners) are required to appoint a Professional Engineer (PE) to design and submit structural plans of buildings and infrastructures (e.g. road, tunnel, railway structure, bridge, viaduct and embankment) to Building Construction Authority (BCA) for approval as required under the current Building Control Act. This requirement is applicable to slope and slope protection structures. It is an offence to carry out building works without first obtaining plan approval and will on conviction be liable to fine, imprisonment or both. In 2004, BCA sent a circular requiring slopes and slope protection structures to be designed to withstand the most onerous ground water conditions. The purpose of this requirement is to enhance their robustness against slope instability as a result of uncertainty in fluctuation of groundwater regime. In addition, the provision of effective drainage systems such as weep holes in earth retaining structures and surface water run-off drain, as well as protection to slope surface are important design considerations.

Upon plan approval, construction works can only commence after obtaining permit from BCA to carry out building works. During construction, the PE appointed by the developer is required to supervise the building works. The builder is required to carry out the works in compliance with the approved plan. Upon completion of the

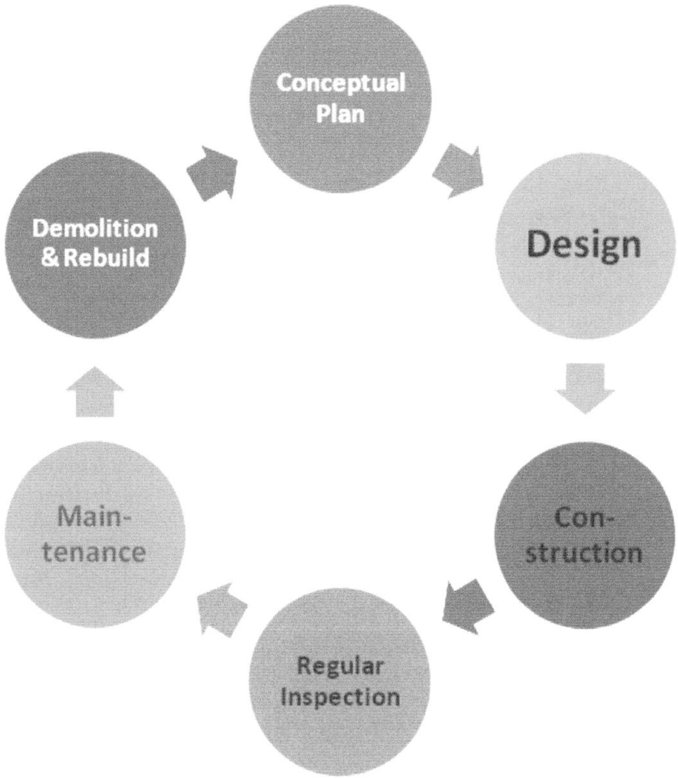

Figure 9.22 Application of Building Control Regulation on the lifecycle of a slope protection structure.

building works, the PE is required to submit as-built plans and a certificate of supervision while the builder is required to submit the certificate of completion of building works. Failure to comply with these requirements is an offence and will on conviction be liable to fine, imprisonment or both. During serviceability stage, all buildings other than landed residential houses are subjected to Periodic Structural Inspection (PSI) under Part V (Inspection of Buildings) of the Building Control Act. Besides building structures, visual inspection would have to cover inspection of earth retaining walls and other slope protection structures. It is an offence for not carrying out PSI as required under the Act and will on conviction be liable to fine. However, PSI is not applicable to landed residential houses. Thus, it is important for the owners (private and government agencies) of these properties to fulfil their duty and responsibility by initiating their own regular inspection and maintenance of the slope or slope protection structures. They are advised to engage a PE to inspect, assess and advise on rectification if they detect structural defects in their buildings. In Singapore, owners are responsible for maintenance of their properties in a good and safe condition. This includes slope and slope protection structures. Therefore, it is in their interest to protect these properties

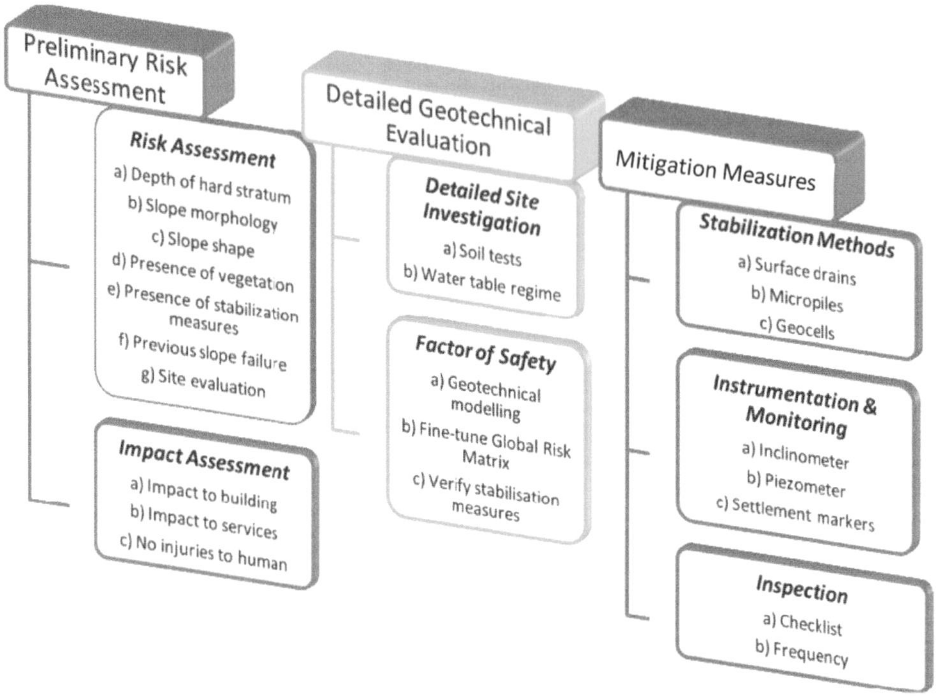

Figure 9.23 Overview of 3-stages slope management system.

to avoid undesirable damage, injury and loss of lives. In this way, it would prevent their financial losses that could be many times more than the routine maintenance cost.

9.2.8 3-Stages slope management framework

The occurrence of landslides arising from exceptional torrential rainfall during 2006 to 2007 occurred in different parts of Singapore served to building and/or land owners as a solemn reminder. Thus, in 2008 government agencies that owned slopes were advised by BCA (Ong, 2008) to take a proactive measure to slope management through adoption of the 3-stage slope management framework (Figure 9.23), which comprises: preliminary risk-impact assessment, detailed geotechnical assessment and mitigation measures.

The purpose of Preliminary Risk-impact Assessment is to identify critical slopes needing further attention on a risk-impact matrix (Figure 9.24). All slopes are first identified and the risk of failure and impact on people and property is assessed across several criteria. Each criterion is given a weightage for different degrees of risk and impact. The summation of the respective risk and impact weightages gives the overall risk/impact level. Then results can be plotted on a risk-impact matrix. Critical slopes are those which fall within the red coloured high-risk zone. They have to be further evaluated to determine the actual safety level and suitable slope strengthening measures

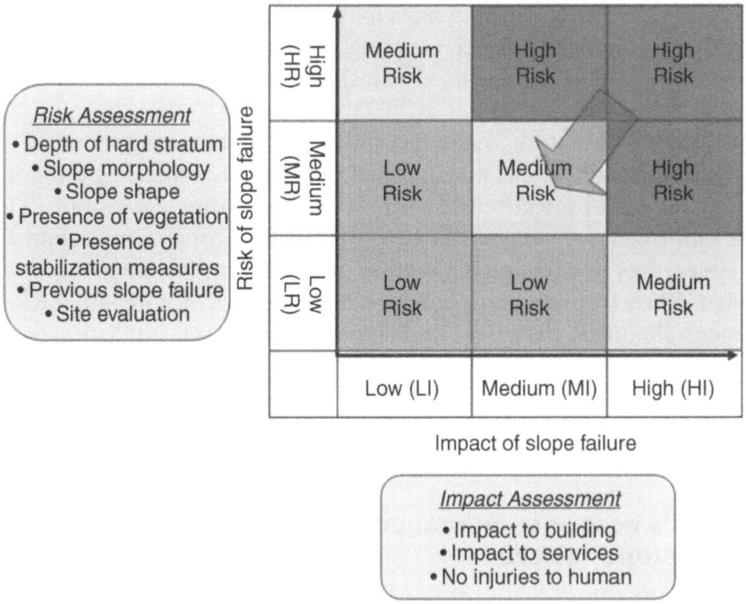

Figure 9.24 Risk-impact matrix.

to be implemented. To do this, investigation needs to be carried out to obtain the actual soil parameters for geotechnical modelling. The objective is to reduce high risk and/or high-impact slopes to a medium level. Generally, medium risk/impact slopes need to be instrumented and monitored while low risk slopes need to be inspected at appropriate intervals corresponding to frequency/amount of rainfall. This initiative kicks off most of the Singapore government agencies to start the slope and slope protection structure program to prepare for the changing climate. BCA has stepped up efforts to work with government agencies and house owners to take proactive actions on slope management via enhancing the Periodic Structural Inspection (PSI) scheme to include inspection and certification of slopes and slope protection structures, conducting dialogues sessions, sending advisory letters and publishing information leaflet to home owners to maintain and protect their slopes.

9.2.9 Periodic Structural Inspection (PSI)

The mandatory periodic structural inspection (PSI) regime is one that is proactive and unique to Singapore. Most exposed slopes in Singapore are composed of unsaturated residual soils which rely on matric suction immediately below the surface for stability. This can be greatly reduced or destroyed by heavy prolonged rainfall, burst underground piping or damaged drainage. The inspection guideline covers both on retaining walls and slope protection structures. Inspecting PEs will be required to give their assessment on the structural integrity and safety of such structures. During inspection the PE can look out for some early warning signs. To the diligent engineer investigation for the above warning signs may involve inspecting the environment outside the site and removing surface covering locally to check for signs of distress. If opportunity allows, it is recommended to inspect just after rainstorm in order to check functionality of drainage system.

Apart from regulation, BCA conduct public seminars to communicate with public highlighting that regular inspection and maintenance are critical as the slope stability and integrity of slope protection structure would over time deteriorate. Therefore, it is prudent that owners, management corporations or property managing agents take preventive measures by carrying out regular inspection and maintenance such as: clear accumulated debris from drainage channels and slope surface, repair cracked or damaged drainage channels, pavement and slope surface cover, unblock weep holes and outlet drainpipes, remove or clear any vegetation causing severe cracking of slope surface, cover and drainage channels, re-turf bare slope surface areas, remove undesirable vegetation from the retaining walls, repair or replace rusted steel for the slope protection structure elements and fill any gap behind the earth retaining structures.

9.2.10 BCA's response management against landslides and slope failure

In the event of landslides and/or slope failure incident occur, BCA will utilize a systematic incident management work flow to ensure an efficient and high productivity response to the incident. Figure 9.25 describes the workflow sequence on how S24A (Dangerous Slope Order) and S25 (Closure Order) are served in the event of landslides and slope failure which endangered public safety.

9.2.11 Summary

Slope failures usually occur during or after severe rainfalls in Singapore, regardless the presence of slope stabilizing features within each slope (i.e. cut-off drains, vegetation or retaining wall). Government agencies in Singapore are prepared to address the climate change and its damaging impacts in Singapore. SLA uses a combination of geospatial tools and Analytical Hierarchy Process (AHP), to create a Slope Hazard Analysis and Repository System (SHARES). The initial vision for SHARES was to simply look into longer term solutions to avoid catastrophic slopes failures that will endanger public safety and damage nearby properties, via the identification of susceptible locations and implementing preventive slope protection works earliest possible, without climate change in mind. By leveraging on SHARES, SLA was able to focus on critical areas, better manage its extensive portfolio of state land and properties and avoid the need to conduct large scale, time-consuming visual inspections, in view of limited resources. It has enabled better and faster decisions to be made in the implementation of crucial preventive measures, before the next catastrophe. As slopes are safer and more resilient after the implementation of preventive measures, usage of SHARES as a tool has proven to be integral to Singapore's sustainable development efforts, as well as provide a good foundation for current efforts in addressing impacts of climate changes.

Predicted future rainfall generated through statistical downscaling and temporal disaggregation from high-resolution climate-change scenarios (Simonovic and Peck 2009), or soil data properties and micro-climatic conditions, could be used as new

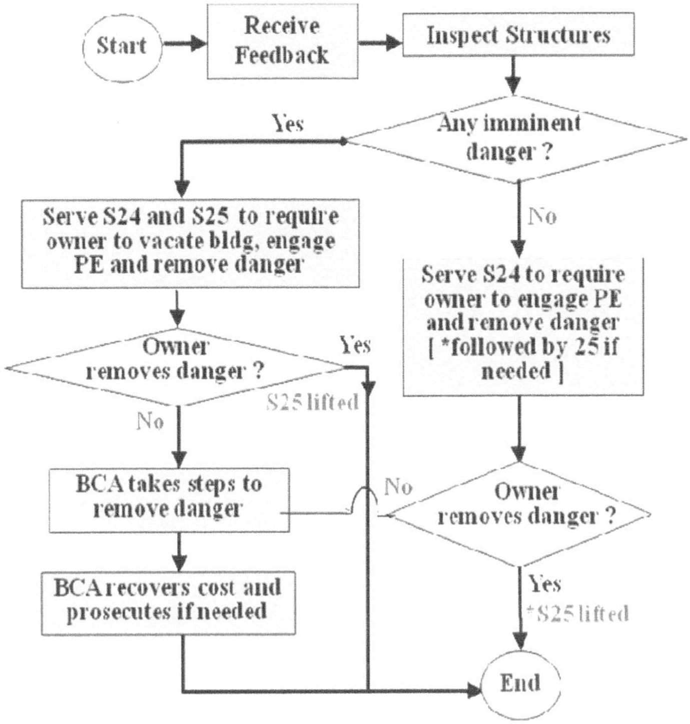

Figure 9.25 BCA work procedure to serve S24A and S25.

factors to help identify locations prone to slope failure more accurately. Collaborative efforts with agencies and academia, shaped by Singapore's current strategy to deal with impacts of climate change, are in the pipeline to further enhance SHARES's analytical capabilities and robustness. The role of citizen engagement and crowd-sourcing data in GIS could be further explored into, as efficient and effective flow of information (accurate, timely, accessible, relevant and transferable) forms an integral part of important operational decisions being made.

9.2.12 Acknowledgements

SLA would like to acknowledge contribution of its partner agencies viz. Ministry of the Environment and Water Resources (MEWR) and Ministry of National Development (MND), who helped in providing the necessary information for this report.

9.3 SLOPE SAFETY PREPAREDNESS FOR EFFECTS OF CLIMATE CHANGE IN MALAYSIA

9.3.1 Slope safety preparedness in Malaysia

Malaysia, as many other countries around the world, has been much affected by the global climate change. According to the Ministry of Science, Technology and

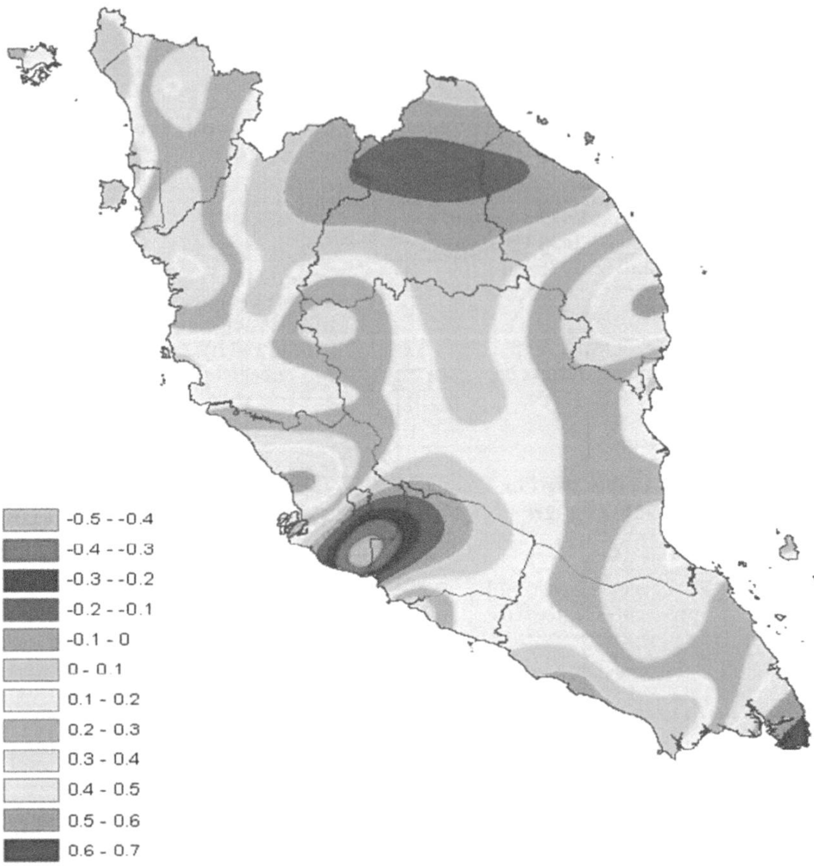

Figure 9.26 Mean annual maximum temperature trend (°C/years) in Malaysia (1951–2012).

Innovation of Malaysia (MOSTI), global warming has affected Malaysia's climate in relation to its temperature, precipitation, wind, lightning and thunder storm. Figure 9.26 shows mean annual maximum temperature trend (°C/10 years) in Malaysia from 1951 to 2012 (MOSTI, 2013).

One of the most important effects of climate change in Malaysia is an increase in rainfall which has resulted in more landslides. Landslides are a big problem that geotechnical engineers are faced with around the world, and especially so in Malaysia. The increase in rainfall has also affected Malaysia's groundwater level. Subsequently, the change in groundwater level has led to a very serious problem in slope stability. The climate condition in Malaysia is characterized by a relatively uniform high temperature, high humidity and high rainfall, with annual rainfall intensity of over 2500 mm. Most of the landslides in Malaysia's two monsoon seasons are caused by heavy rain falls (See-Sew and Siew Wai, 2008).

The most common type of landslide in Malaysia is the shallow slide where the slide surface is usually less than 4 m deep and occurs during or immediately after an

Figure 9.27 Landslide in PayaTerubong.

intense rainfall (Singh, Huat and Jamaludin, 2008). Six major landslides have occurred in Bukit Antarabangsa since 1993, making them significant milestones in the history of landslide disasters in Malaysia, with more than 80 landslides occurring along the road from the main highway to Genting Sempah. Most of these landslides occurred within a few hours after continuous rain for more than 72 hours on 30th June 1995. A total of 21 historical rainfall-induced landslides were reported in the Hulu Kelang area from 1990 to 2011. In 2013, heavy rainfall caused landslide hazards in Bukit Gasing and Paya Terubong. Figure 9.27 shows an image of the landslide in Paya Terubong.

The use of certain species of woody plants to enhance the stability of hill slopes against shallow-seated failures is based primarily on their ability to reinforce the soil they grow in by root binding action and the reduction of soil moisture by evapotranspiration. This particular technique is called the Live Pole. Figure 9.28 shows the Live Pole technique. The basic principle of the Live Pole technique is that the benefits of trees which grow well on hill slopes can be maximized by 'engineering' them. In other words, by planting selected hardy species (short trees or shrubs) in close spacing and deep into the slope, immediate and long term slope stabilization can be achieved (Huat and Mafian, 2011).

Research works have been carried out to review and screen large live cuttings obtained from branches of small trees and shrubs species for their potential ability to propagate. The selected species were then tested for root and stem growth in containers under shade-house condition with irrigation. Ten species were identified and planted in four replicates in a standard medium consisting of crushed well-graded sand and 10% organic matter for about 8 weeks in a shade house. The ten species are: *Hibiscus tiliaceus, Cassia fistula, Dillenia indica, Pterocarpus indicus, Macaranga, Ficus benjamina, Dillenia suffruticosa, Glyricidia sepium, Pajanella longifolia, Erythrina fusca.* Based on screening trials and observations of distribution/location and the shape of

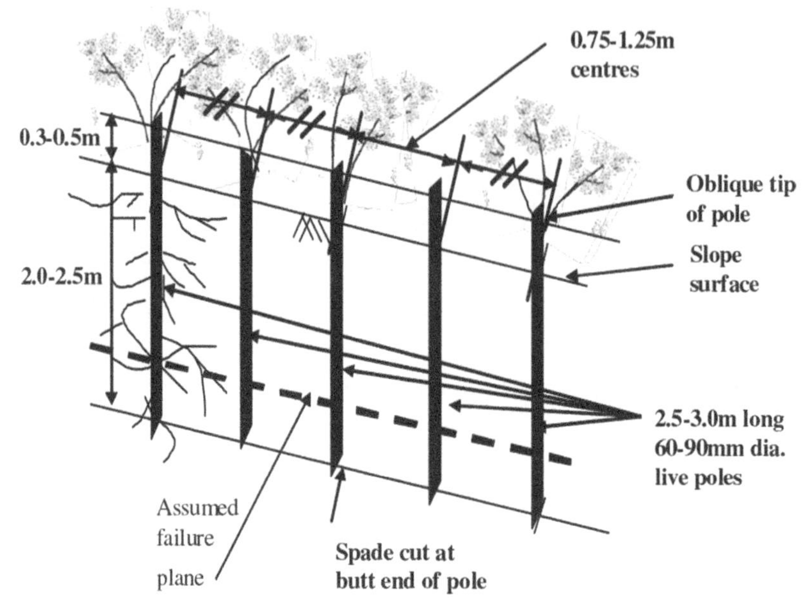

Figure 9.28 Live Pole technique.

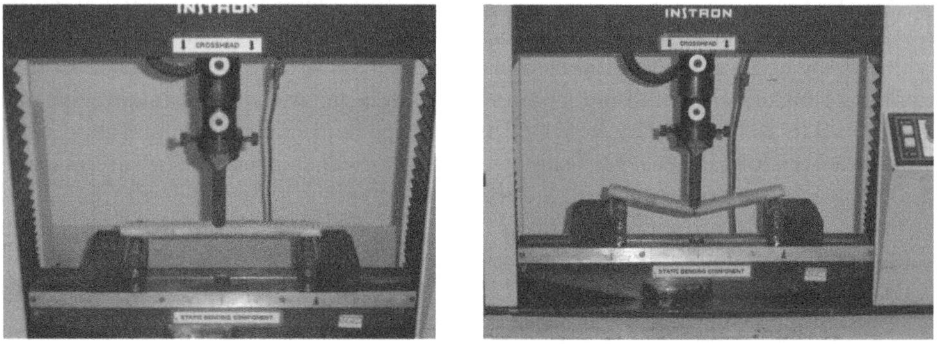

Figure 9.29 Mechanical tests on selected tree species.

growing roots, three species were finally selected as the best candidates for the live pole: *Hibiscus tiliaceus* (*Ht*), *Dillenia indica* (*Di*) and *Dillenia suffruticosa* (*Ds*). Mechanical properties of these species were determined based on BS 373 testing methods, for their compressive, bending and shear properties. Figure 9.29 shows the mechanical tests on the species. Generally, the bending property of *Ht* was almost 26% and 17% better than that of *Di* and *Ds*, respectively. *Ds* however had more strength in compression (nearly 20% and 25% higher than *Di* and *Ht*, respectively). In terms of shear property, *Di* appeared to be better than *Ds* and *Ht* (around 31% and 49% more, respectively). Figure 9.30 shows a comparison of the mechanical properties of selected live poles.

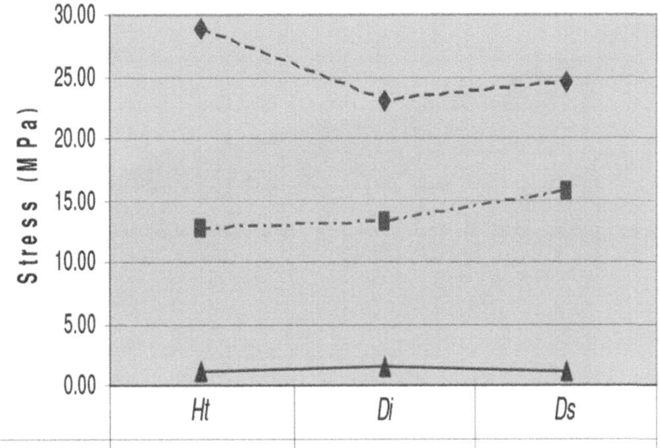

Species	Ht	Di	Ds
- -◆- - Bend. Strength (MPa)	28.925	22.951	24.598
- ·■· - Comp. Strength (MPa)	12.599	13.128	15.696
—▲— Shear Strength (MPa)	1.038	1.549	1.179

Figure 9.30 Comparison of mechanical properties of selected live poles.

For the purpose of the field trial, a meta-stable slope in the vicinity of the Universiti Putra Malaysia (UPM) campus was selected. The slope had a general angle of inclination of about 30°, with shallow failures (0.95–1.5 m deep) at several locations. Most of the failures were rotational failures, but translational failure was also evident in this slope. Two trial strips each of about 2.0 m wide and 7.5 m long along and down the slope were installed with a grid of live poles at 0.5 m and 0.75 m staggered centers across and down the slope. *Hibiscus tiliaceus* and *Dillenia suffruticosa* were selected for the live poles. They were planted as stem cuttings (50–80 mm in diameter) in alternate rows, and to a depth of 1.5 m below the ground surface. The first strip (1) was a re-graded section of an already failed slope. Figure 9.31 shows the trial step.

The planting operation was completed in 2 days. All fresh cuttings of *Ht* and *Ds* had initial lengths of 2.10–2.30 m, diameters between 50–70 mm at upper end and 50–80 mm at butt end; and were almost straight, smoothly tapered, and with no bends or branch points. The poles were installed for most of their length into the slope; they were free of bumps or angularity which might prevent or hinder entry into the prepared holes. Each pole was placed butt end first into the base of a pre-formed hole and driven into the slope using a 14 lb hammer until about 1.5 m of length. After installation, the annulus between the pole and its preformed hole was backfilled with fine dry sand to within 250 mm of the surface of the slope. The final 250 mm was then backfilled with the removed soil and topsoil. On completion of driving, the exposed end of each live pole was trimmed cleanly at an angle of about 60° to the longitudinal axis of the pole. After completing the planting stage, the required details and specification were recorded for each installed pole. The field trial was monitored for the next 6 months.

Figure 9.31 The trial step.

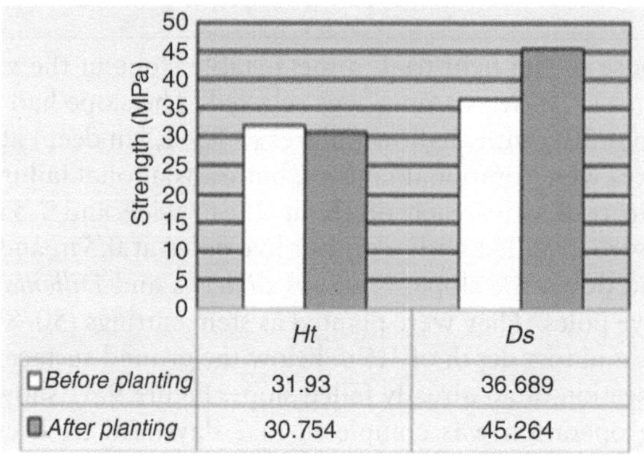

Figure 9.32 Bending strength of Ds and Ht.

Bending strength of *Ht* poles was about the same before and after planting. Figure 9.32 shows the bending strength of the *Ds* poles which had increased by about 25%.

The presence of roots significantly improved the shear strength of the soil, in particular its apparent cohesion. Table 9.2 shows the shear strength of the root system. As a result of incorporating the effect of (*Ht*)/(*Ds*) roots on the slope using construction industry research and information association (CIRIA) Method, to look at the use of vegetation in civil engineering (Coppin & Richards, 1990). According to the study, the influences of *Ht*/*Ds* roots on the slope FOS (Equation 1) increased from 0.93 to

Table 9.2 Shear strength of root system.

	Direct Shear Test			
Sample	σ_n *(kPa)*	φ *(kPa)*	τ *(°)*	*c* *(kPa)*
Jln MARDI SP	306.59	121	17.6	11.2
(Depth 30 to 50 cm)	459.88	131		
	613.17	218		
Jln MARDI SP + Ds roots	306.59	139	16.0	56.5
(Depth 30 to 50 cm)	459.88	199		
	613.17	227		
Jln MARDI SP + Ht roots	306.59	146	17.1	60.2
(Depth 30 to 50 cm)	459.88	220		
	613.17	240		

Figure 9.33 Live pole treated slope in 2014.

1.11 (20%) (Mafian, 2009). Figure 9.33 shows the most recent photo from the site in 2014. The live poles provided a form of vegetated soil nailing or dowelling which offered improvement to slope stability and can be used in the prevention and repair of shallow slips on highway embankment and cut slopes, as well as natural slopes. The stability of the slope is increased over time through the benefits of the live pole, i.e., in the establishment of the root system, the increase in soil suction, and the reduction in soil moisture. Adopting a green solution such as the live pole for slope stabilization demonstrates an example of the role that an engineer can play in helping towards sustainable development and in dealing with the climate change in Malaysia.

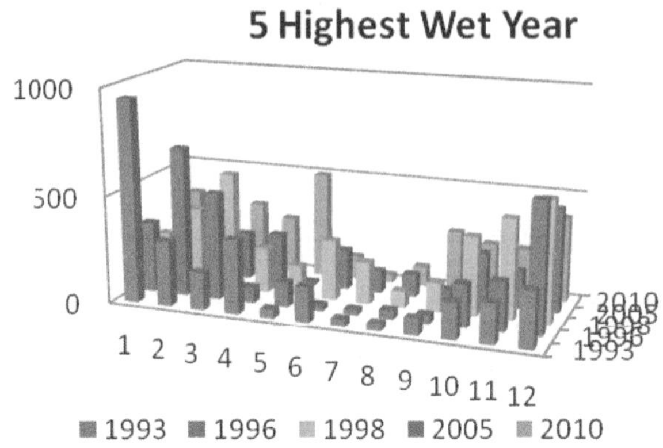

Figure 9.34 Trend of Precipitation in West Java, Indonesia (data from BMKG).

9.4 SLOPE SAFETY PREPAREDNESS FOR EFFECTS OF CLIMATE CHANGE IN INDONESIA

9.4.1 Background on slope stability problems in Indonesia

Groundwater, perched water and ground water pressures, as controlling factors of landslides- are dependent upon climate. Climate variation will affect the hydrological environments governing landslides through changes in antecedent pore-water pressures and trigger landslide event (Rahardjo, 2002; Tohari, 2012). This in turn could lead to a change in the frequency and distribution and mode of landslides. Of particular interest is the existence of water sensitive soils and rocks including clay shales and expansive soils which is believed to be derived from clay shales.

Natural slopes that were originally stable may become unstable after many years due to changes in topography, ground water flow, weathering, change in stresses as well as climate change with changing trends of temperature and rainfall intensity. Figure 9.34 shows five (5) selected rainfall intensities of the highest wet years in West Java, Indonesia where the frequency of landslides is the most severe among other provinces of Indonesia, and may be the most frequent in the world. The change of temperature in South East Asia is of particular concern. According the scenario depicted by IPCC AR4 (2007), the annual change of average temperature in 2080 will be about 2.49°C and the increase of precipitation will be about 5.1% (Figure 9.35). Based on this scenario, Tohari (2012) has shown that the change in the degree of landslide hazard is thus strongly influenced by the characteristics of precipitation in the future. However, the modelling results by Tohari (2012) suggested that climate change is unlikely to have dramatic consequences on the frequency of landslides in West Java region during a wet-month period. Although the climate change signal suggests some changes in periods of instability, it is likely that factors such as land use change and human activity will have a greater impact on landslide frequency than climate change over the next 80 years.

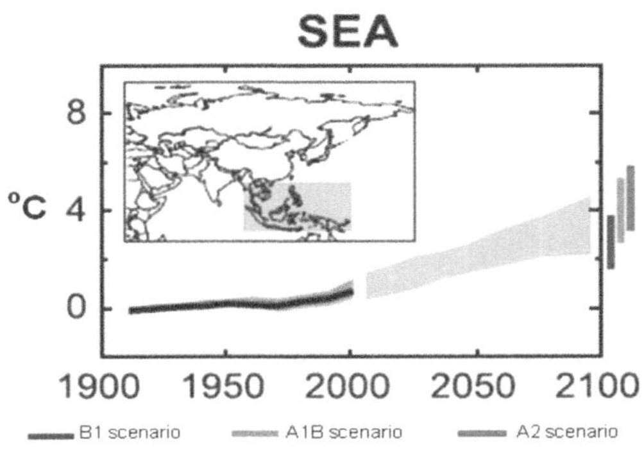

Figure 9.35 Climate Change prediction according to IPCC's AR4, 2007 (Tohari, 2012).

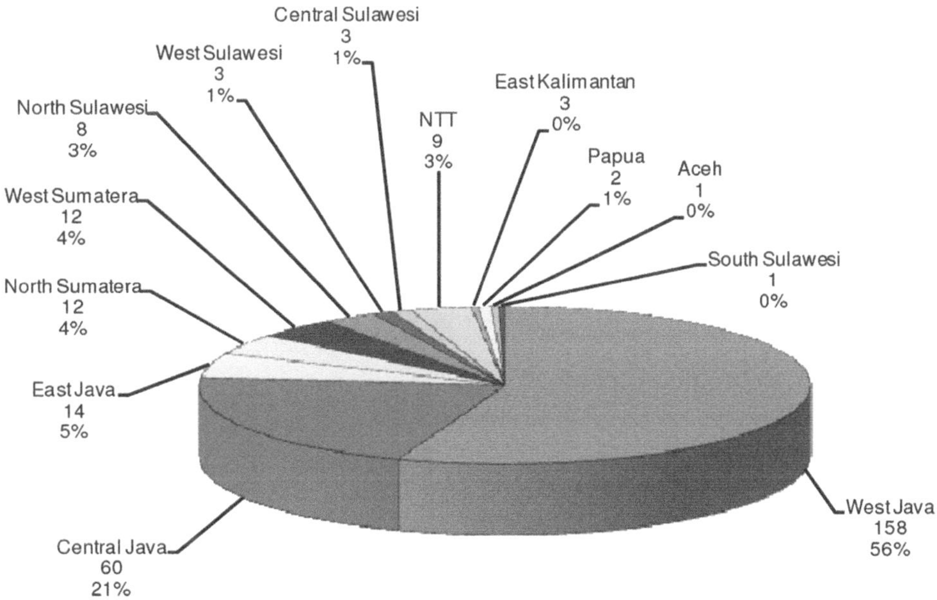

Figure 9.36 Statistics of Landslides in Indonesia from 2003 to 2007 (data from LIPI-Geoteknologi).

9.4.2 Landslides statistics in Indonesia and the existence of clay shales

The landslides statistics in Indonesia between 2003 and 2007 is shown in Figure 9.36. Out of more than 250 landslides occurrences in Indonesia, more than 56% were in West Java. These are attributed to the topographical aspects, rain intensity and the easily degradable materials or water sensitive materials, especially the so called clay shale and expansive soils. Hence it is of interest to discuss our understanding about clay shales. Figure 9.37 shows the distribution of landslides in Indonesia. The extended

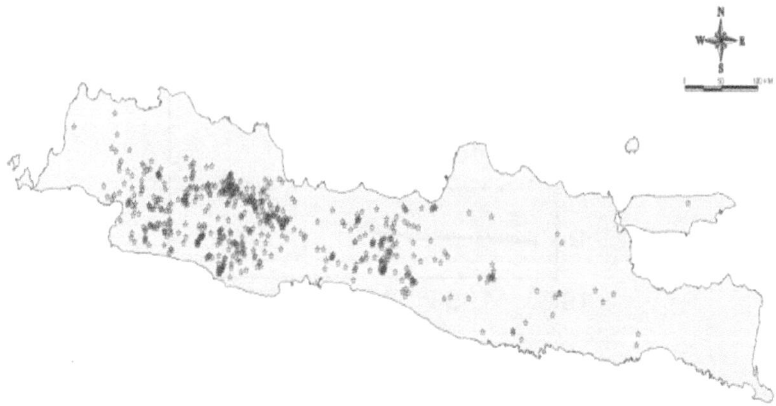

Figure 9.37 Distribution of landslides in Java and its geomorphology.

Figure 9.38 Exposed clayshales and expansive soils in cut and its fissility upon exposure to air or water.

distribution of clay shales and expansive soils deposits are among the reasons why West Java is always prone to landslides.

9.4.3 Clay shales and expansive soils

Clay shale consists of a clastic sediment, originated in the tertiary geological period. It is classified as transition material between Rocks and Soil. The behavior of clayshale is as expansive rock, soften and break upon exposure with water. The main problems of clay shale is its ease of "slaking" and "fissility". The degradation could occur in a matter of hours or days. Expansive soils are those soils that have the capability of adsorption of water, and they swell or shrink depending on their water content. Once it softens, the cohesion is diminished and the friction angle upon shearing is very low, as low as 8–12 degrees. Figure 9.38 shows how clay shale and expansive soils look like upon exposure.

9.4.4 Reactivation of old landslides

Natural slopes are subject to progressive failures where materials tend to be weathered such as clay shales where the chemistry bond degraded gradually and disintegrated.

Figure 9.39 Effect of breccia resliding over clayshale (Rahardjo, 2003).

According to Bjerrum (1966) the disintegration causes release of substantial energy that was stored in its chemical bonding. Consequently, they cause reduction in the shear strength of the materials. Research on landslides of clay shales were started by Bishop (1966), Bjerrum (1966), and Skempton (1964).

Reactivation of breccia re-sliding over clay shale has become common phenomena in the West Java and also Semarang Area. Rahardjo, P.P. (2003) has explained the mechanism as follows: volcanic breccia was deposited on top of clay shale very long time ago as *lahar* flows after volcano eruption. Clay shale is material that is degrading rapidly due to exposure in water or air. In the long run after consolidation, the material might be stable, however, the inclination of clay shale boundary with the breaccia enable creep or even sliding of the breccia over clay shale.

Several landslides accidents occurred in the past and recently, after prolonged rainfall. The breccia is porous material, hence infiltration is possible, while the clay shale is practically impermeable, and hence the water will flow along the boundary between the breccia and clay shale. Creep or sudden landslides could occur along this boundary. An example of its kind was the movement of 20 m thick layer of breccia along the clay shale after three days of rain in Semarang, central java. Figure 9.39 and Figure 9.40 illustrate the landslide occurrence, where the upper layer moved 4 m laterally.

Another case of interest is the movement of bridge foundation located in clayshale. The Penggaron Bridge is located in landslide area. Landslides occurred just after the completion of the bridge causing delay of the commissioning of the bridge for safety reason (Figure 9.41). The causes and the consequences of the landslide of the bridge were investigated and actions for maintaining the safety of the bridge were recommended. The main task of the geotechnical analyses was to ensure (1) whether the ground movement would still occur, (2) location, direction and rate of movement (3) the magnitude of additional loads on piles caused by the ground movement and (4) decision for strengthening of the bridge foundation and, (5) monitoring of the bored pile performance. Due to landslide failure located below the tip of the existing foundation, the strengthening of the foundation was carried out with larger bored piles of 1500 mm diameter and 35 m long surrounding each pile cap of the piers. Inclinometers were installed inside the bored piles and being monitored until present (Figure 9.42). The results of inclinometer measurements can be used to calculate not only pile movement but also the performance of the protection bored piles.

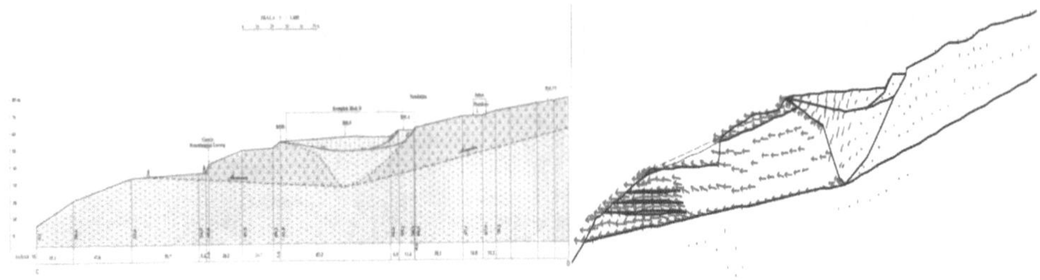

Figure 9.40 Geological Profile and FEM Model (Rahardjo, 2003).

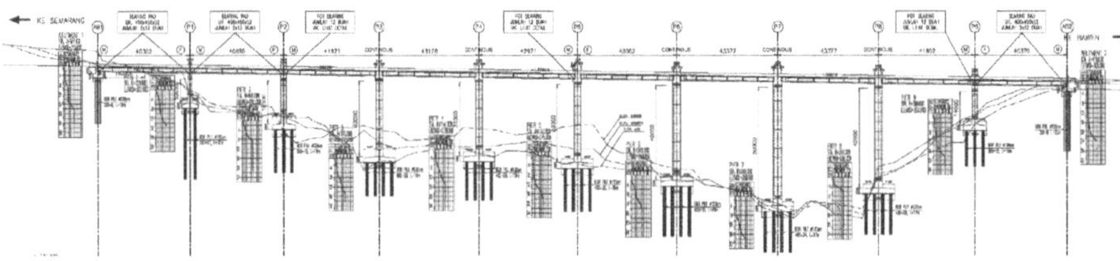

Figure 9.41 The bridge with foundation in the moving ground.

Figure 9.42 Condition of the bridge and movement of foundation as shown by inclinometers.

Based on the results of investigation, it was decided to install bored piles surrounding each pile cap of Pier 2 to pier 8. Each pile cap was strengthened by 16 bored piles with 1.5 m diameter, 35 m deep (Figure 9.43 and Figure 9.44). The pile caps of the new bored piles were connected to the existing pile cap. For each pier, two inclinometers were installed inside the bored piles.

Ten inclinometers were installed inside protection bored piles, however only 8 inclinometers were valid for interpretation. It is of interest to use the results of embedded inclinometers to assess the performance of the bored piles in resisting the soil movement and for calculation of internal forces in the bored piles. Installations of bored piles were completed in mid-November and inclinometer monitoring was started in early December 2011. It was detected that the protection bored piles started to move in the first week of January 2012 and the movement started to slow down about the

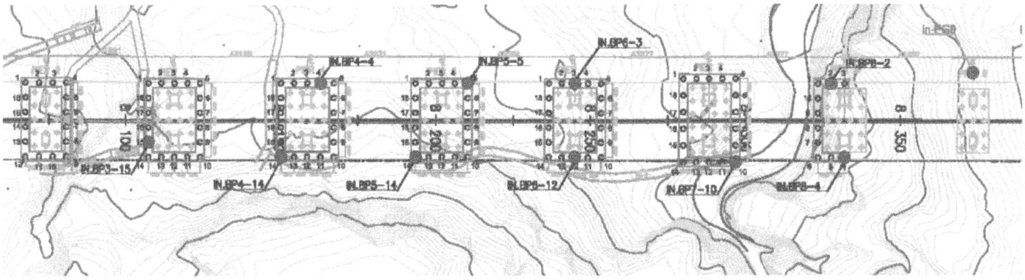

Figure 9.43 Protection of the Bridge Foundation using bored pile "rings".

Figure 9.44 Installation of bored piles for protection of the foundation.

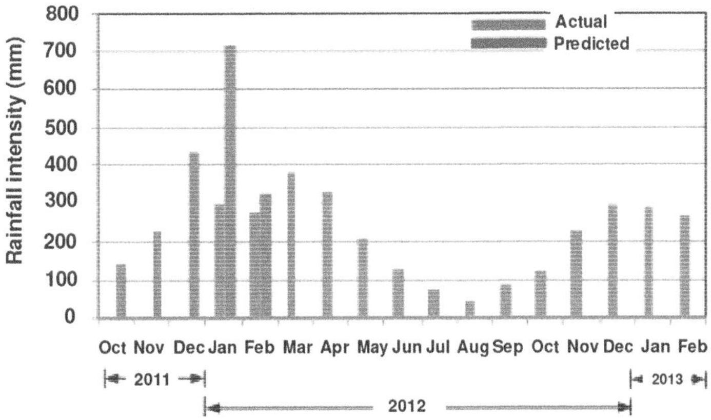

Figure 9.45 Rain intensity comparison between those predicted and measured.

third week of February 2012. Data of rainfall intensity for the same week was very high and this high rainfall intensity was believed to cause the ground movement. The bored piles have apparently been moved to mobilize the passive resistance. In January 2012, rainfall intensity measured near the project site was almost double then predicted value (Figure 9.45).

(a) (b)

Figure 9.46 Slope stability problems in Thailand (a) School building damaged by landslide in Padae, Tak, Northern Thailand (b) Landslide scars in forest area, Natum, Surathani, Southern Thailand.

9.4.5 Summary

Landslides hazards have been a major problem in Indonesia, mainly in Java where rock degradation is very high together with the high intensity of rain, causing the instability of the slopes due to water movement. This section summarizes the increasing number of landslides due to climate change in Indonesia with particular discussion on the existence of clayshale and expansive soils.

9.5 SLOPE SAFETY PREPAREDNESS FOR THAILAND

9.5.1 Background on slope stability problems in Thailand

In Thailand, landslide is one of the most serious natural disasters which can lead to loss of lives and affect various infrastructures, such as highways, life-lines, hillside development, forest land, and highland agriculture (Figure 9.46). While factors such as geology, slope gradient, aspects, vegetation etc., play their important roles in slope stability, rainfall is found to be the final triggering factor of landslide in Thailand. Figure 9.47 shows the rainfall patterns that lead to past major landslide events in the country. Two critical rainfall envelopes (accumulated rainfalls of 150 mm and 300 mm respectively) are empirically drawn in order to categorize localized landslide events and more destructive wide spread landslide/debris flow events. When accumulated rainfall exceeds 300 mm, numerous landslides, normally accompanied by flash flood, are generally to be expected in a wide area within the affected watershed. Even when total rainfall exceeds only about 150 mm, localized slides are frequently observed in slopes affected by human activities such as cut/fill. Though these critical rainfall envelopes provide a useful indicator of when landslides are likely to occur, they do not explicitly take into account the complex interaction between slope and climates. Being empirical in nature, they are only approximate and may not be entirely accurate in all circumstances.

A number of organizations in Thailand have been active in mitigation of landslide risk in various aspects, such as landslide hazard mapping, slope stabilization and

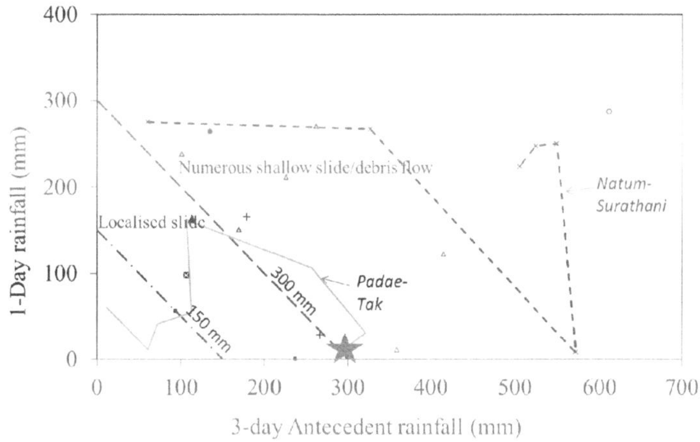

Figure 9.47 Rainfall patterns for past landslide events in Thailand.

development of early warning system, such as Dept. of Highway (DOH), Dept. of Water Resource (DWR), and Dept. of Mineral Resource (DMR), etc. In particular, PTT Public Company Limited (or Petroleum Authority of Thailand) has been instrumental in development of bio-engineering technique used in stabilization of slope along the gas pipeline right-of-way from Myanmar to Thailand (Tansamrit, 2003). As regards basic research, Kasetsart University has been active in areas aimed at understanding the hydro-mechanical properties of unsaturated residual soils and interaction between the climate factors and slope stability by means of laboratory and field studies as well as appropriate modelling (e.g. Jotisankasa & Mairaing, 2010; Mairaing *et al.*, 2012).

This section of the paper highlights some overview of research and practical works carried out as related to slope safety preparedness in Thailand. Particular emphasis would admittedly be placed on the author's own works.

9.5.2 Highlights of research findings on pore water pressure variation with rainfall

Since excessive rainfall is the key triggering factor of landslide in Thailand, much emphasis is placed on understanding the interaction between climate and behaviour of residual soils, particularly the pore water pressure development due to infiltration. Characterization of unsaturated properties has thus been carried out, namely, soil-water characteristic curves, permeability function, unsaturated shear strength (e.g. Mairaing *et al.*, 2012). In particular, a number of representative slopes in various locations in Thailand have been instrumented in order to monitor the variation of pore water pressure and their relation with rainfalls. Some key findings will be presented here.

Figure 9.48 shows the typical contours of pore-water pressure in a granitic fill slope along Highway No.1009, KM 42, Doi-intanon, Chiangmai, and Northern Thailand in dry and rainy season (Jotisankasa *et al.*, 2014). The site is of cloud forest climatic type

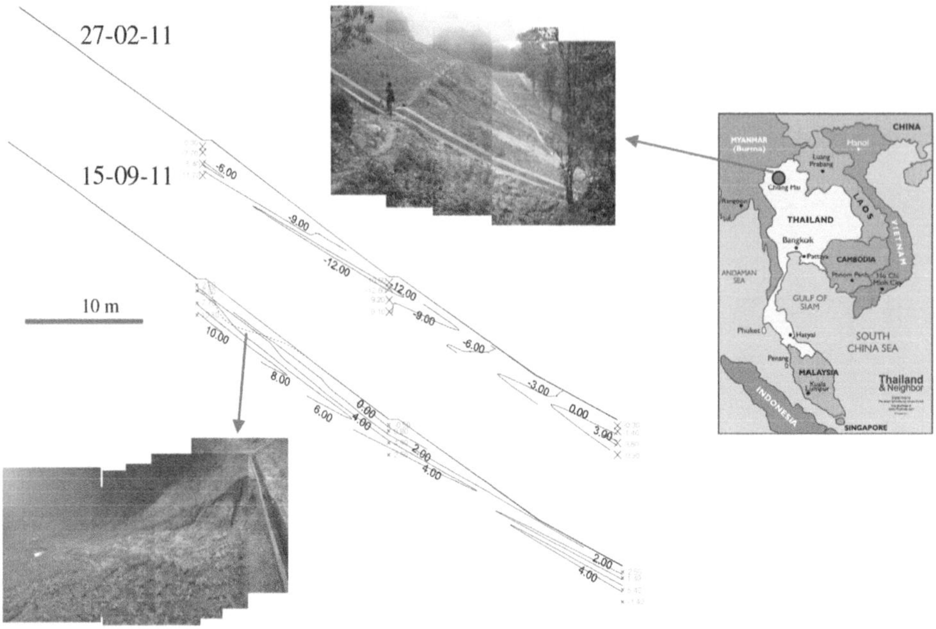

Figure 9.48 Pore-water pressure distribution (kPa) in a fill slope at Doi-intanon site in dry season (27-02-11) and rainy season (15-09-11).

at the elevation of 2,175 m above Mean Sea Level with average high annual rainfall (total annual rain of 3,462 mm in 2011). Therefore, relatively high moisture condition is expected at this site. The slope consists of a decomposed granitic soil fill, which comprises rather heterogeneous alternating layers of two types of soils, i.e. reddish clayey sand and whitish silty sand, classified as SC & SM, mixed with granite boulders. The slope was covered with erosion control blanket vegetated with vetiver and local grasses, together with horizontal drains installed as slope stabilization scheme. The slope was instrumented with 12 tensiometers that are capable of measuring both negative and positive pore water pressure from −90 up to 700 kPa. The tensiometers were installed at the depth of 0.5, 1, 2 and 3 meters at three locations.

In dry season, some suction (5 to 10 kPa) developed in the upper and middle part of the slope while at the slope toe, the ground was still relatively wet with ground water being present at about 1-meter depth which corresponds to the observed spring line. During the instrumentation period on 12 September 2011, the slope suffered from shallow failure and erosion bowl to depth varying between 1 to 3 meters (see Figure 9.48). The incident was triggered by the maximum daily rain of about 100 mm and 3-day antecedent rainfall of up to 180 mm. Distribution of pore-water pressure within the slope about 3 days after the failure is also shown in Figure 9.48 (on 15 September 2011). The location of failed mass (dashed line) falls closely to the zone of high pore-water pressure as observed by the instruments. Due to the heterogeneity of materials at the site, the presence of horizontal drains, and infiltration-induced wetting front, the pore-water pressure regime appeared to be non-uniform and complex.

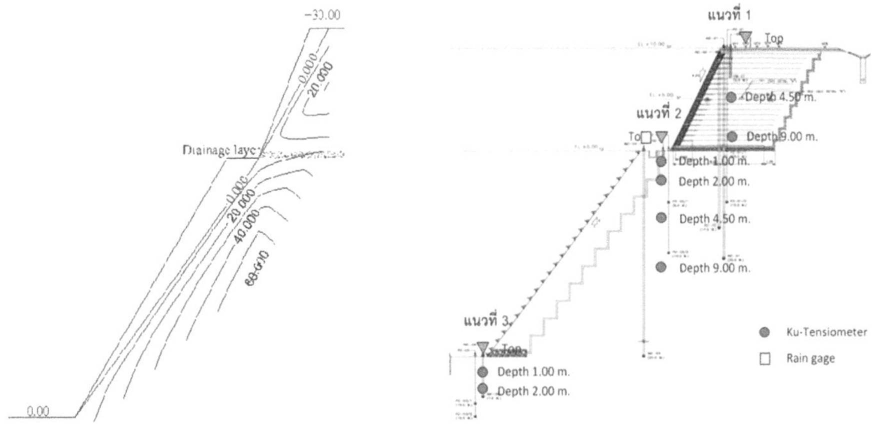

(a) Pore water pressure (in kPa) (b) Location of instruments

Figure 9.49 Pore water pressure distribution for a reinforced slope with underdrain layer (Mae-Lana site, Maehongsorn, Northern Thailand) (Taerakul *et al.*, 2012).

This complexity thus reflects the difficulties in assuming the ground water condition when performing slope stability analysis. In this regards, Jotisankasa *et al.* (2014a) performed back-analysis with known shear strength parameters obtained from laboratory tests and with the ground water condition specified using the pore-water pressure coefficient, $r_u = u/\gamma H$ (u = pore water pressure, γ = soil bulk unit weight, and H = soil height). The pore water pressure coefficient, r_u was employed in order to lump the ground water regime in one simple parameter. The coefficient, $r_u = u/\gamma H$ at failure was about 0.43 or $u = 0.43\gamma H$ and this back-calculated value lies well above the measurement of pore water pressure made a couple of days after the failure. Thus the value of r_u value of about 0.43 may be used as the critical pore-water pressure coefficient for preliminary design and analysis of slope in similar geological and climatic condition. Nevertheless, one should exercise certain cautions since there are still limitations of this finding. For example, Jotisankasa *et al.* (2014a) only performed 2-dimensional stability analysis while the actual failed slope mass and pore-water pressure regime being three-dimensional. More studies are still needed.

Another interesting behaviour of pore-water pressure variation in slope is shown in Figure 9.49 for reinforced embankment slope of highway No.1095 Mae-Hongson Province, also North of Thailand. The slope was constructed of compacted clayey sand (SC) fill derived from weathered argillaceous rock, reinforced by layers of geogrid (0.5 m spacing) and underdrain layer at the base of embankment as shown in Figure 9.49b (Jamnongpipatkul *et al.*, 2008). Tensiometers (capable of measuring both negative and positive pore water pressure) were installed at various depths (from 1 to 9 meter) along the slope. From the results in Figure 9.50, it can be seen that there was some delayed increase in pore water pressure in response to rainfall. Water would take time to infiltrate deeper into the reinforced slope probably due to the soil's lower permeability at greater depth due to its greater density and lower saturation. The pore

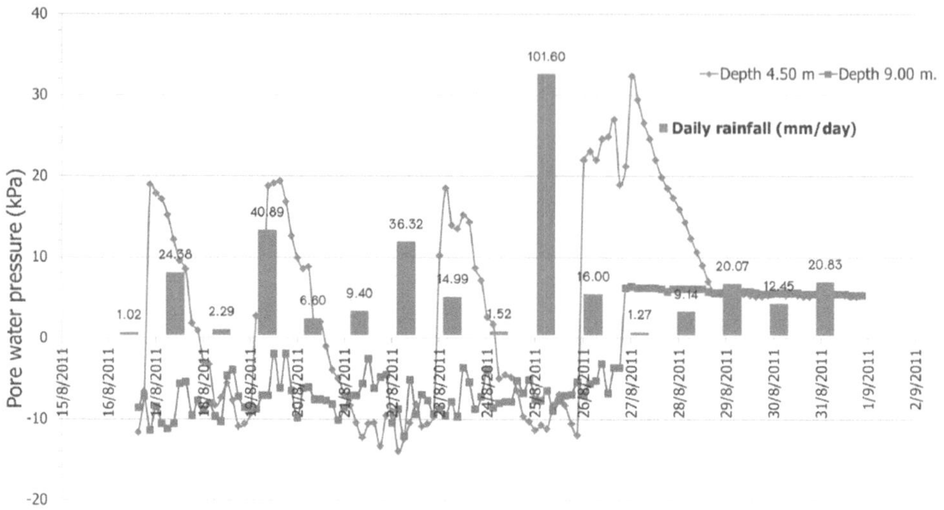

Figure 9.50 Pore-water pressure variation with rainfall within the reinforced slope with underdrain layer at 10 meter depth (Mae-Lana site, Maehongsorn, Northern Thailand) (Taerakul *et al.*, 2012).

water pressure at 4.5 m depth rose to its peak value about 1 day after the rainfall stopped, while the pore water pressure at 9 m depth remained relatively low due to its location being closer to the under drainage layer. One direct implication of this delayed response is that warning based on rainfall alone may not accurately predict the landslide initiation time. This is particular so for soil with lower permeability since infiltration process would take longer time before the pore-water pressure reaches the critical value at critical portion of the slip surface.

9.5.3 Soil bio-engineering

Recently, soil bio-engineering, or the use of vegetation to prevent shallow slides and erosion, has received more attention among practitioners and researchers worldwide due to its relatively low cost, aesthetic and environmental value, as well as sustainability (e.g. Gray & Sotir, 1996; Rahardjo *et al.*, 2012). In Thailand and many other tropical countries, vetiver grass (*Chrysopogon Zizanioides*) have been widely used for shallow slide stabilization and erosion control (e.g. Hengchaovanich, 1998; Truong *et al.*, 2008). Vetiver grass has a very dense fine vertical root system that penetrates as deep as 3–4 meter in some applications, making it very effective for slope stabilization, reduction of runoff erosive energy and sediment trap if planted as hedgerows along the contours. Recently, the author has conducted research aimed at improving the bio-engineering techniques by incorporating other kinds of plants and engineering structures, and also at understanding the hydro-mechanical effects of plants on soil slope (e.g. Jotisankasa, 2013; Jotisankasa *et al.*, 2014b). Figure 9.51 shows example of test site that employs bamboo slope grating with vetiver grass and Jatropha live stake.

Figure 9.51 Example of test sites for bio-slope engineering technique in Thailand: Bamboo slope grating with vetiver grass and Jatropha live stake (a) After installation, 14-Dec-12 (b) 4months, 5-April-13 (c) 8 months, 3-August-13 (d) 1 year, 16-December-13.

9.5.4 Adaptation and emergency preparedness

In order to mitigate landslide hazard effectively and sustainably, the communities at risk must have a responsive understanding of "where" and "when" landslide will take place so that they could plan suitable preventive countermeasures such as early warning system, evacuation protocol and even land management practice for hill slope development. Though various centralized governmental organizations can help develop necessary technologies for that purpose, yet it is the communities themselves who need to react responsibly in case of emergency. By and large, education and community capacity building are the keys to success in adaptation and preparedness to landslide.

In Thailand, several attempts have been made in order to develop technologies for landslide warning such as dynamic landslide hazard map, low-cost landslide warning box, etc, (e.g. Mairaing & Thaiyuenwong, 2010; Soralump, 2013). In practice, however, most local communities still rely on daily rainfall as measured by traditional cylinder rain gauge for warning based on critical rainfall envelope concept, as shown in Figure 9.52. The only drawback associated with manual rain gauge is that its location is usually away from the landslide source uphill where rainfall is usually higher than the measurement point. In response to this situation, several research attempts have been made to develop such telemetry system for landslide warning (Jotisankasa *et al.*, 2010; Jaikaeo *et al.*, 2013)

9.5.5 Summary

- In Thailand, rainfall-induced landslide is considered one of the most serious disasters affecting people's lives and infrastructure. Empirical approach for landslide warning based on critical rainfall envelope has been used.
- On-going research has been carried out at Kasetsart University aimed at understanding at the hydro-mechanical properties of unsaturated residual soils and interaction between the climate factors and slope stability by means of laboratory & field studies as well as appropriate modelling.
- From long-term monitoring of pore water pressure within representative slopes in Thailand, it was found that spatial and temporal variation of pore water pressure within slope can be rather complex due to the heterogeneity of materials at the site,

(a) (b)

Figure 9.52 Technology for landslide warning system (a) Simple cylinder rain gauge (b) Telemetry weather station.

the presence of horizontal drains, and infiltration-induced wetting front. In particular, pore water pressure rise in response to rainfall can be delayed especially for soil with lower permeability.

- Various measures for adaptation and emergency preparedness are employed, such as early warning system, and landslide hazard mapping, etc. In principal, the communities at risk must have a responsive understanding of "where" and "when" landslide will take place and how to mitigate them. Soil-bio engineering or the use of vegetation, notably, vetiver grass, to prevent shallow slides and erosion has been commonly used in Thailand and newer technique involving various plant species in combination with engineering structures are being developed.

9.5.6 Acknowledgements

The author gratefully acknowledges the supports by Chaipattana Foundation, PTT group, Kasetsart University Research and Development Institute (KURDI), Geotechnical Engineering Research and Development Centre of Kasetsart University, and Department of Highways, Thailand.

9.6 SLOPE SAFETY PREPAREDNESS IN VIETNAM FOR EFFECTS OF CLIMATE CHANGE

9.6.1 Background

Global Climate Change – commonly known as global warming and associated sea level rise is one of the most serious challenges facing human beings in the 21st century. Natural disasters and extreme climatic phenomena are increasing in number, affecting the lives of humans across the global (Tai *et al.*, 2009). Unpredictable variability of rainfall patterns could probably be due to global climate change with the increased frequency of extreme weather event, Rainfall is one of the most important triggering factors for shallow and deep – seated landslides in Vietnam (Nawagamuwa *et al.*,

Table 9.3 Temperature trends during the period of 1957–2007.

Climate zone	*Weather station*	*Survey period*	*Increase (°C/year)*
North West	Dien Bien	1958–2007	0.0232
	Moc Chau	1961–2007	0.0121
North East	Sa pa	1958–2007	0.0037
	Ha Giang	1960–2007	0.0176
	Bac Quang	1961–2007	0.0160
North Delta	Phu Lien	1958–2007	0.0111
	Bach Long Vi	1958–2007	0.0106
	Ninh Binh	1969–2007	0.1190
North Central	Vinh	1958–2007	0.0167
	Tuong Duong	1961–2007	0.0177
	Ba To	1980–2007	0.0110
South Central	Quy Nhon	1958–2007	0.0150
	Nha Trang	1958–2007	0.0090
Central Highlands	Buon Ma Thuot	1958–2007	0.0148
	Da Lat	1964–2007	0.0047
South	Can Tho	1958–2007	0.0054
	Rach Gia	1957–2007	0.0045

Source: Institute of Meteorology, Hydrology and Environment.

2011). As a peninsula country located in the tropical monsoon belt of Southeast Asia, Vietnam is one of the nations with high possibility of being the most negatively affected by climate change and global warming (Tai *et al.*, 2009). The country has already been experiencing changes in fundamental climatic elements as well as extreme weather phenomena such as storms, heavy rains, and droughts. They occurred under unusual climatic phenomena like El Niño and La Niña.

9.6.2 Climate change and sea level rise in Vietnam

9.6.2.1 *Temperature*

According to a study reported by the Ministry of Natural Resources and Environment, within 50 years (1951–2000), the annual temperature in Vietnam has increased 0.7°C. The average yearly temperature of four recent decades (1961–2000) is higher than that of the previous three decades (1931–1960). The increasing trend with time is explained more detailed in a report by Tai *et al.* (2009). The temperature in Vietnam is rising by 0.5–0.7°C per 50 years from the South to the North as shown in Table 9.3. A scenario analysis by Nguyen (2009) indicated that the overall temperature in Vietnam in 2100, as compared with the period of 1980–1999 is likely to increase by between 1.1–1.9°C and 2.1–3.6°C. In the high emission scenario, mean annual temperatures in 2020, 2050, 2070 and 2100 are projected to increase by 0.3–0.6°C, 0.8–1.5°C, 1.3–2.2°C and 2.1–3.6°C, respectively, relative to the baseline period (1980–1999) (Table 9.4).

9.6.2.2 *Rainfall and tropical cyclones*

The changing trend of annual rainfall in Vietnam over the nine decade period (1911–2000) is not clear as it depends on the specific time and regions. The annual rainfall

Table 9.4 The temperature changing scenarios for Vietnam.

Climate zones	2020			2050			2070			2100		
	Low	*Medium*	*High*	*Low*	*Medium*	*High*	*Low*	*Medium*	*High*	*Low*	*Medium*	*High*
North West	0.5	0.5	0.5	1.2	1.3	1.3	1.6	1.9	2.0	1.7	2.6	3.3
North East	0.5	0.5	0.5	1.2	1.2	1.3	1.5	1.8	1.9	1.7	2.5	3.2
North Delta	0.5	0.5	0.5	1.2	1.2	1.3	1.5	1.8	1.9	1.6	2.4	3.1
North Central	0.6	0.6	0.6	1.4	1.5	1.5	1.7	2.1	2.2	1.9	2.8	3.6
South Central	0.4	0.4	0.4	0.9	0.9	1.0	1.2	1.4	1.5	1.2	1.9	2.4
Central Highlands	0.3	0.3	0.3	0.8	0.8	0.8	1.0	1.2	1.3	1.1	1.6	2.1
South	0.4	0.4	0.4	1.0	1.0	1.0	1.3	1.6	1.6	1.4	2.0	2.6

Source: Climate Change, Sea Level Rise Scenarios for Vietnam, MONRE, 2009.

Table 9.5 The annual average rainfall in recent decades (mm).

Decade	*Ha Noi*	*Da Nang*	*Tan Son Nhat*
1911–1920	1.521		1.829
1921–1930	1.789		2.063
1931–1940	1.691	1.919	1.924
1941–1950	1.845	2.223	1.926
1951–1960	1.622	1.970	1.805
1961–1970	1.557	2.095	2.005
1971–1980	1.788	2.019	1.828
1981–1990	1.697	1.962	1.813
1991–2000	1.590	2.434	1.850

in Vietnam is within a range of about 700–5000 mm. The most common amount of rainfall recorded is around 1400–2400 mm. Under the influence of climate change, the annual rainfall is likely to increase by between 1.6% and 14.6%. According to a statistic from the institute of Meteorology, hydrology and environment, the annual rainfall has experienced a significant change in the last 100 years (Table 9.5).

The rainfall scenario for the future can be predicted based on the emission scenario according to Quang (2012) as illustrated in Table 9.6. If based on the minimum emission scenario: until the end of 21st century, the precipitation increases popularly by more than 6%, except in Tay Nguyen region, this increase is less than 2%. If based on the medium emission scenario: the precipitation will increase all over the country. The increase is generally around 2%–7%. The range for Tay Nguyen is from 1% to 3%. If the high emission scenario is used: the precipitation will increase all over the country. The increase is generally about 2%–10%. The range for Tay Nguyen is from 1% to 4%.

9.6.2.3 *Sea level rise*

Vietnam is one of the five countries that are likely to be the most affected by global climate change and sea level rise. Several studies have reported these problems. According to the United Nations Environment Program (UNEP, 1993) sea levels around Vietnam

Table 9.6 Rainfall changing scenarios for Vietnam.

Climate zones	*2020*			*2050*			*2070*			*2100*		
	Low	*Medium*	*High*	*Low*	*Medium*	*High*	*Low*	*Medium*	*High*	*Low*	*Medium*	*High*
North West	1.4	1.4	1.6	3.6	3.8	3.7	4.4	5.4	5.6	4.8	7.4	9.3
North East	1.4	1.4	1.7	3.6	3.8	2.8	4.5	5.4	5.7	4.8	7.3	9.3
North Delta	1.6	1.6	1.6	3.9	4.1	3.8	4.8	5.9	6.1	5.2	7.9	10.1
North Central	1.5	1.5	1.8	3.8	4.0	3.7	4.7	5.7	5.9	5.0	7.7	9.7
South Central	0.7	0.7	0.7	1.6	1.7	1.7	2.0	2.4	2.5	2.2	3.2	4.1
Central Highlands	0.3	0.3	0.3	0.7	0.7	0.7	0.9	1.0	1.1	1.0	1.4	1.8
South	0.3	0.3	0.3	0.7	0.8	0.7	0.9	1.1	1.2	1.0	1.5	1.9

Table 9.7 Forecasted sea level rise in Vietnam, 2020–2100, compare to 1980–1999.

Scenario	*Time*								
	2020	*2030*	*2040*	*2050*	*1060*	*2070*	*2080*	*2090*	*2100*
High (cm)	12	17	24	33	44	57	71	86	100
Medium (cm)	12	17	23	39	37	46	54	64	74

increased by 5 cm between the 1960s and 1990s. The Vietnam Hydro-meteorological General Department estimates that the sea level is rising at an average rate of 2 mm per year. According to a research, if the sea level rises by between 0.2 and 0.6 meters, 100–200 thousand hectares of Vietnamese plains is predicted to be submerged. A one-meter rise would result in 0.3 to 0.5 million hectares of the Red River Delta being under water and 90% of the Mekong Delta would be flooded (Danh, 2012). The sea level rise scenarios released by the Ministry of Natural Resources and Environment in 2009 were built with two levels of environmental emergency: a high and medium level. The results showed that compared to the 1980–1999 period, on average, the sea level rise would measure between 30 cm and 33 cm, and between 74 cm and 100 cm by the middle and end of this century, respectively. Table 9.7 presents sea level rise scenarios for Vietnam.

9.6.3 Landslides due to rainfall

Effects of past, present and future climate characteristics on landslide activities can be assessed using historical information, geomorphological evidence, and modelling of slope hydrology and mechanics. However, there are series of problems related to that approach (Malet *et al.*, 2007). First, quantitative assessments of the spatial and temporal relationships between climate and landslide occurrences are often hampered by the inaccuracy and uncertainty of the historical records. Second, the climate-landslide coupling is complex, because climate is related to landslide via the nonlinear soil water system. Process-based models have been proposed to understand this complex interaction. However, landslide triggering systems show complicated response

Figure 9.53 Sliding along the national highway from Hoa Binh to Son la (8/2005).

in relation to geotechnical, hydrological, and climate properties. Third, the uncertainty in future climate parameters is high, especially if the time context is greater than the weather records or because of the low-resolution of the downscaled climate time series (Malet *et al.*, 2007). However, when all is said and done, since climate change brings more rain to Vietnam with higher intensity and longer duration, the landslide possibility would increase. Thus, there should be more research about the relationship between climate change and the occurrence of landslide or critical rainfalls that trigger landslide to be precise.

Vietnam has 75% of the land in the mountainous area where mining, infrastructure development, industrial zones and settlement area takes place. In these areas, especially in rainy seasons, hundreds of both large scale and small scales landslides occur and cause severe damage to lives and properties. Landslide may occur on natural slopes or cut slopes under the influence of its self-weight and some other auxiliary factors, for example: the pressure from the surface water and underground water, seismic forces and other forces. In Vietnam, studies about landslide have been conducted since 1960s. The problems that have received more attention are: landslide at mining areas, sliding along roads, highways, river banks, levees and dams etc. Especially, the high intensity of rainfall in mountainous areas has caused landslide and flash flood in many regions, especially like Son La, Lai Chau, Lao Cai, Dien Bien in the North and Nghe An, Quang Binh, Thua Thien Hue, Quang Nam, Quang Ngai, in the Middle and so on. Over the period of 16 years, from 1990 to 2005, landslide and flash flood have destroyed 13280 houses, damage badly about 115000 houses, 988 people were killed, 628 people were wounded, and a lot of other damage related to farming production and infrastructure (Hanh, 2009). Figure 9.53 shows a slide along a national highway from Hoa Binh to Son La in August 2005 after a long duration of rainfalls. Another severe slope instability incident was the one that happened on September, 2004. After a heavy rain, a huge hilly area of Phin Ngan commune, Bat Xat district, Lao Cai, slided and buried 20 people (Figure 9.54).

Figure 9.54 Landslide in Bat Xat – Lao Cai in 27/09/2005.

Numerous studies have shown a close relationship between rainfall and the occurrences of landslides. Slope sliding often occurs after rainfalls with high intensity and long duration. Many large scale landslides in mountainous areas in Vietnam are related to heavy rains (Table 9.8) and many landslide regions are often in the areas with high annual rainfall (Table 9.9). The reason behind that can be explained by a study of Thu (2012). When the rain comes, the unsaturated area within the soil layer reduces. As a result, the shear strength decreases and leads to the reduction in the factor of safety of the slope. In order to prove this, the slope on talus of the highway from Tam Ky to Tien Phuoc (at 23 km south of Tam Ky town) was chosen for slope stability assessment. The residual soil is silty clay with gravel of the thickness of 7 m. Figure 9.55 shows the geometry and mesh of the example utilized for the steady and transient seepage analyses. In the transient seepage analysis, boundary flux, q, equal to 1.0×10^{-7} m/s is applied to the surface of the soil slope as the influence of rainfall. Evaporation and evapotranspiration are not simulated in this study. The applied boundary flux, q is reviewed by elevation in order to avoid ponding. Figure 9.56 shows the factor of safety obtained from slope stability analysis with time during infiltration and drying processes. The factor of safety was 1.50 before rainfall and decreased as rainwater infiltrated into the slope. After 6 hours of rainfall the factor of safety was 0.98, indicating that failure could have occurred. Subsequently, the factor of safety increased gradually after the rainfall had stopped.

9.6.4 Stability problems in the seadike system in Vietnam

Vietnam covers a land area of 330,000 km^2 and a very long coastline (over 3254 km) which runs through 29 provinces and cities. Approximately 1000 km of sea dike and almost one thousand kilometers estuarine dike and embankment for flood prevention was constructed at Cuu Long river delta. Therefore, the sea dike system is of great importance to Vietnam, one of the most vulnerable countries to climate change. However, this system at the present is only designed to withstand a maximum storm at

Table 9.8 Rainfalls and landslides in Vietnam.

Number	*Time*	*Location of Landslide*	*Rainfall Intensity*
1	11–1964	Que Son, Quang Nam	300–1000 mm
2	10–1969	Quan Cay, Thai Nguyen	600–700 mm
3	12–1986	Son Tra, Quang Ngai	500–1227 mm
4	08–1996	Muong Lay, Lai Chau	300–400 mm
5	11–1999	Phu Loc, Thua Thien Hue	Approximately 1000 mm
6	09–2002	Huong Son, Ha Tinh	500–700 mm
7	09–2005	Van Chan, Yen Bai	Approximately 300 mm

Source: The national centre for Hydro - Meteorological forecasting.

Table 9.9 Landslides occurrence in regions with high annual rainfall.

Number	*Regions occur landslide*	*Annual rainfall (mm)*
1	Bac Quang – Hoang Su Phi (Ha Giang)	2000–4000
2	Bat Sat – Sa Pa (Lao Cai)	2800–3500
3	Muong Lay – Muong Te (Lai Chau)	2000–2500
4	Huong Son (Ha Tinh)	2400–3200
5	A Luoi (Thua Thien Hue)	2400–3000
6	Que Son (Quang Nam)	2500–3500
7	Son Ha (Quang Ngai)	2500–3500
8	Dak Nong (Dak Lak)	2000–2400

Source: The national centre for Hydro - Meteorological forecasting.

level 9. This means that it can only withstand a certain level depending on its significance. According to statistics from the Water Resources University, the sliding situation in Vietnamese sea dike system is at alarming situation, especially in rainy seasons.

- There are 165 points of sliding with a length of 252 km in the North
- In the Middle, 307 sliding points with the total length of over 555 km are found
- The number of sliding points in the South is 265, covering the total length of 450 km.

In addition, in Vietnam, due to the fact that most of sea dike were constructed through many generations by handwork, the location and design of dikes were based on the observations and experience of local inhabitants, where the dike elevation were set based on the highest water levels either experienced directly onsite or suggested by the available history for the area. The crest level and cross section basically do not follow any design standards, so when facing with large storms, rising water level and waves, this dike system becomes too small in comparison to the power of nature. For instance, in the rainy season in 2005, many dikes were broken such as Hai Hau (Nam Dinh) (Figure 9.57), Cat Hai (Hai Phong) (Figure 9.58); Hai Thanh and in some other districts in Thanh Hoa, Ha Tinh, etc., leading to severe in human casualties and property damages.

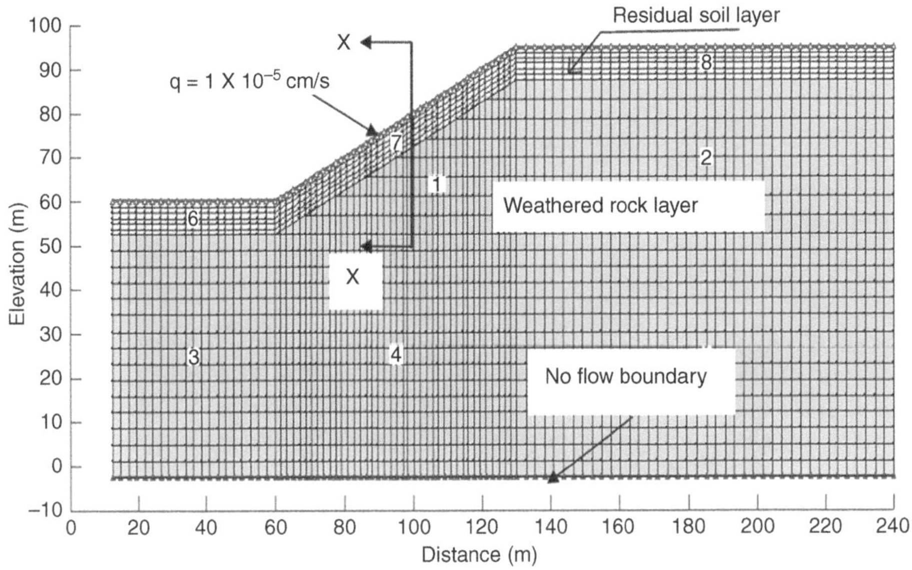

Figure 9.55 Geometry and finite mesh for Seepage analysis.

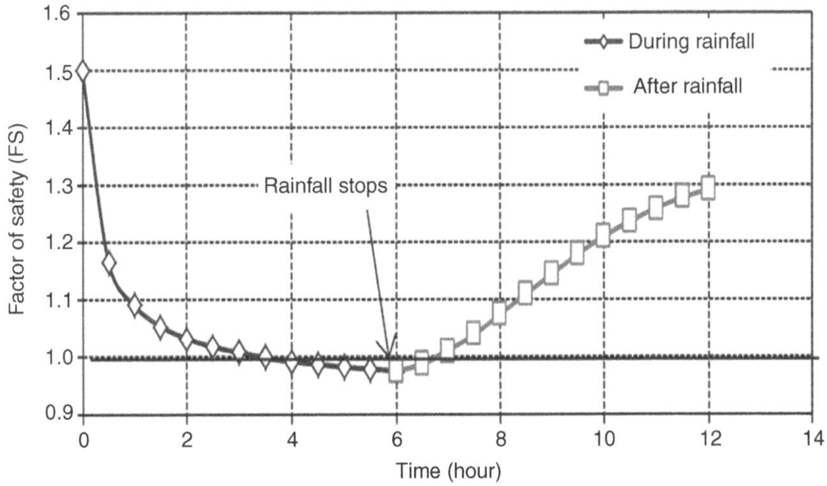

Figure 9.56 Factor of safety versus elapsed time during infiltration and drying processes.

9.6.5 Adaption to sea level rise

Learning from others with similar situations can provide important insights on approaches worth exploring in this region. Currently, these experiences can be monitored on the internet and are being distilled in reports, surveys and guidance materials

Figure 9.57 Hai Hau sea dyke broke due to the Storm No 7 (27/09/2005), source: (Trung 2012).

Figure 9.58 Effect of overtopping at Haiphong dike after Typhoon No. 2 (2005), source: (Cuong 2012).

(Lausche, 2009). According to statistics from Hoa (2012), facing with challenges from sea level rise, in 2006 the Vietnamese government approved a program to upgrade sea dykes from Quang Ninh to Quang Nam province along the country's northern coast. To date, the 13 provinces involved have reinforced 272 km of sea dikes, renovated or built 42 sluices and planted more than 130 ha of wave blocking trees. In Ha Tinh province alone, 100 km of sea dikes have been consolidated. After 3 years of building and upgrading their sea dike system, the 7 Mekong provinces of Tien Giang, Ben Tre, Tra Vinh, Soc Trang, Bac Lieu, Ca Mau and Kien Giang have completed 6 embankment projects. Seven (7) other projects are under construction and 13 more will begin construction soon. However, there are several principles that need to follow:

a) Plan early. Early planning to prevent the bad impacts of sea level rise is essential. Many communities and natural systems have already been at risk. Decisions made today will have consequences for decades. In addition, effective solutions should be made to cope with the gradual changes of the sea level rise problem.
b) Upgrade the existing policy tools: Local governments have many policy tools that have already been available to support local actions for sea level rise preparedness. While new policies will also be needed, existing tools provide the means to start-up the plan.
c) Plans and actions are required to integrate and coordinate across sectors, disciplines and jurisdictions (Lausche, 2009). Also, the government should develop the ability to manage the risk of climate change effectively, and adapt to climate change.

9.6.6 Summary

Slope instability is a very common catastrophe in mountainous regions in Vietnam. It often occurs at a higher frequency in the rainy season and leads to considerable damages. Together with the impact of climate change, storms, and hurricanes that cause heavy rains, the precipitation, rainfall intensity, rainy season, regions with high precipitation are also changing. Thereby, it leads to the change in the scale and development of sliding. So the treatment to lessen slope instability damage is a closed system from information collection, analyzing and assessing the reasons, building the design scheme and the effective plan, executing and monitoring the plan. To manage the bad consequences of landslides and slope instability, it is necessary to have a really good program with multi- participants that can predict, localize, and map future landslides. In addition, the government should educate and improve the knowledge of the entire citizen who live inside the dangerous zone so they can prevent and protect themselves from the threat of slope failure problems.

REFERENCES

Alonso, E.E., Gens, A. & Delahaye, C.H. (2003) Influence of rainfall on the deformation and stability of a slope in over consolidated clays: A case study. *Hydrogeology Journal*, 11, 174–192.

BCA (Building and Construction Authority) (2014) Available from: http://www.bca.gov.sg/AboutUs/about_bca.html.

Bennett, M.R. & Doyle, P. (1997) *Geology and the Human Environment. Environmental Geology*. Chichester, John Wiley & Sons Ltd.

Brabb, E.E. & Harrod, B.L. (eds.) (1989) Landslides: Extent and economic significance. In: *Proceedings of the 28th International Geological Congress: Symposium on Landslides*. Washington, DC, Rotterdam, Balkema. 385 pp.

Brand, E.W. (1992) Slope instability in tropical areas. In: *Proceedings of the 6th International Symposium on Landslides*. Christchurch, New Zealand, Vol. 3. pp. 2031–2051.

Cartier, G. & Virollet, M. (1980) Note technique relative à la surveillance et à l'entretien d'un dispositive de drainage par drains subhorizontaux. France, *Laborataire des Ponts et Chaussées*, Note Technique. 14 pp.

Chigira, M. & Mohamad, Z. (2011) Landslides in weathered granitic rocks in Japan and Malaysia. *Society of Malaysia*, 57, 1–6.

Choi, E.C.C. (1983) Seepage around horizontal drains in hill slopes. *Journal of Hydraulic Engineering, ASCE*, 109, 1363–1368.

Chowdhury, R.K. & Beecham, S. (2010) Australian rainfall trends and their relation to the southern oscillation index. *Hydrological Processes*, 24 (4), 504–514.

Coppin, N. & Richards, I. (1990) *Use of Vegetation in Civil Engineering*. London, Butterworths. pp. 174–180.

Costa Nunes, A.J. (1985) *Written Communication with GCO.*

Core Writing Team, Pachauri, R.K. & Reisinger, A. (eds.) (2007) *Contribution to Working Groups I, II and III to the Fourth Assessment Report of the Intergovernmental Panel on Climate Change*. Geneva, IPCC. pp. 104.

Danh, V.T. (2012) *Adaption to Sea Level Rise in the Vietnamese Mekong River Delta: Should a Sea Dike be built.* Can Tho, Can Tho University.

Derbyshire, E. (1976) *Geomorphology and Climate.* London, John Wiley.

El-Hadji, I.T. & Singh, V.P. (2002) Space-time-frequency analysis of rainfall, runoff and temperature in the Casamance River Basin, South Senegal, West Africa. *Water – South Africa*, 28 (3), 259–270.

Emery, F.E. & Trist, E.L. (1965) The causal texture of organizational environments. *Human Relations*, 18 (1), 21–32.

Fall, M., Azzam, R. & Noubactep, C. (2006) A multi-method approach to study the stability of natural slopes and landslide susceptibility mapping, *Engineering Geology*, 82 (4), 241–263.

Forland, E.J., Alexandersson, H., Drebs, A., Hamssen-Bauer, I., Vedin, H. & Tveito, O.E. (1998) *Trends in Maximum 1-Day Precipitation in the Nordic Region. DNMI Report 14/98, Klima.* Oslo, Norwegian Meteorological Institute. pp. 1–55.

Francis, D. & Hengeveld, H. (1998) *Extreme Weather and Climate Change.* Downsview, Ontario, Environment Canada.

Fredlund, D.G. & Rahardjo, H. (1993) *Soil Mechanics for Unsaturated Soils.* New York, NY, John Wiley & Sons, Inc. 517 pp.

Fredlund, D.G., Rahardjo, H. & Fredlund, M.D. (2012) *Unsaturated Soil Mechanics in Engineering Practice.* New York, NY, John Wiley & Sons, Inc. 926 pp.

Fujibe, F. (2008) Long-term changes in precipitation in Japan. *Journal of Disaster Research*, 3 (1), 51–52.

Fukuoka, M. (1980) Landslides associated with rainfall. *Geotechnical Engineering*, 11, 1–29.

Gallant, A.J.E., Hennessy, K.J. & Risbey, J. (2007) Trends in rainfall indices for six Australian regions: 1910–2005. *Australian Meteorological Magazine*, 56, 223–239.

Geotechnical Engineering Consultant, PT. (2011) *Final Report of Landslides Evaluation of Semarang Bawen Toll Road Section 1, Packet II.* Submitted to PT Trans Marga Jateng.

Glendinning, S., Loveridge, F., Starr-Keddle, R.E., Bransby, M.F. & Hughes, P.N. (2009) Role of vegetation in sustainability of infrastructure slopes. *Proceedings of the Institution of Civil Engineers, Engineering Sustainability*, 162 (ES2), 101–110.

Gray, D.H. & Sotir, R.B. (1996) *Biotechnical and Soil Bioengineering Slope Stabilization A Practical Guide for Erosion Control.* New York, NY, John Wiley & Sons.

Groisman, P.Y., Knight, R.W., Easterling, D.R., Karl, T.R., Hegerl, G.C. & Razuvaev, V.N. (2005) Trends in intense precipitation in the climate record. *Journal of Climate*, 615 (18), 1326–1350.

Hanh, N.H. (2009) Climate change, landslide risks in the mountainous regions of Vietnam. Controlling measured and management. *Vietnam Geotechnical Journal*, 3, 41.

Hengchaovanich, D. (1998) *Vetiver Grass for Slope Stabilization and Erosion Control, with Particular Reference to Engineering Applications. Technical Bulletin No. 1998/2. Pacific*

Rim Vetiver Network. Bangkok, Thailand, Office of the Royal Development Project Board.

Heath, R.C. & Lehr, J.H. (1987) A new approach to the disposal of solid waste on land. *Ground Water*, 25 (3), 258–264.

Hewitt, P.J. & Philip, L.K. (1999) Problems of clay desiccation in composite lining systems. *Engineering Geology*, 53 (1), 107–113.

Hoa, T. (2012) *Sea Dyke – Part of Vietnam's National Development Strategy*. Available from: http://vovworld.vn/en-us/Current-Affairs/Sea-dyke-part-of-Vietnams-national-development-strategy/78709.vov.

Huat, B. & Mafian, S. (2011) Assessment of indigenous plants for live pole applications in slope stability of Malaysia. *Australian Journal of Basic and Applied Sciences* 5 (9), 22–27.

Hutchinson, J.N. (1977) Assessment of the effectiveness of corrective measures in relation to geological conditions and types of slope movement, Theme 3: General report. *Bulletin of Engineering Geology and the Environment*, 16, 131–155.

Intergovernmental Panel on Climate Change (IPCC) (1990) *Climate Change: The IPCC Scientific Assessment*. In: Houghton, J.T., Jenkins, G.J. & Ephraums, J.J. (eds.) Cambridge, Great Britain, New York, and Melbourne, Australia, Cambridge University Press. 410 pp.

Intergovernmental Panel on Climate Change (IPCC) (2007) Climate change: The physical science basis. In: Solomon, S., Qin, D., Manning, M., Chen, Z., Marquis, M., Averyt, K.B., Tignor, M. & Miller, H.L. (eds.) *Contribution of Working Group I to the Fourth Assessment Report of the Intergovernmental Panel on Climate Change*. Cambridge, United Kingdom and New York, NY, Cambridge University Press. 996 pp.

Intergovernmental Panel on Climate Change (IPCC) (2013) Climate change: The physical science basis. In: *Contribution of Working Group I to the Fifth Assessment Report of the Intergovernmental Panel on Climate Change*. Stockholm, Sweden, Final Draft Underlying Scientific-Technical Assessment. 205 pp.

Jaikaeo, C., Phonphoem, A., Jansang, A., Tiwatthanont, P., Tangtrongpaijoj, W., Soralump, S. & Torwiwat, W. (2013) Landslide monitoring and assessment system using low-cost wireless communication. In: *Proceedings of the 6th Conference of the Asia Pacific Association of Hydrology and Water Resources, August 19–21*. Seoul, Korea, Korea University.

Jamnongpipatkul, P., Dechasakulsom, M., Sukolrat, J. & Sawangsuriya, A. (2008) Slope management in Thailand: A case study of reinforced slope from highway route No.1095. In: *International Conference on Slope. Malaysia: Landslide Awareness: Slope Mitigation & Prevention, 4th–6th November 2008*. Malaysia, Istana Hotel Kuala Lumpur.

Jotisankasa, A. (2013) Application of local plant species for live stake as a bio-slope stabilization method in Thailand. In: *Proceedings of the Fourth Tokyo Tech-KU Joint Seminar on Infrastructure Development, October 31–November 1, 2013*. Tokyo, Japan, Tokyo Institute of Technology. pp. 145–150.

Jotisankasa, A. & Mairaing, W. (2010) Suction-monitored direct shear testing of residual soils from landslide-prone areas. *Journal of Geotechnical and Geoenvironmental Engineering, ASCE*, 136 (3), 533–537.

Jotisankasa, A. & Hunsachainan, N., Kwankeo, N., Chunrod, P. & Mancharoen, J. (2010) Development of a wireless landslide monitoring system. In: *Proc. International Conference on Slope. Thailand 2010*. Chiangmai, Geotechnique and Geosynthetics for Slope.

Jotisankasa, A., Mahannopkul, K. & Sawangsuriya, A. (2014a) Evaluating slope stability with respect to pore-water pressure and suction variation with rainfall; a case study of fill granitic slope in northern Thailand. In: *Proceedings of the International Conference on Unsaturated Soils, Unsaturated Soils: Research & Applications. Sydney Australia 2–4 July 2014*.

Jotisankasa, A., Mairaing, W. & Tansamrit, S. (2014b) Infiltration and stability of soil slope with vetiver grass subjected to rainfall from numerical modelling. In: *Proceedings of the*

International Conference on Unsaturated Soils, Unsaturated Soils: Research & Applications. Sydney Australia 2–4 July 2014.

Kilsby, C., Glendinning, S., Hughes, P.N., Parkin, G. & Bransby, M.F. (2009) Climate-change impacts on long-term performance of slopes. *Proceedings of the Institution of Civil Engineers, Engineering Sustainability*, 162 (ES2), 59–66.

Krahn, J., Fredlund, D.G. & Klassen, M.J. (1989) Effect of soil suction on slope stability at Notch Hill. *Canadian Geotechnical Journal*, 26, 269–278.

Krisnanto, S., Rahardjo, H., Fredlund, D.G. & Leong, E.C. (2014) Mapping of cracked soils and lateral water flow characteristics through a network of cracks. *Engineering Geology*, 172, 12–25.

Kuhn, N.J., Schütt, B. & Baumhauer, R. (2011) Managing the impact of climate change on the hydrology of the Gallocanta Basin, NE-Spain. *Journal of Environmental Management*, 92, 275–283.

Kunkel, K.E. (2003) North American trends in extreme precipitation. *Natural Hazards*, 29, 291–305.

Kyoto Protocol to the United Nations Framework Convention on Climate Change (1998) U.N. Doc FCCC/CP/1997/7.

Lausche, B.J. (2009) *Policy Tools for Local Adaptation to Sea Level Rise*. Sarasota, FL, Marine Policy Institute at Mote Marine Laboratory.

Leroueil, S. (2000) Natural slopes and cuts: Movement and failure mechanisms. *Géotechnique*, 51 (3), 197–243.

Li, A.G., Yue, Tham, L.G. & Lee C.F. (2005) Field-monitored variations of soil moisture and matric suction in a saprolite slope. *Canadian Geotechnical Journal*, 42, 13–26.

Li, T. (1989) Landslides: Extent and economic significance in China. In: *Proc. 28th International Geological Congress: Symposium on Landslides*. Washington, DC, A.A. Balkema. pp. 271–287.

Liew, S.S., Liong, C.H. & Low, C.L. (2004) Four landslide investigations in Malaysia. In: *Proc. 15th SEAGC, Bangkok, Thailand.*

Lim, T.T., Rahardjo, H., Chang, M.F. & Fredlund, D.G. (1996) Effect of rainfall on matric suction in a residual soil slope. *Canadian Geotechnical Journal*, 33, 618–628.

Loh, S.Y. & Khoo, V. (2012) Spatially enabled Singapore through Singapore Geospatial Collaborative Environment (SG-SPACE). In: *GSDI 12 World Conference, 19–22 October 2010*. Singapore, Singapore Land Authority.

Mafian, S. (2009) *Slope Stabilization Using Bioengineering (Live Pole) Technique*. MSc Thesis. Selangor, Universiti Putra Malaysia.

Mairaing, W. & Thaiyuenwong, S. (2010) Dynamic landslide warning from rainfall and soil suction measurement. In: *Proceedings of International Conference on Slope 2010: Geotechnique and Geosynthetics for Slopes, 27–30 July, 2010*. Chiangmai, Thailand.

Mairaing, W., Jotisankasa, A. & Soralump, S. (2012) Some applications of unsaturated soil mechanics in Thailand: An appropriate technology approach. *Geotechnical Engineering Journal of the SEAGS & AGSSEA*, 43 (1), 1–11.

Malet, J.P., Remaitre, A., Maquaire, O., Durand, Y., Etchevers, P., Guyomarch, G., Deque, M. & van Beek, L.P.H. (2007) Assessing the influence of climate change on the activity of landslides in the Ubaye Valley. *Landslides and Climate Change: Challenges and Solutions*. London, Taylor & Francis.

Malone, A.W. (1997) Risk management and slope safety in Hong Kong. *Transactions of the Hong Kong Institution of Engineers*, 4 (2–3), 1–90.

Mantovani, F., Soeters, R. & Van Westen, C.J. (1996) Remote sensing techniques for landslide studies and hazard zonation in Europe. *Geomorphology*, 15 (3–4), 213–225.

Marengo, J.A., Rusticucci, M., Penalba, O. & Renom, M. (2010) An inter-comparison of observed and simulated extreme rainfall and temperature events during the last half of the twentieth century: Part 2: Historical trends. *Climatic Change*, 98, 509–529.

Mason, S.J., Waylen, P.R., Mimmack, G.M., Rajaratnam, B. & Harrison, J.M. (1999) Changes in extreme rainfall events in South Africa. *Climatic Change*, 41, 249–257.

McNicholl, D.P., Pump, W.L. & Cho, G.W.F. (1986) Groundwater control in large scale slope excavations – Five case histories from Hong Kong. In: *Proc. 21st Regional Conference of the Engineering Group of the Geological Society, Sheffield.*

Mendes, J. (2011) *Assessment of the Impact of Climate Change on an Instrumented Embankment: An Unsaturated Soil Mechanics Approach*. PhD Thesis. Durham, Durham University.

Mirhosseini, G., Srivastava, P., Stefanova, L. (2013) The impact of climate change on rainfall Intensity–Duration–Frequency (IDF) curves in Alabama. *Regional Environmental Change*, 13 (1), 25–33.

Montgomery, D.R., Schmidt, K.M., Dietrich, W.E. & Greenberg H.M. (2000) Forest clearing and regional landsliding in the Pacific Northwest. *Geology*, 28, 311–314.

Morris, P.H., Graham, J. & Williams, D.J. (1992) Cracking in drying soils. *Canadian Geotechnical Journal*, 29 (2), 263–277.

MOSTI (2013) *General Climate Information*. Malaysian Meteorological Department. Retrieved from: http://www.met.gov.my/index.php?option=com_content&task=view&id=1731&Itemid=1595.

National Climate Change Secretariat (NCCS) (2012) *National Climate Change Strategy*.

National Environment Agency (NEA) (2013) *Press Release – IPCC's Findings Suggest Future Increases in Global Sea-Level Rise and More Extreme Rainfall in the Region.*

National Climate Change Secretariat (NCCS) (2012) *Speech on Climate Change by Mr Teo Chee Hean, Deputy Prime Minister, Coordinating Minister for National Security and Minister for Home Affairs, at the Committee of Supply Debate.*

National Environment Agency (NEA) (2010) *Singapore's Second National Communication – Under the United Nations Framework Convention on Climate Change.*

Nawagamuwa, U.P., Rathnaweera, T.D., Rangana, L.L. & Palihawadana, M.P. (2011) *Effects of Global Warming on Landslide Frequencies in Ratnapura District, Sri Lanka*. Society for Social Management Systems.

Ng, C.W.W., Springman, S.M. & Alonso, E.E. (2008) Monitoring the performance of unsaturated soil slopes. *Geotechnical and Geological Engineering*, 26 (6), 799–816.

Ng, S.Y., Mukherjee, A. & Foo, K.Y. (2009) Using geospatial model for managing risk of slope failure. In: *ESRI Asia Pacific UC 2009 – Industrial Track Papers and Presentations, 20–21 January 2009*. Singapore, Singapore Land Authority (SLA).

Nguyen, P.K. (2009) *Climate Change, Sea Level Rise Scenarios*. M. o. N. R. a. Environment. Ha Noi. Vietnam Publishing House of Natural Resources. Environment and Cartography.

Norris, J.E. & Greenwood, J.R. (2000) In-situ shear and pull-out testing to demonstrate the enhanced shear strength of root reinforced soil. In: *Proc. 8th International Symposium on Landslides*. Cardiff, Thomas Telford.

Oberthur, S. & Ott, H.E. (1999) *The Kyoto Protocol: International Climate Policy for The 21st Century (International and European Environmental Policy Series)*. Paris, Springer. 359 pp.

Osborn, T.J., Hulme, M., Jones, P.D. & Basnett, T.A. (2000) Observed trends in the daily intensity of United Kingdom precipitation. *International Journal of Climatology*, 20, 347–364.

Penman, H.L. (1948) Natural evapotranspiration from open water, bare soil and grass. *Proceedings of the Royal Society of London Series*, 193, 120–145.

Pitts, J. (1985) *An Investigation of Slope Stability on the NTU Campus, Singapore, Applied Research Project RPI/83*. Singapore, Nanyang Technological Institute. 54 pp.

Pricewaterhouse Coopers (PWC) (2009) *Transportation & Logistics 2030 – Volume 1: How Will Supply Chains Evolve in an Energy-Constrained, Low-Carbon World?*

Potts, D.M., Kovacevic, N. & Vaughan, P.R. (1997) Delayed collapse of cut slopes in stiff clay. *Geotechnique*, 47 (5), 953–982.

Quang, N.M. (2012) *Climate Change, Sea Level Rise Scenarios for Vietnam*. M. o. N. R. a. Environment. Ha Noi. Vietnam Publishing House of Natural Resources. Environment and Cartography.

Rahardjo, H., Leong, E.C., Gasmo, J.M. & Tang, S.K. (1998) Assessment of rainfall effects on stability of residual soil slopes. In: *Proceedings 2nd International Conference on Unsaturated Soils, Beijing, China*, Vol. 27–30 (2). pp. 280–285.

Rahardjo, H., Hritzuk, K.J., Leong, E.C. & Rezaur, R.B. (2003) Effectiveness of horizontal drains of slope stability. *Journal of Engineering Geology*, 69 (3–4), 295–308.

Rahardjo, H., Lee, T.T., Leong, E.C. & Rezaur, R.B. (2005) Response of a residual soil slope to rainfall. *Canadian Geotechnical Journal*, 42, 340–351.

Rahardjo, H., Satyanaga, A., Leong, E.C. & Ng, Y.S. (2010) Effects of groundwater table position and soil properties on stability of slope during rainfall. *Journal of Geotechnical and Geoenvironmental Engineering, ASCE*, 136 (11), 1555–1564.

Rahardjo, H., Santoso, V.A., Leong, E.C., Ng, Y.S. & Hua, C.J. (2011) Performance of horizontal drains in residual soil slope. *Soils and Foundations, Japanese Geotechnical Society*, 51 (3), 437–447.

Rahardjo, H., Satyanaga, A. & Leong, E.C. (2012a) Unsaturated soil mechanics for slope stabilization. *Geotechnical Engineering Journal of the SEAGS & AGSSEA*, 43 (1).

Rahardjo, H., Santoso, V.A., Leong, E.C. & Ng, Y.S. (2012b) Variability of residual soil properties. *Journal of Engineering Geology*, 141–142, 124–140.

Rahardjo, H., Satyanaga, A. & Leong, E.C. (2013a) Effects of flux boundary conditions on pore-water pressure distribution in slope. *Journal of Engineering Geology, Special Issue on Unsaturated Soils: Theory and Applications*, 165, 133–142.

Rahardjo, H., Santoso, V.A., Leong, E.C., Ng, Y.S., Tam, C.P.H. & Satyanaga, A. (2013b) Use of recycled crushed concrete and secudrain in capillary barrier for slope stabilization. *Canadian Geotechnical Journal*, 50, 1–12.

Rahardjo, H., Leong, E.C., Satyanaga, A., Ng, Y.S., Tan, H.T. & Hua, C.J. (2014a) *Rainfall-induced Slope Failures and Preventive Measures in Singapore. Geotechnical Engineering Monograph*. Singapore, NTU-HDB Research Collaboration Project, Nanyang Technological University. 84 pp.

Rahardjo, H., Satyanaga, A., Leong, E.C. & Wang, J.-Y. (2014b) Comprehensive instrumentation for real-time monitoring of flux boundary conditions in slope. *Procedia Earth and Planetary Science*, 9, 23–43.

Rahardjo, H., Satyanaga, A., Leong, E.C., Santoso, V.A. & Ng, Y.S. (2014c) Performance of an instrumented slope covered with shrubs and deep rooted grass. *Soils and Foundations, Japanese Geotechnical Society*, 54 (3), 417–425.

Rezaur, R.B., Rahardjo, H., Leong, E.C. & Lee, T.T. (2003) Hydrologic behaviour of residual soil slopes in Singapore. *ASCE Journal of Hydrologic Engineering*, 8 (3), 133–144.

Rahardjo, P.P. (2002) Landslides of natural and man made slope, case histories and their counter measures in Indonesia. In: *Proceeding of Seminar "Slope-2002"*. Bandung, Parahyangan Catholic University.

Rahardjo, P.P. (2003) Reactivation of breccia resliding over clay shale. In: *Proceeding of Pan American Conference*. Soil and Rock, Boston.

Rahardjo, P.P., Halim, Y. & Wisanto, Y. (2012) Measurement of bored piles' performance for protection of bridge foundation in landslides area. In: *Proceeding of International Conference on Transportation Geotechnics*. Japan, Hokkaido University.

Rouainia, M., Davies, O., O'Brien, T. & Glendinning, S. (2009) Numerical modelling of climate effects on slope stability. In: *Proceedings of the Institution of Civil Engineers, Engineering Sustainability*, 162 (ES2), 81–89.

Sebastian, O. & Ott, H.E. (1999) *The Kyoto Protocol: International Climate Policy for the 21st Century (International and European Environmental Policy Series)*. Paris, France, Springer. 359 pp.

See-Sew, G., & Siew-Wai, C. (n.d.) *Geotechnical Challenges in Slope Engineering of Infrastructures*. Gnpgeo.com.my. pp. 1–20. Retrieved from: http://www.gnpgeo.com.my/download/publication/2008_04.pdf.

Sherif, M.M., Mohamed, M., Shetty, A. & Almulla, M. (2011) Rainfall-runoff modeling of three wadis in the northern area of UAE. *Journal Hydrological Engineering, ASCE*, 16 (1), 10–20.

Simonovic, S.P. & Peck, A. (2009) *Updated Rainfall Intensity Duration Frequency Curves For the City of London Under The Changing Climate. Water Resources Research Report, Book 29*. Department of Civil & Environment Engineering, University of Western Ontario, London, ON.

Singapore Land Authority (SLA) (2011) *Geospace Garners International Awards*.

Singapore Land Authority (SLA) (2012) Country report on SDI activities in Singapore. In: *Nineteenth United Nations Regional Cartographic Conference for Asia and the Pacific (UNRCC-AP) 2009, E/CONF.100/CRP.9, 29 October–1 November 2012*. Bangkok, Thailand.

Singh, H., Huat, B.K., & Jamaludin, S. (2008) Slope assessment systems: A review and evaluation of current techniques used for cut slopes in the mountainous terrain of West Malaysia. *Electronic Journal of Geotechnical Engineering*, 13. 1–24.

Smith, D.D. (1980) Long-term performance of horizontal drains. *Transportation Research Record*, 783, 39–45.

Snowy Mountains Hydro-Electric Authority (1983) *Written Communication with GCO*.

Soralump, S. (2013) Multi-way landslide and debris flow warning for resilience and sustainable community. In: *Proceedings of the International Symposium, Hanoi Geo-Engineering 2013, Natural Resources Engineering and Disaster Mitigation for Infrastructure Development*. Hanoi, Vietnam National University Publisher.

Springman, S.M., Jommi, C. & Teysserie, P. (2003) Instabilities on moraine slopes induced by loss of suction: A case history. *Geotechnique*, 53 (1), 3–10.

Stern, N. (2007) *The Economics of Climate Change: The Stern Review*. Cambridge, Cambridge University Press. 712 pp.

Suppiah, R. & Hennessy, K.J. (1998) Trends in total rainfall, heavy rain events and number of dry days in Australia. *International Journal of Climatology*, 18, 1141–1164.

Taerakul, A., Jotisankasa, A., Soralump, S. & Sukolrat, J. (2012) Behaviour of pore-water pressure and rainfall: A case study of reinforced soil slope, highway route No.1095 Mae-Hongson province. In: *Proceedings of the 17th National Convention in Civil Engineering (NCCE17), 9 to 11 May, 2012, Udonthani* (in Thai).

Tansamrit, S. (2003) The progress of the use of vetiver grass system for erosion control and slope stabilization along the Yadana gas pipeline right-of-way, the Yadana gas pipe-line project. In: *Proceedings of 3rd International Vetiver Conference, Guangzhou, China, October 6–9* 2003.

Temesgen, B., Mohammed, M.U. & Korme, T. (2001) Natural hazard assessment using GIS and remote sensing methods, with particular reference to the landslides in the Wondogenet Area, Ethiopia. *Physics and Chemistry of the Earth, Part C: Solar, Terrestrial & Planetary Science*, 26 (9), 665–675.

Thu, T.M. (2012) *Rainfall Induced Landslides in Vietnam*. Hanoi, Water Resources University.

Tohari, A. (2012) Impact of climate change on landslides hazard in West Java. In: *Proceeding of Workshop on Slope Stability*. Bandung, Parahyangan Catholic University.

Toll, D.G., Rahardjo, H. & Leong, E.C. (1999) Landslides in Singapore. In: *Proceedings of the 2nd International Conference on Landslides, Slope Stability and the Safety of Infra-Structures, Singapore, July 27–28*. pp. 269–276.

Truong, P., Van, T.T. & Pinners, E. (2008) *The Vetiver System for Slope Stabilization an Engineer's Handbook*. 1st edition. Published by the Vetiver Network International.
Uhl-Bien, M.R. & McKelvey, B. (2007) Complexity leadership theory: Shifting leadership from the industrial age to the knowledge era. *The Leadership Quarterly*, 18 (4), 298–318
UNEP (1993) *Vietnam and Climate Change. Fact Sheet Producted by the Information Unit on Climate Change*. Geneva, United Nations Environment Programme.
United Nations (1993) United Nations framework convention on climate change. *Treaty Series*, 1771, 107 pp.
Veder, Ch. & Lackner, K. (1984) Economical slide stabilization by slightly upward inclined horizontal drainage borings. In: *Proceedings of the 4th International Symposium on Landslides, Toronto, Canada*, Vol. 1. pp. 723–728.
Wilson, G.W., Fredlund, D.G. & Barbour, S.L. (1997) The effect of soil suction on evaporative fluxes from soil surfaces. *Canadian Geotechnical Journal*, 34, 145–155.
Wilson, G.W., Machibroda, R.T., Barbour, L.S. & Woyshner, M.R. (1993) Modelling of soil evaporation from waste disposal sites. In: *Proceedings of the Joint CSCE-ASCE National Conference on Environmental Engineering, July 12–14*. pp. 281–288.
Zhai, P., Zhang, X., Wan, H. & Pan, X. (2005) Trends in total precipitation and frequency of daily precipitation extremes over China. *Journal of Climate*, 18, 1096–1108.

Chapter 10

Review of landslide occurrence in Spain and its relation to climate

Jordi Corominas
Division of Geotechnical Engineering and Geosciences, Department of Civil and Environmental Engineering, Universitat Politècnica de Catalunya-BarcelonaTech, Barcelona, Spain

Rosa María Mateos
Geohazard Department, Geological Survey of Spain, Granada, Spain

Juan Remondo
Department of Earth Sciences and Condensed Matter Physics, University of Cantabria, Santander, Spain

FOREWORD

In Spain, landslides are phenomena mostly concentrated in the mountainous areas and coastal cliffs. Even though many of them have a natural origin, the activity in some locations has increased as a result of changes in land use and due to excavations. The increasing use of mountainous areas for tourism, has led to an unusual frequentation in clearly unstable areas. New roads and human settlements are spreading towards places in which slides, rockfalls and other type of movements occur relatively frequently, thus increasing the risk to population. Climate change is expected to produce the increase of temperatures in the Iberian Peninsula. The future rainfall pattern is, however, still uncertain. Despite of this, all models predict an increase of the intense precipitation events across the whole country with the subsequent occurrence of debris flows and shallow landslides.

10.1 CONTEXT OF LANDSLIDE OCCURRENCE IN SPAIN

The Iberian Peninsula is shared by Spain and Portugal. The western and central part of the Peninsula is a plateau ("Meseta") composed of crystalline and metamorphic rocks that form the Hercinian basement. The plateau has a mean height of 660 m and bounded in its Southern and Northeastern edges by the Guadalquivir and Ebro river basins, respectively. The plateau itself contains two river basins, the Duero and Tagus, filled with sediments of Cenozoic age and separated by the Central range (Figure 10.1).

Surrounding the plateau, several alpine ranges extend: in the North, the Pyrenees and its extension to the West, the Cantabrian range; the Baetic ranges to the South; and the Iberian range in between both. Finally, two archipelagos are of interest for landslides. In the Mediterranean Sea, the Balearic islands which are the extension of the Baetic system towards the East. In the Atlantic Ocean are the Canary Islands

Table 10.1 Unstable lithologies in Spain, associated types of failure and geographical distribution (synthesis based on data from Corominas, 1989; Araña *et al.*, 1992; Díaz de Terán *et al.*, 1997).

Lithology	*Age*	*Type of failure*	*Context*
Black shales	Carboniferous Silurian	Slides, earthflows	Pyrenees and Cantabrian Range
Marls and schists	Mesozoic and Paleozoic	Slides	Baetic ranges (Nevado Filábride and Alpujarrides Complex)
Claystones, marls and evaporites	Keuper	Rotational and translational slides, earthflows	Pyrenees, Coastal Ranges of Catalonia, Cantabrian Range, Iberian Range, Tramuntana Range, Baetic ranges
Red-lilaceous clays, marls and siltstones (Weald Facies)	Lower Cretaceous	Rotational and traslational slides	Cantabrian and Iberian Ranges
Alternances of blue marls with limestones	Aptian	Rotational and translational slides and earthflows	Iberian Range
Alternances of lutites, red sandstones, lignites (Garum Facies)	Upper Cretaceous	Rotational and translational slides and earthflows	Pyrenees
Marly clays	Lower Eocene – Lutecian	Rotational slides and earthflows	Pre-Pyrenees, Pre-Baetic Ranges
Claystones and alternances of sandstones, clays and limestones (Flysch)	Cretaceous-Lower Eocene	Earthflows, slides	Pyrenees, Cantabrian coast
Massive gypsum	Oligocene	Rockfalls and topples	Ebro Basin
Clays, claystones, siltstones	Miocene	Rotational slides and earthflows	Cenozoic basins (Duero, Tagus). Intra-mountainous depressions (Vallès-Penedès)
Boulders and gravels with sandy-silty or clayey matrix (glacial till)	Pleistocene	Debris flows and avalanches. Rotational slides	Pyrenees, Cantabrian Range,
Gravels, sand, silt and clays (colluvium)	Pleistocene-Holocene	Shallow slides and debris flow	All the mountain ranges
Basalts and volcanic ashes	Miocene, Pliocene, Pleistocene	Large slides Rockfalls	Canary Isles

whose most prominent peak is Mount Teide, the highest peak of Spain. The archipelago consists of seven large and several minor islands of volcanic origin whose formation began in the Oligocene and continues today.

The slope failures are strongly controlled by the presence of susceptible lithologies (Table 10.1). A synthesis of the most susceptible ones is found in Corominas (1989), Araña *et al.* (1992) or Díaz de Terán *et al.* (1997). Vegetation and land use cover may also exert some influence on the stability of the slopes, but to a lesser extent. The

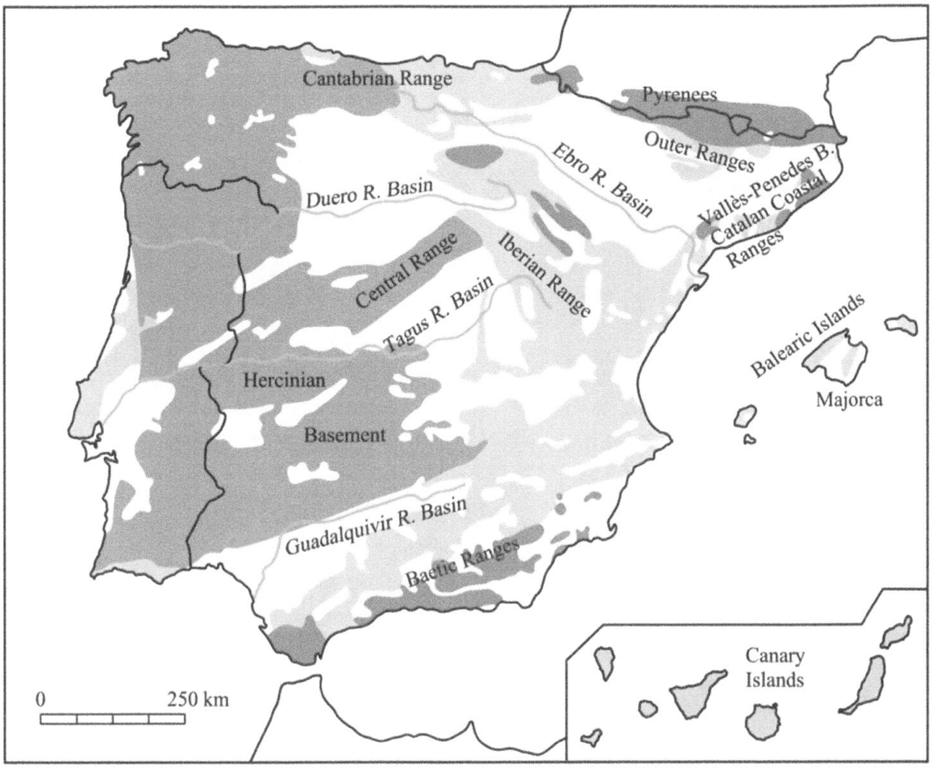

Figure 10.1 Main morphostructural units of Spain: Hercinian Basement (grey), Outer Alpine Ranges (light green) Inner Alpine Ranges (dark green), Cenozoic basins (yellow), and volcanic islands (orange).

climatic regime, sea and river erosion, and earthquakes are, in this order, the most significant natural triggering mechanisms in the country.

The Hercinian Basement occupies the westernmost part of the plateau and is mostly stable. This is due to the resistant characteristics of the materials (plutonic rocks, gneiss, quartzites and schists) and to the gentle relief. Landslides in Spain occur mostly in three main geomorphological contexts (Corominas 2005, 2006): the Alpine ranges associated to the steep relief, the coastal cliffs subjected to the sea erosion, and the Cenozoic basins filled with hard soils and weak rocks which are susceptible to landsliding, particularly in locations affected by fluvial erosion (Figure 10.1).

10.1.1 Alpine ranges

The Pyrenees and Baetic ranges contain an inner core of igneous and metamorphic rocks shaped to a great extent by the glacial and periglacial activity. The erosive action of the Pleistocene glaciers and the incision of the present fluvial network, locally exacerbated by the orogenic uplift (the Baetic Ranges), have resulted in a steep relief, which

Figure 10.2 Large earthflow reactivated in June 2013 in the Cantabrian Range, during an abnormal wet episode, which coincides with a decreasing trend of the annual precipitation during the last decades in Northern Spain (photo taken by A. González-Díez).

constitutes the main predisposing factor for instability. Outcrops of susceptible lithologies (e.g. shales, phyllites) (Figure 10.2) and a favourable rainfall regime, especially in the Mediterranean area, are other required ingredients for the occurrence of the slope failures (Corominas and Baeza, 1992).

The Iberian and Cantabrian ranges as well as the outer ranges of the Pyrenees and the Baetic ranges are composed of thick sedimentary sequences made of strong (limestones, conglomerates) and weak argillaceous rocks. The calcareous nature of the outer ranges provides to large compartments of the ranges a relative stability with the only exception of the locations showing structural weaknesses (unfavorable dipping bedding planes, joints, faults, schistosity planes) through which large landslides can develop. Instead, the argillaceous formations (marls and gypsum of Keuper, flysch facies, Weald facies) are highly unstable and are affected by earthflows and both rotational and complex slides.

The most frequent types of failures in the alpine ranges are shallow landslides and debris flows, generated in surficial formations, particularly till deposits and colluvium. The mobilized volumes usually range between a few hundred to a few thousand

Figure 10.3 Earthflow at Zubiza, Navarra, Ebro Basin triggered after an abnormally rainy winter in early March 2013 (photo taken by J.R. Mendióroz).

cubic meters. However, failures in till may reach up to hundreds of thousands of cubic meters. Earthflows occur in locations where argillaceous formations and low grade metamorphic rocks outcrop, particularly in the outer Pyrenean ranges, and in both the Cantabrian and Baetic ranges.

10.1.2 Cenozoic basins

The Cenozoic basins nest the main rivers draining the plateau. Other basins are a few intramontaneous basins within the alpine ranges. The Cenozoic basins are filled with detritic sedimentary sequences of continental and marine origin. They are mostly composed of argillaceous rocks (claystones, siltstones, marls, shales) interbedded with sandstones and conglomerates. The sub-horizontal arrangement of the strata in the Duero and Ebro basins often results in tabular reliefs capped with layers of sandstones or limestones, showing frequent rockfalls and topples. The river meanders undermine the banks composed of argillaceous rocks causing translational and rotational slides (Berganza and Modrano, 1978). Although landslides generated are usually of small size, some of them can reach several million cubic meters (Figure 10.3). The marine sediments of the Guadalquivir and Vallès-Penedès basins often contain sequences of high-plasticity clays that display an expansive behavior (Tsige *et al.*, 1994) and generate both translational slides and earthflow even on gentle slopes.

Figure 10.4 Sa Caixota coastal landslide (Ibiza, Balearic islands) on 18th September 2005, triggered after intense rains (>150 mm/24 h). Numerous dwellings were affected and people evacuated (Photo taken by R.M. Mateos).

10.1.3 Coastal cliffs

Spain has a coastline of 7.880 kilometers which houses about 65% Spanish population. Almost 50% of coastline consists of hard and soft cliffs, which are particularly abundant in the Atlantic area, in both Northern and Southern edges of the Mediterranean coast, and in the Canary and Balearic Islands. In the locations where the alpine ranges meet the sea the landscape is characterized by a rugged coast as well. Although rock fall is the predominant mechanism of failure, large landslides may develop through weak lithological contacts.

The whole Cantabrian coast is dotted with landslides caused by the erosion and undermining of the cliffs. Are particularly susceptible the outcrops of Cretaceous and Eocene flysch in the Basque Country due to the unfavorable arrangement of the layers and the low strength marls of facies Keuper in Asturias. Similarly, large instabilities have been generated in the northern coast of Majorca, Balearic Islands (Figure 10.4) (Mateos, 2012; Mateos, *et al.*, 2013a).

The Canary Isles were formed by piling of successive lava flows and pyroclastic products that have built large volcanic edifices. The erosion and retreat of the stacked lava flows generates imposing cliffs with frequent rockfalls. Canary slopes have produced the largest landslides known in Spain, of around several cubic kilometres, like in the Orotava valley and Teguise, Tenerife Island or in el Golfo, El Hierro Island (Ancochea *et al.*, 1994; Watts and Masson, 1995; Martí *et al.*, 1997; Carracedo *et al.*, 1999). These landslides are prehistoric and the most accepted hypotheses for failure indicate an origin associated with the accumulation of volcanic material, dyke intrusions and the associated seismicity, and marine erosion (Hürlimann *et al.*, 1999). Despite of this, the influence of climate and, indirectly, the drop in sea level associated with glacial episodes cannot be ruled out (Carracedo *et al.*, 1999; Ablay and Hürlimann, 2000). In Gran Canaria island, the surface of failure of the large slides in the Tirajana Depression have made use of the presence of weak layers (tuff, ashes and ash flows) interbedded in the lava flows (Lomoschitz *et al.*, 2002).

Landslides in Spain are the geohazard that generate greater economic losses after floods. Ferrer (1995) estimated economic losses due to landslides for the period spanning 1986–2016 of about 160 million euros per year. The greatest damage was recorded in the most populated and urbanized regions of the country, and especially in the coastal fringe.

According to Ayala *et al.* (2004), the economic losses due to landslides during a ten year period (1990–2000) amount to €42 million per year. An analysis in the Basque Country for the period 1982–1995 estimates a damage of €10 million per year (Del Val *et al.*, 1996).

Several economic analyses have been carried out at local scale. Remondo *et al.* (2005) estimated direct damage to the road network in a small municipality in northern Spain (Deba, Guipúzcoa Province) of about 50 km^2, at 20,000 €/year for the next 50 years. Bonachea *et al.* (2009) calculated € 1.1–3.3 million for a period of ten years in a larger area of the same province. Herrera *et al.* (2013) estimated €15 million damage for the last decade in the Tena valley (central Pyrenees).

Landslides have also caused numerous casualties. Ayala (1995) quoted 17 deaths between 1991 and 1993. The Statistical Yearbook of the Ministry of Food, Agriculture and Environment (MAGRAMA, 2013) reported 36 deaths for the period 1995–2011.

10.2 CLIMATIC TRIGGERS OF LANDSLIDES IN SPAIN

According to the geographical situation and orographic conditions, three main climatic zones, and therefore precipitation regimes, can be distinguished in Spain:

The Atlantic (oceanic) climate located in the northern part of the country, especially in the region called "Green Spain" (Northern coast and Cantabrian Range). It is characterized by relatively mild winters and warm summers. The climate is determined by the Atlantic Ocean winds whose moisture gets trapped by the mountains circumventing the Spanish Atlantic coast. Because of the Foëhn effect, the southern slopes fall inside the rain shadow zone and so Green Spain contrasts starkly with the rest of Spain. But the summers remain warmer than many comparable North-west European regions. The rainfall is generally abundant, exceeding 1,100 mm and is fairly evenly spread out over the year. Temperatures vary only slightly, both on a diurnal and a seasonal basis, and averages range from 10°C in January to 23°C in July.

The Mediterranean climate is the predominant climate in Spain, covering most of the Iberian Peninsula and the Balearic islands, and different subtypes can be distinguished: typical Mediterranean located in the Mediterranean coastal region. It is characterized by irregular precipitations (annual rainfall around 400–700 mm), especially during autumn and spring. Winters are short and mild and summers dry and hot. A particular subtype of this Mediterranean Climate occur in the inland areas of the Iberian Peninsula, on the Central Plateau, where there is a continental-Mediterranean climate with very hot summers (regularly exceeding 35°C) and very low temperatures in winter (under −5°C regularly occurring). Although it rarely rains during summer, there is often heavy rainfall in spring and autumn (annual rainfall exceeds 500 mm).

The Semiarid climate can be found in south eastern Spain and in certain parts of the Ebro Valley. It is very hot during the summer and the drought usually extends into the autumn. It is sub-desert climate, with annual rainfall about 300 mm.

Apart from the three main climate zones, other sub-zones can be found, such as the Alpine climate in the highest mountain ranges (Cantabrian range, Pyrenees, Baetic system, Iberian range and Central System) and a typical subtropical climate in the Canary Islands.

In general, within the year, precipitations tend to concentrate in autumn and spring, but there are areas (north and west coast) where rains are frequent and persistent throughout the year. On the contrary, in other areas, such as the SE coast, precipitations are very scarce, short-lasting and intense. Generally speaking, in the Iberian Peninsula precipitations decrease southwards and eastwards.

Rainfall is the main landslide-triggering mechanism for landslides in Spain (Ferrer and Ayala, 1997; Corominas, 2005). The rainfall infiltration reduces the strength of the slopes by increasing water pressure in pores and joints of the ground. The duration and intensity of rainfall episode and the composition of the slope materials are the main factors controlling the type of failure mechanism.

Heavy rains in Spain are associated to low pressures connected with polar front fluctuations or with the influx of subtropical air masses. The 40°N parallel, which divides the Iberian Peninsula into two halves, delimit the influence between winds coming from N and NW and S and SW.

Different patterns exist in the seasonal distribution of slope failures, particularly between the Mediterranean environment and the rest of the Peninsula. In the Coastal Mountain Ranges of Catalonia, the Eastern Pyrenees and the Iberian Range, landsliding events are mostly concentrated in Autumn, October and November, although there have been sporadic episodes distributed throughout the other seasons. In the Baetic ranges and in the Cenozoic Basins, there is a predominance of failures in winter and spring. In the Cantabrian Range activity develops through the year, but especially in three periods: late summer-early autumn, late autumn-early winter and late spring-early summer. On the Canary Isles, slides and rockfalls are mostly concentrated in the winter months. Referring to the Balearic Islands, landslides and rockfalls are concentrated in autumn and winter. In this region, the maximum precipitation takes place during the autumn months due to the arrival of the first high cold air masses, which contrast with the high temperature retained by the sea. This phenomenon is known as "gota fría" (cold air pool), in which heavy storms are accompanied by intense rainfall with episodes of up to 300 mm in 24 hours (Mateos, 2012).

The analysis of series of dated landslide reactivation episodes back up to the end of nineteenth century shows certain cyclicity. Wet periods of greater landslide activity (1905–1930 and 1958–1987) alternate with periods of relative quiescence such as between the 30s and the 50s, and in the last third of the last century. Such cyclicity was also observed in regions of Europe, although it did not occur simultaneously (Corominas, 2005).

In any case, it should be remembered that anthropic disturbances may produce significant changes in the frequency of the slope failures. In Spain, the number of

Figure 10.5 Diezma landslide in the A-92 motorway (South Spain) which took place in March 2001 blocking the motorway for several days. Photo by J.M Azañón.

human-induced landslides has also increased during the past decades due to vast urban and infrastructure development. Human activities have disturbed large volumes of geologic materials in housing development, transport network and facilities, dams and reservoirs, and communication systems. For example, during the past 20 years, a surface area of 44 hectares has been urbanized in Spain every day, and they have expanded into landslide-prone areas. In this sense, urban development has mainly occurred in coastal areas, as the main source of income in the country is tourism (10.2% Spanish GDP); numerous vacation residences and tourist resorts have been built. Subsequently, coastal development has been a major factor in the recent increase in damaging slope failures in Spain. Regarding the transportation network, road and railway cuts caused not only shallow failures but also large landslides. This is the case of the Almodóvar del Río landslide (Guadalquivir Basin) related to the high-speed railway line works in 1992 and the Diezma landslide (Baetic Range) (Figure 10.5), which took place in March 2001, blocking the A-92 motorway for several days (Azañón *et al.*, 2010). But the most catastrophic human-induced event was the failure of the Aznalcóllar (Guadalquivir basin, SW Andalusia) tailings dam in 1998. A landslide developed through claystones located underneath causing a breach in the dam through which water and tailings flushed out towards the valleys of nearby rivers. This event created great concern, as the Doñana National Park- an emblematic and protected natural environment- was at risk (Alonso *et al.*, 2010).

Analyses carried out in the Basque Country, western Cantabrian range (Remondo, 2001; Remondo *et al.*, 2005) have shown that the frequency of slides and the mobilized volume increased almost one order of magnitude between 1954 and 1997, but it has not been possible to correlate this noticeable increase with comparable increases in total rainfall, number of storms or the annual number of rainy days above certain thresholds, which show no significant changes. On the other hand, it appears that there exists a relationship between the intensity of human intervention in the territory, through very different actions which modify the strength of the soil layer and therefore reducing, the rainfall threshold needed to trigger slides.

Begueria (2006) and García-Ruiz *et al.* (2010) studied the effects of land-use changes on landslide activity in the Pyrenees. Gutiérrez (2009) found that the slides affecting the escarpments of the Ebro river in the surroundings of Zaragoza are caused by leaks in the sanitation network. According to Domínguez *et al.* (1999) 20% of the mass movements in the Carboniferous terrain in central Cantabrian range were caused by human activity. Remondo *et al.* (2005) found that 7% of the landslides occurred in the second half of the 20th century in the lower Deba valley (Eastern Cantabrian range) were directly triggered by human activity and a further 30% showed evidence of a possible relationship. Forest logging, changes in slope drainage, excavations and clearings are among the main causes of instability (Corominas, 2005). Failures are frequent in road cuts and embankments. For instance, the rainy 1995–96 and 1996–97 winters in the South of Spain (Baetic ranges) caused hundreds of mass movements along the main roads. The 10 km long Ardales-Campillo road sector (Malaga) experienced over 100 failures (González *et al.*, 1997).

Indirect evidences of the human-induced landslide activity have been observed at global scale (Remondo *et al.*, 2005; Cendrero *et al.*, 2006). Evidence of such possible relationship in northern Spain has been described (González-Díez *et al.*, 1996, 1999; Remondo *et al.*, 2008), and it appears to be linked to increasing sedimentation rates in estuaries (Remondo *et al.*, 2005; Bruschi *et al.*, 2013).

10.2.1 Landslides-climatic scenarios in Spain

Three meteorological situations have been identified for triggering landslides (Corominas, 2000; Corominas *et al.*, 2002): (a) short duration high-intensity rainfall which cause widespread shallow slides, debris flow and rockfalls; (b) rainfall episodes of moderate to low intensity lasting for several days or weeks reactivating rotational and translational slides, and earthflows; (c) abnormally either seasonal or interannual rainfall episodes that reactivate large-scale slides. Furthermore, in the alpine range freeze-thaw cycles are found to generate rock fall activity although a quantitative relationship cannot be established.

10.2.1.1 High intensity short-lasting rain storms

High intensity storms typically occur in the Mediterranean region. They trigger widespread shallow landslides and debris flows in slopes covered with surficial deposits (i.e. colluvium, till) and weathered argillaceous rocks. For example, in June 2000 an intense precipitation of 171 mm in 19 hours caused shallow landslides, debris flows and rockfalls in Montserrat, Coastal ranges (Marquès *et al.*, 2001). The Basque Country has also seen episodes of superficial landslides caused by heavy rains (Remondo *et al.*, 2005). In the Eastern Pyrenees, precipitations above 185 mm/day normally trigger shallow landslides (Gallart and Clotet 1988; Corominas and Moya, 1999). However, in road cuts where the excavation procedure might have disturbed the rock mass, thresholds are lower (Domínguez *et al.*, 1999; Moya and Corominas, 1997). In this type of failures, antecedent rainfall is not usually required (Corominas, 2000; González-Diez *et al.*, 2005). In the last decade, however, the availability of rain gauges with higher time resolution has shown that rainfall storms over 60 mm, may also trigger shallow landslides provided that hourly rainfall exceeds 15 mm/hr (Hürlimann *et al.*, 2003).

Figure 10.6 Shallow landslides in the vicinity of Berga, Eastern Pyrenees triggered by heavy rains in December 1997 (photo taken by J. Corominas).

Similar thresholds have been obtained in the Cantabrian range (Domínguez *et al.*, 1999).

10.2.1.2 Low to moderate intensity rainfall episodes

These episodes are related to precipitations associated to Atlantic depressions. Rainfall episodes of moderate intensity (over 40 mm in 24 h) are found to trigger or more likely, reactivate earthflows and mid-size rotational and translational landslides (few tens to hundreds of thousands of cubic meters) provided that 90 mm or more of rainfall had accumulated in the preceding 3–10 days (Corominas and Moya, 1999). This type of slide occurs in low-permeability clayey and silty-clayey formations.

10.2.1.3 Abnormal seasonal or annual rainfall

This type of abnormal amount of precipitation is related to fluctuations of the global atmospheric circulation, leading to a southward shift of polar fronts. Despite the fact that historical records show that most first-time large slope failures were triggered by non-climatic factors (Corominas, 2000; González-Diez *et al.*, 2005), rainfall is the most frequent cause of reactivation for dormant slides or the acceleration of those already active. However, the definition of a relationship between rainfall and slide activity for large landslides is difficult. The reason is that we do not yet avail of sufficient knowledge of the hydrological behavior of large slides. Advances in this field require complex mechanical-hydrological modelling, which needs a great deal of data on the terrain and instrumental ones, which are rarely available. In general, long rainy periods (at seasonal, annual or ten-year scale) appear to have a certain influence in the reactivation of large landslides although the relationship can often only be established qualitatively (Corominas *et al.*, 2002). In some particular geomorphological contexts, as in the case of extraordinary amounts of groundwater (e.g. contact with karstic massifs) or due to brusque topographic changes (e.g. toe erosion), large landslides can be reactivated by very intense, short-lived rainfall episodes (Corominas and Alonso, 1990).

Other types of mass movements, such as falls, happen as a consequence of freeze-thaw cycles, but they are also linked to intense precipitations like those in February and March 2013 in the island of La Gomera (Canary Islands), which caused several rockfall episodes and two fatalities. In high mountain areas, particularly the Pyrenees and Cantabrian Range, periglacial conditions cause episodes of gelifluction and solifluction, as well as landslides affecting the moraine deposits.

Although less frequently, fluvial undercutting is also a cause of slope instability. Examples of multiple rotational slides due to lateral migration in the Ebro River have been described by Gutiérrez *et al.* (2009) in the scarps near Zaragoza, mainly when the river floods. Basal erosion is also the cause of numerous mass movements in coastal cliffs. Almost every year landslides affect the flysch-composed coast of Guipúzcoa (N Spain), and frequently damage roads.

10.3 HISTORICAL EXTREME LANDSLIDING EVENTS

Based on the analysis of 20 landslide events spread throughout Spain, Ferrer and Ayala (1997) found that large slope failures take place associated to abnormal precipitation events ranging between 15 and 120% of the mean annual precipitation.

Multiple-Occurrence of Regional Landslide Events or MORLEs is a special type of landslide phenomenon. One event alone can involve thousands to ten thousands of individual landslides occurring almost simultaneously, over areas extending up to 20,000 km^2 (Crozier, 2005). In the last hundred years, several extreme climatic episodes have been identified producing widespread landslide activity in Spain. The most significant took place in the Pyrenees in October 1937 and 1940 and in November 1982, in the Basque country in the Cantabrian range in August 1983, in the Baetic ranges in the winter of 1996–1997, all of them due to abnormal precipitations. Finally, other abnormal situation took place in the period of 2008 and 2010 a combination of persistent precipitations and low temperature caused an unusual number of slope failures in Majorca (Balearic Islands).

Heavy rains and subsequent floods have been reported several times in the recent past in the Eastern Pyrenees. On 7–8th November 1982, heavy rains lasting for 36 hours had catastrophic consequences. The recorded rains ranged between 80 and 350 mm, with a maximum daily record of 342 mm at the ski resort of La Molina (Novoa, 1984). These rains caused extensive flooding and thousands of landslides which produced a toll of 20 lives and about 600 million of Euros in direct damages. The most striking geomorphic effects were concentrated in first-order watersheds. In a 1,250 km^2 area, more than 1,800 shallow landslides, debris flows, earthflows and rockfalls were reported (Gallart and Clotet, 1988). The spatial distribution of the failures were strongly controlled by the lithology and, particularly, by the availability of surficial formation (colluvium). In this event a few large dormant landslides were also reactivated such as Pont de Bar and Gòsol, mobilizing 20 and 5 million cubic meters respectively (Corominas and Alonso, 1990). The reactivation cannot adequately be explained without considering the geological setting which favors the groundwater flow to the landslides. Both landslides are in contact with limestones that have developed a karstic network. The infiltrated water in a large drainage basin is collected and transferred rapidly to the slide mass, causing the sharp increase in pore pressures and destabilizing the whole. Another landslide with a contrasting behavior was the Vallcebre landslide with an estimated volume of 20 million of cubic meters. This landslide affects a thick layer of claystones and shales with some gypsum lenses. Despite the impervious nature of these materials the landslide reacted immediately to the rain because the presence of well-developed graben and tension cracks (Corominas *et al.*, 2005). Omitting these types of features may lead to a misleading perception of the role of the climate in triggering landslides.

Between 2008 and 2010, the island of Mallorca in the Balearic Islands experienced the coldest and wettest winters of the last 40 years. Accumulated rainfall was twice the average and values of intense rainfall up to 296 mm/24 h were recorded in the heart of the Tramuntana range, the main mountainous alignment of the island. Additionally, high precipitation coincided with anomalous, low temperatures and freezing in the highest zones of the range. As a result, 34 mass movements were recorded on this region (Mateos *et al.*, 2012). Most of the rockfalls started in rocky massifs when there were several freeze-thaw cycles in the days prior to the failure, when the temperature hovered at around 0°C, regardless of whether the maximum daily rainfall was overly intense (around 30 mm). The largest event was the Son Cocó rock avalanche, which mobilized around 300,000 m^3 (Mateos *et al.*, 2013a). The landslides recorded took place after episodes of moderate to extreme intense rainfall, with very high accumulated rainfall totals (>800 mm/3 months). They were shallow landslides and earth flows. The largest one - Estellencs- took place on 8th March 2010, involving a volume of 30,000 m^3, which blocked the main road crossing the range for three and a half months. Fortunately, there were no deaths but there were numerous cases of damage to dwellings, holiday apartment blocks, barns and power stations, and especially the road network in the range, most significantly the numerous blockages on the Ma-10 road (the main transportation corridor of this region), cutting off the access to several locations and villages. These mass-movements caused significant economic losses in the different tourist resorts. Total economic losses (direct and indirect) were valued at approximately 11M Euro (Mateos *et al.*, 2013b).

During the period spanning from October 1996 to March 1997 the rainfall was so heavy in the southern and eastern part of the Iberian Peninsula (Andalusia) that the average of the previous years was more than doubled. In some parts of Andalusia, the rainfall was 50–80% of the previous all-time record, indicating the extraordinary intensity of the rainfall in the region during that period (Irigaray *et al.*, 1999). These heavy rains led to partial reactivations of many old landslides and to the generation of many new landslides, mostly corresponding to shallow earth flows and rock/soil falls, although several slides and complex movements were also recorded (Lamas *et al.*, 1998).

In the Cantabrian Range, a relationship seems to have existed between periods of rainfall increase and landslide frequency for the last 100.000 years (González-Díez *et al.*, 1996, 1999). In recent times, episodes of slope instability linked to rainy periods took place in 1983, 1988, 1992, 1994, 1997, 2006 and 2013 in Asturias, Cantabria and the Basque Country, being the rain episode of 1983 one of the biggest events ever recorded in northern Spain (Figure 10.7). In the Bajo Deba, the heavy rains

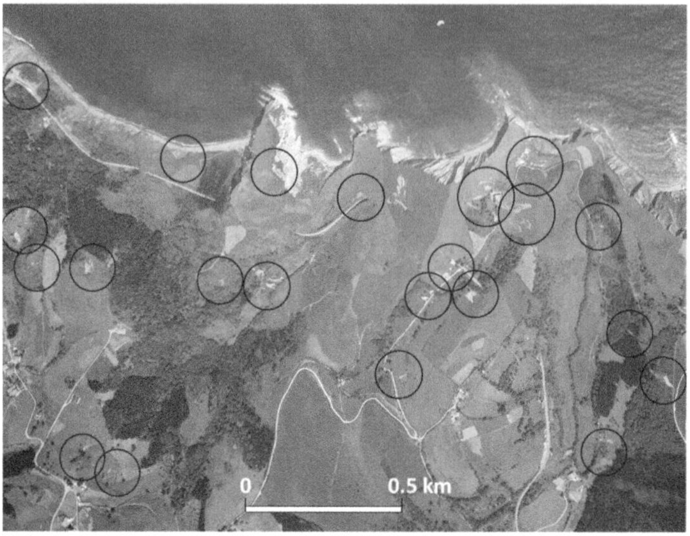

Figure 10.7 Ortophoto showing shallow landslides triggered by the Multiple-Occurrence of Regional Landslide Event (MORLE) occurred in the Cantabrian range in August 1983, the most intense landsliding event registered in the area.

occurred in August 1983 (113 and 128 mm the 25th and the 26th, respectively) and July 1988 (70 mm/24 h), as well as the precipitation of October and December 1992 (140 mm/day and 130 mm/day, respectively), triggered a great number of slope movements (more than two thousand shallow slides and debris flows) and floods, causing extensive damage and several fatalities. Estimates indicate that around 40% of those shallow landslides, occurred within the period 1954–1997, were triggered by those single events. Furthermore, in most cases those weather conditions coincided with high tidal coefficient and storm surges, producing large landslides in the Flysch coastal cliffs.

More recently, the 2012–2013 winter and the 2013 spring were abnormally wet periods with many intense rainstorms (persistent Atlantic depressions) in the Cantabrian Range, which produced many slope instability processes, especially reactivations of large ancient landslides. For instance, a large deep-seated landslide of about half a million cubic meters developed in Palaeozoic rocks affecting the village of Sebrango (Picos de Europa, Cantabria). The landslide was noticed in June 2013 and led to the evacuation of the village. This abnormal wet episode coincides with a decrease trend of annual precipitation observed for the last decades in northern Spain.

10.4 PROJECTED CHANGES IN WEATHER PATTERN UPON CLIMATE CHANGE

There is much uncertainty with regard to the occurrence of heavy rainfall episodes in the future. Christensen and Christensen (2003) predict an increase in the frequency of torrential rains during summer months in Europe, although the results for the Iberian Peninsula are highly uncertain.

According to the IPCC climate change future scenarios (A2 and B2), the Spanish Meteorological Agency (AEMET, 2009) has developed the regional scenarios for Spain and for the XXI century. The results can be summarized as follows:

- All model simulations show warming in the future across the whole country and in all seasons, with the consequent reduction of snow coverage and altitudinal displacement of vegetation. In the central area of the Iberian Peninsula, the temperature increase may range from 5°C to 7°C in summer and from 3°C to 4°C in winter (A2 scenario).
- A marked contrast between winter and summer patterns of precipitation. Intensities of summer heat waves will increase throughout the country.
- Intense precipitation events will increase across the whole country in frequency, especially in winter.
- An increase likelihood of drought events, specifically on the Mediterranean region and southern Spain.
- A reduction of annual precipitation, <10% in northern Spain and between 20 and 30% in the South.

These scenarios are consistent with those obtained by De Castro *et al.* (2004) that may be summarized as follows: (a) an increase in winter rainfall on the Cantabrian coast and the northern basin of the Duero river (b) a decrease in rainfall in absolute terms and a possible increase in the irregularity of rainfall in the Mediterranean Arch; (c) the rise in sea level and size of the sea waves; and (d) higher temperatures with the consequent permafrost melting, reduction of the snow coverage and altitudinal displacement of vegetation.

The Figure 10.8 shows the change in average annual precipitation (expressed in percentage terms) for the period 2011–2040 with respect to the period 1961–1990, on the basis of different global models and utilizing different technique for statistical regionalization. The trend is not as consistent as for temperature predictions. However, there is a slight reduction of precipitation in the western half of the Iberian Peninsula and an insignificant increase in the eastern part (AEMET, 2009). The seasonal distribution of precipitation does not show any pattern of behavior, with the exception of a greater reduction in summer (AEMET, 2014). This indicates the relatively low reliability of the predictions for this variable. This fact is a consequence of the error introduced by regionalization techniques when applied to precipitation and also, a consequence of the geographical location of the Iberian Peninsula, between high latitudes, where precipitation will increase, and subtropical areas, where a reduction of precipitation is expected. The figure also shows that there will be a decrease in the number of days with precipitation, which can be estimated in 10–15 days at the end of the century. Droughts duration tends to increase, mainly in the southern part of the Peninsula. The intensity and frequency of extreme precipitations will be higher but with no particular trend and with greater uncertainty values in southern Spain (Costa *et al.*, 2012; Herting *et al.*, 2013; AEMET, 2014).

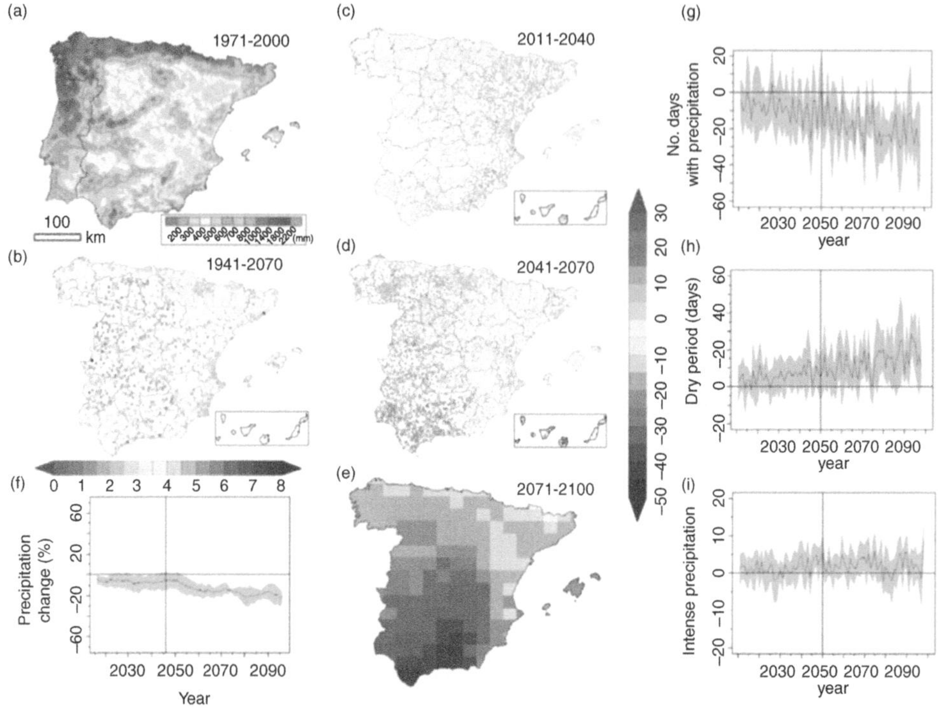

Figure 10.8 Projected 21st century changes in precipitation and temperature. A) Average total annual precipitation in the Iberian Peninsula and the Balearic Islands (1971–2000). B) Annual maximum temperature changes in 2041–2070 based on the global prediction model ECHAM4, which considers the scenario SRES A2, and analogues method for downscaling. C, D and E, examples of maps of total annual precipitation changes for Spain for different time periods with respect to 1961–1990 data: C) precipitation changes in 2011–2040 according to the global prediction model CGCM2, on the basis of emission scenario SRES A2, and analogues method (Anal_FIC) for downscaling; D) precipitation changes in 2041–2070 based on the global prediction model ECHAM4, which considers the scenario SRES A2, and analogues method for downscaling; E precipitation changes in 2071–2100 according to the global prediction model CGCM2 (scenario A2) and utilizing a grid (Anal_INM) for downscaling. C and D include the projected changes for Canary Islands. Time series of simulated annual precipitation and extreme precipitation indices obtained by dynamic regional downscaling and smoothed with a 10-year running mean; green lines show time series ensemble mean and green shading shows uncertainty: F) total annual precipitation change, G) Changes in the number of days with precipitation per year; H) dry periods duration; I) Intense precipitation occurrence rate. Sources: AEMET, 2009, 2011, 2014.

Finally, IPCC Report (2007) refers that sea level is projected to rise at an even greater rate in this century of around 3 mm/yr. Wind-driven waves and storms are seen as the primary drivers of short-terms coastal processes on many European coast (Smith *et al.*, 2000), specifically in the north-eastern Atlantic during at least the early part of the 21st century (2010–2030), with a shift of storm centre maxima closer to

European coast (Lozano *et al.*, 2004). These works also show a decline in storminess and wind intensity eastwards into the Mediterranean.

10.5 EXPECTED CHANGES IN LANDSLIDE PATTERN GIVEN CLIMATE CHANGE

It is unlikely the occurrence of new first-time large failures due to the reduction of the number of abnormal seasonal rainfall episodes and to the fact that a rise in sea level could be a factor that reduces the potential for an incision of the fluvial network. In the Baetic ranges, torrential flooding could favor new failures by slope undermining mainly in slopes made of shales and phyllites.

The immediate consequence of greater frequency of intense rainfall will be an increase in shallow slides, debris flow and rockfalls. The increase may also be due to anthropic disturbances and to a lesser protection of slopes by vegetation, which will have more adverse climatic conditions with regard to development and which will be affected increased forest fires. Substitution of plant species favors slope failures, especially in those areas in which the autochthonous vegetation is substituted by plants with shallower roots, which provide poorer retention of the surficial formation. This effect has been already observed in the Cantabrian range (Remondo *et al.,* 2008). In SE Spain, the semi-arid environment also determines a scarce vegetal cover and unprotected soil. Consequently, heavy rainfall events result in very intense erosive processes, shallow landslides and significant bedload contribution to the drainage network (Millares *et al.*, 2013). Nevertheless, some current works in central Spain (Duero River Basin) and the Balearic Islands show a generalized negative trend in annual runoff which is the consequence of a socioeconomic process with the abandonment of rural activities and the resulting changes in land-use/land cover (Morán Tejeda *et al.*, 2010). In these areas, forest and shrub areas have considerably increased due to abandonment of agricultural and traditional practices. In the medium-long term, however, the rate of appearance of failures will be restricted by the availability of movable material on the slope (Marqués *et al.,* 2001). Infill of the hollows with new material and weathering may require decades or hundreds of years in the Mediterranean context. Only slopes with unlimited sediment supply (i.e. till deposits and thick layers of weathered shales and slates) might increase the frequency of debris flows.

The occurrence of shallow landslides involves the direct supply to the water course of the material mobilised and the erosion of the landslide scarps made up of colluvium and clay formations which may degrade to badlands. This causes a significant increase in solid load in suspension in the water course, with the resulting decrease in quality and possible reservoir clogging downstream. For example, in the upper Llobregat river basin (Eastern Pyrenees), 50% of the solid load is supplied by the badlands, which take up 3.7% of the area, and 32% of the load results from the erosion of spoil tips which take up 4% of the area. The mean annual erosion is 1,000 t/km^2/year (Clotet and Gallart, 1983).

Increased winter rainfall in the Cantabrian Range and the northern Duero Basin would favour the reactivation of some big rotational slides and earthflows, especially if the increased rainfall is accompanied by river floods capable of sustaining the erosive action of the meanders on the river banks. In the rest of the Peninsula, the loss of seasonal and interannual rainfall will mean that some large movements will become dormant landslides. The exceptions will involve large landslides associated with particular geological context such as landslides fed with large amounts of groundwater in the event of intense downpours or those situated on riverbanks subjected to extraordinary floods or coastal cliffs.

As a consequence of what has been indicated in the previous sections, large slides in the Cantabrian Range are the most susceptible to reactivation. In the remaining mountain ranges, the reactivation of large movements will only occur in particular contexts (areas with an extraordinary supply of groundwater, areas of river erosion).

In case heavy rainfall becomes more frequent, the increase in shallow slides, debris flows and rockfalls will take place virtually in all the mountain ranges, including the Cantabrian range. However, in the Central Range and in the Mediterranean sector of the Iberian and the outer Baetic ranges, one should not expect a significant increase of soil slope failures due to the calcareous nature of the rock formations and the scant soil cover.

Even though the changes in rainfall pattern are subjected to a high uncertainty, what seems likely is the increase of exposure and that human activity itself is today the main geodynamic agent, which directly and indirectly introduces changes that destabilize the slopes. How landslide activity will evolve in the future will depend more on this than on the precipitation (changes from human activities are huge compared to those due to rainfall). We entered a new geological era, the Anthropocene, characterized by the dominance of men in the processes of erosion and sedimentation. The changes in vegetation will probably have a limited impact on slope instability and, in any case, are not easy to predict (e.g. the root systems of Mediterranean species tend to deepen, except pines, however they maintain less soil moisture and tend to have greater biological activity).

The rise in sea level, along with the frequency of sea storms, will cause undermining, fall and sliding of the coastal cliff, especially in those made up of soft rocks, such as the Triassic and Miocene formations of the northern coast of Majorca and Southwestern coast of Ibiza (Balearic Islands), the Cantabrian coast (flysch in Zumaya, Triassic in Asturias), lava flows stacks on the Canary Islands and, to a lesser degree, the fractured rocky massifs on the Costa Brava and the Costa del Sol.

10.6 CLIMATE CHANGE ADAPTATION STRATEGIES FOR CHANGING LANDSLIDE PATTERN

The impact of increased shallow slides and debris flow can be partly mitigated through policies aimed at the reforestation of slopes and the maintenance of species better adapted to the existing environment. The forest expansion also constitutes a clearly sustainable element for protection against rockfalls (protection forest). Some regional administrations as in the Basque Country have already given priority to the implementation of reforestation and fire-fighting policies.

The best adaptation tool involves both urban and land-use planning with the aim of avoiding, as far as possible, the development in areas most susceptible to slope instability, particularly for sensitive activities and facilities (schools, hospitals, etc.). It is the less expensive measure which promotes a sustainable development while reconciling the interest of conservation with urban growth.

Since 2004 Spain has taken some steps towards the definition of a coherent set of public policies to deal with climate change. One of the cornerstones of this institutional response is the National Climate Change Adaptation Plan (PNACC in Spanish). This program was adopted in October 2006 after endorsement by the Cabinet of Ministers. The PNACC is the reference framework tool for the coordination of Public Administrations' efforts dealing with the assessment of impacts, vulnerability, and adaptation to climate change in the Spanish sectors acknowledged as potentially affected (water management, agriculture, forests, biodiversity, coasts, health, tourism, etc.). The first action lines drawn in the PNACC for mountain areas, coastal areas, soils, transport, urban planning and constructions among others, do not take into account landslides at all. Landslide frequency/magnitude is not deemed to be an indicator of climate change.

The management of natural hazards in Spain is carried out by the regional administrations and, specifically by Civil Protection and Emergency Services (Mateos, 2007). In 1992 the Basic Civil Protection Regulations were approved, establishing the framework of emergency plans for each region, which would constitute an organisational and administrative network dealing with emergencies. In addition to the regional territorial plans, the Basic Regulations also established the possibility of preparing thematic plans focusing on "*particularly significant*" hazards in each region. Nevertheless, the thematic plans are mainly oriented to flood and seismic hazard and they do not take into account landslide-hazard. There are some exceptions, such as the 153/1997 decree-law in the Basque Country, which describes the natural risks in the community including mass movements. The decree includes a Civil Protection Plan and emergency strategies. This plan has the rank of a law and was approved by the Regional Basque Parliament. It includes rules and protocols to face a natural disaster and establishes that every municipality with more than 20,000 inhabitants must have a specific emergency plan. Nevertheless, these plans are more or less factual and do not refer to any particular type of hazard.

The main challenge facing Spain is to extend civil protection against natural hazards to risk management policies, such as urban and land-use planning. For this, section 15.2 of the new Spanish Land Bill (approved in 2008), included the requirement to draw up natural-risk maps within the ambit of land-use planning. In this regard, some regional administrations have already taken a first step by including in their urban planning laws the requirement to consider landslide hazard, among other hazards, as criterion to be taken into account when authorising development. These regions are encouraging efforts to improve detailed maps to enhance land-use location and authorisation decision-making.

The Catalonian Geological Survey has initiated the systematic hazard prevention mapping for different processes (rockfalls, landslides, snow avalanches, torrential floods). In 2005, the Catalonian regional administration approved a new Urban Planning Law establishing that land development must provide the adequate level of protection to the population. This law states that the development should be avoided in areas threatened by natural hazards if countermeasures are not feasible. Furthermore, the law states that urban settlements cannot develop on slopes over 20% unless no alternative land exists. The assessment of both natural and geological hazards for urban planning purposes must take into account the official maps of the Catalonian Geological Institute (CGI). All the local development plans are nowadays sent to the CGI for providing reports on the potential hazards. These reports consider specifically all kind of natural hazards and provide recommendations to the municipalities. Between year 2000 and 2007, more than 250 reports were issued, which represents approximately 25% of the municipalities in Catalonia. In 2007 the CGI has started an ambitious plan of landslide hazard mapping for the whole Catalonia at 1: 25,000 scale that it is expected to be completed within this decade. At present Catalonia region has only 24 official landslide maps to depict at an appropriate scale the landslide hazardous areas. Because of this, many reports issued by the CGI have yet no legal binding and the policies to restrict the development in the landslide threatened areas cannot be fully implemented, depending largely on the interest and pro-active behavior of the local communities. This situation will change in few years once the plan for mapping geological hazards at 1:25,000 had been completed.

Since 2003, the Balearic Islands´ regional legislation has required all local authorities to draw up 1:25,000-scale landslide susceptibility maps. For any land use plan for an area with a medium or high degree of susceptibility, a binding expert report on the real hazard must be commissioned. In case a development potentially involves a major risk (e.g. marinas, hotel facilities, etc.), a positive suitability report must be also obtained from the regional Emergency Service.

In the case of Andalusia, the Regional Land Use Plan (2006) includes a chapter focused on natural hazard prevention, including landslides and erosion. The strategic items for this Plan are: (1) the necessity to include landslide- hazard at all levels of land-use planning. (2) establish territorial criteria to prevent this natural hazard, (3) support studies focused on a better knowledge of landslide- hazard and to draw up a detailed risk-map to be able to take decisions relating to land use, and (4) the planning of works and infrastructures necessary for the reduction of risk as well as the establishment of observation and alerting networks.

Damages caused by natural risks in Spain are covered by the Insurance Compensation Consortium, which covers the consequences of extraordinary risks (Royal Act 2022/1986) and includes: flooding, earthquakes, tsunamis, volcanic eruptions, storms of certain intensity, and falls of meteorites or other planetary objects. Landslides are not specifically included, except when they are linked to high intensity storms. When the situation of natural disaster is officially declared, the uncovered damages caused by landslides on infrastructure and crops, are compensable, as occurred after the heavy rains of 1983 in northern Spain.

In the case of landslide hazard in coastal areas, the main framework is based on two legal documents, the Coastal Act and the Spanish Strategy for Coastal Zone Management (MAGRAMA, 2013). The integrated strategy aims at coordinating policies affecting the coast and its economic development, through sustainable management of resources, conservation and hazard protection actions. The lack of a global management strategy to *fight coastal erosion and the short-term urgency to protect specific stretches has* resulted in a reactive approach based initially on hard engineering structures (especially groins, but also seawalls and detached breakwaters). The net long-term effect is the degradation of many coastal areas and sediment starvation, which is enhanced by the retention of sediment in river dams and associated river regulation policies (Gracia *et al.*, 2013). Major activities that have been conducted in some coastal

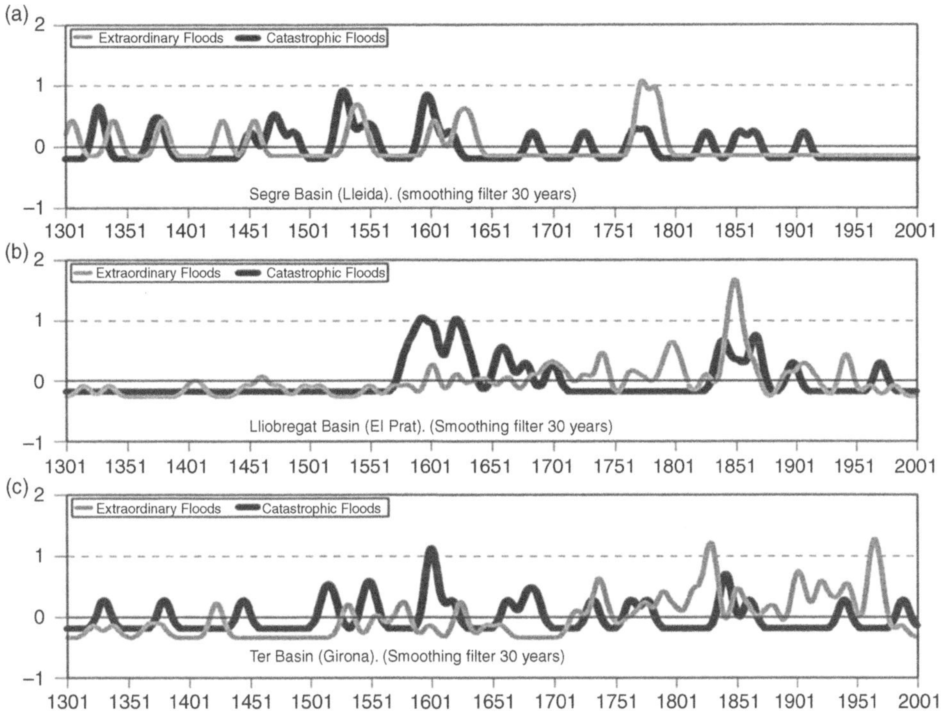

Figure 10.9 Historical flash floods in Catalonia, North Eastern Spain between years 1300 and 2000 (Llasat *et al.*, 2005).

sand flats are the restoration and preservation of dune areas (for example, in Galicia, Balearic and Canary Islands), through initiatives by the Coastal Authorities, where attempts have been made to effectively combine tourist activities with environmental protection. As a general conclusion, it will be very difficult to establish protection measures against erosion and undermining of coastal cliffs, except in special cases for which measures are economically feasible. Likewise, only those large slides threatening property and infrastructures of value could be corrected and contained.

The Geological Survey of Spain (IGME) is currently preparing a National Landslide Catalogue, according to the INSPIRE criteria. This catalogue may become a useful tool for future evaluations and to establish a reliable relationship between climate change and landslide occurrence.

10.7 FINAL REMARKS

Predictions of the future behaviour of slopes are based on the scenarios derived from the available climate change models. At present, these scenarios show high uncertainty with regard both to the areal distribution and the frequency of the intense rainfall events in the Iberian Peninsula. In this respect, in spite of the fact that increased sea temperature should favour storms in the Mediterranean environment, it cannot be ascertained that torrential downpours will be more frequent.

There is still great uncertainty with regard to the response of both shallow and large landslides to rainfall episodes. Despite many landslide-triggering rainfall thresholds have been proposed for shallow landslides in the scientific literature, these vary greatly depending on the geologic, morphological and climatic conditions of each region. Critical rainfall thresholds have been proposed in Spain for the Eastern Pyrenees, but they still need to be established for the rest of the territory. On the other hand, there is no available knowledge on the hydrological response of most active or dormant large landslides distributed throughout the main mountain ranges. The increase in temperature and rainfall in the Cantabrian range did not always accompanies the highest activity of the landslides in the past. Indeed, tectonism, channel incision and human intervention might have played a significant role as well (González-Díez *et al.*, 1999). As climate models allow for better definition of the rainfall regime on the Iberian Peninsula, the current hypotheses will require confirmation.

In the Mediterranean area, the changing scenario should be evidenced by the increased frequency of high-intensity rainfall and, consequently, a higher number of flash floods, debris flows, shallow slides and rockfalls. During the second half of the last century it was observed an increase of extreme meteorological events and of landslide occurrence. However, the analysis of the flash flood frequency during the last 700 years in Catalonia, NE Spain (Figure 10.9), does not show any clear trend.

Moreover, it should be recalled that during the second half of 20th century, a trend of precipitation reduction was registered in many parts of Spain, particularly in the Cantabrian Range (4.8 mm/year in Santander) and in the SW, being the period 2000–2010 the decade with the lowest values of precipitation since 1950 (OECC, 2012).

The increase of the frequency of wintertime reactivation of earthflows and large slides along with the increase of both sea storms and the instability of sensitive coastal cliffs, would be indicators as well.

Finally, it must be recalled that climate is not the only driver of the landslide activity. The changes in the precipitation regime and the sea level rise may exacerbate the occurrence of landsides in the susceptible regions. However, what may produce higher damages is the combination of the latter with the disturbance of the slopes caused by human intervention and the increase of exposure of the elements at risk.

10.8 KEY CONCLUSIONS/RECOMMENDATIONS ON THE WAY FORWARD

- Specific reports at national scale must be prepared on the results of the evaluations and projects regarding landslides.
- It is required a follow-up and update of the Landslide National Inventory.
- Climate change parameters must be processed and considered in the updating of the existing hazard maps
- Improve the communication, training and public awareness.

- The implementation of good practices and land use planning is strongly recommended, among them: proper and enforced building codes and construction practices. A revision of the design standards (see Eurocodes) might be necessary.
- Civil engineering structures and protection measures in landslide zones have to be evaluated and if necessary, redesigned in order to fulfill new safety conditions and possible increased loads in relation with climate change effects. Tasks must be shared by all the stakeholders which includes the municipalities, autonomous regions, national authorities, research institutions, owners and insurance companies.

REFERENCES

Ablay, G. & y Hürlimann, M. (2000) Evolution of the north flank of Tenerife by recurrent giant landslides. *Journal of Volcanology and Geothermal Research*, 103, 135–159.

AEMET (2009) *Generación de Escenarios Regionalizados de Cambio Climático para España.* Ministerio de Medio Ambiente, Medio Rural y Marino. 158 pp.

AEMET (2011) *Iberian Climate Atlas.* Ministerio de Medio Ambiente, Medio Rural y Marino. 80 pp.

AEMET (2014) *Guía de escenarios regionalizados de cambio climático sobre España a partir de los resultados del IPCC-AR4.* Madrid, Ministerio de Agricultura, Alimentación y Medio Ambiente. 198 pp.

Alonso, E., Pinyol, N.M. & Puzrin, A.M. (2010) *Earth Dam Sliding Failure: Aznalcóllar Dam. Spain. Geomechanics of Failures. Advanced Topics.* Dordrecht, Netherlands, Springer. pp. 129–167.

Ancochea, E., Hernán, F., Cendrero, A., Cantagrel, J.M., Fuster, J.M., Ibarrola, E. & Coello, J. (1994) Constructive and destructive episodes in the building of a Young ocean island, La Palma, Canary Islands, and genesis of the Caldera de Taburiente. *Journal of Volcanology and Geothermal Research*, 60, 243–262.

Araña, V., Badiola, E.R., Berga, L., Carracedo, J.C., Cendrero, A., Coello, J., Corominas, J., Dabrio, C., Díaz de Terán, J.R., Durán, J.J., Elízaga, E., Ferrer, M., García, M., Garzón, M.G., Goy, J.L., López, J., Martínez-Goytre, J., Mezcua, J., de la Nuez, J., Salinas, J.L., Soler, V., del Val, J. & y Zazo, C. (1992) Riesgos geológicos en España. Estado de la cuestión. In: *III Congreso Geológico de España y VIII Congreso Latinoamericano de Geología.* Tomo 2. pp. 671–745.

Ayala, F.J. (1995) Probabilidad y vulnerabilidad en movimientos de ladera. In: *Reducción de riesgos geológicos en España, ITGE.* Madrid. pp. 95–113.

Ayala, F.J., Olcina, J. & Vilaplana, J.M. (2004) Impacto social de los riesgos naturales en España en el periodo 1990–2000 (II). *Gerencia de riesgos*, XXI, 17–29.

Azañón, J.M., Azora, A., Yesares, J., Tsige, M., Mateos, R.M., Nietoe, F., Delgadof, J., López-Chicano, M., Martín, W. & Rodríguez-Fernández, J. (2010) Regional-scale high-plasticity clay-bearing formation as controlling factor on landslides in Southeast Spain. *Geomorphology*, 120, 26–37.

Beguería, S. (2006) Changes in land cover and shallow landslide activity: A case study in the Spanish Pyrenees. *Geomorphology*, 74, 196–206.

Berganza, F. & Modrano, R. (1978) The instability phenomenon of the banks of the river Duero in its passage through Aranda (Burgos) in Tertiary soils. In: *III International Congress IAEG. Sect. I(1).* Madrid. pp. 197–208.

Bonachea, J., Remondo, J., González-Díez, A., Díaz de Terán, J.R. & Cendrero, A. (2009) Landslide risk models for decision making. *Risk Analysis*, 29 (11), 1629–1643.

Bruschi, V.M., Bonachea, J., Remondo, J., Gómez-Arozamena, J., Rivas, V., Méndez, G., Naredo, J.M. & Cendrero, A. (2013) Analysis of geomorphic systems' response to natural and human drivers in northern Spain; implications for global geomorphic change. *Geomorphology*, 196, 267–279.

Carracedo, J.C., Day, S.J., Guillou, H. & y Torrado, F.J.P. (1999) Giant Quaternary landslides in the evolution of La Palma and El Hierro, Canary Islands. *Journal of Volcanology and Geothermal Research*, 94, 169–190.

Cendrero, A., Remondo, J., Bonachea, J., Rivas, V. & Soto, J. (2006) Sensitivity of landscape evolution and geomorphic processes to direct and indirect human influence. *Geografia Fisica e Geodinamica Quaternaria*, 29 (2), 125–137.

Christensen, O.B. & Christensenm, J.H. (2004) Intensification of extreme European summer precipitation in a warmer climate. *Global and Planetary Change*, 44, 107–117.

Clotet, N. & Gallart, F. (1983) Sediment yield in a mountainous basin under high Mediterranean climate. *Zeitschrift für Geomophologie*, 60, 205–216.

Corominas, J. (1989) *Litologías inestables. Monografía nº 3.* Zaragoza, Sociedad Española de Geomorfología. pp. 81–96.

Corominas, J. (2000) Landslides and climate. In: Bromhead, E.N. (ed.) *VIII International Symposium on Landslides.* Cardiff, UK, Keynote Lectures, CD_ROM.

Corominas, J. (2005) Impacto sobre los riesgos naturales deorigen climático: inestabilidad de laderas. In: Moreno, J.M. (coordinator) *Proyecto ECCE. Evaluación Preliminar de los impactos en España por efecto del Cambio Climático.* Madrid, Ministerio de Medio Ambiente. pp. 549–579.

Corominas, J. (2006) El clima y sus consecuencias sobre la actividad de los movimientos de ladera en España. *Cuaternario y Geomorfología*, 20 (3–4), 89–113.

Corominas, J. & Alonso, E.E. (1990) Geomorphological effects of extreme floods (November, 1982) in the southern Pyrenees. In: Sinniger, R.O. & Monbaron, M. (eds.) *Hydrology in Mountainous Regions II. Artificial Reservoirs, Water and Slopes.* Wallingford, UK, IAHS, Publ. 194. pp. 295–302.

Corominas, J. & Baeza, C. (1992) Landslides in the Eastern Pyrenees-Spain. In: del Prete, M. (ed.) *Movimenti franosie metodi di stabilizzazione.* Italia, Consiglio Nazionale delle Ricerche. Pub. no 481. pp. 25–42.

Corominas, J. & Moya, J. (1999) Reconstructing recent landslide activity in relation to rainfall in the Llobregat River basin, Eastern Pyrenees, Spain. *Geomorphology*, 30, 79–93.

Corominas, J., Moya, J. & Hürlimann, M. (2002) Landslide rainfall triggers in the Spanish Eastern Pyrenees. In: *Proceedings 4th EGS Conference on Mediterranean Storms.* Mallorca. CD-ROM.

Corominas, J., Moya, J., Ledesma, A., Lloret, A. & Gili, J.A. (2005) Prediction of ground displacements and velocities from groundwater level changes at the Vallcebre landslide (Eastern Pyrenees, Spain). *Landslides*, 2, 83–96.

Costa, A.C., Santos, J.A. & Pinto, J.G. (2012) Climate change scenarios for precipitation extremes in Portugal. *Theoretical Applied Climatology*, 108 (1–2), 217.

Crozier, M.J. (2005) Multiple occurrence regional landslide events in New Zealand: Hazard management issues. *Landslides*, 2, 247–256.

De Castro, M., Martín-Vide, J. & y Alonso, S. (2004) *El clima de España: pasado, presente y escenarios de clima para el siglo 21.* Evaluación de los Impactos del Cambio Climático en España (ECCE).

Del Val, J., Delgado, J., Eguiguren, E., García, I. & Garrote, R. (1996) Ejemplo de análisis de inestabilidades del terreno para la atención de emergencias, el caso del País Vasco. In: *Actas del VI Congreso Nacional y Conferencia Internacional de Geología Ambiental y Ordenación del Territorio.* Granada.

Díaz de Terán, J.R., Cendrero, A., Araña, V., Badiola, E.R., Berga, L., Carracedo, J.C., Chacón, J., Coello, J., Corominas, J., Dabrio, C., Durán, J.J., Elízaga, E., Ferrer, M., García, M., Garzón, G., González, A., Goy, J.L., López, J., Martínez-Goytre, J., Mezcua, J., Nuez, J. de la, Salinas, J.L., Soler, V., Val, J. & del Zazo, C. (1997) Geomorphological hazards in Spain. In: Embleton, C. & Embleton-Hamann, Ch. (eds.) *Geomorphological Hazards of Europe*. Amsterdam, Elsevier Science Publishers. pp. 429–456.

Domínguez, M.J., Jiménez, M. & y Rodríguez, A. (1999) Press archives as temporal records of landslides in the North of Spain: Relationships between rainfall and instability slope events. *Geomorphology*, 30, 125–132.

Ferrer, M., (1995) Los movimientos de ladera en España. In: Instituto Tecnológico Geominero de España (ed.) *Reducción de Riesgos Geológicos en España*. Madrid. pp. 69–82.

Ferrer, M. & y Ayala, F.J. (1997) Relaciones entre desencadenamiento de movimientos y condiciones meteorológicas para algunos deslizamientos en España. In: *IV Simposio Taludes y Laderas*. Granada, Vol. 1. pp. 185–197.

Gallart, F. & Clotet, N. (1988) Some aspects of the geomorphic process triggered by an extreme rainfall event: The November 1982 flood in the Eastern Pyrenees. *Catena Supplement*, 13, 75–95.

García-Ruiz, J.M., Beguería, S., Alatorre, L.C., Constante, A. & Puigdefábregas, J. (2010) Land cover changes, shallow landsliding in the Flysch sector of the Spanish Pyrenees. *Geomorphology*, 124, 250–259.

González, B., Recchi, C., Oliveros, M.A. & Ortuño, L. (1997) Adaptación de dos proyectos de carreteras a las condiciones geológico-geotécnicas reales deducidas de la experiencia en tramos cercanos. In: *IV Simposio Nacional sobre Taludes y Laderas Inestables*. Granada, Vol. 1. pp. 157–169.

González-Díez, A., Salas, L., Díaz de Terán, J.R. & Cendrero, A. (1996) Late Quaternary climate changes and mass movement frequency and magnitude in the Cantabrian region, Spain. *Geomorphology*, 15, 291–309.

González-Díez, A., Remondo, J., Díaz de Terán, J.R. & Cendrero, A. (1999) A methodological approach for the analysis of the temporal occurrence and triggering factors of landslides. *Geomorphology*, 30, 95–113.

González-Díez, A., Remondo, J. & Cendrero, A. (2005) Consideraciones sobre la relación entre movimientos de ladera y clima. In: *VI Simposio Nacional sobre Taludes y Laderas Inestables*. Valencia, Spain, Vol. 3. pp. 1103–1130.

Gracia, V., Sánchez-Arcilla, A. & Anfuso, G. (2013) Spain. In: Pranzini, E. *et al.* (ed.) *Coastal Erosion and Protection in Europe*. Abingdon, Routledge. pp. 254–274.

Herrera, G., Gutiérrez, F., García-Davalillo, J.C., Guerrero, J., Notti, D., Galve, J.P., Fernández-Merodo, J.A. & Cooksley, G. (2013) Multi-sensor advanced DInSAR monitoring of very slow landslides: The Tena Valley case study (Central Spanish Pyrenees). *Remote Sensing of Environment*, 128, 31–43.

Hertig, E., Seubert, S., Paxian, A., Vogt, G., Paeth, H. & Jacobeit, J. (2013) Changes of total versus extreme precipitation and dry periods until the end of the twenty-first century: Statistical assessments for the Mediterranean area. *Theoretical Applied Climatology*, 111, 1–20.

Hürlimann, M., Corominas, J., Moya, J. & Copons, R. (2003) Debris-flow events in the Eastern Pyrenees. Preliminary study on initiation and propagation. In: Rickenmann, D. & Chen, C. (eds.) *3rd International Conference on Debris-Flow Hazards Mitigation*. Davos, Millpress. pp. 115–126.

Hürlimann, M., Ledesma, A. & y Martí, J. (1999) Conditions favouring catastrophic landslides on Tenerife (Canary Islands). *Terra Nova*, 11, 106–111.

IPCC (2007) *Climate Change, 2007. The Physical Science Basics*. Summary for Policymakers. 21 pp.

Irigaray, C., Fernández, T., El Hamdouni, R. & Chacón, J. (1999) Verification of landslide susceptibility mapping. A case study. *Earth Surface Processes and Landforms*, 24, 537–544.

Lamas, F., El Hamdouni, R., Irigaray, C., El Amrani, N. & Chacón, J. (1998) Rainfall triggered landslides in the Betic Cordillera: October 1996 to April 1997. In: *Proceedings 8th International Congress of the International Association for Engineering Geology and the Environment*, Vancouver, Canada, Vol. 1–5. pp. 2001–2008.

Llasat, M.C., Barriendos, M., Barrera, A. & Rigo, T. (2005) Floods in Catalonia (NE Spain) since the 14th century. In: Benito, G., Ouarda, T.B.M.J. & Bárdossy, A. (eds.) *Palaeofloods, Historical Data & Climate Variability: Applications in Flood Risk Assessment. Journal of Hydrology*, 313 (1–2), 32–47.

Lomoschitz, A., Meco, J. & y Corominas, J. (2002) The Barranco de Tirajana basin, Gran Canaria (Spain). A major erosive landform caused by large landslides. *Geomorphology*, 42, 117–130.

Lozano, I., Devoy, R.J., May, W. & Anderson, U. (2004) Storminess and vulnerability along the Atlantic coastlines of Europe: Analysis of storm records and of greenhouse gases induced climate scenarios. *Marine Geology*, 210, 205–225.

MAGRAMA (2013) *Ministerio de Agricultura, Alimentación y Medio Ambiente*. Available from: http://www.magrama.gob.es/es/.

Marquès, M.A., Martín, E. & Gascón, M. (2001) Episodio extremo y distribución espacial de movimientos de laderas en Montserrat (Barcelona, España). In: *V Simposio Nacional sobre Taludes y Laderas Inestables*. Madrid, Vol. 3. pp. 1315–1326.

Martí, J., Hürlimann, M., Ablay, G.J. & Gudmundsson, A. (1997) Vertical and lateral collapses on Tenerife (Canary Islands) and other volcanic ocean islands. *Geology*, 25 (10), 879–882.

Mateos, R.M. (2007) Natural hazard and land-use planning in Spain. *Journal of the European Federation of Geologists*, 23.

Mateos, R.M., García-Moreno, I. & Azañón, J.M. (2012) Freeze-thaw cycles and rainfall as triggering factors of mass movements in a warm Mediterranean region: The case of the Tramuntana range (Majorca, Spain). *Landslides*, 9, 417–432.

Mateos, R.M., García-Moreno, I., Herrera, G. & Mulas, J. (2013a) Recent mass movements in the Tramuntana range (Mallorca, Spain). In: Margottini, C., Canuti, P. & Sassa, K. (eds.) *Landslide Science and Practice*. Springer. Vol. 4: Global Environmental Change. pp. 27–37.

Mateos, R.M., García-Moreno, I., Herrera, G. & Mulas, J. (2013b) Damage caused by recent mass-movements in Majorca (Spain), a region with a high risk due to tourism. In: Margottini, C., Canuti, P. & Sassa, K. (eds.) *Landslide Science and Practice*. Berlin, Springer. Vol. 7: Social and Economic Impact and Policies. pp. 105–113.

Millares, A., Polo, M.J., Moñino, A., Herrero, J. & Losada, M.A. (2013) Bedload dynamics and associated snowmelt influence in mountainous and semiarid alluvial rivers. *Geomorphology*, 206, 330–342.

Morán-Tejeda, E., Ceballos-Barbancho, A. & Llorente-Pinto, J.M. (2010) Hydrological response of Mediterranean headwaters to climate oscillations and land-cover changes: The mountains of Duero River basin (Central Spain). *Global and Planetary Change*, 77, 62–76.

Moreno, J.M. (Coord.) (2005) *Evaluación Preliminar de los Impactos en España por Efecto del Cambio Climático. Proyecto ECCE – INFORME FINAL*. Madrid, Ministerio de Medio Ambiente.

Moya, J. & y Corominas, J. (1997) Condiciones pluviométricas desencadenantes de deslizamientos en el Pirineo Oriental. In: *IV Simposio Taludes y Laderas*. Granada, Vol. 1. pp. 199–212.

Novoa, M. (1984) Precipitaciones y avenidas extraordinarias en Cataluña. In: *Jornadas sobre Inestabilidad de Laderas en el Pirineo*. Barcelona. pp. I.1.1–I.1.15.

PSMSL. Available from: http://www.psmsl.org/products/trends/ [Accessed 20th May 2014].

Remondo, J. (2001) Elaboración y validación de mapas de susceptibilidad de deslizamientos mediante técnicas de análisis espacial. Tesis Doctoral. Oviedo, Universidad de Oviedo.
Remondo, J., Bonachea, J. & Cendrero, A. (2008) Quantitative landslide risk assessment and mapping on the basis of recent occurrences. *Geomorphology*, 94, 496–507.
Remondo, J., Soto., J., González-Díez, A., Díaz de Terán, J.R. & Cendrero, A. (2005) Human impact on geomorphic processes and hazards in mountain areas. *Geomorphology*, 66, 69–84.
Smith, B., Burton, I., Klein, R. & Wandel, J. (2010) An anatomy of adaptation to climate change and variability. *Climate Change*, 45, 223–251.
Tsige, M., González de Vallejo, L., Doval, M. & Barba, C. (1994) Microfabric of Guadalquivir "Blue Marls" and its engineering geological significance. In: *Proceedings 7th International IAEG Congress.*
Watts, A.B. & Masson, D.G. (1995) A giant landslide on the north flank of Tenerife, Canary Islands. *Journal of Geophysical Research: Solid Earth*, 100, B12, 24487–24498.

Chapter 11

TR3 project: slope safety preparedness for effects of climate change contribution for Switzerland

C. Bonnard
PBBG SA, Lausanne, Switzerland

C. Crot & E. Prina Howald
School of Business and Engineering Vaud, Yverdon-les-Bains, Switzerland

L. Laloui
Swiss Federal Institute of Technology, Lausanne, Switzerland

ABSTRACT

Due to the situation of Switzerland, located mainly in the Alpine range in which the effect of climate change conditions is quite significant, various consequences with respect to landslide frequency and intensity are expected. These impacts are important as a large development of constructions for infrastructures and holiday residential buildings has been observed during the last 20 years and has needed the implementation of limitations within the town and country planning procedures. In parallel to a long phase of production of hazard maps for the different dangers, produced at the cantonal level, that is nearly achieved, several actions have been taken to reduce the potential increase of landslide activity, namely either the velocity of slide zones, by drainage schemes, or the importance of debris flows by protection works in the torrents using steel nets. Within the regulation city plans limitations are also foreseen to reduce the potential damage of buildings located in exposed areas. But the notion of risk specifically due to climate change is not really considered and quantified in the regulations produced by the federal state.

11.1 PREFACE

Answering the request of the Joint Technical Committee on Landslides JTC-1 within the TR3 Project, several scientists from Switzerland have set up a task force to prepare a contribution on the various actions put forward in this country in order to mitigate the effects of climate change, with respect to landslide safety, through prevention and protection actions. Indeed the development of this strategy had already begun in the 1990s, at the national level as well as at the cantonal levels, so as to cope with the serious increase of damage caused by landslides of all kinds. Different administrative units and committees have worked to produce regulations, recommendations and codes of practice, aiming at a better consideration of landslide hazards in town and country planning and construction projects.

Given the complex administrative structure in Switzerland, in which the state, the cantons and the communes have various fields of competence and responsibilities, it is not easy to outline a global concept for slope safety preparedness when facing effects of climate change; nevertheless it is worthwhile to present several standards, documents and researches that have been published during these last years and that contribute to the objectives of the present TR3 Project. This contribution cannot be regarded however as an official position by the authorities, but rather as a prospective vision on this theme. It is therefore a compilation of information presently available and published recently, and does not result from a specific research program. It allows the comparison of Swiss advances in this field with other countries.

This paper is prepared at the very moment when most of landslide hazard maps in Switzerland are available, after a long process that lasted nearly 20 years. It is now the task of all cantons and communes to incorporate such information in their legal planning documents. It is also illustrated with several examples of landslide zones or events that appear to be somehow influenced by the effects of climate change and that prove the necessity to undertake a slope safety preparedness coordinated policy, in order to reduce the impact of such hazards in the long term.

11.2 GEOGRAPHICAL FRAMEWORK

Switzerland is a small country extending over an area of 41,285 km^2 (www.swissworld.org), located in the center of Europe between Germany, to the North, Italy to the South, Austria to the East and France to the West. It includes three distinct geographical regions, namely the Alps (60% of its area), the Plateau (or flat zone, 30% of its area) and the Jura range (10% of the area of the country) (Figure 11.1). Although the economic heart of Switzerland is located in the Plateau area, as well as the majority of its inhabitants, the sector of the Alps, in which most of gravity-induced natural hazards occur, is far from being a deserted zone. 10% of the population lives there, and it is above all an important leisure activity and rest area, in which 60% of tourist trade is concentrated, despite of a rugged topography at an average elevation of 1,700 m asl. The highest summit is Dufour peak in Monte Rosa Massif which is 4,634 m asl high. Moreover all mountain ranges have been deeply carved by several glacial episodes in the recent Quaternary phases, so that prevention and management of all natural hazards and in particular landslide hazards appear as important topics for the safety of the society; however hazards related to floods and snow avalanches constitute a more serious problem to manage, all over Switzerland.

At the institutional level, Switzerland includes three levels of political and administrative authorities, extending from the level of the commune (or district council) to the level of the state (federal government), with the cantonal level (or regional authority) in between, including 26 cantons that are all responsible for the management of natural hazards. Each of these three political levels includes an executive body (government level) and a legislative body (parliament level). It has to be pointed out that a lower entity must always abide to the decisions and laws of the higher authorities.

As far as the management strategy of natural hazards is concerned, it is established by the state, through different recommendations or standards, but it is implemented at the cantonal level. This explains why the hazard maps have needed quite a long time to be carried out (indeed some 20 years), despite of the promise of a state financial support if completed within a fixed deadline, that has been extended several times (see chapter below on Federal prevention strategy). The methodology to be applied has been set up in a federal recommendation, published in 1997 as far as landslide hazards are concerned, and that will be updated within one year or two, in order to give more details on the requests regarding the methodologies to be used depending on the level of investigation (OFAT, OFEE, OFEFP, 1997).

11.3 LANDSLIDE TYPES IN SWITZERLAND

The complex geological conditions of the Alps and the Prealps, as well as the last glacier episode (Würm) covering more than 2/3 of the country have created numerous predisposition situations favoring the development of all types of landslides in Switzerland. Indeed, due to the weak geotechnical characteristics of the mollassic formations, covering most of the Plateau zone, and to the intense erosion caused by numerous rivers, landslide phenomena are also quite frequently met in the low level areas of Switzerland, but display smaller sizes (see Figure 11.1).

The main landslide type, as far as its extent is concerned, includes permanently moving slides, that are mainly translational, like the cases of La Frasse and Braunwald described below. The largest slide area, called Heizenberg, is located in the Canton of Graubünden and extends over an area of 45 km^2, about one third of this area is still active (Noverraz *et al.*, 1998). Similar large slides are met in the eastern part of Switzerland, because of the presence of extensive friable black schists (Bündnerschiefer).

Another main factor influencing the large number of slides in the Prealps, a zone that is located in the north of the Alps, is the presence of numerous Flysch formations, even leading to an areal proportion of 30% of the whole Canton of Schwyz that is affected by slides (see upper part of Figure 11.1).

Apart from these mostly permanently moving slide zones, that may record fairly long periods of marginal stability, a few slides display occasional severe crisis periods, that may last for several weeks or months: before its stabilization, La Frasse landslide that is presented below experienced critical periods of movements in its lower part, i.e. some six acceleration phases during the last 150 years (Bonnard, 1984, and Figure 11.12). The highest recorded velocity affecting a large slide was observed at Falli Hölli Landslide, in the Canton of Freiburg, in 1994, implying a total volume of some 30 million m^3, with a peak value of 6 m/day that led to a total displacement of 250 m for the central sliding mass, in four months (Bonnard *et al.*, 1995).

As far as sudden large landslide events are concerned, the compilation of past information allows the statement that approximately one major event per century occurred in Switzerland, as shown by the following list of well-known events: Yvorne, 1584; Derborence, 1714 and 1749; Goldau, 1806; Elm, 1881; Randa, 1991.This list of major landslide events that led to serious destructions includes translational slides, as well as rockfalls. The last large event in Switzerland occurred in 1991 in Randa, in the Canton of Wallis, when a total volume of 30 million m^3 of gneiss and micaschists fell in two main stages that each lasted for several hours, and that were separated by a period of three weeks. This special circumstance allowed a precise monitoring of the movements before the second rockfall event (Noverraz *et al.*, 1992).

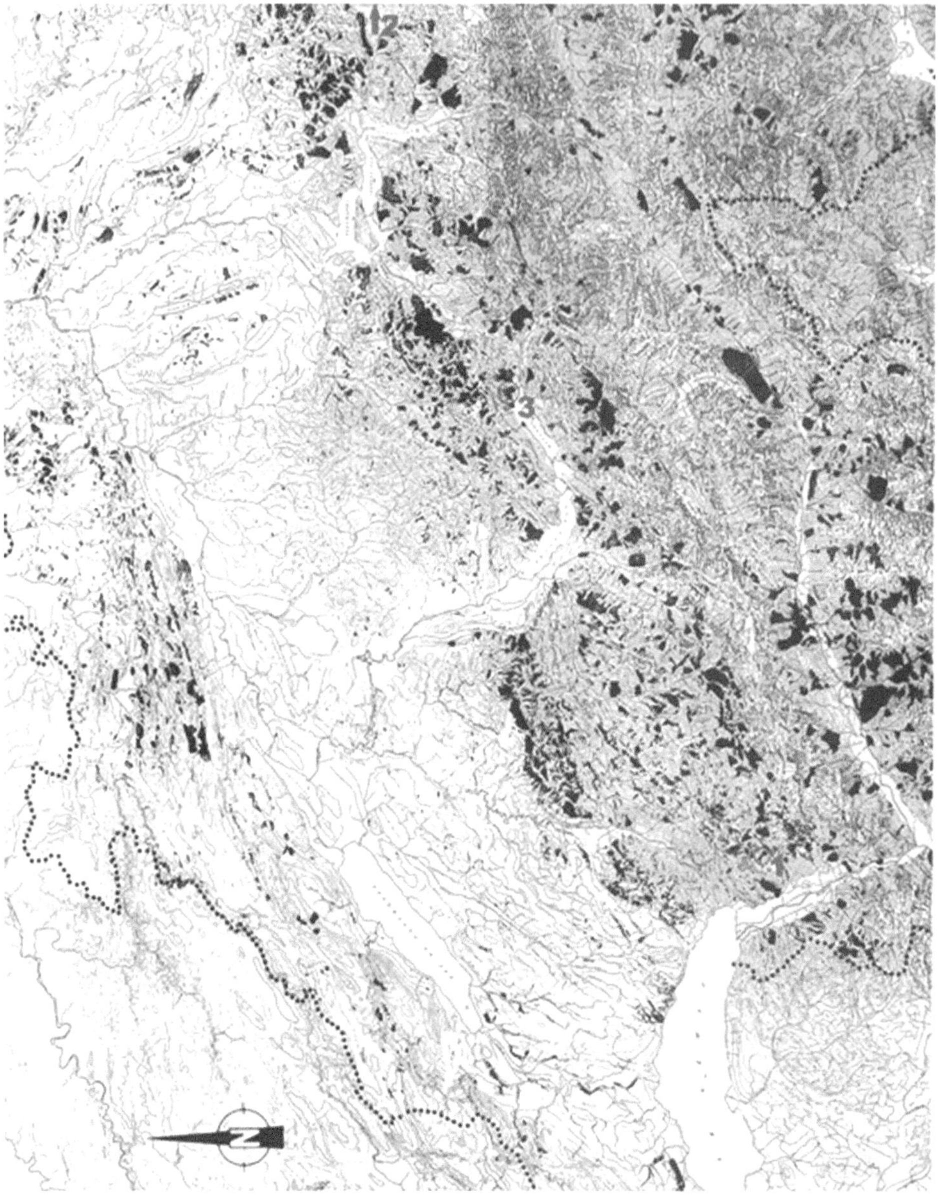

Figure 11.1 Map of past and present large unstable zones in the western and central parts of Switzerland; this map has been established by a visual interpretation of topographic maps at a scale of 1:25,000, considering the geomorphological features of the contour lines and cross-checking this approach in a large number of well-known landslide cases. The slide and rockfall zones, shown in black, represent thus more than 6% of the Swiss territory and are mainly observed in the Alps and Prealps. The location of the different case studies commented at the end of this paper, La Frasse landslide (1), Braunwald landslide (2) and Brienz debris flow (3) is also indicated (Noverraz *et al.*, 1990).

Of course the largest landslide event in Switzerland ever occurred is clearly the well-known Flims case, some 8,000 years ago, which caused a major damming of the river Rhine during a long period. This damming process is however not rare, as shown in Figure 11.2, in which it appears that prehistoric events are quite numerous, but historic events are also fairly frequent.

Apart from large slides, rockfall events are frequently met in the Alps, mostly in uninhabited areas, but some major phenomena have affected developed zones; in most of the cases rainfall events were the triggering cause, but sometimes also earthquakes (Bonnard, 1994); for instance, the village of Grächen in the Canton of Wallis was partly destroyed twice, in 1755 and 1855, as a consequence of earthquakes that triggered a rockfall. It has to be noted that the last major rockfall in Randa, the triggering mechanism of which was not clearly identified, did not cause any victim (except some animals), thanks to the intuitive prevention action of the mayor who evacuated a hamlet a few days before the first fall phase.

Finally the largest number of landslide events that affect inhabited and agricultural zones includes debris flows and mud flows, even if the volume of each case is not particularly high (a large part of them implies volumes between 10,000 and 30,000 m^3). The largest recent recorded event implying a massive debris flow occurred at the village of Fully, in the Canton of Wallis, in October, 2000. Because of specific circumstances (spill of the outlet of a dam through a gallery that led to the failure of a penstock), combined with a critical rainfall period, the debris flow mass reaching a total volume of 100,000 m^3, extended in a zone just next to the village, but happily did not cause any victim nor destruction of houses.

In numerous areas of Switzerland, especially in its central part, a large number of debris flows occurred the last 10 years, mainly in 2005, 2007 and 2014, so that one can note that this type of hazard is clearly increasing in frequency. It also occurs in some slopes in which no previous event had been recorded.

Due to the location of numerous houses or farms, as well as infrastructures, at the toe of steep slopes, or on debris fans, these debris flows often cause serious damage, but the number of victims has been very low, thanks to warning systems. One of the main debris flow zones in Switzerland is the area around the river Glyssibach in which the town of Brienz, in the Canton of Bern, has extended its residential area; this case is presented in one of the last chapters of this report. It is important to note that, if this danger area was clearly identified in the 19th century, following several major debris flow events, it did not incur major damage in the 20th century, so that a large tourist development took place near the shore of the lake, in parallel with the construction of protection works. Unfortunately the last severe debris flow of August, 2005 took the authorities by surprise and caused two fatalities.

11.4 CLIMATE CHANGE TREND IN SWITZERLAND

The most exhaustive source of information regarding future climate change is published within the Intergovernmental Panel on Climate Change reports (IPCC). According to the latest scenarios, the world global mean temperature over the 21st century may rise by some 2–4°C depending on the different assumptions. Climate model computations show that the change in temperature will be stronger in the high latitudes compared

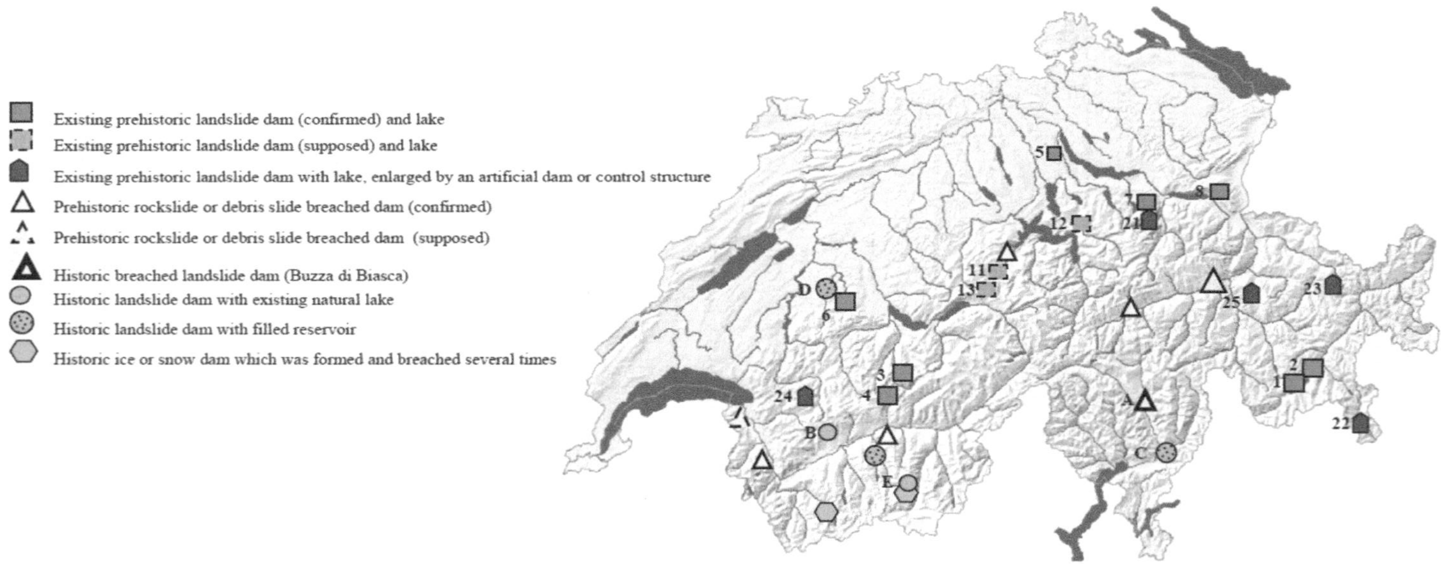

Figure 11.2 Map of all landslide dams identified in Switzerland (confirmed or supposed), that are still existing or have been breached; this list also includes a few historic events, some of them formed by ice or snow avalanches (Bonnard, 2011, including the detailed list of cases).

to the equatorial belt. Temperature change will be greater over the continents, which concerns somehow the area where Switzerland lies in Europe. As far as precipitation is concerned, it will not increase everywhere, but in mountainous regions storm events may be more violent.

In the Alps in general, wintertime temperatures will rise by 3 to 4°C in the 21st century and even more in summer, due to the reduction of the snow and ice cover (Beniston *et al.*, 2006). Therefore, whatever the emission scenario that is considered, the freezing level will rise by some 600 m in winter and to more than 1,000 m in summer. This will definitely accelerate the glacier retreat and contribute to the melting of the upper layers of the permafrost zones, which will certainly increase the landslide hazards, in particular debris flows – that might even occur without precipitation, as it was already observed in the case of the Ritigraben in the canton of Wallis (Zimmerman *et al.*, 1997). The progressive melting of morainic deposits in high altitudes, reducing their cohesion, will increase the number and importance of debris flows that will reach the lower part of valleys in which numerous villages are expanding.

In Switzerland, climate simulations for the period 2071–2100 with respect to conditions one century before show a noticeable shift of the seasonality of mean precipitation, with increases in winter and spring, and reduction in summer and fall (Bonnard *et al.*, 2008). The main cause of these patterns may be explained by the strong summer warming and drying in the Mediterranean zone; in winter, the precipitation will fall rather as rain than as snow. As a result of such seasonal changes, the frequency of extreme rainfall events may also increase, in particular as the warmer atmosphere in summer will bring more energy to these storms. The effect of such evolution appears to be different between the Northern and the Southern Alps (Figure 11.3).

In the future climate conditions, the freezing level associated with heavy rainfall will be on average some 500 m lower than the present level in summer (i.e. 3500 m asl), so that more intense floods will occur in torrents, causing debris flows, as well as mud flows on slopes. One must observe that these mechanisms do appear as more frequent in the last 10 years, although this time perspective is not the one considered in climate change studies.

It is worthwhile observing that Switzerland has signed the Kyoto protocol aiming at limiting the increase of CO_2 in the atmosphere, which is one of the major factors justifying the increase in temperature in the atmosphere and the consecutive climate change. Thanks to severe regulation measures, our country has been able to comply with the Kyoto objectives for the period 2008–2012, despite of the continuous economic and demographic development. However this limitation (reduction of greenhouse gases by 8% with respect to 1990 level during the considered period) is not sufficient to modify the trend in climate change, but just slows the process (OFEV, 2014).

11.5 FEDERAL PREVENTION STRATEGY

Historically the need to carry out natural hazard maps derives from the problems caused by snow avalanches. Indeed in January 1951, a peculiar meteorological event caused numerous snow avalanches in the Alps. During the night of January 16, 1951, a north-west active wind drift occurred carrying humid maritime masses of air from the North Sea towards the Swiss Alps, which caused important snow falls on the northern

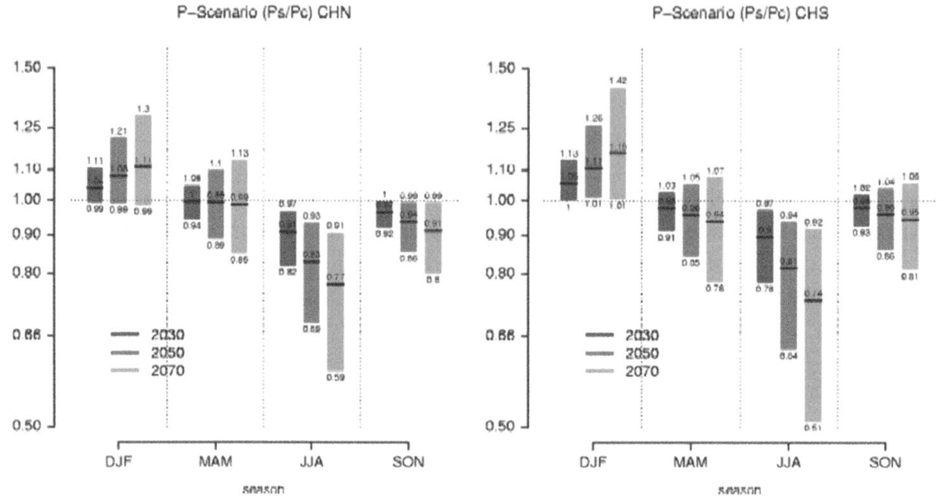

Figure 11.3 Seasonal rainfall predictions in Switzerland for Northern Alps (left graph) and Southern Alps (right graph) regions. The changes are expressed for the years 2030, 2050 and 2070 as a ratio of the present state (1990). Data from OcCC, Switzerland, 2007.

slopes of the Alps. More than the double of the average precipitation during the month of January was recorded. During some spells the intensity of snow fall reached 10 to 15 cm/ hour. Therefore a large part of the Swiss Alps was impacted by catastrophic snow avalanches; some 1500 events caused 98 victims and numerous destructions. They are considered as 100 year return period events (www.planat.ch).

As a consequence of this disaster, the population as well as the authorities realized the importance to consider natural hazards in land planning, which was until then hardly focussed in economic development. During the 1970s, two major steps were taken, first the preparation of indicative hazard maps covering the whole of the Swiss territory and second the coming into effect of the federal law on town and country planning. This law dating from 1979 (LAT) is the first legal text that obliges the authorities to take natural hazards into account and specifically to produce hazard maps (LAT art. 6, al. 2: In order to set up their master plans, the cantons develop basic studies in which they designate the parts of their territory that: [...] are severely threatened by natural disasters or pollution events).

As far as the indicative hazard maps are concerned they have been prepared at large scales (1:100,000) for different types of hazards: slides, rockfall, floods and snow avalanches. As they were generated in an automatic way (mainly with respect to slope), they did not consider the geological conditions of the surface and the local topographical details. For instance in the case of flooding, a uniform standard buffer zone along all the main river beds has been determined to display the endangered zones. This methodology has put forward the excessive expression of danger zones and the need of a better landform analysis. However these indicative maps have allowed the

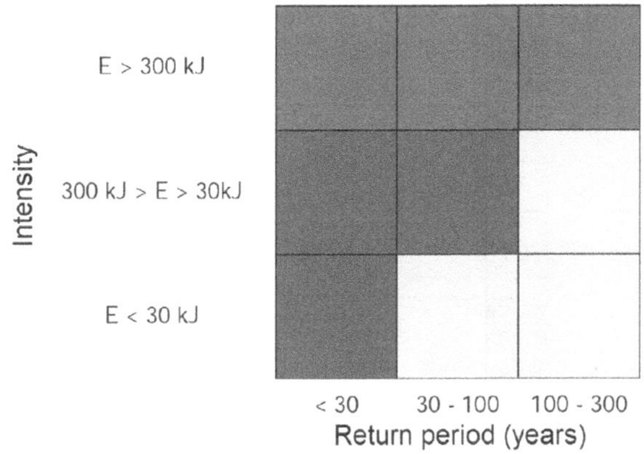

Figure 11.4 Hazard matrix including a qualitative assessment of intensity and probability of occurrence; the reference values for the categories of probability of occurrence are the same for all hazards, namely 30 and 300 year return periods, whereas the reference values for the categories of intensity vary according to the type of danger. The example shown applies to the rockfall of single blocks. Some changes to this matrix have been introduced according to the cantons and to the hazards; indeed some squares of this initial matrix now display two colors, so as to reflect a progressive hazard level change.

population and the authorities to be made aware of the presence of natural hazards, as well as to feel the need to take this aspect into account in town and country planning.

Following the production of these indicative or preliminary maps, a second generation of maps was developed through several research projects: hazard maps (called "danger maps" in an inappropriate way, because of the complex meaning of this term in German). A federal recommendation for the production of these hazard maps was published and the cantons organized their establishment. These maps established for each type of hazard require a more detailed analysis with field investigations (especially for the slides), so as to qualitatively assess the possible intensity of events as well as their probability of occurrence, as shown in Figure 11.4 (OFAT, OFEE, OFEFP, 1997; Lateltin *et al.*, 2005).

The determination of hazard intensity as well as its probability of occurrence is based on one hand on the analysis of past and present events and on the other hand on the field observation. No consideration of climate change impact is included in this process and therefore a possible progressive evolution of natural hazards is not contemplated. However these hazard maps are not supposed to be valid forever and a possibility of updating them is foreseen, in relation with the revision of the town and country planning documents, which is planned every 10 years.

The temporal development of the political and social consideration of hazards explains the difference noted in the progression of snow avalanche mapping with respect to other natural hazards, as shown on Figure 11.5. Due to the early development of snow avalanche hazard maps, the values of return periods selected for the

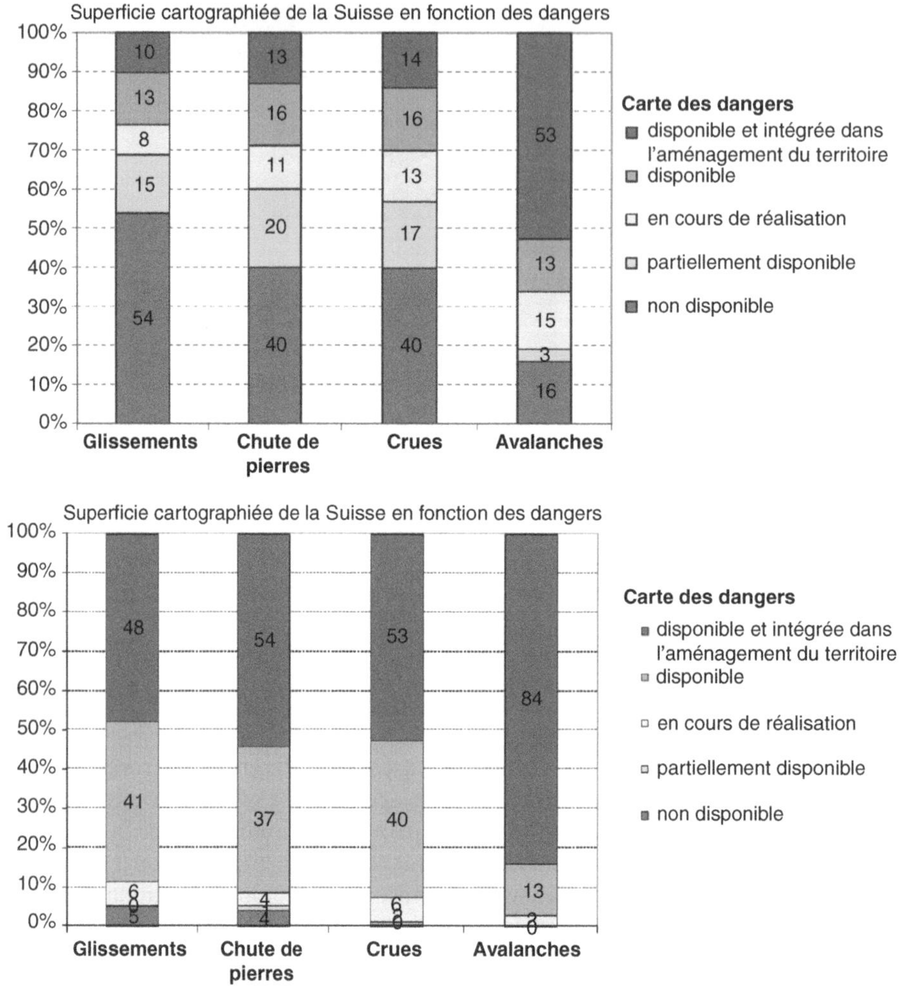

Figure 11.5 and 11.6 State of hazard mapping in Switzerland in January 2006 (upper graph) and January 2014 (lower graph), for slides, rockfall, floods and snow avalanches (from left to right). The displayed colors have the following meaning:

- Dark green: available and integrated in town and country planning
- Light green: available
- Light yellow: mapping in process
- Dark yellow: partially available
- Red: not available

The figures given in the graphs are areal percentages.

different categories of probability of occurrence of such maps have also been selected for all types of hazard maps.

In January 2006 the development in the production of hazard maps could be synthetized by the graph shown in Figure 11.5, based on the relative area mapped for different types of hazards (www.bafu.admin.ch).

Presently, i.e. in January 2014, as shown in Figure 11.6, most of the Swiss territory (at least the areas where people live or work) is mapped. However an important effort still needs to be done as far as the integration of these maps in town and country planning. This last stage is crucial as it allows a major step from hazard knowledge to risk prevention and protection of the population and infrastructure, in order to insure an adequate security which is a human need and one of the key requirements of the welfare of society.

11.6 THE DIFFERENT LEVELS OF PREVENTION ACTION

The natural hazard management strategy, as developed by PLANAT, the platform on natural hazards (www.planat.ch), as well as the consecutive legal framework, is produced by the federal state. It also supplies application regulations as far as the hazard maps are concerned as well as for the construction of protection measures within the framework of the town and country planning.

The production of hazard maps is carried out under the responsibility of the cantonal authorities that are entitled to a certain degree of freedom to organize this mapping and adapt the hazard categories to some local specific conditions. That particular situation explains why very different degrees of completion according to the cantons are observed, as shown in Figure 11.7.

Once these maps are produced for the whole canton and approved by the cantonal authorities (and endorsed by the federal authorities, which is a condition to receive a federal financial support), these maps have to be integrated by the communes or local authorities in their local planning documents; these local authorities are also requested to carry out eventual local protection works (as flood hazard protection works are concerned they still depend on the cantonal authorities as they often imply numerous communes).

This approach depending on three distinct decision levels is a Swiss peculiarity. It presents the advantage to produce planning documents that are close to the reality of the natural conditions and that correspond to the real needs of the population. However the corresponding disadvantage lies in the fact that each canton or commune is somehow free to adapt some criteria to its needs, so that some discrepancies in maps and in their use appear.

For example the canton of Freiburg was one of the main leaders in the production of landslide hazard maps, following the incentive caused by the crisis of Falli Hölli landslide in 1994, which was characterized by a 6 month crisis with a peak velocity of the slide mass of 6 m/day, causing the destruction of a holiday village (Bonnard *et al.*, 1995). This circumstance led to a quite severe assessment of numerous areas in which tourist development was underway, as shown by the landslide hazard map in the area of the Hohberg slide near the Black Lake (Schwarzsee) presented in Figure 11.8, in which Flysch formations are mainly met. Numerous holiday houses located in red zones, like in the hamlet of Rohr, implying a high level of hazard, were thus affected by a prohibition of any modification in the corresponding buildings. This critical situation led to some legal actions in court by the owners to reduce the severe limitations in the use of their chalets.

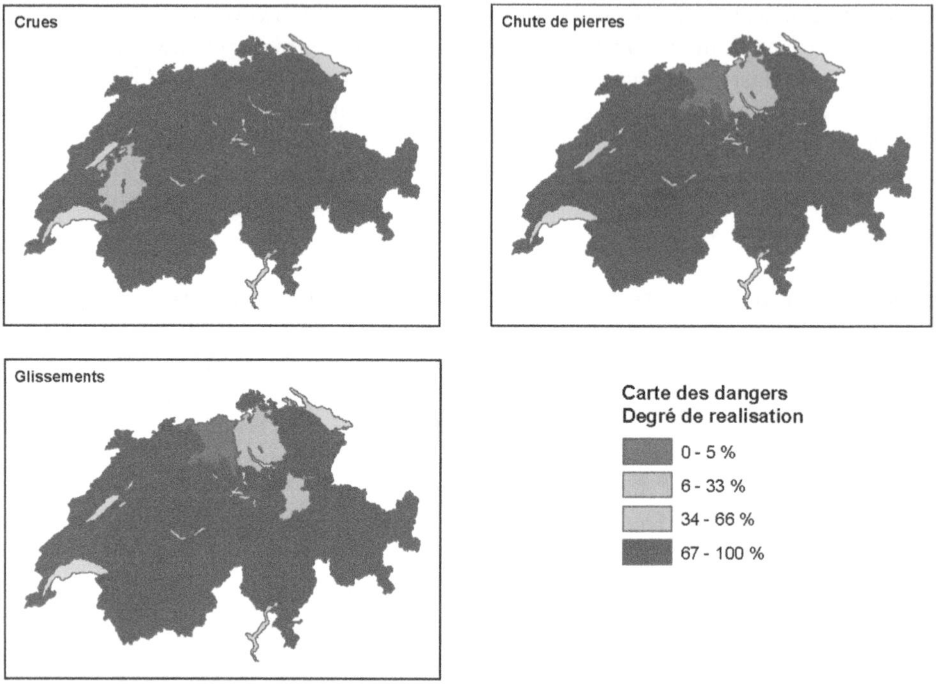

Figure 11.7 Present state of hazard mapping according to the cantons and to the type of hazard (top left: floods; top right: rockfall; bottom: slides). The degree of completion of hazard maps in percentage is expressed by four colors (www.bafu.admin.ch).

11.7 PROTECTION OBJECTIVES

In order to take natural hazards into account in town and country planning, Switzerland has not aimed at developing risk maps that are often a source of controversy, in particular in the financial assessment of the potential cost of disaster events. Indeed it is for example very difficult to assess the economic losses induced by the destruction of a road; these costs are however much higher than the direct damage to the road, in many circumstances.

Therefore the federal town and country planning office has developed a matrix, shown in Figure 11.9, aiming at a first qualitative definition of protection objectives, depending on the type of infrastructure exposed to hazards, on the return period considered and on level of protection desired. In this matrix, a complete protection required (degree 0) means that no hazard intensity is admitted in the corresponding zone or for a specific infrastructure, for frequent or rare events, but not for very rare events; an efficient protection against events of medium or high intensity (degree1) means that a reduced hazard intensity can then be accepted, in particular during very rare events for villages for instance; an efficient protection against events of high intensity (degree 2) means that damage can be accepted, for instance for uninhabited buildings, minor roads and protection forests even for frequent events; finally no protection (degree 3) means that whatever is the return period, some damage may occur: it corresponds to

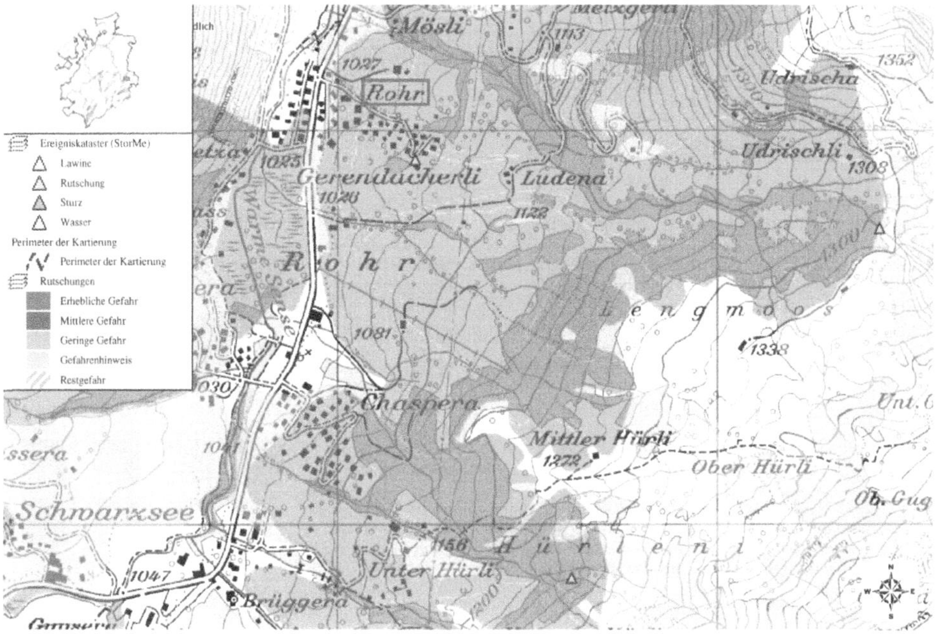

Figure 11.8 Hazard map in a tourist area of the canton of Freiburg near Schwarzsee where numerous holiday houses or chalets are built. All the red zones imply a prohibition to build new constructions and limit the right of the owners of existing buildings to modify their dwelling. In the blue zones specific prescriptions are imposed, whereas in yellow zones, only warnings are given to the owners of individual houses and some limitations are imposed to important or strategic buildings like hotels or hospitals (www.geo.fr.ch).

undeveloped zones in which tourist go skiing or climbing at their own risk. Indeed the cantons often use the limit of individual lethal risk (1×10^{-5} dead/year) to assess if protection measures are needed; the use of a utility/cost coefficient is also considered to check if a protection measure is efficient.

This procedure insures reaching a certain level of safety but maintaining it in the long term is a real challenge!

11.8 TOWN AND COUNTRY PLANNING ACTIONS

Rather than producing risk maps, the Swiss cantonal authorities directly integrate the hazard maps in town and country planning procedures with the help of the concept of protection objectives. Thus principles are set at the federal level so that the application of these rules is as uniform as possible over the whole country, in the 26 cantons. These principles are detailed hereafter according to 4 categories (OFDT, OFEG, OFEFP, 2005); this iterative process requires a regular reassessment:

1. To identify hazards: this first phase is realized by the production of hazard maps and by the application of the protection objective matrix presented above

Legend

= Complete protection	= no intensity is allowed	= 0
= Protection against medium and high intensities	= a low intensity is allowed	= 1
= Protection against high intensities	= a medium intensity is allowed	= 2
= No protection	= even a high intensity is allowed	= 3

Categories of objects				Protection objectives			
Nr	Buildings	Infrastructures	Environmental patrimony	Return period [years]			
				1–30 Frequent	30–100 Rare	100–300 Very rare	>300 Extremely rare
1		Walking trails or skiing tracks in mountain areas	Natural environment	3	3	3	3
2.1		Pedestrian footpaths and commercial cross-country skiing tracks, agricultural roads,communal water network		2	3	3	3
2.2	Uninhabited buildings	Communal roads, cantonal water network	Protection forest, agricultural land	2	2	3	3
2.3	Isolated buildings and hamlets that are temporarily or permanently occupied, cowsheds	Cantonal or main communal roads, national water or gas network, mountain railway lines, skiing areas	Protection forest as far as it protects grouped dwellings	1	1	2	3
3.1		National or main cantonal roads and railway networks, cable cars and t-bar lifts		0	1	2	3
3.2	Villages with grouped dwellings, industrial and craftmanship devoted areas, future building zones, camping grounds, sport and leisure campgrounds	Railway stations and traffic control stations		0	0	1	2
3.3	Special risks, specific vulnerability or indirect damage	Special risks, specific vulnerability or indirect damage		Determination according to the specific case			

Figure 11.9 Protection objective matrix, drawn from federal recommendations for town and country planning and natural hazards, 2005 (OFDT, OFEG, OFEFP, 2005).

(see Figure 11.9). This uniform process insures a comparable level of safety over the whole country and guarantees the development of society.

2. To avoid hazards: this second phase tends to limit and avoid hazards as much as possible. Town and country planning is important in this phase, especially at the level of local planning by duly considering natural hazards; the residual risks can also be limited.

   Presently the Swiss authorities rather privilege an appropriate use of the land rather than the construction of important protection infrastructures, except in the zones where specific public works need to be built (like motorways). A long term assessment of benefits and financial charges including the consideration of high maintenance costs may lead to the prediction of high risk situations. This vision also leads to determine buffer zones in which the hazard processes can be contained, by-passed or slowed down.

3. To manage risks: As a large number of built infrastructures and buildings are already located in hazard zones, a risk management procedure must be developed sometimes through the protection objectives presented above; it includes the application of the principle of proportionality and the consideration of a cost/efficiency ratio, as well as the consideration of a level of accepted risk.

   These technical and financial assessments are completed with prevention and information measures. All partners (administrations, insurance companies, owners) need to work jointly so that the population be duly informed on these hazard problems. In case of disasters, the population will be able to react in an appropriate way so as to avoid a worse issue.

   This third aspect of risk management also implies the due coordination of the protection works carried out with the domains concerned by the environment and nature protection, like agriculture, forestry, hydraulic energy management and tourist industry.
4. To control safety conditions periodically: Not only hazard maps need to be reassessed at regular intervals, but risk level must also be revised with the changes in use of land, urban development, changes in population needs, decaying of protection works and evolution of technique and scientific knowledge. However, as the initial version of hazard maps has been recently carried out in Switzerland, that updating phase has not been discussed yet.

11.9 GENERAL ADAPTATION MEASURES TO CLIMATE CHANGE IN SWITZERLAND

In Switzerland, the adopted policy to cope with climate change is set up by the federal government. Therefore, it is uniform for the whole territory on a legislative point of view. It is aiming at exploiting all opportunities created by these modifications as well as minimizing the possible induced risks. In order to reach these objectives, a multisectorial collaboration and above all a national coordination between all cantons that are responsible for implementing such measures is indispensable.

The analysis of the general situation in Switzerland has pointed out several potential changes that will have an impact on nature and population, in numerous domains (OFEV, 2012):

- Increase of heat waves in towns
- Increase of summer droughts affecting agriculture
- Aggravation of flood risks, effect on water management
- Higher sensitivity of slopes and increase of landslides
- Rising of snow fall limit in winter affecting tourism
- Degradation of water, soil and air quality
- Modification of natural environment, landscapes and distribution of animal and vegetal species
- Propagation of harmful living organisms, illnesses and exotic species.

This global vision sheds some light on the relative importance of the problem of gravity-induced natural hazards. Indeed the adaptation measures to face this new situation imply modifications in numerous domains. Therefore the concern related to landslide hazards is relatively weak because the Swiss population is definitely not affected as a whole; the flood hazards are much more significant as well as more visible, and also induce much more risks, so that the sensitivity of the population as well as of the authorities is quite marked for such hazards.

The plan of action adopted by the federal government in spring 2014, as a second phase of its strategy in order to adapt the economic conditions to the effects of climate change, includes measures already planned in a first phase and in development within nine sectors; one of them deals with the management of natural hazards. Apart from the improvement of transversal knowledge and coordination between cantons and communes, it mainly includes the elaboration of more reliable climatic and hydrological scenarios, so as to assess the risks, but also the opportunities arising from the climate change. This plan of action constitutes a framework for the period 2014–2019. The evolution observed during this period will be monitored and new adaptation actions will be decided at the end of this period.

11.10 CONTRIBUTION OF RESEARCH TO SLOPE SAFETY PREPAREDNESS

In order to understand the development of all the actions undertaken in Switzerland to mitigate the hazards related to climate change, it is important to step back and first recall the serious flood events of 1987 that led to a change of paradigm, i.e. a transition from a pure defence against natural hazards to integrated risk management. Then it is worthwhile mentioning the National Research Project PNR 31, one of the largest multidisciplinary research projects, that was launched by the Swiss Scientific Research Fund from 1991 to 1997, called "Climate Change and Natural Disasters". 53 research projects were carried out in most Swiss universities and institutes of technology in four main fields, namely climate change and meteorological studies, natural disasters and related climatic conditions, effect of climate change on the biodiversity and the environment and socio-economical impacts of climate change and natural disasters. This very large investigation led not only to significant and somehow surprising results in many fields, but above all developed the conscience among the authorities and the population that the future evolution of climate, mainly related to anthropic actions, could and might have serious consequences on the whole society. The importance of this project depended a lot on the fact that many unexpected interactions between climate change and natural disasters were investigated and that regular coordination meetings with all research teams were organized which contributed to the coherence of this remarkable effort (PNR31, 1998).

As far as landslides of all kinds are concerned, six projects dealt with different aspects related to debris flow hazard, sensitivity of Flysch formations in specific climatic conditions, stability of permafrost zones and rockfall zones in specific geological conditions, and effect of annual rainfall variation on large slides over the 20th century (Noverraz *et al.*, 1998). If an increase of debris flow hazard was anticipated in relation to more intense rainfall events in mountain areas mainly in summer, the evolution of the movements of large slides (covering areas of more than 1 km^2) with the slight increasing trend of annual rainfall was not worrying. In the mountain valleys influenced by the Atlantic depressions running from West to East, the increase in slide velocity was slightly more perceptible than in the valleys oriented South to North in which rainfall shadow effects were recorded. It is interesting to note that the largest rockfall event in the period of this research project, Randa rockfall in the canton of Wallis, that implied the movement of some 30 million m^3 in 2 major stages, was not triggered by climatic related conditions.

Following this large coordinated research effort, numerous national and international projects on landslide mechanisms, behaviour and risks were developed, for instance the European projects IMIRILAND and Mountain Risks in which research teams from Switzerland were participating (Bonnard *et al.*, 2004). A large effort was also dedicated to shallow slides and their triggering mechanisms in relation to storms, as a non-saturated approach is needed in order to model their behaviour. An important step was reached in 2013 in the investigation of shallow slide triggering, through the thesis by John Eichenberger, entitled Geomechanical modelling of rainfall-induced landslides in partially saturated slopes, presented at the Swiss Federal Institute of Technology, Lausanne; it was thus possible to trigger a real slide by watering it and to model its failure in the exact reality conditions (Eichenberger, 2012).

Based on this unsaturated slope modelling, an early-warning system was designed to foresee the triggering of shallow slides considering the progressive saturation of the soil mass with a given rainfall time distribution (Eichenberger *et al.*, 2010). This model considers a finite element and infinite slope approach and is based on the computation of changes in capillary stresses (variation in matric suction and degree of saturation) in a soil column. An analytical consideration of equilibrium conditions allows the determination of a capillary threshold, depending on the soil water retention curve, for which the instability phenomenon begins.

Another applied research effort has been devoted to protection systems for debris flows based on the use of nets similar to rockfall nets, in particular by the firm Geobrugg. This technique as shown in Figure 11.10, allows the retention of large debris flow volumes as they occur for instance in the Illgraben Valley in the canton of Wallis; once the net is filled, it is able to sustain more debris flows and thus replaces the need of concrete sills.

This design with protection nets is implemented more and more in valleys displaying debris flow hazard and can reach unexpected dimensions, like in the case of Oberwil im Simmental, where the suspended net held by prestressed cables extends over a length of 42 m for a height of 20 m (see Figure 11.11).

It is finally worthwhile mentioning that another system of rockfall net is presently developed, so as to stop the rock blocks at the top of a cliff and then release them along the cliff behind a safety net, called attenuating system, so as to collect the blocks at the toe of the slope where the access is easier (www.wsl.ch). This system has been successfully tested by the Swiss federal research institute WSL in St-Leonard, in the Canton of Wallis, in 2012.

Finally large research and development efforts have been undertaken to improve the landslide warning systems and in particular the continuous monitoring of slope movements, as shown below in the case of La Frasse landslide. A system called

Figure 11.10 Geobrugg protection net installed in 2006 in the riverbed of the Illgraben; a first debris flow of some 1,000 m^3 was first withheld, and then five more debris flow events occurred on this structure without damage (www.geobrugg.com).

Figure 11.11 Debris flow protection net above Oberwil im Simmental achieved in 2013, designed by Emch + Berger, Bern and Geotechnisches Institut, Brienz (Steiger, 2014).

ROBOVEC (Manetti *et al.*, 2007) including a laser distance meter placed on a mobile plate can measure distances according to various preset orientations. The precision reaches some of +/– 2 mm for a distance of 140 m. This system results from a research project between both Swiss Federal Institutes of Technology of Zurich and Lausanne as well as the Superior Technical Institute of Ticino (SUPSI) with the contribution of industrial partners.

All these new approaches of risk management thus aim at determining which level of safety can be reached at which cost, when facing frequent as well as rare events, in an interdisciplinary planning procedure implying a solidarity with future generations.

11.11 SPECIFIC ADAPTATION MEASURES TO CLIMATE CHANGE IN SWITZERLAND WITH RESPECT TO LANDSLIDES

It is indeed difficult to focus on specific measures taken by the authorities or the federal and cantonal administrations to adapt to climate change conditions with respect to landslides. The main actions have been promoted since the 1990s, in order to improve the consideration of natural hazards and in particular the different landslide hazards in country and town planning. This essential step (OFAT, OFEE, OFEFP, 1997) was developed as a federal recommendation with a similar formal structure and scope for different hazards (floods, landslides, snow avalanches). The requirement, after the identification of hazards, to assess the hazards and then to propose protection or management measures, with a regular updating of these last two steps, implies that the climate change parameters will be processed and considered in the updating and adjusting of the hazard maps. However, it is not possible to demonstrate the impact of such procedure, as it is only the end of the first iteration of landslide hazard maps (and even not everywhere, as shown in the chapter on the different levels of prevention actions).

What is important is that most of the involved authorities dealing directly or indirectly with landslides hazards begin to realize that the past protection measures are not necessarily appropriate to the future conditions and that gaps in safety level following climate change conditions must be closed, implying a sharing of tasks between all the responsible entities (federal, cantonal and communal authorities, research institutions, owners and insurance companies). In parallel with a revision of the design standards, as inspired by the philosophy of the Eurocodes, the civil engineering structures along the motorways, and in particular in landslide zones, are assessed and often redesigned, in order to take into account new safety conditions and possible increased loads. In particular, it appears that the safety of anchored walls stabilizing slowly moving landslide zones, if they are not regularly monitored, cannot be demonstrated; therefore rather than adding some anchors to gain additional resisting forces, these walls are rebuilt with the consideration of more severe loads in relation with climate change effects.

As far as the stabilization of large landslides by drainage tunnels or boreholes is concerned, the hydrogeological parameters considered are generally not greatly influenced by climate change impacts. Indeed the increase of temperature will rather lead consecutively to an increase of evapotranspiration, especially in spring. It is interesting to note that in different large slides that have been drained during the last decade or so, the variation of annual rainfall does not appear in the evolution of residual movements, as the case of La Frasse landslide shows below.

11.12 STABILIZATION OF LA FRASSE LANDSLIDE

This large slide covering an area of about 1 km^2 with a total volume of some 42 million m^3, and extending in the Prealps of the Canton of Vaud, not far from the lake of Geneva, is quite active since the last glacier retreat, i.e. since more than 10,000 years (Bonnard, 1984). The average velocity of its central part during the 20th century was 13 cm/year and in the toe zone, average values of 60 cm/year were reached. As this toe zone is crossed by a major road, it was decided to build a drainage gallery, 720 m long, once the hydrogeological and geomechanical modelling could prove that the reduction

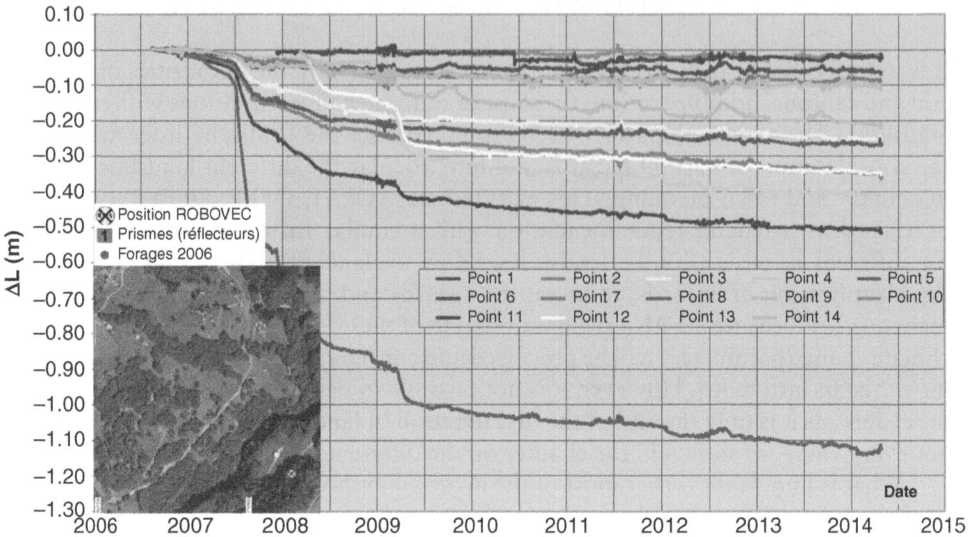

Figure 11.12 Evolution of movements in the toe zone of La Frasse landslide that was stabilized by a drainage gallery between 2007 and 2009. In the most active part of the slide, the velocity as measured at the surface of the ground by a continuous monitoring was reduced from 60 cm/year on an average to 1.5–2.0 cm/year. During the period 2009–2014, the temperature and rainfall conditions were quite variable, but the landslide velocity of point 8 (the most active one) was nearly constant. Measurement provided by Swiss GeoTesting Sàrl (Steinmann, 2014).

of velocity of the slide would be substantial, i.e. a 95% decrease (Laloui *et al.*, 2004). This hydrogeological modelling considered severe rainfall episodes of several days, as they can be expected in the future.

From the gallery 30 to 50 m long vertical drainage boreholes were carried out and allowed a significant drawdown of the groundwater level, so that the landslide movements in the toe zone, continuously monitored by a distance meter system, have shown more than 95% decrease in velocity during the last 5 years, as shown in Figure 11.12. It is interesting to observe that during this period, after the achievement of the drainage gallery, a very important rainfall amount was recorded in the months of December, 2011 (367 mm) and in the month of July, 2014 (322 mm), but no acceleration trend was monitored.

This example proves that a well-designed drainage system is able to be quite efficient, even for large slides, provided the geological and hydrogeological situation is well known. The main criteria to consider for such a design have been collected and published in 2010 by the Swiss federal Office of the Environment (Parriaux *et al.*, 2010). Such instructions are considered as a useful tool to reduce the potential impact of climate change on unstable slopes. Indeed during the last decades several large landslides in Switzerland implying villages have been partially or totally stabilized, like in Campo Valle Maggia in Canton Ticino or Braunwald in Canton Glarus (see chapter below). The results are however not always sustainable.

For La Frasse landslide, it is planned that in the coming years the road that is severely affected by numerous bumps and deformations (see photo in Figure 11.12) will be reshaped as the residual movements are all below 2 cm/year. It could be finally checked, after the very serious rainfall episode of May, 2015, that this stabilizing trend was nearly unaffected by this rare event, which confirms the expected sustainability of the future repair works.

11.13 MANAGEMENT OF THE VILLAGE OF BRAUNWALD LOCATED ON A SLOWLY MOVING LANDSLIDE

Located in the Canton of Glarus, in the centre of Switzerland, this very large slide covering an area of 3.5 km^2, with depths varying between 10 and 100 m, is moving slowly, between 1 and 3 cm/year, but experiences occasional crisis events during which the velocity may reach some 10 cm/day in specific zones. This long lasting movement since the end of the glaciation period, monitored for more than a century, is mainly caused by the infiltration of precipitation (rain and snow) of about 1 m/year on average; indeed no river flows on this slide with a modest slope of about 15°, but which ends up by a cliff limiting the west side of the main valley. The main reason of this slide can be attributed to the geological conditions as the slope is formed by impervious schists with some layers of sandstone and limestone, as well as thin blue clay layers that cause several superimposed slip surfaces. The specific pattern of this slide is that its movements are mainly cyclic, with main movements in winter and spring and residual movements in summer and autumn.

This alpine area has developed into a tourist resort since the end of the 19th century with a few peculiarities: no car traffic exists in this region as the only access is provided by a cable driven funicular. Even no combustion engine is allowed even for the construction of the drainage system! A large hotel was built on the lower part of the slope at the end of the 19th century, in order to offer the best view on the surrounding mountains; already in February 1899 the first minor damage was observed. Indeed several crisis episodes occurred during the 20th century, but the major event took place in February, 1999, when major damage was caused to buildings and a large debris flow (100,000 m^3) originated from the top of the cliff and flew in the valley (Frank *et al.*, 2000).

After some attempts to improve the safety of the slope by superficial drainage, a major drainage scheme was built in the years 1983–85 including a series of deep sequent wells filled with stone rubble reaching the slip surface and a drainage gallery made of precast elements introduced by jacking. This system led to some improvements but did not stabilize the whole endangered area. That is why new investigation boreholes were carried out in 2012 in order to supply more reliable characteristics for the geomechanical modelling that is under way. In parallel the monitoring of the slope movement by satellite observations was developed.

As far as town and country planning is concerned, a major change occurred in 2011 in the management of the commune that had gained an autonomy only in 1939: the canton has decided to regroup all communes into three entities, and now Braunwald is included in the commune called South Glarus. On the other hand as far as natural hazards are concerned, the city plan foresees that in the most active zones new buildings are prohibited, despite of the fairly limited slide velocities; in the areas where the movements do not display critical accelerations, new buildings are allowed provided that they respect some requirements in relation with their foundations and structures (Figure 11.13). Thus the commune takes quite concrete steps in order to ensure safe conditions even in the perspective of climate change effects. However the present measures are probably not sufficient and the geomechanical modelling underway will determine the most appropriate measures to be taken, in order to guarantee a sufficient preparedness level.

11.14 PROTECTION ACTION FOLLOWING A LARGE DEBRIS FLOW AFFECTING THE TOWN OF BRIENZ

As far as the debris flow phenomena are concerned, that are quite frequent in Switzerland, there exists a large documentation on the long-term past activity of some of them, which helps to assess to which extent these phenomena can be affected by climate change conditions. One of them affects the town of Brienz in the canton of Bern, in which 3000 inhabitants live, but which also includes quite numerous secondary dwellings (holiday houses) as it is located on the shore of the Lake of Brienz. This area has experienced a severe disaster on August 23, 2005. After several days of intense rainfall, landslides occurred in the steep mountainous region above the town. These earth and rock masses formed a severe debris flow that devastated a part of the town and caused some 2 fatalities, destroying numerous houses (happily many were not occupied on that day), as shown in Figure 11.14.

Numerous similar events occurred in the past as shown by the archive data and as stated by the development of protection works. Indeed several stone walls were built in the 18th century in and along the river bed to stop and limit the impact of debris flows. However, as the high mountain range (nearly 2000 m asl) above Brienz displays poor geological conditions, a large quantity of sediments as well as large debris flows are carried downward by the torrent Glyssibach, so that the check dams were regularly filled despite of the frequent raising of the dams.

In the 20th century, the number of severe debris flows has been decreasing, so that a large area of permanent and holiday residences has been built quite near to the torrent Glyssibach. Following the disaster of August 2005, the authorities became aware to take some actions, like the construction of a reservoir upstream of the village, in order to store a large part of the debris flow materials; moreover a larger space has been foreseen along the lower part of the torrent channel without constructions, but the general planning of the area has not been drastically modified (Figure 11.15). Thus the stream bed still crosses a quite populated zone. It is foreseen the one hundred year return period floods will be controlled by the engineering measures, but the possibility of obstruction of the channel by fallen trees may still cause serious damage (Figure 11.16). This significant example demonstrates that the possible effects of climate change are not quite fully integrated in local land planning, as they are very difficult to prove and to quantify; the past debris flow events are also quite connected to the change in the use the mountain land (forest clearance, over exploitation of pasture land).

The protection works carried out have indeed seriously modified the hazard map; thus a scenario with a return period of 300 years only affects permanent residences

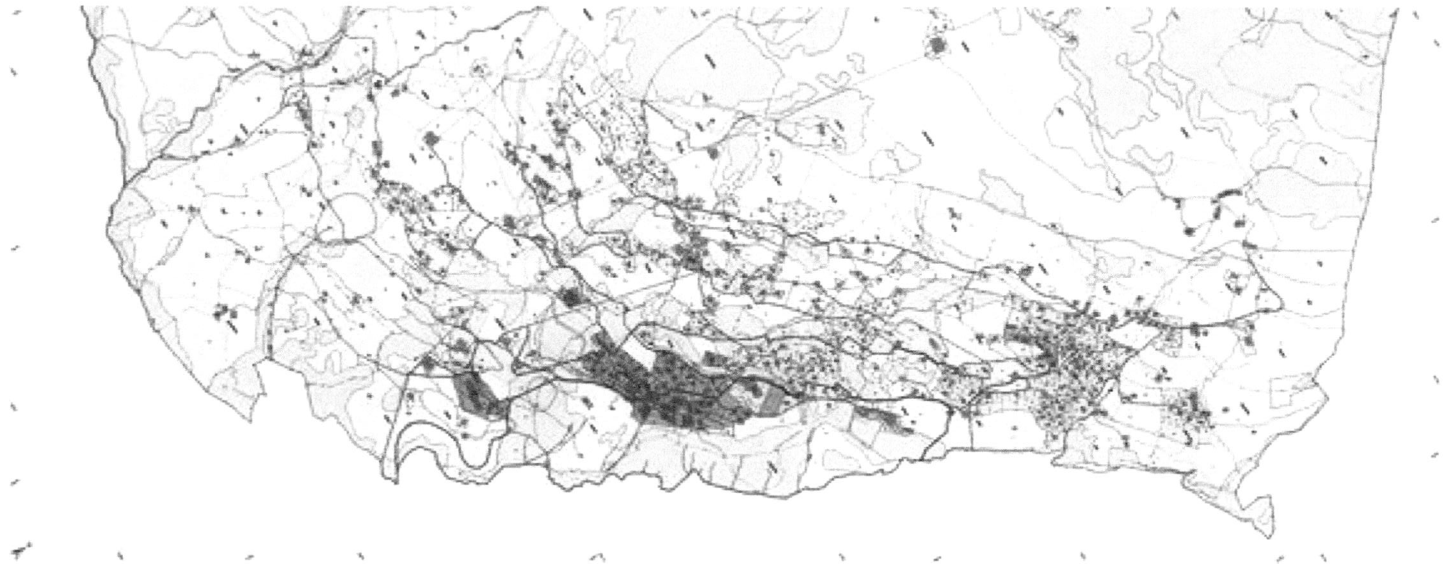

Figure 11.13 Braunwald city planning, in the canton of Glarus, specifying the zones in which no new building is allowed (brown zones) and zones in which specific conditions have to be met for the foundations and the structures of the buildings (yellow zones).

Figure 11.14 Partial view of the town of Brienz in the canton of Bern, following the debris flow disaster of August 23, 2005. The buildings in red will be fully demolished, those in blue sustained medium to important damage and will be properly repaired and those in yellow sustained minor damage. The distance from the river channel is thus not always an appropriate criteria to assess the effective hazard level (Ryter, 2007).

Figure 11.15 Detailed view of some chalets that were damaged as well as of the Glyssibach river bed (Ryter, 2007).

with an average intensity. The zone with a red colour is more or less limited to the river bed and only in a rare scenario (300 years), some low intensity overflooding may occur.

It is true that alarm systems have been foreseen to announce the occurrence of a large debris flow, which is useful for the local population, but the presence of some

Figure 11.16 Part of the landslide hazard map for the town of Brienz; it will be modified once the protection is achieved (Zimmermann, 2014).

tourists that are not prepared to escape according to planned evacuation scenarios may not guarantee a full efficiency of the warning actions.

11.15 FUTURE EVOLUTION OF RISK SITUATION IN SWITZERLAND

In order to try to give a prospective global view on the landslide risk evolution in Switzerland, three aspects have to be considered, which cannot be all quantified in a timely correlated way depending on the rhythm of evolution of these factors:

- It is certain that the impact of climate change conditions will be more severe in the long term (even if some persons pretend the contrary, like Dr. Ueli Gruner, 2012), in particular in summer for debris flow events, in terms of intensity and frequency increase. It is even possible that new debris flows will occur in torrents where they had not been observed in the past, in relation with the progressive elevation change of the permafrost level; this represents a major challenge for the hazard mapping as it is mostly based on the observation of past events, for debris flows. For example, the Spreitgraben torrent near Guttannen, in the canton of Bern, has experienced important debris flows related to permafrost melting, which was not anticipated in the initial hazard maps. On the contrary, as far as large slides are concerned, it can be predicted that their activity will slightly decrease or maybe remain stable, as the increase in temperature will cause a higher evapotranspiration and regulate the annual water balance in the watershed (Briceño *et al.*, 2007). Some other factors related to technical developments may also reduce the activity of large slides, like hydroelectric schemes reducing the flow of a river at the toe of a slide.

- The persons, buildings and infrastructures exposed to landslide hazards will certainly increase in the future, both in terms of number and intrinsic value, given the notable increase of population in Switzerland (about 1% per year), the increase of tourist development in the Alps and the larger and larger investments carried out even in exposed areas, despite of the progressive implementation of hazard maps in town and country planning. This trend is however difficult to assess and to predict in the far future as many economic and political parameters may evolve in an unexpected way in the 21st century; in particular contradictory measures are taken to control the expansion of tourist areas, as recently voted by the Swiss population, which may lead to a higher concentration of exposed elements in some hazardous zones. The evolution of such zones is quite difficult to predict, as shown by the spectacular case of Vallamand, in the Canton of Vaud, in which some 15 owners had to be expelled from their houses and could not rebuild another house during 10 years (Fretz *et al.*, 2013).

  The protection actions taken by all administrative authorities in order to increase the level of safety and thus contribute to a better preparedness against landslide risks are presently quite impressive all over Switzerland and may still develop in the future, as the conscience of the threat caused by natural hazards increases and as the general economy is able to finance such works. It even includes the protection of historical monuments, like the Chillon Castle along the lake of Geneva (Privat *et al.*, 2014). It is therefore possible that a large number of danger situations will be prevented or decrease their impact thanks to the different protection works underway or that are planned in a near future; the example of the region of Brienz mentioned above is quite significant of a future clear decrease in landslide potential impact.

  In particular the federal railway and motorway administrations are developing the implementation of large programs to identify all hazards for their infrastructures and to build appropriate protection works, like the installation of nets for rockfall and debris flows, the construction of tunnels and so on. The ASTRA (roads federal office) wants to establish a platform which allows the evaluation of the gravitational natural hazards which are threatening the Swiss roads. This project is monitored by the ASTRA, the OFEV (environment federal office), the PLANAT (national platform of natural hazards), some of the High Schools and some of the cantons. Its objectives are to evaluate the threat of the gravitational natural hazards, taking the intensity and the probability into account, to quantify and compare the consequences, to establish priority for the planning of countermeasures and to privilege and find the best measures depending on their costs (ASTRA; 2014).
- This policy also has major effects on specific flood risks, by developing buffer areas in which inundation may occasionally take place, so that the riverbed erosion phenomena in narrow stretches of rivers that are likely to cause landside dams may also be less significant.

Considering on the other hand that the level of information on hazardous events is definitely increasing through the huge development of communication media, it can be assessed that the level of safety when facing landslide hazards will be probably higher than now, provided that hazard maps are duly implemented, but it is not sure that the cost of the unforeseen disasters will decrease. Finally, within the integrated risk management policy, the key elements imply an improved knowledge of hazards and risks, a dialogue on risks between the authorities and the population, a holistic planning of protection and management measures, the consideration of appropriate structures to accommodate an "excess load (residual risks)", an efficient emergency preparedness and the timely identification of exceptional hazard events; none of these elements has to be understated.

11.16 CONCLUSIONS

Despite of numerous actions carried out at various political levels in Switzerland and of the presence of a large civil protection organisation, the state of preparedness of this country to face the landslide hazards caused by the consequences of climate change is not totally satisfactory. Of course the official authorities claim significant results, and the engineers and geologists, as well as some cantonal environment offices, are deeply involved in hazard assessment and protection works required by the numerous landslides that affect more and more the development of the country. It can also be mentioned that, in a UN global assessment report dated 2011, called Living with Risk, Switzerland is classified among the safest countries, based on a risk reduction index. What is missing indeed is a more efficient coordination between all the persons and institutions that deal with this long-term problem, which will be reached only if a stronger political will is expressed by the authorities at all levels. It has to be observed that this will is alas not really permanent and focussed on the protection of the population.

Indeed, the way of assessing landslide hazards, at the cantonal and communal levels, are sometimes differing and the local management plans sometimes do not reflect the importance of this constraint. In a few cases, on the contrary, the preliminary or indicative maps are used in the local management and prohibit the development of possible buildings, without analysing the real impact of hazards of mere importance.

It is hoped that in the near future, a new energy will be developed to really increase the level of safety and avoid future disasters related to all types of landslides, in particular through the implementation of monitoring data and research results in risk reduction activities. This seems to be the case with the new strategy of infrastructure reinforcement developed for the national motorways, but this program may last quite some years to complete its objectives. One of the difficulties is related to the quantification of the effects of climate change on natural hazards, especially in the perspective of the modification of hazard maps. The uncertainties affecting these maps and the corresponding scenarios are of such magnitude that the effects of climate change appear to be relatively marginal.

Indeed hazard maps are a dynamic tool that evolves with the progress of knowledge, the new recorded events and the protection works carried out; all these changes have to be analysed within the revision process of the local management plans, every 10 to 15 years, and the local authorities fear that the level of hazard will be increased, implying less possibilities of development of their communities. In some cantons, it is noted that all local management plans do not reflect the changes in the hazard maps or fail to implement these documents in the analysis of new projects.

This fairly pessimistic vision, mainly expressed by the first author who has been working in landslide investigation and mitigation for nearly 35 years, but which is not shared by all the persons contacted (indeed the situation may differ for other natural hazards), may derive from a certain impatience to see fulfilled the objectives of the international decade for disaster reduction (IDNDR). It is also a challenge for the other co-authors to contribute in the future to a safer country.

ACKNOWLEDGEMENTS

The authors wish to thank many persons, colleagues as well as federal and cantonal offices for answering their questions and supplying interesting information for this project. A special mention has to be made for the contribution of Mr. Andreas Götz, former vice-director of the Swiss Federal Office for the Environment president of PLANAT, and of Mr. Jörg Häberle, responsible for natural hazards in the canton of Bern, as well as Philippe Arnold, Stéphane Commend, Stefano Coraducci, Bernard Loup, Helmut Steiger, Gilbert Steinmann and Markus Zimmermann.

REFERENCES

ASTRA (2014) Gestion des dangers naturels sur les routes nationales. Rapport 19003, Berne.

BAFU. Available from: http://www.bafu.admin.ch/naturgefahren/11421/11423/11911/index.html?lang=fr.

Beniston, M., Stephenson, D.B., Christensen, O.B., Ferro, C.A., Frei, C., Goyette, S., *et al.* (2006) Future extreme events in European climate: An exploration of regional climate model predictions. *Climatic Change*, 81, 71–95.

Bonnard, Ch. (1984) Determination of slow landslide activity by multi-disciplinary measurement techniques. In: *Proceedings of the Int. Symp. on Field Measurements in Geomechanics, Zürich, September 1983*, Vol. 1. pp. 619–638.

Bonnard, Ch. (1994) Los deslizamientos de tierra: Fenómeno natural o fenómeno inducido por el hombre? In: *Proceedings 1st Panam. Symp. on Landslides. Guayaquil, Ecuador* Vol. 2. pp. 1–15.

Bonnard, Ch. (2011) Technical and human aspects of historic rockslide-dammed lakes and landslide dam breaches. In: *Natural and Artificial Rockslide Dams*. Heidelberg, Springer. pp. 101–122.

Bonnard, Ch., Noverraz, F., Lateltin, O. & Raetzo, H. (1995) Large landslides and possibilities of sudden reactivation. *Geomechanics Colloquy, Salzburg*. Felsbau No. 6/95, Essen. pp. 401–407.

Bonnard, Ch., Forlati, F. & Scavia, C. (eds.) (2004) Identification and mitigation of large landslide risks in Europe: Advances in risk assessment. In: *IMIRILAND Project, European Commission, Fifth Framework Programme*. Rotterdam, Balkema. 325 pp.

Bonnard, Ch., Tacher, L. & Beniston, M. (2008) Prediction of landslide movements caused by climate change: Modelling the behaviour of a mean elevation large slide in the Alps and assessing its uncertainties. In: *Proceedings of the 10th Int. Symp. on Landslides and Engineered Slopes. X'ian, China*, Vol. 1. pp. 217–227.

Briceno, S., Basabe, P. & Bonnard, Ch. (2007) Landslides and climate change: A world perspective, but a complex question. In: *Proceedings of Int. Conf. on Landslides and Climate Change*. Ventnor, Isle of Wight, UK.

Canton de Fribourg. *Geoportail*. Available from: http://www.geo.fr.ch/.

Eichenberger, J. (2012) *Geomechanical Modelling of Rainfall-Induced Landslides in Partially Saturated Slopes* MSc Thesis. Lausanne, Laboratoire de mécanique des sols EPFL. 246 pp.

Eichenberger, J., Nuth, A. & Laloui, L. (2010) Modelling landslides in partially saturated slopes subjected to rainfall infiltration. In: Laloui, L. (ed.) *Mechanics of Unsaturated Geomaterials*. Hoboken, NJ, John Wiley & Sons pp. 235–249.

Frank, S. & Zimmermann, M. (2000) Rutschung "Bätschen", Murgänge in der Wüechtenrus (Braunwald und Rüti, GL) – Eine Grossrutschung und ihre Folgeprozesse. *Bulletin angewandte Geologie* 5 (1), 131–138.

Fretz, J., Prina Howald, E., Bonnard, Ch. & Berchten, J.-D. (2013) Risk assessment of a unstable cliff above a housing estate in Vully-les-Lacs, Switzerland – Management of residual risks for the cantonal road RC 502c. *8 Simposio Nacional sobre Taludes y Laderas Inestables*, Palma de Mallorca.

Geobrugg. Available from: http://www.geobrugg.com/contento/Portals/35/media/Download-Brochures/Geobrugg-AG_Laves-torrentielles_fr.pdf.

Gruner, U. (2012) Available from: http://www.k-h.ch/bilder/pdf_publikationen/2004_12_Gruner-Klima-Sturz.pdf.

Laloui, L., Tacher, L, Moreni, M. & Bonnard, Ch. (2004) Hydro-mechanical modeling of crises of large landslides: Application to the La Frasse Landslide. In: *IX Int. Symp. on Landslides, Rio Janeiro, Brasil* pp. 1103–1110.

Lateltin, O., Haemming, Ch., Raetzo, H. & Bonnard, Ch. (2005) Landslide risk management in Switzerland. *Journal of the International Consortium on Landslides* 2 (4), 313–320.

Manetti, L. & Steinmann, G. (2007) 3DeMoN ROBOVEC – Integration of a new measuring instrument in an existing generic remote monitoring platform. In: *Proceedings of 7th Int. Symposium on Field Measurements in Geomechanics, 2007*. 12 pp.

Noverraz, F. & Bonnard, Ch. (1990) Mapping methodology of landslides and rockfalls in Switzerland. In: *Proceedings VIth Int. Conf. and Field Workshop on Landslides, Milan, Italy* pp. 43–53.

Noverraz, F. & Bonnard, Ch. (1992) L'écroulement rocheux de Randa, près de Zermatt. In: *Proceedings VIth Int. Symp. on Landslides, Christchurch, New Zealand* Vol. 1. pp. 165–170.

Noverraz, F., Bonnard, Ch., Huguenin, L. & Dupraz, H.(1998) *Grands glissements de versants et climat. Projet VERSINCLIM, Comportement passé, présent et futur des grands versants instables subactifs en fonction de l'évolution climatique, et évolution en continu des mouvements en profondeur.* Rapport final PNR 31. Berne, Zürich, FNRS. 314 pp.

OcCC (2007) Klimaänderung und die Schweiz 2050; Erwartete Auswirkungen auf Umwelt, Gesellschaft und Wirtschaft. Organe consultatif sur les changements climatiques. ProClim.

OFAT, OFEE, OFEFP (1997) *Prise en compte des dangers dus aux mouvements de terrain dans le cadre des activités de l'aménagement du territoire*. Berne, Confédération Suisse. 42 pp.

OFDT, OFEG, OFEFP (2005) *Recommandation: Aménagement du territoire et dangers naturels*. Berne, Switzerland. 48 pp.

OFEV (2012) *Adaptation aux changements climatiques en Suisse*. Berne, Confédération Suisse. 64 pp.

OFEV (2014) Protocole de Kyoto: la Suisse a rempli l'engagement pour 2008–2012. *Proclim-Flash* 60 5.

Parriaux, A., Bonnard, Ch. & Tacher, L. (2010) *Glissements de terrain: Hydrogéologie et techniques d'assainissement par drainage – Guide pratique*. Berne, Office Federal de l'environement (eds.). Nr 1023. 128 pp.

Planat. Available from: http:// www.planat.ch/fr/bon-a-savoir/chronique.

PNR 31 (1998) *Le regard de Janus: Changements climatiques et catastrophes naturelles*. Genève, Georg Editeur (ed.). 82 pp.
Privat, G., Prina Howald, E. & Feuz, R. (2014) Chutes des blocs sur la Commune de Veytaux, Suisse. In: *Proceedings of the Rock Slope Stability, 2–4 April 2014*. Marrakech.
Ryter, U. (2007) *Brienzer Wildbäche*. Available from: http:// www.be.ch/naturgefahren.
Steiger, H. (2014) Personal communication.
Steinmann, G. (2014) Personal communication.
Swissworld, *Geography*. Available from: http://www.swissworld.org/en/geography/.
WSL. *Site d'essais de chutes de blocs à St-Léonard (VS)*. Available from: http://www.wsl.ch/info/organisation/versuchsanlagen/stleonhard/index_FR.
Zimmermann, M. (2014) Personal communication.
Zimmermann, M., Mani, P., Gamma, P., Gsteiger, P., Heiniger, O. & Hunziker, G. (1997) Murgangefahr und Klimaänderung – ein GIS-basierter Ansatz. *Schlussbericht NFP 31*. v/d/f, Zürich. 161 pp.

Chapter 12

Review of landslide occurrence and climate change in Taiwan

Meei-Ling Lin
National Taiwan University, Taiwan

Sheng-Chi Lin and Yu-Ching Lin
National Science and Technology Center for Disaster Reduction, Taiwan

ABSTRACT

In Taiwan, the landslide is one of the most severe hazards in mountainous areas, especially for those nearby urban areas and roads. More than 75% of Taiwan is mountainous, forcing human activity to meet the expectations of land development. In addition, the frequency of typhoons and rainfall are predicted to intensify as a result of climate change. The continuation of global warming will likely lead to more landslides. Taiwan's government agencies have adapted measures to ease the effects of climate change, measures that need not only be implemented as swiftly as possible but also be formulated to minimize risks. In this report, we introduce government agencies' climate projections, trend analysis of landslide frequency variations, and the construction of the landslide risk map. These agencies report that the total precipitation and the maximum rainfall intensity of the central and southern Taiwan will increase. As per its population density and human development index, central Taiwan has a high social vulnerability. The landslide risk map incorporates rainfall triggering effects, geological environmental conditions, and social vulnerability. It suggests high levels of risk in central and southern Taiwan, and efforts should be made to strengthen adaptive capacity, including measures for lowering social vulnerability and establishing an integrated operating system.

12.1 INTRODUCTION

The three main issues related to the global warming (Figure 12.1) are science, mitigation, and the combined issue of impact and adaptation. The science issue emphasizes the importance of climate change observations, scientific question clarifications, and future climate change projections. Mitigation focuses on controlling or reducing the emission of greenhouse gasses through policymaking, industrial transformation, and advances in scientific techniques. Impact and adaptation assesses the extent of climate change effects and evaluates preparation methods and adaptive measures for reducing global warming; the keyword is "reduction," as we will still experience the impact of persistent greenhouse gasses.

The current status of these three issues and the background of this report are described in the following sections. Regarding mitigation, although Taiwan is neither a member of the UN Framework Convention on Climate Change nor a signatory to

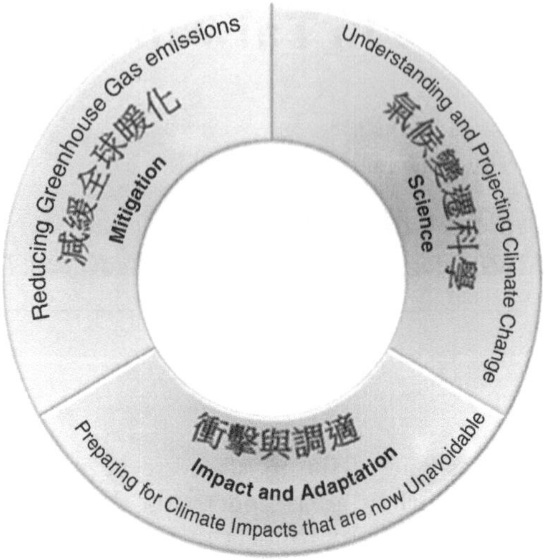

Figure 12.1 The three main global warming issues from the IPCC Climate Change Report (National Science Council, 2011).

the Kyoto Protocol, it has taken consistent actions to reduce global warming. National carbon emission reduction targets were set, the Sustainable Energy Policy Convention was promulgated, and the Greenhouse Gas Reduction Act was drafted to account for economic developments and environmental sustainability while achieving energy conservation and carbon reduction goals.

To adapt to climate change, after the 2010 implementation plan for the National Adaptation Policy Framework and Adaptation Programs of Action for Climate Change, the Council for Economic Planning and Development of Taiwan developed a national adaptation strategy framework and implemented a substantial action plan that accounted for various aspects of climate change, including natural disasters; water resources; coastal conditions; agricultural production and biodiversity; health; infrastructure; energy supply and industrial economics; as well as land-use planning and management.

The National Science Council (NSC) of Taiwan has summarized the research on long-term local climate change projections and its impact on scientific issues. By the end of 2009, the Sustainable Development Research Committee of NSC launched the Taiwan Climate Change Projection and Information Platform Project, the duration of which was 2010 to 2016. The project promoted climate change research and integrated climate change impact information applications. It aimed to enrich the climate change research in Taiwan, to consolidate the capacity for climate change research and projection, as well as to implement climate change information applications and services. "The Science Report of Taiwan Climate Change 2011" published by the National Science Council (NSC) reviewed and integrated relevant research results and scientific advances to provide up-to-date information on climate change on both global

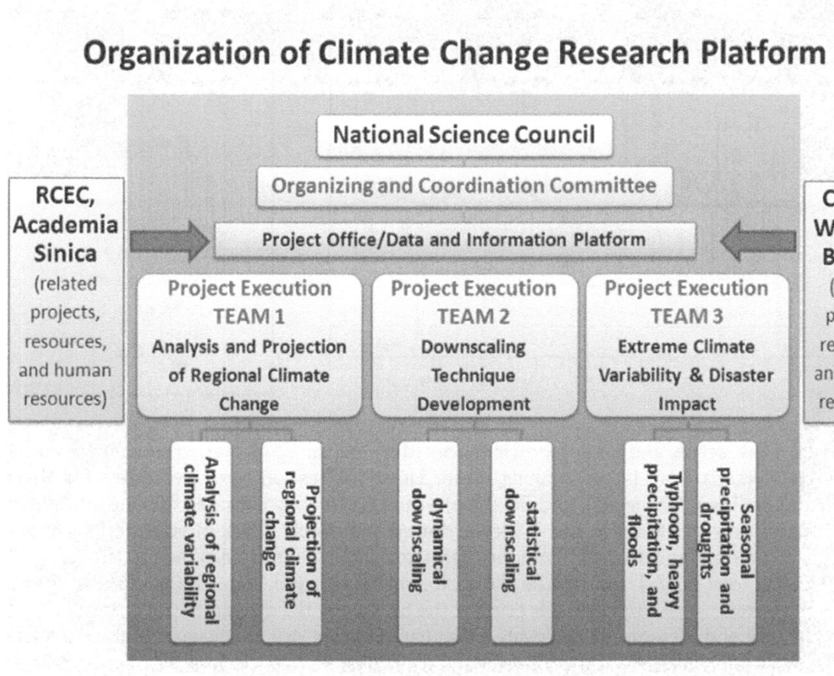

Figure 12.2 TCCIP framework and organizations (National Science Council, 2011).

and Taiwanese scales, thereby serving as a scientific base for further research and government climate-change policy-making.

A 3-year Taiwan Climate Change Projection and Information Platform Project (TCCIP) was planned and implemented by the National Science and Technology Center for Disaster Reduction (NCDR). The project integrated Taiwanese research resources to project future climate change and assess the impacts of related disasters in Taiwan. Figure 12.2 shows the TCCIP framework and organizations.

12.2 PROJECTION OF CLIMATE CHANGE IN TAIWAN

Many Taiwanese research agencies have long been analyzing climate change and projecting future climate scenarios. One such analysis is “The Science Report of Taiwan Climate Change 2011” published by the National Science Council (NSC) in 2011 (NSC, 2011). This comprehensive report includes all the latest scientific advances and technological developments on climate change research in Taiwan and is used by TCCIP framework as the base-line reference for projecting Taiwan’s future climate change scenario. Statistically significant linear trends in various meteorological variables related to landslides, including the annual precipitation, number of rainy days, as well as typhoons over the last 100 years (1911 to 2009), 50 years (1960 to 2009),

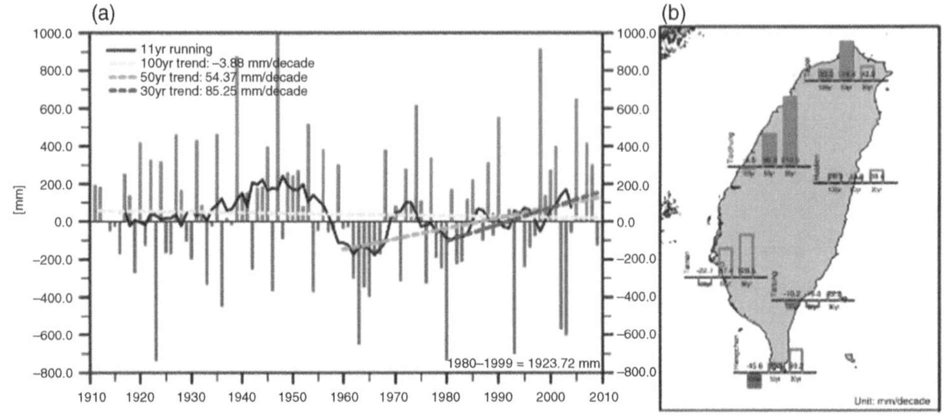

Figure 12.3 (a) Time series and annual precipitation (daily rainfall ≧0.1 mm) trends in Taiwan. The black line is the 11-year running mean. Linear regression lines are shown for the last 100 (yellow), 50 (green), and 30 (blue) years. Solid lines indicate that linear trends are statistically significant at the 5% level; dotted lines imply a lack of statistical significance. Trend variations calculated from the regression slopes are shown in the top left corner. The annual total rainfall mean of the climatological base period from 1980 to 1999 is shown in the bottom right. (b) From left to right, the bars show the magnitudes of change in the last 100, 50, and 30 years, respectively. Filled bars indicate that the linear trends pass tests of statistical significance at the 5% level, and unfilled bars show a lack of statistical significance. Blue indicates a decrease, and red indicates an increase (National Science Council, 2011).

and 30 years (1980 to 2009) are analyzed to illustrate the climate change trends in Taiwan.

12.2.1 Rainfall

No significant statistic trend on Taiwan's annual rainfall was found over the past century, Fig. 12.3 shows, as per a 5% alpha level. Although the annual rainfall does not have a statistically significant long-term variation trend, the annual precipitation has exhibited a decreasing trend of 3.88 mm per decade over the past one hundred years. Yet it exhibited significant increases of 54.37 mm and 85.25 mm per decade over the past fifty and thirty years, respectively.

One of the most noticeable facts is that the average annual total number of rainy days in Taiwan has decreased significantly. The number of rainy day has exhibited a decreasing trend of four days per decade over the past one hundred years, and the rate of decrease has accelerated to six days per decade over the past 30 years. This indicates that the decrease in rainy days has recently become more discernible, as shown in Fig. 12.4. The three years from 2002 to 2004 were those with the minimum numbers of rainy days over the past one hundred years. In terms of seasonal characteristics, the numbers of rainy day in all four seasons exhibit decreasing trends, with summer showing the fastest decreasing rate. Furthermore, number of days of torrential rain (defined as daily rainfall larger than 200 mm) has exhibited a significant increasing

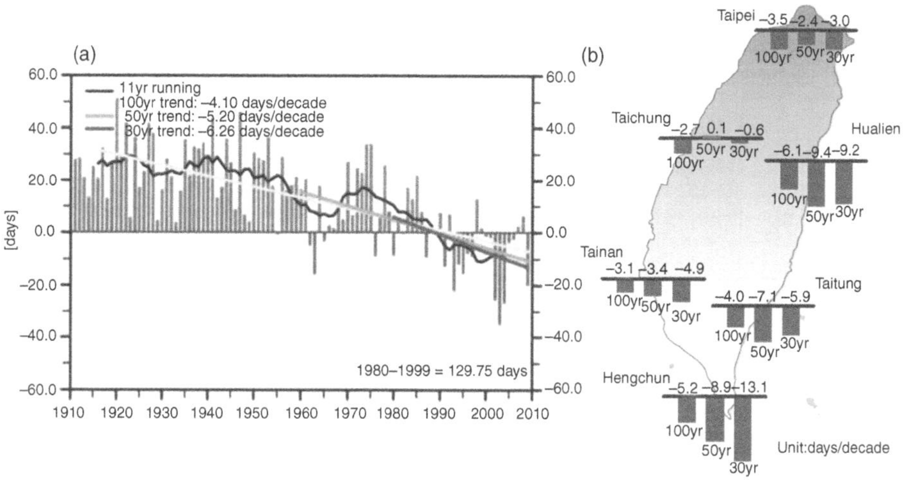

Figure 12.4 (a) Time series and annual total number of rainy day (daily rainfall ≧0.1 mm) trends in Taiwan. The black line is the 11-year running mean. Linear regression lines are shown for the last 100 (yellow), 50 (green), and 30 (purple) years. Solid lines indicate that linear trends statistically significant at the 5% level; dotted lines indicate a lack of statistical significance. Trends calculated from the regression slopes are shown in the top left corner. The mean of annual total number of rainy day with base period from 1980 to 1999 is shown in the bottom right. (b) From left to right, the bars show the magnitudes of change in the last 100, 50, and 30 years, respectively. Filled bars indicate that the linear trends at a 5% significance level; unfilled bars are statistically insignificant. Blue indicates decreases, while red indicates increases (National Science Council, 2011).

trend over the past 50 years and past 30 years. In addition, the number of typhoons with extremely intensive rainfall has increased over the past decade.

Additionally, number of days with torrential rain (daily rainfall ≧200 mm) in Taiwan has a significant increasing trend in the last 50 years and 30 years. However, the 50–60 year variation was also significant. Whether this trend will continue in the future is unknown. Days with light rain (daily rainfall <1 mm) have decreased substantially, with a decrease of two days per decade in the last 100 years and four days per decade in the last 30 years. The number of days with torrential rain shows fluctuations for periods of 10–20 years. There were fewer days with torrential rain from 1920 to 1935 and from 1965 to 1985, but more days from 1940 to 1960 and after 1985 (Fig. 12.5).

12.2.2 Typhoons

To objectively define the recurring period of typhoons affecting Taiwan, the NSC (2011) analyzed the characteristics of past typhoons, standardizing the range as within 300 km from the Taiwanese coast line as the boundary of the typhoon centers. It was found that more typhoons affected Taiwan after 1990 than from 1961 to 1989, with a sharp increase in 2000 (Fig. 12.6). This change was probably caused by the

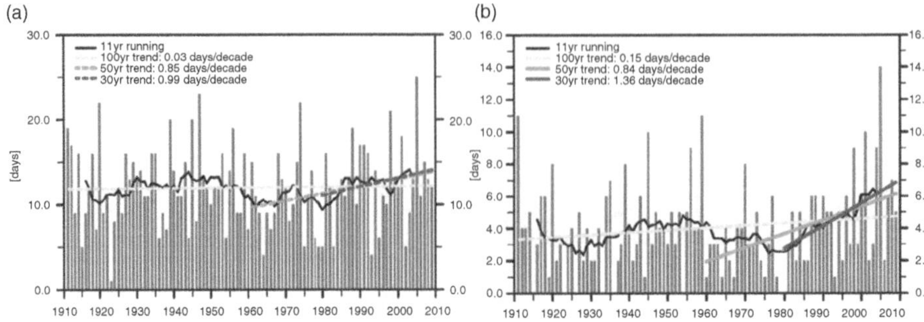

Figure 12.5 Time series and trends for the annual number of days with (a) extremely heavy rain (daily rainfall ≧130 mm) and (b) torrential rain (daily rainfall ≧200 mm) in Taiwan from 1911 to 2009. The black line shows the 11-year running mean. Linear regressions lines are shown for the last 100 (yellow), 50 (green), and 30 (purple) years. Solid lines indicate that linear trends satisfying significance tests at a 5% alpha level; dotted lines imply a lack of statistical significance. Trends calculated from regression slopes are shown in the top left corner (National Science Council, 2011).

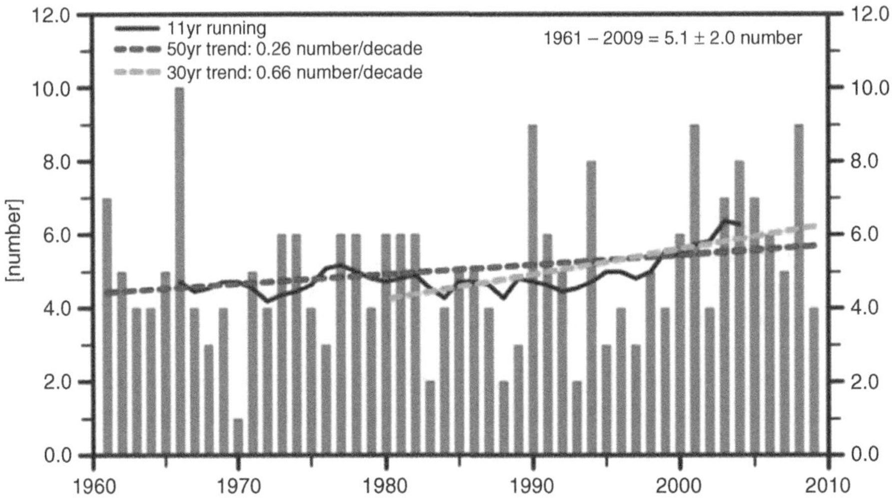

Figure 12.6 Annual number of typhoon affecting Taiwan from 1961 to 2009. The black line is the 11-year running mean. Linear regression lines are shown for the last 50 (blue) and 30 (green) years. Solid lines indicate that linear trends satisfy significance tests at the 5% level; dotted lines indicate a lack of statistical significance. Trends calculated from regression slopes are shown in the top left corner (National Science Council, 2011).

northward shift of typhoon tracks over the Northwestern Pacific Ocean. The proportion of typhoons within the "intense" category increased after 1980 (Fig. 12.7). These changes in typhoon intensity and track are closely related to inter-decadal variations in the Northwestern Pacific Ocean sea surface temperature and tropical cyclones.

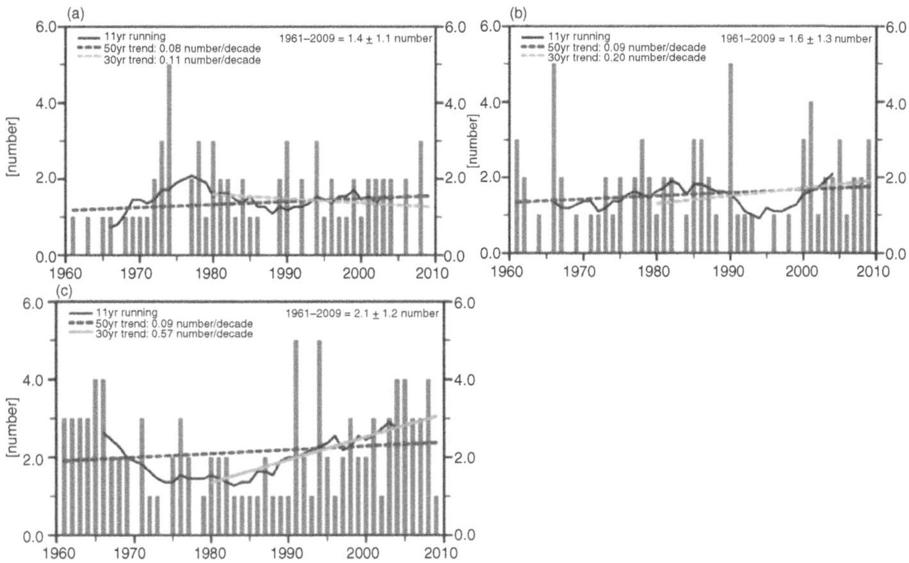

Figure 12.7 Annual number of typhoons affecting Taiwan from 1961 to 2009 with the maximal intensity reaching the (a) minor, (b) moderate, or (c) intense categories (The Science Report of Taiwan Climate Change, 2011).

12.2.3 Projection of future climate change in Taiwan

Using the IPCC-developed scenario of A1B, which the international science community considers to be the most likely future outcome, the NSC has created a list of climate change projections for Taiwan as per a spatial downscale. Results are as shown in Table 12.1 and described as follows.

12.2.3.1 Temperature

The majority of climate models project that the increases of the near-surface temperature of Taiwan at the end of the 21st century will range between 2°C and 3°C relative to the end of the 20th century. These temperature increases are slightly lower than the averaged global temperature rises. In terms of regional and seasonal changes, northern Taiwan will likely warm faster than southern Taiwan, and spring will be least affected by climate change.

12.2.3.2 Precipitation

The global trend of greater disparities between wet and dry seasons will continue in the 21st century. The mean winter precipitations in northern, central, southern, and eastern Taiwan are projected to decrease in the future, with about half of the models projecting decreases between −3% and −22%. In southern Taiwan, the change in the mean precipitation in spring is similar to that in winter. In summer, with the exception of northern Taiwan, over three-quarters of the models project an increase

Table 12.1 The average projected changes of ground temperature and precipitation in four regions of Taiwan under the SRES scenario of A1B (National Science Council, 2011).

		Change in Ground Temperatures (%)							Change in Precipitation (%)						
Regions	Seasons	Minimum	10.0	25.0	50.0	75.0	90.0	Maximum	Minimum	10.0	25.0	50.0	75.0	90.0	Maximum
Northern Taiwan	Winter (DJF)	1.9	1.9	2.1	2.7	3.1	3.4	3.8	-44.0	-29.0	-20.0	-13.0	-3.0	7.0	33.0
	Spring (MAM)	1.7	1.8	2.2	2.6	2.9	3.4	3.9	-31.0	-24.0	-14.0	-8.0	6.0	17.0	36.0
	Summer (JJA)	1.7	1.8	2.2	2.7	3.0	3.3	4.0	-15.0	-12.0	-1.0	14.0	19.0	46.0	64.0
	Autumn (SON)	1.6	1.8	2.2	2.6	3.1	3.3	3.8	-33.0	-25.0	-10.0	8.0	21.0	28.0	34.0
Central Taiwan	Winter (DJF)	1.8	1.9	2.0	2.6	3.0	3.3	3.7	-49.0	-33.0	-22.0	-15.0	-4.0	6.0	22.0
	Spring (MAM)	1.6	1.8	2.1	2.6	2.9	3.3	3.8	-36.0	-25.0	-16.0	-10.0	3.0	17.0	41.0
	Summer (JJA)	1.8	1.8	2.2	2.7	3.0	3.2	4.0	-15.0	13.0	-2.0	-14.0	26.0	64.0	69.0
	Autumn (SON)	1.6	1.8	2.2	2.6	3.0	3.3	3.7	-34.0	-23.0	-7.0	11.0	25.0	31.0	45.0
Southern Taiwan	Winter (DJF)	1.7	1.7	2.0	2.5	2.9	3.2	3.4	-47.0	-34.0	-22.0	-13.0	-5.0	5.0	8.0
	Spring (MAM)	1.5	1.8	2.0	2.5	2.8	3.0	3.6	-41.0	-26.0	-21.0	-14.0	-5.0	22.0	34.0
	Summer (JJA)	1.7	1.7	2.2	2.5	2.9	3.2	4.0	-20.0	-19.0	7.0	16.0	26.0	69.0	76.0
	Autumn (SON)	1.5	1.7	2.1	2.6	2.9	3.1	3.6	-28.0	21.0	-8.0	13.0	25.0	36.0	55.0
Eastern Taiwan	Winter (DJF)	1.8	1.8	2.0	2.6	3.0	3.3	3.7	-44.0	-31.0	-20.0	-12.0	-3.0	5.0	17.0
	Spring (MAM)	1.5	1.8	2.1	2.5	2.8	3.3	3.8	-37.0	-25.0	-18.0	-11.0	1.0	20.0	36.0
	Summer (JJA)	1.7	1.7	2.2	2.6	2.9	3.2	4.0	-17.0	-15.0	3.0	15.0	26.0	57.0	64.0
	Autumn (SON)	1.6	1.8	2.1	2.6	3.0	3.2	3.7	-30.0	-23.0	-10.0	10.0	23.0	33.0	43.0

in the seasonal mean precipitation, and about half of the models project the change to range between +2% and +26%. This projected scenario of increasing precipitation during rainy seasons and decreasing precipitation during dry seasons will pose a great challenge to the management of water resources.

12.2.3.3 Typhoons

Historical data show that both the frequency and the intensity of the Western North Pacific typhoons and typhoons afflicting Taiwan are largely influenced by inter-decadal variability. The direct effect of global warming on the linear trend's variation is not significant. Nevertheless, most climate models project the number of typhoons to decrease and the intensity as well as associated extreme rainfall of the remaining typhoons to increase under the global warming effect.

12.3 THE RAINFALL FACTOR OF CLIMATE CHANGE ON SLOPE STABILITY AND LANDSLIDE RISK IN TAIWAN

12.3.1 Landslide events and triggering factors

The mountain area takes up to about 75% of total area in Taiwan, as shown in Fig. 12.8. The population distribution, patterns of living, and industry development are

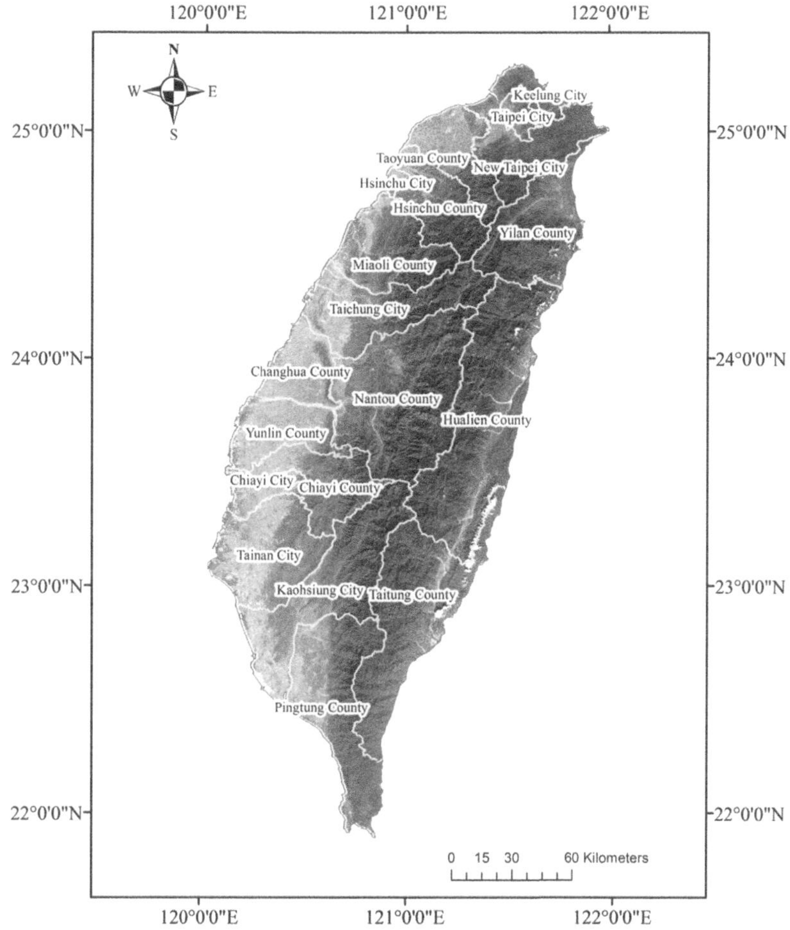

Figure 12.8 Morphological map of Taiwan.

closely tied to the mountainous environment. Additionally, Taiwan is at the convergence zone of the Philippine sea-plate and the Euro-Asian plate, where geological hazards induced by earthquakes frequently occur (e.g., the Chi-Chi Earthquake). The occurrence frequency of landslides and debris flow in the central Taiwan increased after the Chi-Chi Earthquake in 1999. The distribution of the landslide events from 1999 to 2010 versus the peak ground acceleration distribution (PGA) of the Chi-Chi Earthquake is as shown in Fig. 12.9, in which the distribution of landslides appears to be more concentrated in a zone with a PGA larger than 80 gal and increases with PGA.

In addition, Taiwan is in a sub-tropical area. Here tropical cyclones can produce typhoons or heavy rainfall, which induces high accumulations of (as high as 3,000 mm per year) and/or high intensities of rainfall. The torrential rainfall can speed up the weathering process of geological materials, and mass movements often occur in Taiwan. According to the landslide disaster database of the National Science and

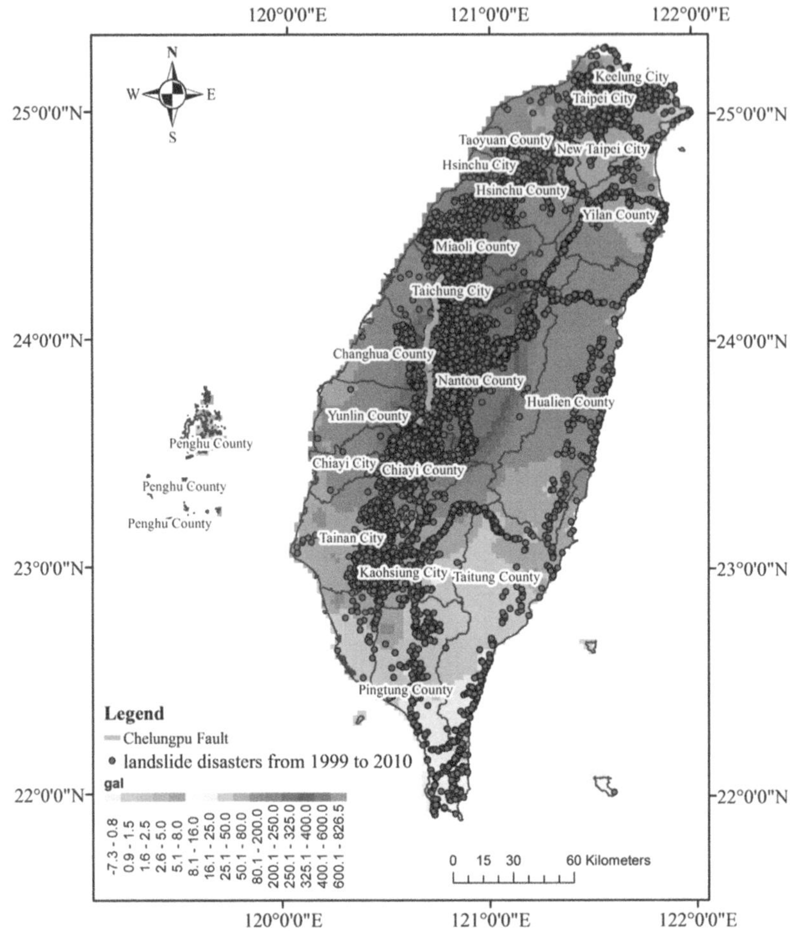

Figure 12.9 Landslide distribution from 1999 to 2010 versus peak ground acceleration distribution of the Chi-Chi Earthquake, 1999.

Technology Center for Disaster Reduction (NCDR), most of the landslide disasters from 1989 to 2013 are located along the mountainous roads and hills nearby urban areas. Approximately 77%, 15%, and 8% of landslide disasters were triggered by typhoons, torrential rainfall, and earthquakes, respectively, as shown in Fig. 12.10.

12.3.2 Extreme landslide events and extreme rainfall from typhoons

Almost all historic landslide disasters with severe casualties were induced by typhoons, such as the devastating typhoon Morakot in 2009. The Shiaolin landslide that caused more than four hundred deaths was as shown in Plate 1. Fig. 12.11 illustrates the distribution of historical extreme landslide disaster events triggered by typhoons. In

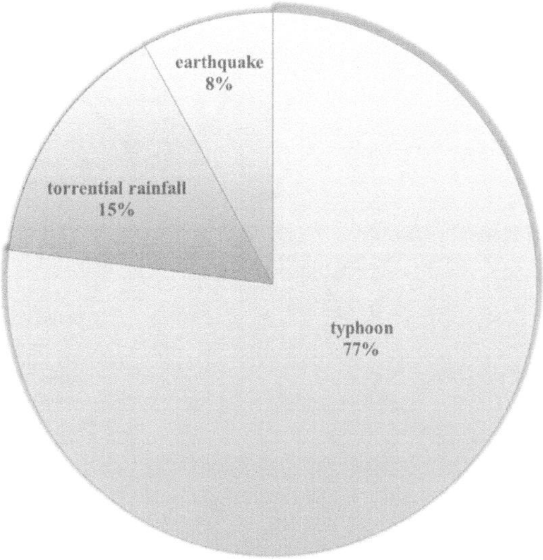

Figure 12.10 Triggering events of landslide disasters from 1989 to 2013.

Plate 12.1 The extreme landslide event induced by Typhoon Morakot in 2009 devastated Shiaolin Village, leading to more than four hundred deaths or missing people.

Fig. 12.11, the distribution of extreme landslide events did not have a specific pattern. In the last 40 years, typhoons with intense precipitation (typhoons with precipitation in the top 10%) often caused severe disasters (e.g., Typhoon Herb, Nari, and Morakot). The frequency of these typhoons is as shown in Fig. 12.12. For the period of 1970 to 1999, these typhoons occurred once every three to four years. However, after year

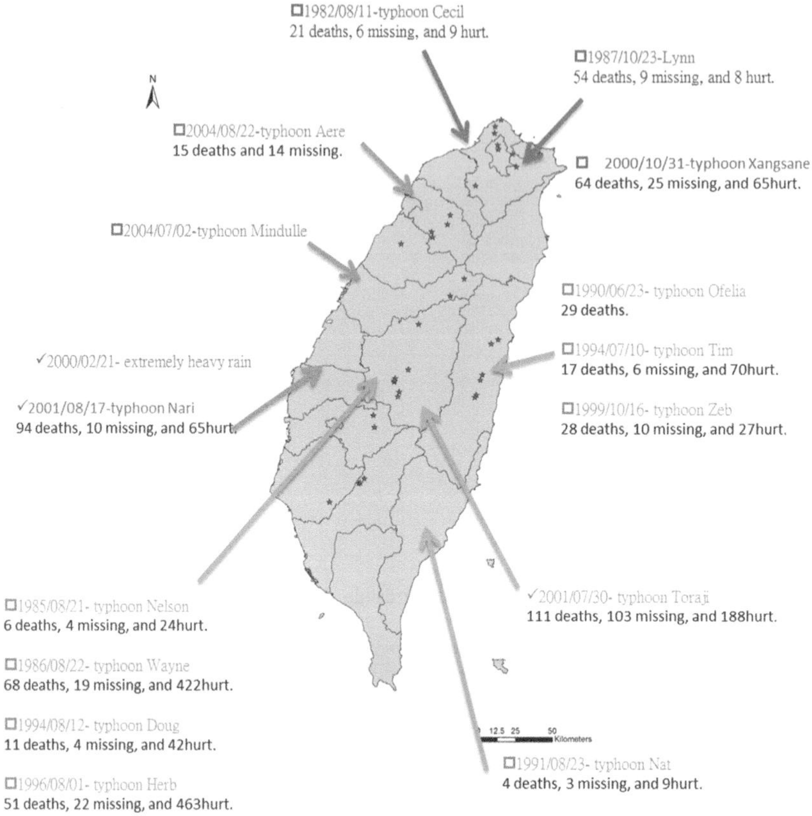

Figure 12.11 Distribution of historical extreme landslide events in Taiwan.

2000, these typhoons occurred annually, on average (NSC, 2011). The number of high-precipitation typhoons has increased substantially in the last decade. Whether these variations will continue in the future requires further research. Despite the uncertainty about the future, government agencies must acknowledge the possible effects of climate change. Although annual rainfall in the last 40 years has not changed significantly, the ratio of typhoon precipitation to total precipitation has increased from 15% in the 1970s to 30% in the 2000s. Precipitation increases in the wet season and decreases in the dry season, leading to uneven seasonal precipitation. The rainfall is the major triggering factor of landslide occurrence, and the trend of increasing rainfall carried by typhoons in the future will likely affect the frequency of landslides.

12.3.3 The correlation of rainfall and landslides

To identify the effects of rainfall on landslide occurrence, a logistic regression statistical method was used to establish the rainfall threshold for landslides (NCDR, 2010), based on the landslide events induced by typhoons from 1989 to 2009. A daily rainfall

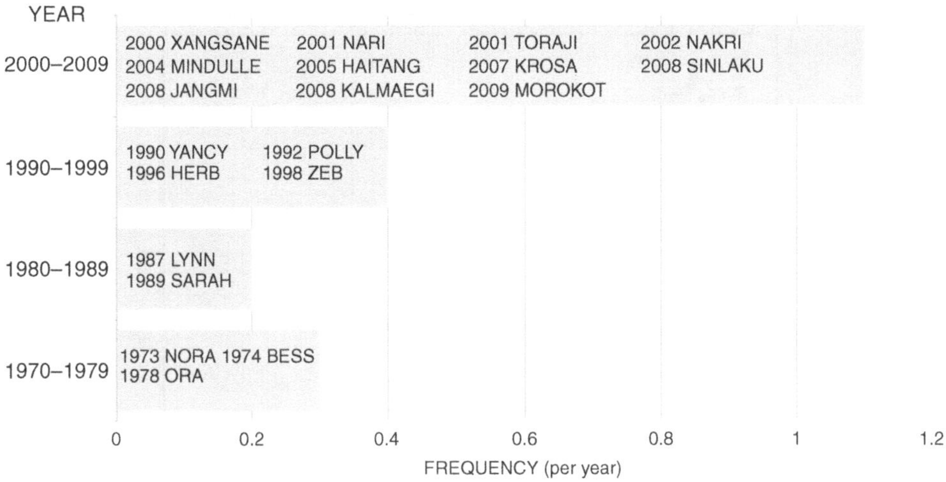

Figure 12.12 Variation in the frequency of typhoons with intense precipitation. A typhoon with intense precipitation is defined as one of the 20 top-ranking typhoons on the integrated typhoon precipitation index (approximately 10% of the total). The typhoon precipitation index is taken as the precipitation capacity of typhoon; this accounts for the maximum total accumulated rainfall, average rainfall intensity, average top five rainfall intensity, average daily rainfall, and the maximum hourly rainfall (National Science Council, 2011).

threshold of each county was determined as the rainfall corresponding to probability of landslides occurrence larger than 50%. These thresholds range from 280 mm to 550 mm as shown in Fig. 12.13.

12.3.4 Effects of climatic changes on landslide occurrence

The number of annual landslide occurrences in Northern, Central, Southern, and Eastern Taiwan in the last 20 years (1989 to 2009) was analyzed in Fig. 12.14. The overall number of landslide occurrences has increased significantly since 2004. In Central Taiwan, the number increased significantly since 1999. Some studies have suggested the high number of landslides to be associated with the Chi-Chi Earthquake (Lin *et al.*, 2000; Lin and Jeng, 2000; Wang *et al.*, 2003 and Lin *et al.*, 2004). The number of landslides in Southern Taiwan increased significantly after 2008, especially in 2009. The number of annual landslide occurrences in Northern Taiwan was different, as it decreased after 2001, possibly due to the implementation of landslide control policies by local governments. The changes of the annual landslide occurrences in Eastern Taiwan were not significant.

One analysis on the frequency of daily rainfall exceeding the rainfall threshold, as shown in Fig. 12.13 for the last 20 years (1989 to 2009), considered the relation between landslides and rainfall. Its results show that the average frequency was about 0.5% from 1989 to 1999 and increased to 0.8% by 2009 as shown in Fig. 12.15. This increasing trend is consistent in all areas of Taiwan, with the most significant increase being in Southern Taiwan as illustrated in Fig. 12.16. The frequency of rainfall

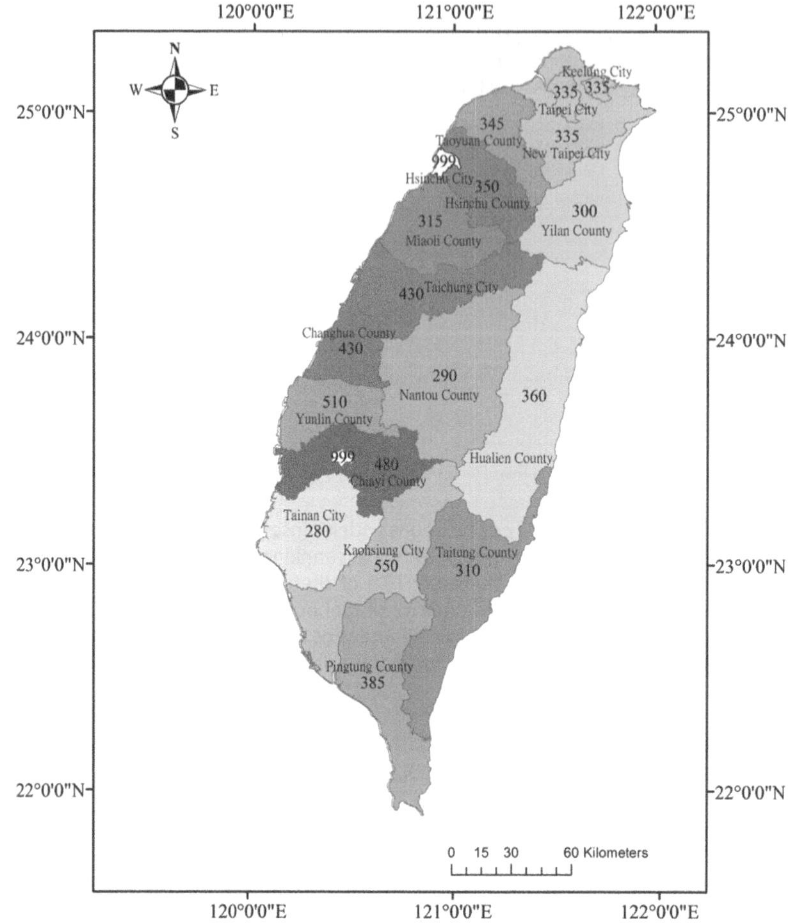

Figure 12.13 The landslide rainfall threshold of each county in Taiwan.

exceeding the rainfall threshold increased more to the west of the Central Mountain Range than to the east of the Central Mountain Range. This result might be due to intensive rainfall-induced landslide disasters that have increased from 1989 to 2009. The frequency of daily rainfall exceeding the threshold and the annual occurrence of landslides appear to be positively correlated in Central and Southern Taiwan, but no such relationship was found for Northern and Eastern Taiwan.

Furthermore, landslide scars were mapped from SPOT and FORMOSAT-2 satellite images taken from 2002 to 2012. Landslide frequency, taken as the number of scar occurrences per $0.5 \times 0.5\,km^2$ area being greater than six is plotted versus the probability of daily rainfall exceeding 350mm, the warning threshold, from 1989 to 2009 in Fig. 12.17. The mapping found three rainfall concentrated zones and

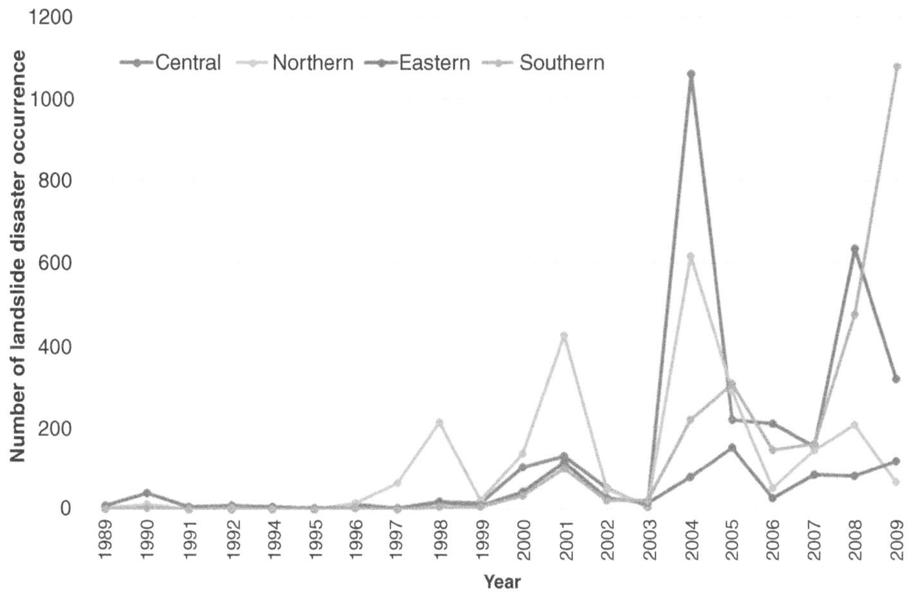

Figure 12.14 The number of annual landslide occurrences from 1989 to 2009.

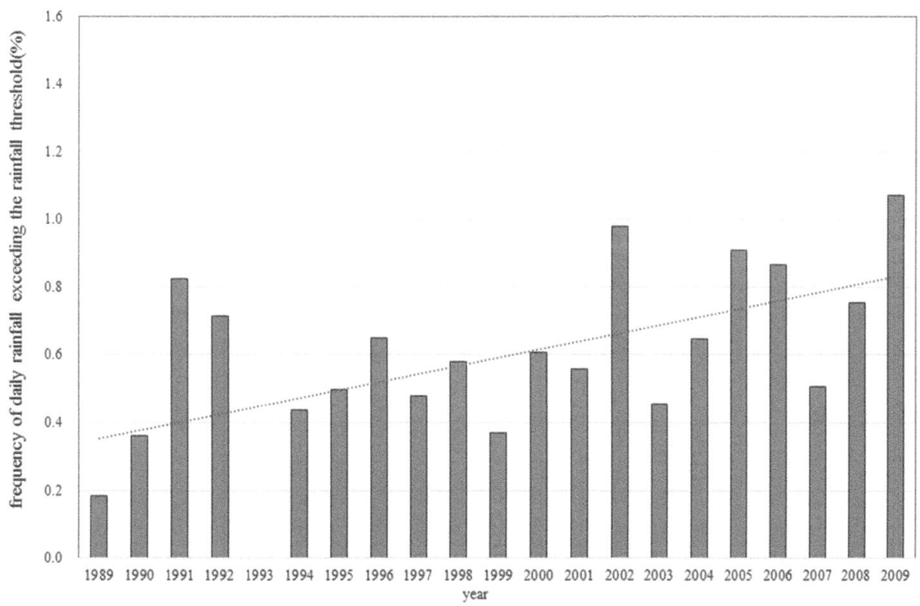

Figure 12.15 The frequency of daily rainfall exceeding the daily rainfall threshold based on landslide occurrences from 1989 to 2009. The red dashed line is the linear regression curve.

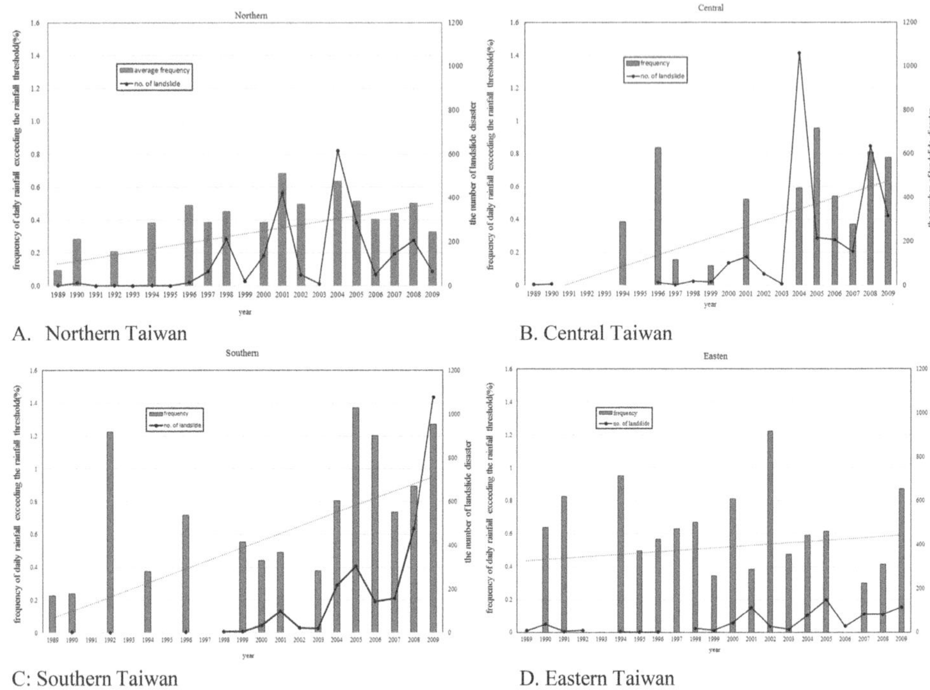

Figure 12.16 The frequency of daily rainfall exceeding the daily rainfall threshold of landslide occurrences in (A) Northern, (B) Central, (C) Southern, and (D) Eastern Taiwan from 1989 to 2009. The solid line is the number of landslides, and the red dashed line is the linear regression curve.

their occurrence probability of rainfall threshold is larger than 0.6%. The distribution of landslide frequencies greater than six is consistent with the contour of the 0.6% rainfall occurrence probability. The mapping shows a strong spatial relationship between the landslide frequency and the daily rainfall exceeding the threshold, suggesting that the landslides in Taiwan are highly related to rainfall. Thus, future rainfall trends can affect landslide occurrences directly. The climate change projections from NSC Report using the IPCC-developed scenario A1B, were considered by the international science community as the most likely future outcomes. The results follow:

(a) In summer, over three-quarters of the models project an increase in the seasonal mean precipitation, except in Northern Taiwan. About half of the models project the change to range from +2% to +26%. This projected scenario of increasing precipitation during rainy seasons can lead to a higher number of landslides.
(b) Most climate models project that, under global warming effects, the number of typhoons will decrease but the intensity and associated extreme rainfall of typhoons will likely increase.

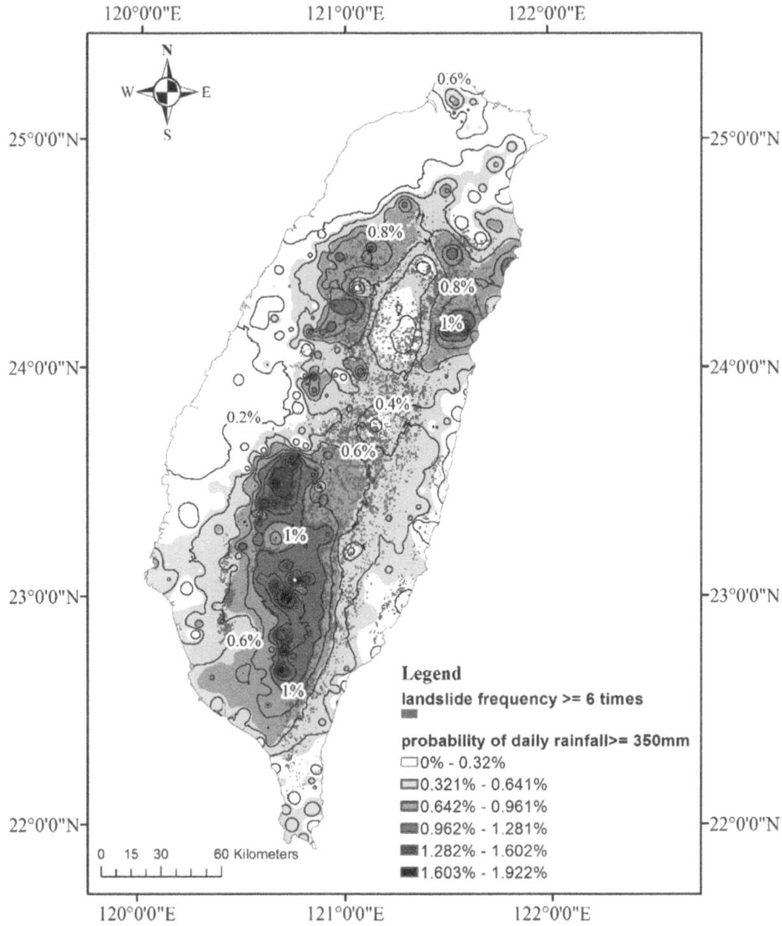

Figure 12.17 Landslide occurrence frequency from satellite images versus the probability of daily rainfall exceeding 350 mm from 1989 to 2009.

In summary, these findings suggest that the frequency and scale of future landslides are likely to increase.

12.4 IMPLICATIONS ON LANDSLIDE RISK MANAGEMENT AND ADAPTATION STRATEGIES

12.4.1 Framework for the development of a landslide risk map

After the 2010 implementation plan for the National Adaptation Policy Framework and Adaptation Programs of Action for Climate Change, the Council for Economic Planning and Development (National Development Council) of Taiwan developed a

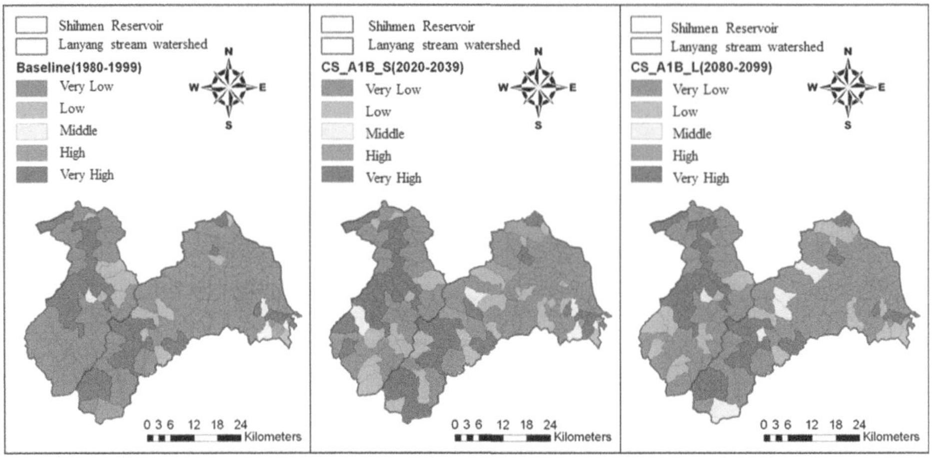

(a) The CSMK35-A1B scenario for landslides

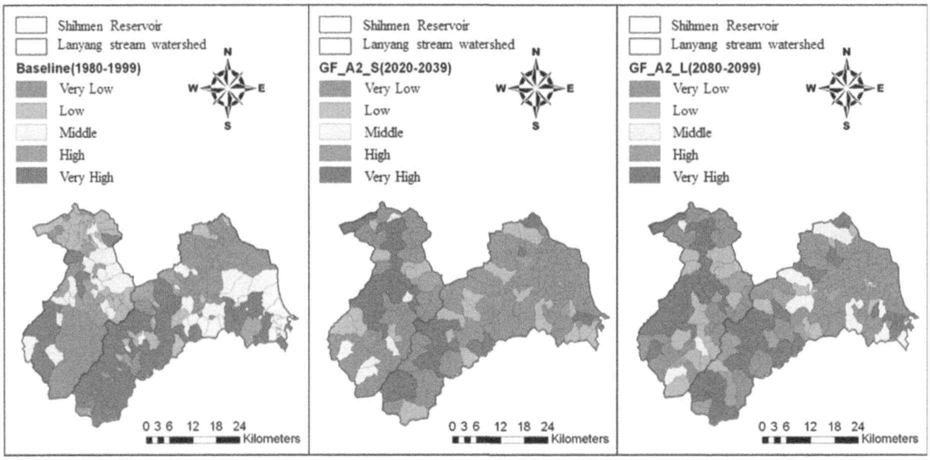

(b) The GFCM21-A2 scenario for debris flow

Figure 12.18 Landslide and debris flow hazard maps in the upstream area of Shihmen Reservoir and Lanyang Stream Watershed, based on the CSMK35-A1B and GFCM21-A2 scenarios, respectively, as per projected condition using baseline as 1980 to 1999; projected as 2020 to 2039 (SWCB, 2012).

national adaptation strategy framework and implemented a substantial action plan. The national adaption strategy framework implemented a substantial action plan that accounted for various aspects affected by climate change, including natural disasters, water resources, coast lines, agricultural production, biodiversity, public health, infrastructures, energy supplies, industrial economics, and land-use planning and managements.

The Soil and Water Conservation Bureau (SWCB) (2012) accounted for the projected rainfall based on the CSMK35-A1B, and GFCM21-A2 scenarios. The hazard

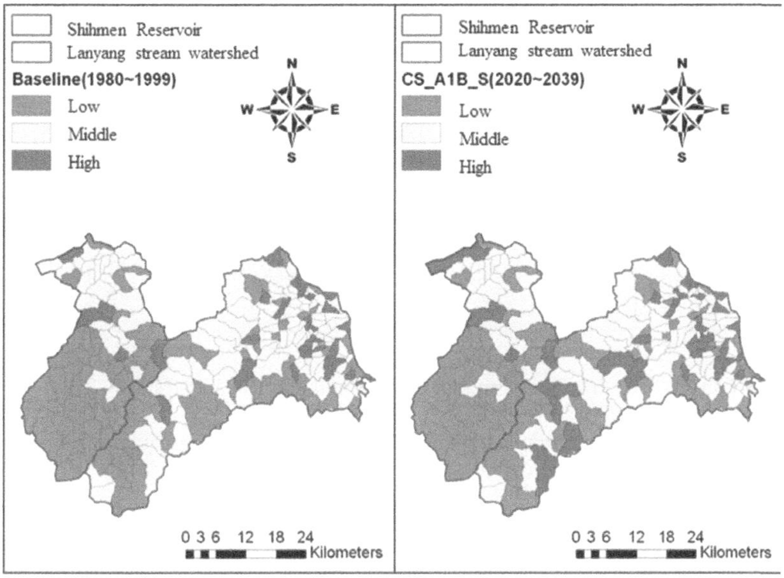

(a) The CSMK35-A1B scenario for landslides

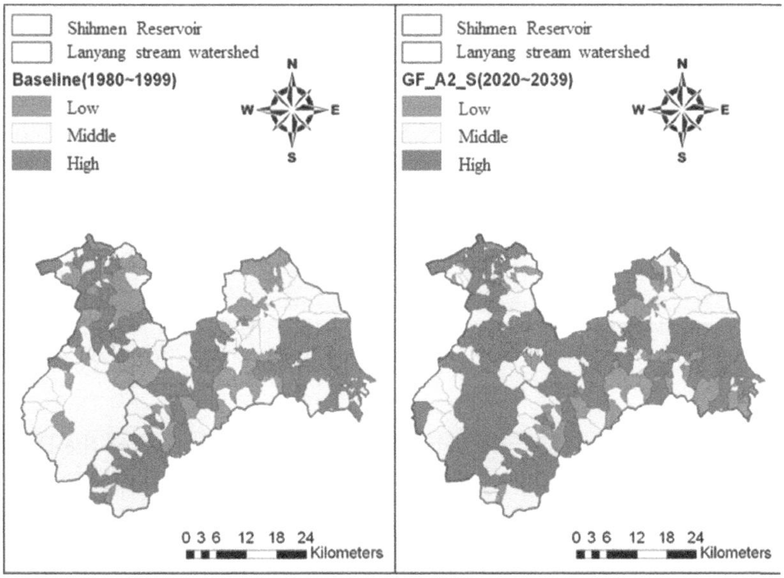

(b) The GFCM21-A2 scenario for debris flow

Figure 12.19 Landslide and debris flow risk maps in the upstream area of Shihmen Reservoir and Lanyang Stream Watershed, based on the CSMK35-A1B and GFCM21-A2 scenarios, respectively, as per projected condition using baseline as 1980 to 1999; projected as 2020 to 2039 (SWCB, 2012).

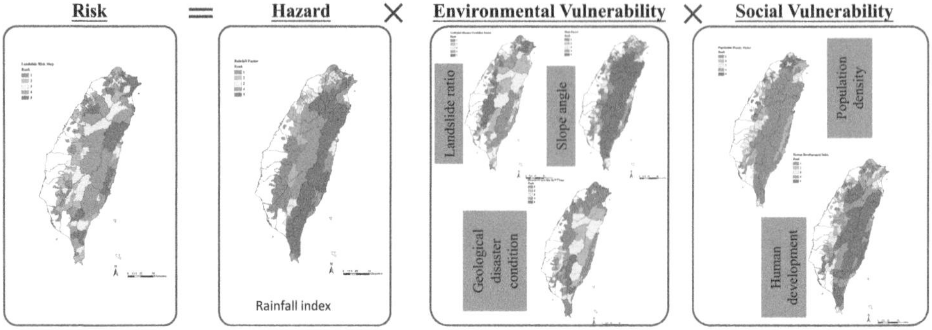

Figure 12.20 Proposed framework for a landslide risk map (NCDR, 2014).

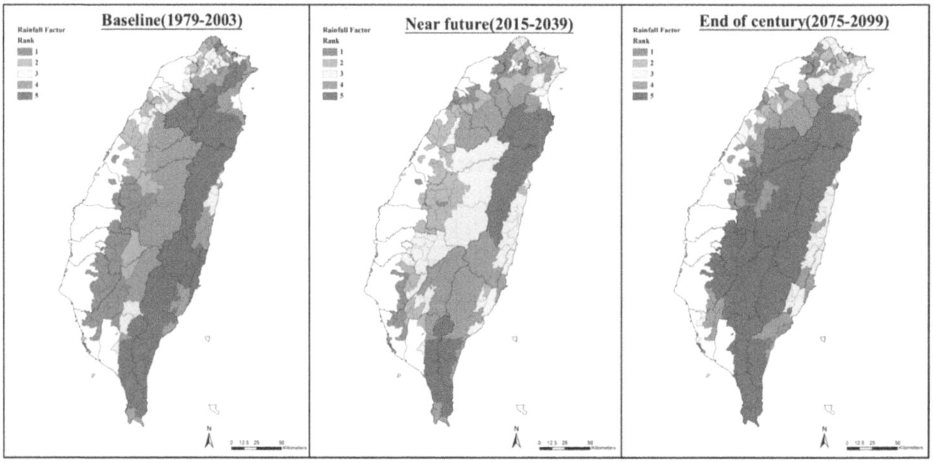

Figure 12.21 Landslide hazard map as per projected rainfall using the baseline as 1979 to 2003; the near future as 2015 to 2039; and the end of the century as 2075 to 2099 (NCDR, 2014).

maps (Fig. 12.18) of landslide and debris flow were constructed for the Dahan River Watershed upstream of Shihmen Reservoir and the Lanyang River Watershed located in the northern and northeastern parts of Taiwan, respectively. The resulting landslide risk maps are shown in Fig. 12.19. From these results, number of sub-watersheds with higher landslide and debris flow risk will increase in both watersheds in the future, and generally the trend of increase is more significant for the period of 2020 to 2039.

NCDR (2014) constructed the projected landslide risk maps, accounting for the future trends of population and rainfall due to climate change. The proposed framework of the landslide risk map is shown in Fig. 12.20. The risk index was composed of the following:

(a) Hazard factor: rainfall index, probability of 24-hour rainfall over 350 mm, dependent on future climate change as shown in Fig. 12.21;

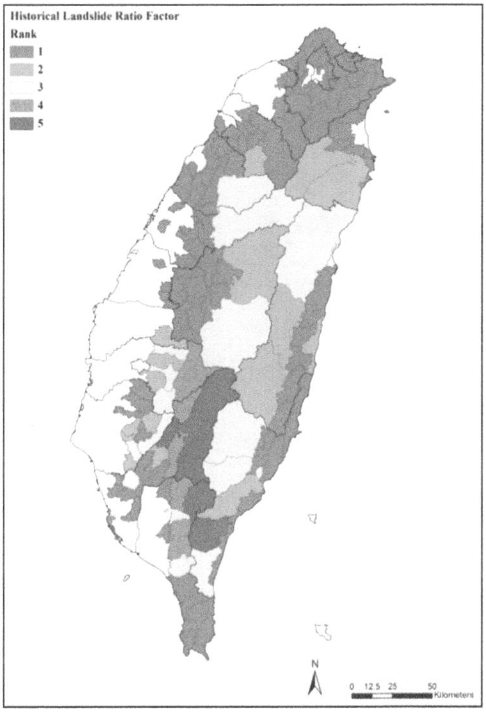

Figure 12.22 Distribution of the historical landslide ratio for the environmental vulnerability factor (NCDR, 2014).

(b) Environmental vulnerability factors: historical landslide ratio (Fig. 12.22), geological disaster condition (Fig. 12.23), and slope angle (Fig. 12.24);
(c) Social vulnerability factors: population density of baseline and future projections, (Fig. 12.25) and human development index (Fig. 12.26), which change with time.

The hazard factor map, using an occurrence probability of 24-hours rainfall over 350mm, is a projection of the A1B scenario using the baseline of 1979 to 2003 for the near future (2015~2039) and the end of the century (2075~2099). The projected rainfall frequency predicts an increase in the mountain area at the end of century.

The environmental vulnerability factors were developed based on the landslide ratio, geological disaster condition, and slope angle. A lack of clear variation trends of the environmental index in the future prompted an assumption of no change in the geological condition over time. Therefore, the environmental vulnerability map is assumed to be the same as the baseline map, as shown in Fig. 12.27. The map shows a higher environmental vulnerability in the central and southern areas.

The social vulnerability was evaluated based on the population distribution and projected human development index, as shown in Fig. 12.28. The population density variation trend in the near future was projected, and the population density at the end of century was assumed to be the same as that of the near future. The population density

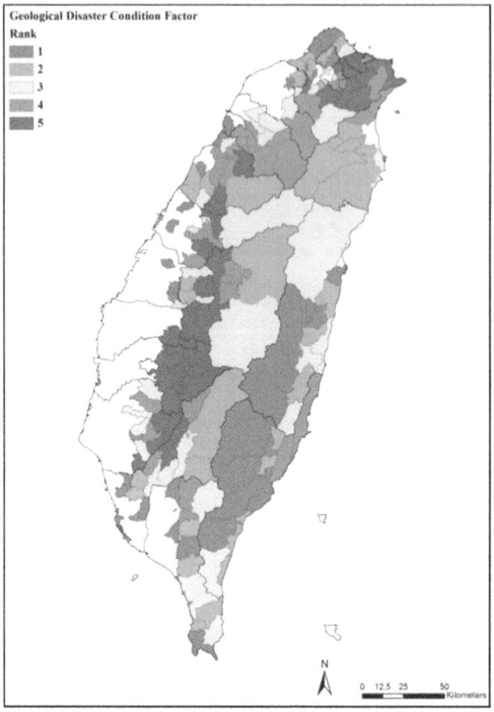

Figure 12.23 Distribution of the geological disaster condition for the environmental vulnerability factor (NCDR, 2014).

in some towns in the central area will decrease, and thus the social vulnerability will be lower than that of the baseline. The human development index and the population density index in combination imply a higher social vulnerability around urban areas.

The combined landslide risk map based on the A1B scenario of future climate change for the baseline, the near future, and the end of the century is shown in Fig. 12.29. In Fig. 12.29, the landslide risk slightly decreases in the Northern and Southeastern part of Taiwan due to a decrease in projected rainfall. However, it shows a significant increase of future landslide risk in Central, Southern, and Eastern Taiwan because of the trends of increasing rainfall in both the near future and the end of the century. Furthermore, the result for the present and that for the end of the century, as shown in Fig. 12.17 and Fig. 12.29, show a high relative landslide risk in the near-future and the end-of-century associated with the central and southern towns. This is consistent with the historical landslide disaster occurrence frequency. In addition, the sustained human development trend and high probability of daily rainfall exceeding 350 mm in the central and southern towns near urban areas in both the past and future imply a future increase in the landslide risk. Consequently, the Taiwanese government faces more challenges, and land-use planning or management in the central and southern mountainous areas should be adopted in consideration of future climate changes.

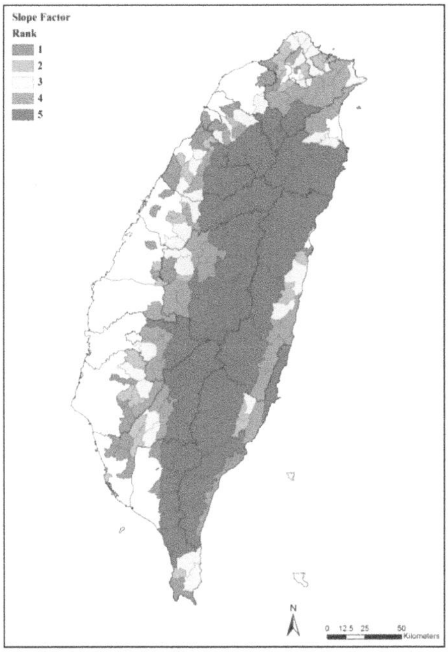

Figure 12.24 Distribution of the slope angle for the environmental vulnerability factor (NCDR, 2014).

However, the current evaluation only accounts for the daily accumulated rainfall. Previous studies have indicated the rainfall intensity to be a significant factor for causing landslide hazards. According to the climate change science report (NSC, 2011), the projected trends of climate change also suggest an increase in rainfall intensity, which is one of the landslide triggering factors. Therefore, the current estimations of the risk map could be underestimating the general risk of landslides.

12.4.2 Landslide disaster: impacts and challenges

As discussed previously, the increase of rainfall will lead to a higher risk of landslides in most parts of Taiwan and is likely to cause tremendous compound soil and water disasters. As the intensity of typhoons increases, Taiwan faces an increasing threat on its contingency and recovery capacities of its disaster prevention systems. Moreover, the critical infrastructure (e.g., bridges, roads, and mitigation facilities, etc.) will suffer from various levels of damage. For an example, Typhoon Morakot carried a huge amount of rainfall and caused a catastrophic landslide that buried Shaolin Village. Prior to Typhoon Morakot, such a catastrophe has never occurred before in Taiwan. Case studies of such events have led to the following possible impacts of future climate change:

(A) Increasing magnitudes of landslide disasters in the future;
(B) Increasing risk of cascading disasters and increasing difficulties of recovery;

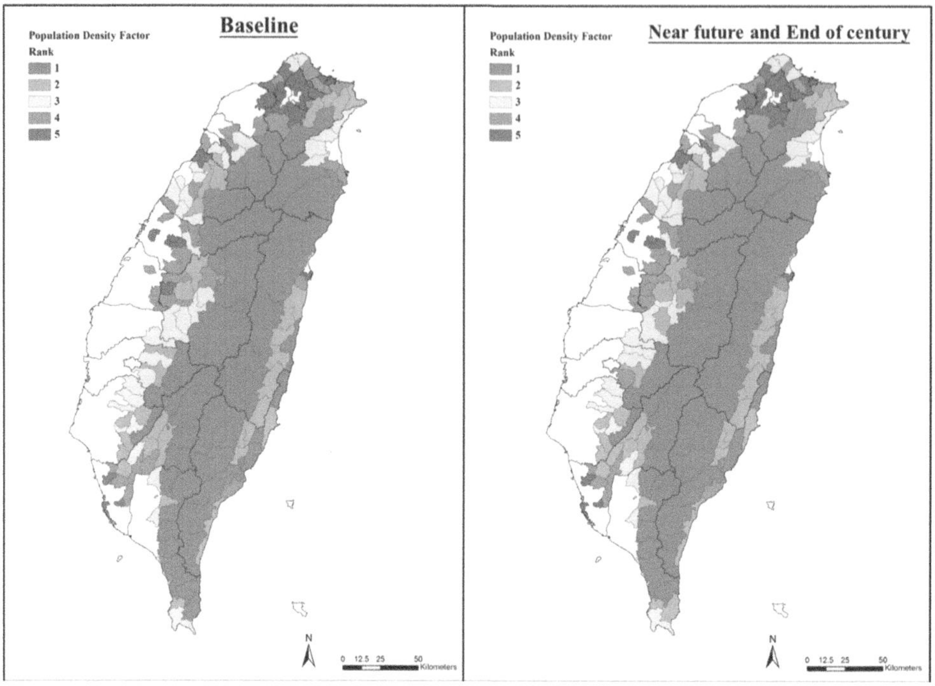

Figure 12.25 Distribution of the population density of baseline (left) and projected near-future and end-of-century (right) for the social vulnerability factor (NCDR, 2014).

(C) Serious compound disasters;
(D) An insufficient disaster prevention level for design code.

12.4.3 Landslide risk management, adaptation strategies, and policy considerations in Taiwan

Adaptation strategies are formulated based on two fundamental concepts: risk-avoidance and risk reduction. The prior concept involves avoiding high-risk regions whereas the latter involves increasing regional resilience to withstand the impacts of climate change. Overall adaptation strategies are interdisciplinary and the top priorities regarding all sectors, which include (Council for Economic Planning and Development, 2012):

(A) Implementation of regional planning and management;
(B) Enhancing disaster prevention and avoidance capacities in environmental, social, and economic systems;
(C) Executing comprehensive river basin management;
(D) Prioritizing high-risk regions against climate change hazards;
(E) Enhancing prevention and protection capacities for adaptation in urban areas.

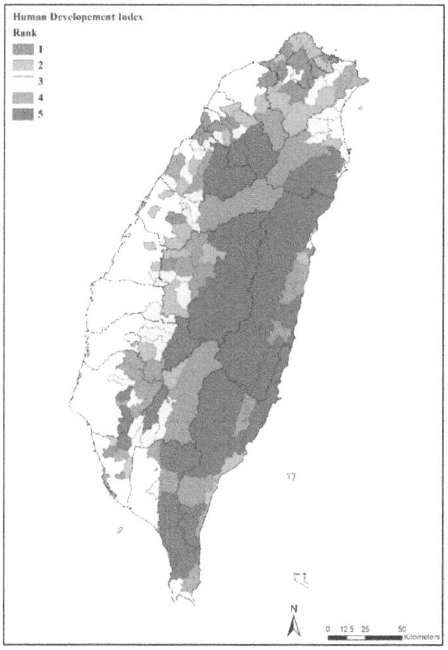

Figure 12.26 The human development index (baseline) for the social vulnerability factor (NCDR, 2014).

Therefore, the governmental adaptation to climate change is to implement disaster risk evaluation and comprehensive policy adaptation, thereby reducing the dangers of climate change. The strategies of disaster prevention and avoidance include:

(A) Surveying and evaluating climate change disaster risk and identifying the high risk areas.
(B) Enhancing the integration of environmental monitoring and disaster warning systems; creating a platform for information exchange, thus enhancing the capacities in response to the impacts of climate change.
(C) Reviewing and evaluating the vulnerabilities and prevention capacities of current critical infrastructures as well as reinforcing disaster prevention and protection plans.
(D) Emphasizing the impact of climate change and increasing disaster resilience in construction and development plans.
(E) Executing comprehensive river basin management in view of reducing the overall risk of climate change.
(F) Strengthening capacities for responding to the impact of extreme climatic events, publicizing the risk information of impacted and endangered areas, as well as carrying out disaster-reduction education, early warning system, and drills for high risk areas.

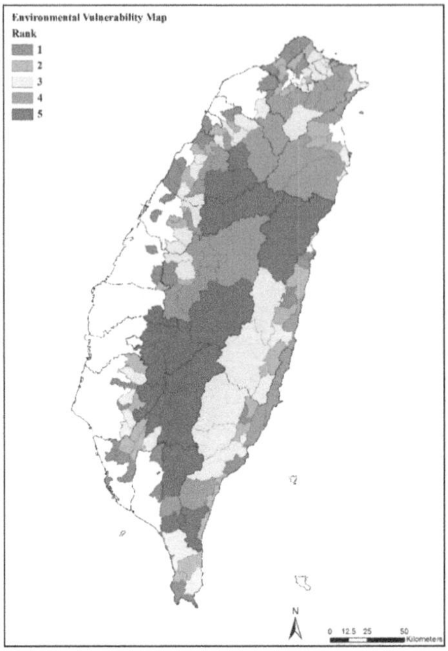

Figure 12.27 The environmental vulnerability map (baseline) (NCDR, 2014).

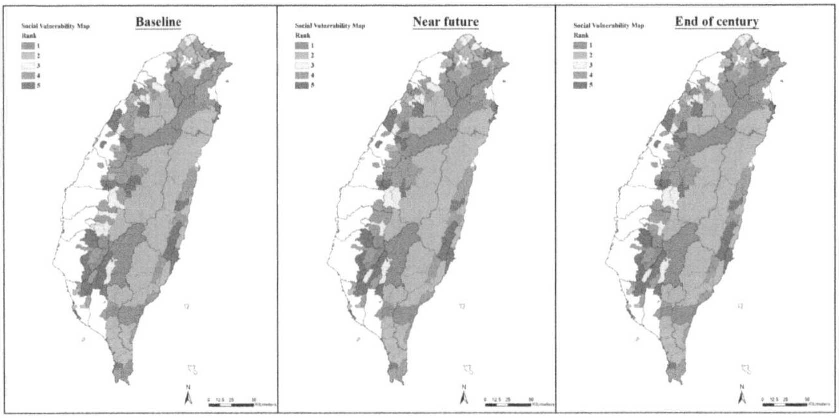

Figure 12.28 The social vulnerability map for the baseline, near future, and end of the century (from left to right).

12.5 CONCLUSIONS

The landslide risk maps were developed based on science reports with climate change projections for the near-future and the end-of-century scenarios of A1B from IPCC TR4, which is considered by the international science community to be the most likely

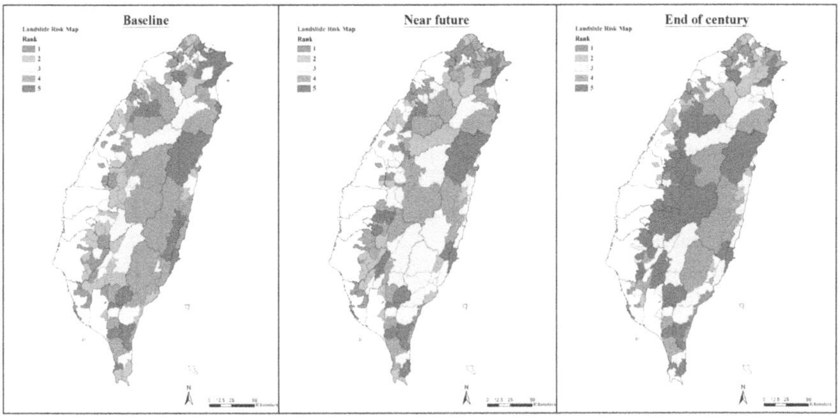

Figure 12.29 Landslide risk map based on the A1B scenario for future climate change for the baseline, near future, and end of the century (NCDR, 2014).

future outcome. Based on these projections, the increasing trend of typhoon precipitation in the last 20 years will continue at a snowballing magnitude, and the average rainfall frequency exceeding the landslide threshold rainfall, which was about 0.5%, will increase to 0.8%, especially in Central and Southern Taiwan. Thus the landslide hazard will become an increasing trend, even without considering human activity. Science, technology, land-use planning and management, as well as policy enforcement should be integrated for implementing a comprehensive strategy. Together, these measures can reduce the effects of global climate change.

ACKNOWLEDGEMENTS

Authors would like to acknowledge the help of the Ministry of Science and Technology, Council for Economic Planning and Development, National Science and Technology Center for Disaster Reduction, and Soil and Water Conservation Bureau, Council of Agriculture in supporting this report.

REFERENCES

Council for Economic Planning and Development (2012) *Adaptation Strategy to Climate Change in Taiwan* Taipei, Taiwan, ROC, Council for Economic Planning and Development.

Lin, M. L., & Jeng, F. S. (2000). *Characteristics of hazards induced by extremely heavy rainfall in Central Taiwan—Typhoon Herb.* Engineering Geology, 58(2), 191–207.

Lin, M. L., Wang, K. L. and Chen, T. C. (2000) *Characteristics of the slope failure caused by the Chi-Chi earthquake.* Int. Workshop Annual Commemoration of Chi-Chi Earthquake, Taipei, pp. 199–209.

Lin, M. L., & Tung, C. C. (2004). *A GIS-based potential analysis of the landslides induced by the Chi-Chi earthquake*. Engineering Geology, 71(1), 63–77.

National Science Council (2011) *Climate Change in Taiwan: Scientific Report 2011 (Summary)*. Taipei, Taiwan, ROC, National Science Council. 67 pp.

National Science and Technology Center for Disaster Reduction (2010) *Risk assessment method for typhoon-induced disasters*. New Taipei, Taiwan, ROC, National Science and Technology Center for Disaster Reduction (in Chinese).

National Science and Technology Center for Disaster Reduction (2014) *Risk Maps of Disaster Under Climate Change*. New Taipei, Taiwan, ROC, National Science and Technology Center for Disaster Reduction (in Chinese).

Soil and Water Conservation Bureau Council of Agriculture (2012) *Disaster Potential Evaluation of the Slopelands and Its Adjusting Plan Under Climate Change*. Nantou, Taiwan, ROC, Soil and Water Conservation Bureau, Council of Agriculture (in Chinese).

Wang, W. N., Chigira, M., & Furuya, T. (2003). *Geological and geomorphological precursors of the Chiu-fen-erh-shan landslide triggered by the Chi-chi earthquake in central Taiwan*. Engineering geology, 69(1), 1–13.

Chapter 13

Landslides and climate change in the United Kingdom

T.A. Dijkstra, G.O. Jenkins, D. Gunn and C. Dashwood
British Geological Survey, Keyworth, Nottingham, UK

R. Dankers
Met Office, Exeter, UK

N. Dixon
Loughborough University, Loughborough, UK

D.N. Petley
University of East Anglia, Norwich, UK

A. Gibson
University of Portsmouth, Portsmouth, UK

M.G. Winter
Transport Research Laboratory, Edinburgh, UK

13.1 INTRODUCTION

This report is a contribution to the Joint Technical Committee (JTC) 1 'Natural Slopes and Landslides'. The objectives of this UK country report are (1) to provide an overview of landslide activity, (2) to sketch the climate change context and highlight examples of research into the possible consequences for landslide activity, and (3) to discuss the relevant regulatory framework and multi-disciplinary hazard assessment.

13.2 LANDSLIDING IN THE UK

The maritime climate and relatively low relief of the United Kingdom limit opportunities for large scale and destructive landslides to occur. Neotectonic activity is also limited to low intensity events that are not known to have triggered landslides. The distribution of landsliding is not uniform (Figure 13.1); it is characterised by a landscape containing a wide variety of unstable slopes, ranging from very large ancient, relict landslides that were formed under very different climate conditions (see, for example, the Mark Nab landslide case study below) to small translational failures (see, for example, the Hollin Hill landslide case study below). Different types of bedrock lithologies, a variety of superficial deposits and both long-term and recent exposures to different climate conditions have resulted in a patchwork of different types of landslides. This complex geology has produced many locations where debris can accumulate. This debris becomes destabilised with prolonged or intense rainfall events. Such accumulations are a particular problem as they are not well represented on maps and therefore difficult to include in regional assessments of landslide hazard. They are often subject to rapid onset and high mobility upon failure. Modern landsliding receives increasing attention through numerous relatively small events that affect the lives and livelihoods of the population as these interact with infrastructure (for example in Glen Ogle and Rest and Be Thankful, Scotland, Winter *et al.*, 2009; Straidkilly, Northern Ireland, Hughes *et al.*, 2013, Creighton 2006; Hatfield Colliery landslide in 2013), can cause large scale damage to property (Cayton Bay coastal landslides in 2013) and can cause fatalities. On 2nd January 2001, the first recorded landslide fatality since 1986 occurred as a consequence of a rainfall induced event at Nefyn Bay, N Wales (Statham, 2002). An unusually intense, prolonged rainstorm generated a small failure in the cliff disturbing the debris apron on the lower slope. This resulting mass movement engulfed a car park and pushed 2 vehicles over a cliff, fatally wounding one occupant (Statham, 2002). Subsequent investigations identified the hazard posed by these accumulations as a far more widespread problem in the region than previously appreciated (Gibson *et al.*, 2002; Jenkins *et al.*, 2007). Recently, the very wet period for most of England and Wales during the summer of 2012 and into early 2013 led to a substantial increase in landslide activity resulting in much disruption of transport infrastructure and a further four fatalities (at the coastal cliffs of Burton Bradstock in 2012; at a road tunnel entrance in Beaminster in 2012; and following a slope failure leading to a house collapsing in Looe in 2013). For relevant case studies please visit www.bgs.ac.uk/landslides. All landslide locations mentioned in this report are shown in Figure 13.1.

Many parts of the coastline around the UK are susceptible to landsliding. The official UK mapping agency, the Ordnance Survey, estimates the length of the UK coastline to be about 17,820 km. The coastline comprises a wide variety of geological and geomorphological settings. Much of it, particularly in the north and the west, consists of hard, crystalline rocks that form near-vertical cliffs. Along the South Coast and in the east there are large sections of weak sedimentary rocks, including glacial deposits, which are readily eroded. The UK coastal zone is affected by a stormy climate, with winter storms producing strong winds and large wave heights, particularly when driven by (south-)westerlies from the Atlantic affecting coastlines in the south and west of the country. There is active coastal erosion in most areas, but the mechanism of landsliding varies widely. The hard rock cliffs are mainly affected by events ranging from individual rockfalls through to larger rock slope collapses. In the softer materials, failure occurs primarily through translational and rotational landslides, and undercutting of soft cliffs. Baseline rates of erosion can be high. For example, Lee (2002) reported long term erosion rates as high as 3 m/year for soft sections of the UK coastline, whilst the hard rock cliffs typically retreat at rates of less than 0.05 m/year. However, retreat rates are highly variable and episodic, especially where the primary retreat mechanism consists of low frequency, high magnitude collapse events (e.g. Foster *et al.*, 2008; Lee, 2008; Quinn *et al.*, 2009; Dawson *et al.*, 2009).

Many landslides in the UK can be associated with 'wet-year' sequences when effective rainfall is consistently above mean. On the Isle of Wight, in Southern England, such phenomena have been well described, with significant correlations between wet-year

Mark Nab landslide, North York Moors

The Mark Nab landslide lies within the North York Moors National Park. It was recently mapped and investigated by the British Geological Survey. The landslide has developed in Jurassic strata and forms a striking landscape with high ridges of back rotated blocks that extend down the valley side. At the rear of the escarpment the thick cap of well-jointed sedimentary rocks, comprising: the Dogger Formation, Saltwick Formation, Eller Beck Formation and Cloughton Formation, effectively provides a static load on the Whitby Mudstone Formation and supplies ground water into the slope. Dilation of existing joints due to stress-relief increases the transmission of water into the slope. The maximum depth of rupture is 50 m, with the slide plane exploiting the full depth of the weak mudstones of the Whitby Mudstone Formation. Towards the base of the landslide, the dominant style of movement is preserved in flows of less competent material. Radiocarbon dating of organic-rich sediments sampled from near the toe of the landslide suggest that the last movements occurred between approximately 2240 and 2340 years BP. Large, relict landslides form a large proportion of the landslides recorded in the BGS National Landslide Database. The residual stability of the majority of relict landslides in the UK is largely unknown and although they are generally considered dormant or inactive, there is the potential for them to reactivate and cause damage.

sequences and landslide activity in 1958–60 (Evans Grant CCS, 1994), and 2000-1 (Hutchinson and Bromhead, 2002) and more recently this correlation has been addressed for SW UK and S Wales for the period from 2006 to 2013 (Pennington *et al.*, 2014; Figure 13.2). Some prominent large landslides associated with wet-year sequences (although also affected by anthropogenic influences) include the 1993 Holbeck Hall Landslide in Scarborough, N England (West 1994; Lee *et al.*, 1998; Lee, 1999; Lee and Clark, 2000) and the 1994 St Dogmaels Landslide in S Wales (Fletcher and Siddle, 1998). Engineering designs currently in consideration are required to take into account not only the anticipated change in annual rainfall, but the influence of the wet-year periods (Mott MacDonald, 2013).

Hollin Hill, North Yorkshire

The Hollin Hill landslide observatory was established by the British Geological Survey in 2006 (Jenkins *et al.*, 2006) as an instrumented slope investigating the relationship between water content and slope failure in the Whitby Mudstone Formation. Landsliding at Hollin Hill is complex and includes shallow, retrogressive rotational failures on the upper slopes with translation and flow from the base of the Whitby Mudstone Formation over the scarp slope of the Staithes Sandstone Formation on the lower part of the slope (Gunn *et al.*, 2013). Permanent monitoring systems have been installed, including electrical resistivity tomography (ERT) arrays, which have mapped the distribution of the Whitby Mudstone and Staithes Sandstone formations. These systems are also being used in time lapse mode to image the near-surface moisture movement driving the landslide processes. The ERT array installations include a large area, low spatial resolution grid to investigate the potential coupling between the upper and lower slope hydrogeological processes and a small area, high spatial resolution grid to investigate the hydrogeological processes causing the slope movements.

13.2.1 National Landslide Database (NLD)

The Landslides Team at the British Geological Survey (BGS) catalogue landslide information in the National Landslides Database (NLD). It is used for a wide range of applications including their national landslide susceptibility map GeoSure (e.g. Foster *et al.*, 2012; Pennington *et al.*, 2009; Figure 13.1a). The BGS NLD is the most comprehensive source of information on landslides in Great Britain and currently holds records of more than 16,500 landslide events that are continually updated and added to as information is reported (Foster *et al.*, 2012). As well as routinely collecting data from ongoing regional geological surveys and the published scientific literature, the online press has been monitored for information about landslides through various Internet search engines since 2006, and since spring 2012 twitter feeds are monitored on a daily basis. Each of the landslide event records can hold information on over 35 attributes, including location, dimensions, landslide type, trigger mechanism, damage caused, slope aspect, material, movement date, vegetation, hydrogeology, age, development and a full bibliographic reference. The information within the database is corporately maintained and held in a digital format that can be adapted and updated (see Foster *et al.* (2012).

13.2.2 Landslide potential

Landslides occur ultimately due to the effect of gravity, although other factors such as geology, topography, weathering, drainage and man-made construction can all contribute to the overall stability of a slope. Landslides rarely comprise a single type of movement but are often the result of a combination of several types. Many of the landslides in the National landslide Database are ancient and occurred under different climatic conditions to those of the present day (e.g. Pleistocene). If left undisturbed these ancient mass movement deposits may remain stable for many years, however poorly planned development can sometimes reactivate these ancient slides. The potential for landsliding (slope instability) to be a hazard has been assessed using 1:50,000 scale digital maps of superficial and bedrock deposits. These have been combined with information from the BGS National Landslide Database and scientific and engineering reports. The detailed digital data illustrated in the map are available as attributed vector polygons, as raster grids and in spreadsheet format (Figure 13.1b).

13.2.3 Engineered slopes

Engineered slopes (cuttings and embankments) deserve special attention as these form a large proportion of UK transport networks (there are 20,000 km of earth structures on the UK highway and rail networks; £20B of the £60B asset value of major highway infrastructure is earthworks, Glendinning *et al.*, 2009a, b) and there are many thousands of km of flood embankments and thousands of water retaining earth structures. Most were not built to modern engineering standards and their gradual deterioration over time means that many fail due to serious instability and/or serviceability problems. With a focus of historical development in the south east of England, many embankments and cuttings are constructed of or within the London Clay Formation, Lias Group and Gault Formation, with many of the older earthworks such as railway

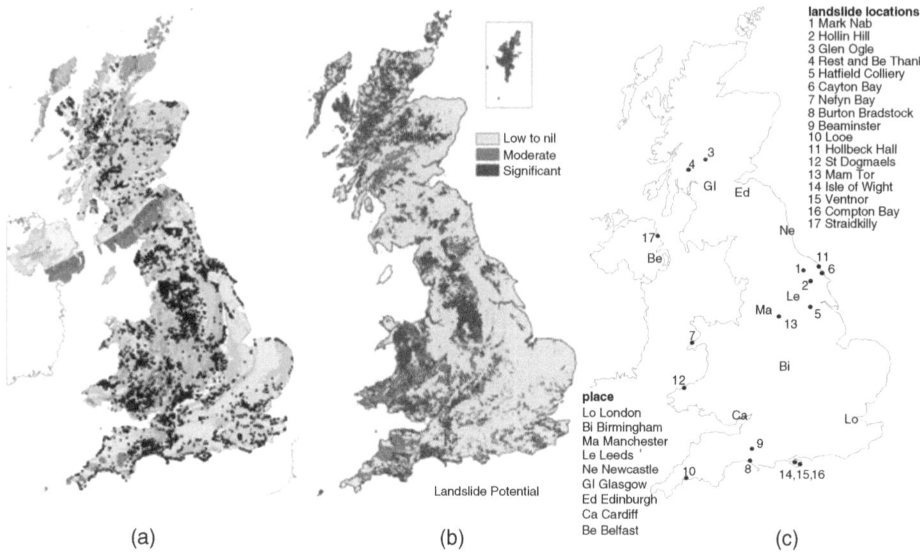

Figure 13.1 (a) Landslide distribution (black dots) in the UK superimposed on the 625k scale bedrock geology map. There is some clustering of landslides where favourable bedrock conditions exist. The coastal fringe is also affected by widespread landsliding. (b) Landslide potential summarised into three main categories. (c) location of landslides mentioned in the text. BGS © NERC 2015. All rights reserved. Contains Ordnance Survey data (coastline) © Crown copyright and database right 2015.

embankments constructed from loosely tipped clay and cut slopes formed at steeper angles than would be used in modern designs. These fine-grained, plastic materials are involved in a significant percentage of engineered slope failures, although other materials distributed across the UK, such as glacial till deposits, are also susceptible to slope failure. Slope instability poses a major safety hazard; with derailment from slope failures the greatest risk faced by the operational railway and the possibility of catastrophic collapse of water retaining earthworks a growing concern. Instability and serviceability problems lead to the imposition of rail speed restrictions; highway slope repairs often lead to lane closures, both leading to severe travel delays. The continuing maintenance and remediation of earthworks is a major engineering and cost constraint for UK infrastructure owners. As more operational equipment is located within the geotechnical asset such as signalling, telecoms, power and noise barriers, even minor slope failures cause major service disruption and incur significant repair costs. With the cost of emergency repair ten times greater than the cost of planned maintenance works, the economic loss and social cost in the UK caused by delays is enormous (Glendinning *et al.*, 2009a, b; Kilsby *et al.*, 2009; Loveridge *et al.*, 2010; Toll *et al.*, 2012; Dijkstra *et al.*, 2014).

Instability of engineered slopes is caused by prolonged and/or intense periods of rainfall (Davies *et al.*, 2014; Toll *et al.*, 2012; Loveridge *et al.*, 2010, Dijkstra and Dixon, 2010). Pore water pressures within engineered slopes are influenced by

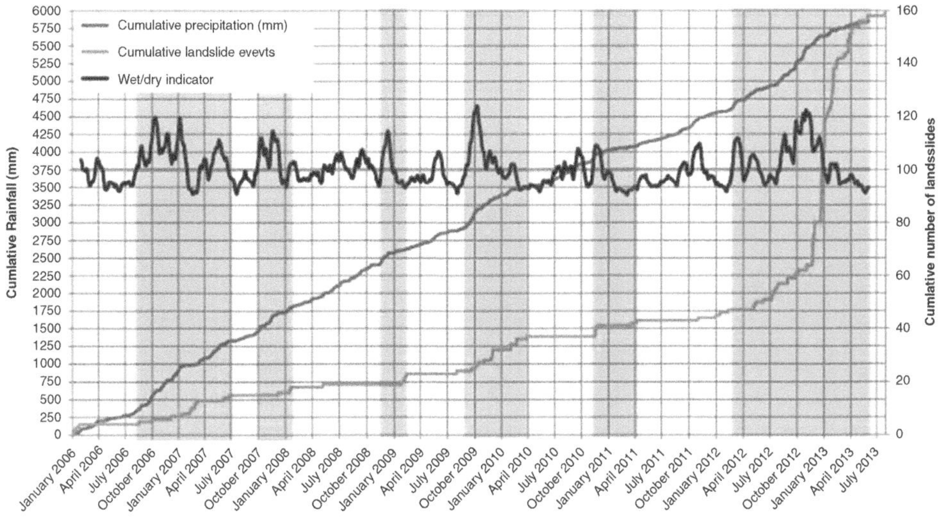

Figure 13.2 A correlation between landslides and effective precipitation for SW England/S Wales. The black line indicates the proportional difference between actual and long-term average (LTA) precipitation where 100 represents equality and values >100 show conditions wetter than LTA (Landslides information from BGS NLD; weather data from Met Office and Wunderground.com; Pennington *et al.*, 2014).

seasonal weather patterns and the vegetation on the slope (Smethurst *et al.*, 2012; Briggs *et al.*, 2014; Chambers *et al.*, 2014). Negative pore water pressures typically occur following dry periods during the summer months when evaporation and transpiration remove soil water. Positive pore water pressures occur following wet periods as water infiltrates the soil, and in the UK are typically highest between March and May. In the UK, rainfall is generally of low intensity and high frequency, although intensive rainfall events do occur. Therefore periods of prolonged, higher than average intensity rainfall are considered extreme events. Increases in pore water pressure during extreme wet periods trigger slope instability. In recent times, wet periods in the winter of 2000-2001, summer and winter 2012 and winter 2013 have resulted in significant and increased numbers of landslides in both embankment and cutting slopes. These have caused major disruption to the transport network. There is evidence that these engineered slopes are deteriorating over time. The slopes are subjected to cycles of summer drying and winter wetting driven pore water pressure changes, and hence cycles of effective stresses. These cause strains within the slope that can lead to strain softening and formation of shear zones, and this can ultimately cause progressive failure. The importance of this mechanism is that slopes often fail following rainfall events that they have survived in the past. The deterioration of engineered slopes in the UK is a significant concern. This is exacerbated by the changing climate, which for the UK is forecast to result in concentration of rainfall into winter months and for more intensive rainfall events, and hence more frequent extreme events. Research is being undertaken by a consortium of UK universities to investigate the impact of these changes in weather patterns on the frequency, magnitude and distribution of slope failures in engineered slopes (Bouch *et al.*, 2012; Dijkstra *et al.*, 2014).

Barrow upon Soar railway slope failure, December 2012

The need for high-resolution capture of spatial heterogeneity, an understanding of construction history and the interaction of different processes is clearly shown in the example of the derailment of a locomotive pulling twenty loaded stone hopper wagons just north of Barrow-upon-Soar, Leicestershire (UK). This derailment took place on December 27, 2012 and was caused by a single rotational slip in the earthwork embankment failure causing a depression of about 0.5 m along a length of some 25 m at the top of the embankment. The depression in the railway line resulted in derailment of seven wagons and closed the 'slow' line for one and a half weeks. As a consequence of persistently higher than average rainfall in the preceding months, soil moisture levels in the embankment and in the adjacent slopes were very high, with widespread inundation of the toe of the railway embankment. The embankment was constructed in Victorian times in the mid-19th century with further widening in the 1870s by placing ash layers on the side slopes. This created local changes in permeability facilitating seepage flows and contributing to rotational failure. Records show that small pre-failure strains had been developing in the previous year. Exceptionally wet conditions thus provided an exacerbation and acceleration of an already marginally stable situation. This failure shows that detailed parameterisation and an understanding of the interaction of different processes are required to fully appreciate the distribution of landslide hazard in these engineered slopes (RAIB, 2013).

13.3 THE CLIMATE CHANGE CONTEXT

The climate of the UK, located between 50° to 60°N, has a large maritime influence dominated by a westerly flow from the Atlantic between the Azores High to the south and the Icelandic Low to the north, and is influenced by the moderating effects of the Gulf Stream. As a consequence winters are relatively mild for its latitude, and the weather is highly variable with rainfall occurring at all times of the year. Although the potential impacts of climate change are generally thought to be much greater in other parts of the world, the UK is already vulnerable to extreme weather, including flooding and heatwaves (CCRA, 2012). In recent years, the UK annual average temperature has generally been between about 0.5°C and 1°C warmer than the 1961–1990 average, albeit with significant year-to-year variability (Jenkins *et al.*, 2009a, b). There are indications that annual rainfall has been increasing, but the signal in the frequency of prolonged wet and dry spells is unclear (Met Office, 2011). The latest climate projections for the UK suggest increases in summer and winter temperatures, increases in winter rainfall, decreases in summer rainfall (although small increases are also possible), more days of heavy rainfall and rising sea levels. Historic trends and future projections of rainfall and heavy rainfall events are discussed in more detail in the following sections.

13.3.1 Historical and future trends in rainfall

In recent years, the UK annual average temperature has generally been between about 0.5°C and 1°C warmer than the 1961–1990 average, albeit with significant year-to-year variability (Jenkins *et al.*, 2009a, b). Based on CRUTEM3 data (Brohan *et al.*, 2006) the average temperature trends from 1960 to 2010 are 0.23 (0.04–0.42) °C per decade in winter and 0.28 (0.16–0.40) °C per decade in summer (numbers in brackets indicate the 5th to 95th percentiles of regression slopes) (Met Office, 2011). Over the same period (but based on the HadEX (Alexander *et al.*, 2006) and European Climate Assessment & Dataset (Klein Tank *et al.*, 2002) datasets) there has been a uniform signal for increased annual rainfall over the UK, although only in the northern parts there is high confidence that this trend is significantly different from zero. The signal in the frequency of prolonged (greater than 6 days) wet spells is rather mixed, and there has been no clear change in the number of dry spells (Met Office, 2011).

Global climate projections show a strong contrast in precipitation changes across Europe, with decreases in the south and increases in the north. The UK is in the northern region with projected increases of up to 10%. There is, however, considerable uncertainty in the position of the transition zone between increasing and decreasing precipitation over Europe, meaning that some southern parts of the UK may actually experience decreases in mean precipitation. Further north the agreement between the models is better, indicating higher confidence in the projections (Met Office, 2011).

13.3.2 Regional projections for the UK

The following discussion of projected changes in precipitation and rainfall events makes use of the UKCP09 climate scenarios that represent the current state-of-the art assessment of future climate changes in the UK over the 21st Century (Jenkins *et al.*, 2009b; Murphy *et al.*, 2009). These climate projections are probabilistic in the sense that they represent a range of possible future changes in UK climate. The range of possibilities is necessarily wide to take account of uncertainties in natural climate variability, including how the UK's climate may respond to global warming, the future trajectory of emissions, and how these might magnify any regional climate change effects. UKCP09 used a complex methodology to combine information from a number of different sources, including both global and regional climate models, in order to provide estimated probabilities of changes in key climate variables. The estimated probabilities do not include uncertainties in processes which are not captured by any current climate models, particularly affecting natural variations from year to year, the location of storm tracks and the causes of more extreme flood and drought conditions, and in this sense they are only partial representations of the overall uncertainty.

Projecting future climate change and variability is fraught with uncertainty, especially at the relatively small scales of the UK. The climate varies naturally from year to year and decade to decade, and confidently projecting the details of these fluctuations is beyond current scientific capabilities. In the longer term, for a decade and beyond into the future, the rise in greenhouse gas concentrations is expected to continue to cause a global warming trend over and above year-to-year natural variability. However, it is still extremely hard to forecast how regional climates in the UK will respond, especially for changes in precipitation (CCRA, 2012).

13.3.3 General overview of UKCP09 projections

UKCP09 projects a range of changes in UK climate over the 21st century, accounting for both natural variability and the uncertain response of regional climates to global warming. In the near-term (next one to two decades), most of the uncertainty range is associated with natural variability, and while the long-term average temperatures are expected to be warmer than present, periods of cooler temperatures are also possible at times. As for precipitation, season-to-season variability means that there could be wetter or drier periods, once again reflective of the UK's variable climate (CCRA, 2012). In the medium term (mid-century), the build-up of greenhouse gases is expected to be significant enough for systematic shifts in the UK climate to start to become evident even though year-to-year variability will still occur. This implies a clear shift toward warmer temperatures, although with natural variability e.g. cold winters may still occur. A shift towards generally wetter winters is expected, but the future response of UK summer rainfall is less well-understood. UKCP09 suggests that there is a greater likelihood that summers will be drier, but wetter summers are not ruled out. Regardless of what happens to seasonal averages, a greater fraction of precipitation is expected to fall during heavy rainfall events (CCRA, 2012). The future trajectory of emissions is also uncertain, but the range of plausible emissions scenarios does not significantly affect the range of possible UK climate changes until the latter part of the 21st century, because the different emissions pathways do not start to diverge until mid-century in terms of their effects on temperature rise. We know that higher emissions lead to greater warming, but it is harder to predict what may happen to annual and seasonal rainfall patterns (CCRA, 2012). UKCP09 considered three emission scenarios (High, Medium and Low). In the following sections the focus is on changes under the Medium scenario (relative to a 1961–1990 baseline), details of the other scenarios can be found on the UKCP09 website (ukclimateprojections.metoffice.gov.uk) and in Murphy *et al.* (2009).

13.3.4 Projection of precipitation and heavy rainfall events

Results from the probabilistic UKCP09 scenarios are typically presented for three probability levels: the 50% level is the central estimate; the 10% probability level is level of change that is very likely to be exceeded; the 90% is the level of change is very likely not to be exceeded. In other words, according to the UKCP09 methodology changes are likely to be higher than the 10% probability level, but lower than the 90% level. Under the Medium emission scenario, the central estimate (50% probability) of changes in annual precipitation by the 2080s shows relatively little change (between −3% and +2%) for most parts of the UK. Changes range from −16% in some places at the 10% probability level, to +14% in some places at the 90% probability level, with no clear geographic pattern. The biggest increases in average precipitation are seen in winter, with changes up to +33% (central estimate) along the western side of the UK, with 10% and 90% probability levels of +9 and +70%, respectively. Decreases of a few percent (−11% to +7%) are seen over parts of the Scottish Highlands. In summer the largest decreases are seen in the far south of England, down to about −40% (likely range between −65% to −6%), while changes over parts of Northern Scotland are close to zero (−8 to +10%) (Jenkins *et al.*, 2009b).

As an indicator of changes in heavy rainfall events, UKCP09 provides estimates of changes in the amount of precipitation on the wettest day of winter and summer (99th percentile of daily precipitation in a season). In winter, these changes range from zero (likely range −12% to +13%) in parts of Scotland to +25% (likely range +7% to +56%) in parts of England (Figure 13.3). In summer, the amount of rainfall on the wettest day is generally projected to decrease in parts of Southern England (central estimate-12%), but the likely range (−38% to +9%) encompasses the zero point, meaning that a (small) increase is still possible. In parts of Scotland the rainfall on the wettest day of summer is projected to increase by +12% (likely range −1% to +51%; Figure 13.4).

UKCP09 also provides estimates of changes in daily weather based on synthetic daily time series of a number of climate variables produced by a weather generator. As it is not yet possible to project changes in storm track and anticyclones with confidence, these synthetic time series allow making a plausible estimate of how the characteristics of daily weather could be affected by changes in the more basic aspects of a future climate (such as monthly mean temperature and precipitation) that can be projected with more confidence. The results suggest an increase in the frequency of 10-day dry spells, particularly in Southern England and Wales, associated with summer drying (Jenkins *et al.*, 2009b). At the same time the number of heavy rain days is projected to increase as well over most of the lowland UK, with multiple-fold increases during the winter months and up to two-fold increases in summer (CCRA, 2012).

13.3.5 Projections of mid-latitude depressions and storms

Severe windstorms have become more frequent across the UK in the past few decades, though not above a level seen in the 1920s (Jenkins *et al.*, 2009b). It is however not yet possible to make projections of changes in mid-latitude storm tracks into the future. The UKCP09 project found that Met Office Hadley Centre models have a tendency to project storm tracks weakening slightly and moving further south, while other models generally show less change in the position of the track, but a wider range of changes in strength. Furthermore, the models' storm tracks do not compare well with observations. UKCP09 concluded that there is no consistent signal of change in either mid-latitude storms or blocking near the UK, and that the changes that are seen are relatively modest, which suggest that the potential for substantial changes might be small (Murphy *et al.*, 2009).

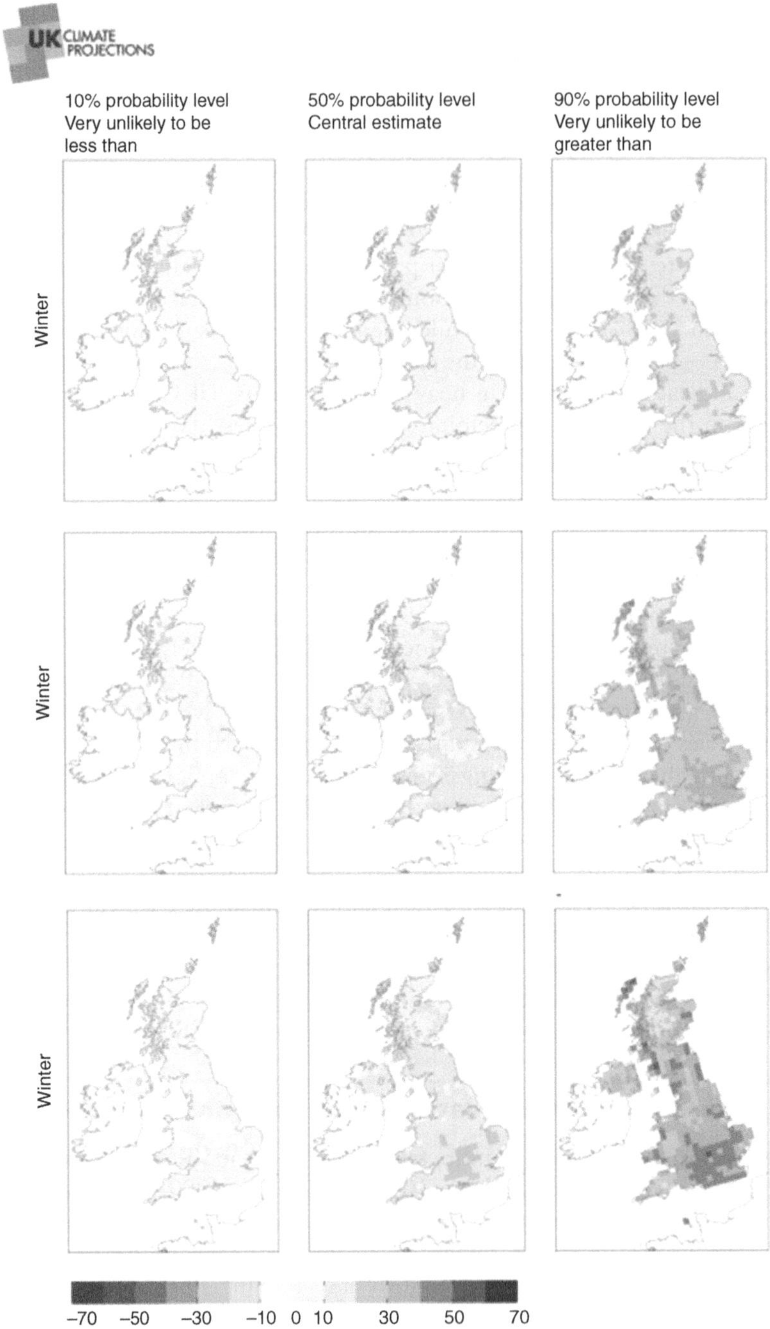

Figure 13.3 Change in precipitation (%) on the wettest day of winter for the 2020s (top row), 2050s (middle row), and 2080s (bottom row), UKCP09 Medium emissions scenario. Source: Murphy *et al.* (2009)/UKCP09 website.

Figure 13.4 Change in precipitation (%) on the wettest day of summer for the 2020s (top row), 2050s (middle row), and 2080s (bottom row), UKCP09 Medium emissions scenario. Source: Murphy *et al.* (2009)/UKCP09 website.

13.3.6 Sea level rise and storm surges

Sea level around the UK rose by about 1 mm/yr in the 20th century, corrected for land movement. The rate of rise in the 1990s and 2000s has been higher than this (Jenkins *et al.*, 2009b). UKCP09 also provides projections of sea level rise around the UK. Under the Medium emissions scenario, the range of absolute sea level rise (i.e. before land movements are included) is projected to be between 12 and 76 cm during the period 1990–2095 (Jenkins *et al.*, 2009b). Taking vertical land movement into account gives slightly larger sea level rise projections relative to the land in the more southern parts of the UK where land is subsiding, and somewhat lower increases in relative sea level for the north. The land movements are typically between −10 and +10 cm over a century. A wider range of scenarios was also developed for sea level rise and storm surges to be used for contingency planning and sensitivity analysis. Under the highest scenario sea level in London was found to increase by 90 to 190 cm (Lowe *et al.*, 2009). However the top of this range is considered very unlikely to be realised during the 21st century (Jenkins *et al.*, 2009b).

Future projected trends in storm surge height were found to be small everywhere around the UK, and in many places can be accounted for by natural variability. This means that changes in extreme sea level by 2100 will likely be dominated by increases in local mean sea level (Jenkins *et al.*, 2009b). Seasonal mean and extreme wave heights were generally projected to increase to the south west of the UK, reduce to the north of the UK and experience little change in the southern North Sea. Changes in the annual maxima are typically in the range −1.5 to +1 m (Jenkins *et al.*, 2009b).

13.4 CLIMATE CHANGE AND SLOPE STABILITY

It is generally accepted that creating slope stability models based on static information such as fixed recharge boundary conditions calculated from past behaviour is no longer acceptable for the long-term forecasting of responses of slope systems to climate change. Model development needs to incorporate observed seasonal and short-term behaviour of pore water pressures in slopes to create non-steady-state systems that can then be simulated with sequences of future climate scenarios relevant to the long-term responses of slope systems. Once a better, formalized, set of scenarios of climate-vegetation-slope interactions has been developed, it should be possible to develop flexible models based on UKCP09 probabilistic outputs of climate change. In turn, this could then lead to the development of solutions for the optimum design and management of slopes. The probabilistic forecasts delivered as part of UKCP09 are an exceptional resource, but to implement these data on a large scale into the modelling of long-term slope stability still requires much further research (Dijkstra and Dixon, 2010).

During a very wet period from November 2012 to January 2013, the UK experienced a large number of landslides (see Figures 13.2, 13.5 and 13.6; Pennington and Harrison 2013). Regional differences in deviations from the long term average are apparent on Figure 13.5 for annual rainfall for 2012; during this period Scotland was drier than average, while central and southern UK received in many places more than 135% of the long term average rainfall. A large proportion of the landslides

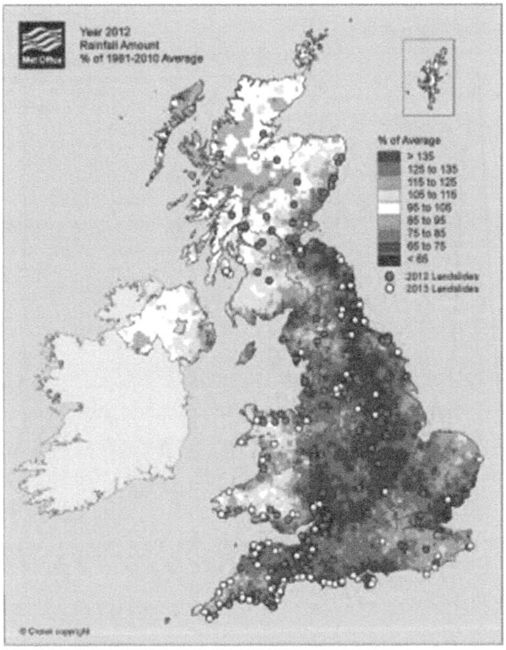

Figure 13.5 Landslides reported in 2012 (red) and Jan–Jul 2013 (yellow) and precipitation as a percentage of LTA [source: Met Office, BGS © NERC 2015].

(43%) reported during this time took place on engineered slopes such as road and railway embankments and cuttings. These slope failures are usually small-scale slumps or flows. Reported observations have shown that these are generally triggered by heavy precipitation and occur within a short period of time after prolonged heavy rain. This period occurs at the end of a very wet summer and autumn and a continuous period of above-average precipitation, which resulted in further incidents of unstable slopes.

Regional analysis of the antecedent precipitation signature required to trigger landslides in the SW England and S Wales highlighted two significant trends – a long-term gradual accumulation from 7 to 90-days before the events occurred and a higher intensity rainfall accumulation rate in the 7 days leading to failure, suggesting that a long period of precipitation is required to prepare the landscape for instability and that a final period of more intense precipitation is necessary to trigger landslides as has been clearly articulated by others in the UK context (e.g. Dixon and Brook, 2007; Collison *et al.*, 2000; Dijkstra and Dixon, 2010). The trends can be interpreted as a threshold envelope where antecedent precipitation plotted above the envelope is a signature for conditions 'wetter than usual' that can lead to multiple landslide events, and that signatures plotting below this trend represent conditions 'drier than normal', which would not likely result in precipitation-triggered landslides. Using a Poisson link function a subset of predictors was selected involving antecedent precipitation over 1, 7 and 90 days for the period from March 2006 until April 2013, and this in turn was used to compute the probability of at least one landslide occurring. In Figure 13.7 these

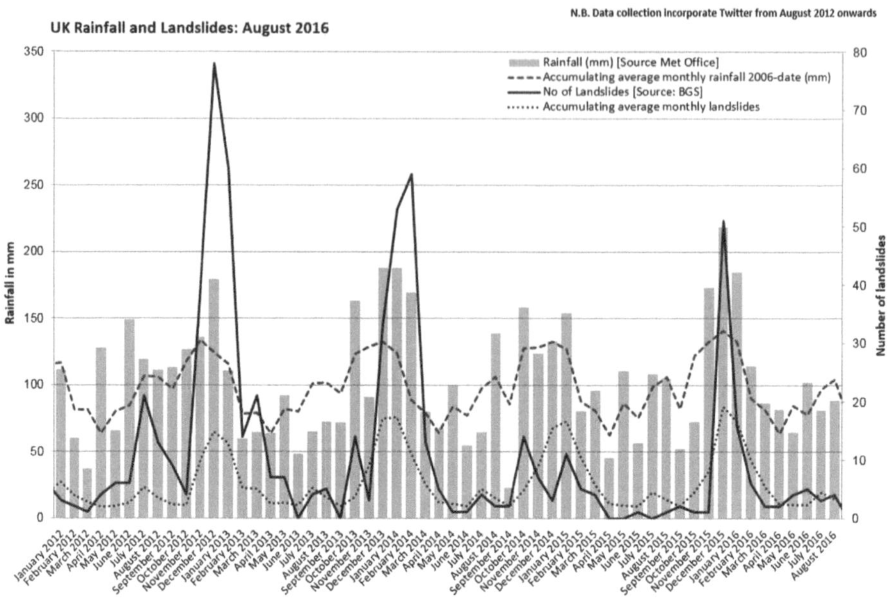

Figure 13.6 Precipitation and landslide incidence in the UK [sources: Met Office, BGS].

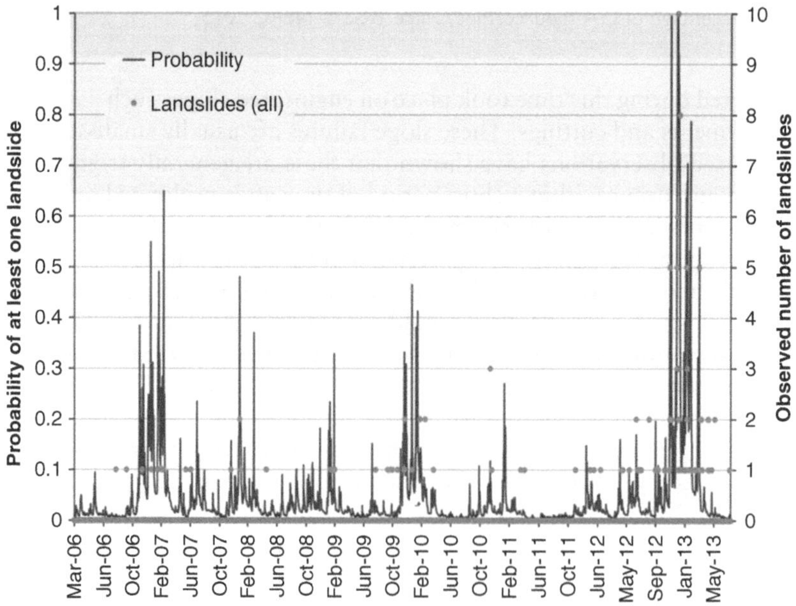

Figure 13.7 The probability of at least one occurrence of any type of landslide on a given date taking into account antecedent precipitation conditions (after Pennington *et al.*, 2014).

probabilities are plotted along with the observed number of landslides on each day. This analysis provides an opportunity to evaluate the changing probability of landslides occurring in response to antecedent effective rainfall. This analysis needs further work as it presently; (i) incorporates different modes of failure (although dominated by shallow translational events, larger slides are also included), (ii) covers a large area represented by regional rainfall characteristics and (iii) reflects historical patterns and is therefore based on the assumption that there is continuity in the modes of failure and the precipitation pathways and thresholds that trigger these events.

Mam Tor landslide, Derbyshire

Research on the deformation phases observed at the Mam Tor landslide in Derbyshire provides an example of a site-based investigation into climate change and landslide response. This study is one of only a small number that have attempted to quantify the possible impact of the forecasted changing climate on the incidence of landsliding in the UK (Dixon and Brook, 2007; Skempton *et al.*, 1989; Waltham and Dixon, 2000; Rutter *et al.*, 2003; Walstra *et al.*, 2007; Hodgetts and Rutter, 2013). The Mam Tor landslide is a kilometre-long feature in shales and sandstones with a backscarp height of some 70m. The initial slope failure occurred over 4000 years ago as a rotational landslide that has developed into a large debris slide at its toe. A road built across it nearly 200 years ago, and now closed, provides evidence of continued movement of the slide mass. Since the 1960s, the slope movements have been studied by several university research groups using a range of techniques including direct surveying and photogrammetry using aerial photographs. Current mean annual movement is up to 0.25 m; this increases greatly when winter rainfalls exceed thresholds of both 210 mm/month and 750 mm in the preceding 6 months. Such movements have a return period of 4 years, calculated since 1900. Dixon and Brook (2007) applied climate change forecasts produced by UK Climate Impact Programme (UKCIP 2002) to a statistically based rainfall threshold model and used to investigate the future stability of the Mam Tor landslide. For predicted changes in precipitation, it is shown that return periods for the exceedence of the combined current 1-month trigger and 6-month antecedent total rainfall thresholds could fall from 4 to 3.5 years by the 2080s based on the UKCIP medium-high scenario. All four climate change scenarios give a reduction in return periods. There is also likely to be a shift from instances of instability in late Autumn to Winter.

Debris flows are a particularly good example of events being triggered by excess rainfall and following a number of large disruptive events in 2004 further research was commissioned into this relationship. While the evidence of a general trend of global warming and increased climate instability is overwhelming, climate change models are subject to a number of uncertainties (Winter *et al.*, 2010a). It follows that the full range of possible climate change impacts on slope stability, both negative and positive, needs to be considered. It is thus helpful to draw conclusions from an analysis of a number of complementary data sets (Winter *et al.* 2010b; Winter & Shearer, 2013), including current climate, recent climate trends and climate change predictions. The observed recent trends in the climate of the recent past are relevant as well as the predictions of future climate change. Work by Barnett *et al.* (2006) described trends in climate across Scotland during the last century. The figures for 1914 to 2004 indicate a definite, but small, increase in the running annual average rainfall for the north and, particularly, the west of Scotland while figures for the east of Scotland are broadly stable.

The deterministic UKCIP02 (Hulme *et al.*, 2002) climate change predictions for Scotland show little predicted change in annual mean precipitation over the next few decades, with any change being within the range that can be attributed to natural variability. However, Galbraith *et al.* (2005) noted that even by the 2020s a distinct seasonal pattern can be discerned; while little significant change was predicted for the spring or autumn precipitation amounts, changes were found to be likely for the winter and summer. Winter precipitation was predicted to increase by between 10% and 15% in the east, with changes elsewhere being within the range of that which can be accounted for by natural variability. In the summer, decreases in the average precipitation were predicted to be widespread with only the far northwest seeing little change. By the 2020s Galbraith *et al.* (2005) indicated that the decreases are likely to be greatest in the south east of Scotland and could be as much as 20% lower than present day modelled levels.

The most recent UKCP09 climate change predictions (Jenkins *et al.*, 2007) allow probabilistic estimates of climate change (see Section 13.3). These probabilities represent the relative degree to which each climate outcome is supported by the evidence available (Anon., 2011; Walking-the-Talk, 2011) and it is possible to build-up a picture of the range of possible climate changes and the associated probabilities of occurrence (Winter & Shearer, 2013). Rainfall intensity is clearly an important issue for landslides and, particularly, debris flows.

UKCP09 is the most detailed data set and it suggests that there is a general propensity for rainfall events of greater intensity (storms) which are intrinsically linked with the triggering of debris flows, and that these rainfall events may become more frequent. Notwithstanding this it should be noted that the prediction of high intensity rainfall events is neither the purpose of climate models nor are such models best-suited to the prediction of such phenomena. Changes in rainfall coupled with increased potential evapotranspiration (particularly in the summer) and a longer growing season (leading to increased root water uptake) are expected to substantially affect soil moisture.

The soils that form the slopes subject to debris flow comprise a wide range of materials, albeit that periglacial, glacial, and post-glacial (including paraglacial) processes dominate their formation and subsequent modification. In terms of the composition of debris flows, the particle sizes range from coarse granular (including boulders) to fine cohesive with most sizes between potentially represented (e.g. McMillan *et al.*, 2005); observations of the characteristics of the materials involved in the August 2004 debris flow events in Scotland indicate such a diverse composition (Winter *et al.*, 2006). Both Winter *et al.* (2009) and Milne *et al.* (2009) acknowledged the importance of water-bearing soils, particularly peat, located on high, relatively flat ground as trigger materials for gulley-constrained debris flows. The materials that are later eroded, entrained and transported in the body of the flow may be of entirely different characteristics; often coarser-grained materials from morainic and related deposits are critical to the development of such flows.

Increased rainfall during the winter months seems likely to increase the prevalence of debris flow in Scotland. This is particularly so when considered in the context of the likelihood of more intense rainfall events.

The reduced soil moisture, as a result of both predicted temperature increases and rainfall decreases, during the summer and autumn may mean that the short-term stability of some slopes, particularly those formed from granular materials, may be enhanced by the suction pressures that may develop in such conditions. However, although climate models are not best-suited to predicting the frequency of storm events, the available evidence for climate change does point to an increase in short duration, high intensity rainfall events, even in summer. Soils under high levels of suction are vulnerable to rapid inundation (Toll, 2001), and a consequent reduction in the stabilising suction pressures, under precisely the conditions that tend to be created by short duration, localised summer storms.

In addition, non-granular soils may form low permeability crusts during extended dry periods as a result of desiccation. If these do not experience excessive cracking due to shrinkage, then runoff to areas of vulnerable granular deposits may be increased. However, the formation of drying cracks could lead to the rapid development of instabilities in soil deposits, potentially creating conditions for the formation of debris flows. It is thus clear that there are a number of climate sensitive potential failure mechanisms that may lead to the initiation of a debris flow.

Vegetation will also be affected by climate change. Lower overall levels and changed patterns of rainfall might be expected to increase the pressure on vegetation and thus to reduce its beneficial effect upon slope stability. Additionally, extended periods of exceptionally dry weather could potentially lead to wildfires and associated debris flows such as those described by Cannon *et al.* (2008). This seems less likely to be a factor in the relatively wet climate of Scotland; it seems more likely that an extension to the growing season will predominate in terms of the effects of vegetation – a factor that is most likely to have a positive effect upon instability. However, the possibility of vegetation desiccation cannot be dismissed during prolonged summer dry spells.

The available climate change predictions present a picture that tends to suggest that debris flow frequency and magnitude may well increase in Scotland, at least in the winter months. A rather more complex picture emerges for the summer months, but one possible outcome is that the frequency of events may decrease whilst the magnitude of those that do occur may increase.

The debris flow risks relate primarily, but by no means exclusively, to transport infrastructure. The effects of climate change seem likely to increase the frequency and/or the magnitude of debris flow hazards. This increases the elements at risk (i.e. the infrastructure) and their vulnerability is likely to increase in line with road traffic and rail passenger growth. Debris flow risk is thus likely to increase as a result of these factors. However, this must be set against planned network upgrades that afford the opportunity to reduce such hazards and risks at the outset. This approach of infrastructure upgrade seems to be unlikely to be viable across the network; it certainly is likely to have only limited applicability much beyond the strategic road network for example. In such cases route-based exposure management schemes and localised hazard mitigation schemes will be necessary. The complex interplay between changing land use and the effects of emerging technology on work patterns will change the way in which society addresses such infrastructure and may also introduce new infrastructure that may itself be at risk from debris flow hazards.

A83, Rest and be Thankful landslides

Fig. A

Observations of the south-westerly facing slopes of Beinn Luibhean (Figure A) since 1989 previously led to the simple conclusion that, typically and on average, one debris flow reached the A83 strategic road in the area each year. However, that rate of erosion has increased substantially in recent years and they have now become much more frequent. The area is known as the Rest and be Thankful and records indicate that 10 events (or closely spaced groups of events) occurred between October 2007 and March 2014, a period of around 6.5 years, and that each of these events had an impact on the strategic road and was reported by the company that operates the road. The most recent event in March 2014 (Figure B), closed the road for several days. The management and mitigation at this site has focussed on three main areas (a) warning during periods of higher hazard and risk (Figure C; Winter *et al.*, 2013b); (b) the use of debris flow fences to reduce the likelihood of material reaching the road (Figure D); and (c) the diversion of traffic onto a temporary route, the Old Military Road (OMR; Figure E), which is located at a lower level on the hillside and for which the likelihood of material reaching the road in areas both with and without debris fences is lower.

Fig. B

Fig. C

Fig. D

Fig. E

13.4.1 Climate change and coastal slopes

The nature of the UK coastline is such that climate change might be expected to cause increases in the rate of mass movements. In particular, sea level rise could be accompanied with higher levels of basal under-cutting, which could promote instability, as could increased wave heights (which will increase the area affected by wave erosion and wetting) or wave energies. In addition, higher rainfall intensities could lead to elevated levels of surface erosion and to higher pore water pressures, whilst higher temperatures may increase rates of weathering.

Most of the work on UK coastal landslide processes in the context of climate change has occurred in soft rock areas. A number of approaches have been taken, of which statistical models of the effects of changing the wave energy climate have been the most popular. Thus, for example, Hackney *et al.* (2013) used global climate models to examine potential changes to erosion rates associated with sea level rise and increased wave energy over the next century. On the Isle of Wight, on the south coast, rates of erosion were found to increase by about 0.50 m/year above background rates of 0.30 to 0.70 m/year, whilst on the east coast rates of erosion were found to increase by about 0.30 m/year above background rates of 0.80 to 1.90 m/year. This suggests that climate change will dramatically increase the rate of coastal landsliding in soft rock coastal areas in the UK, but that the effects will vary greatly according to the wind, wave and tide environment, and in some areas can be offset by glacio-isostatic uplift (e.g. Lambeck, 1995).

An alternative approach to modelling the behaviour of soft cliff systems is to use a slope stability approach, in which the inputs are varied to simulate the climate changing. Thus, Brooks *et al.* (2012) modelled the behaviour of soft cliffs in eastern England using an idealised combined hydrological and stability model of a section of cliff that had been investigated in detail. Analysis of historical data suggested that the rate of retreat has increased in this area in recent years; the modelling allowed attribution of the effects to the complex interaction of environmental forcings. The study identified the key environmental thresholds that lead to increased rates of instability, and demonstrated that anticipated changes in climate will lead to an increase in the frequency at which these thresholds will be exceeded, thus driving more landslide events.

A third approach has been to measure directly the rates of retreat of soft rock cliffs, and to try to relate changes to variations in climatic conditions over the same period to infer future behaviour. Thus, Brooks and Spencer (2012) found that the rate of retreat of soft rock cliffs in eastern England was highly sensitive to the rate of sea-level rise, with higher sea-level rise rates being associated with much higher rates of erosion. The dataset was modelled, allowing future rates of landsliding (in this case measured by the rate of sediment production) to be simulated. This suggested that the process rate might rise by as much as an order of magnitude by the end of the 21st Century when compared with early 20th Century rates.

Whilst most models have focused on the role of increased wave energy, Castedo *et al.* (2012) demonstrated that increases in groundwater level may also play a crucial role in promoting instability in coastal cliffs. This suggests that further work is needed on the combined effects of changes to sea level, wave heights and groundwater levels in coastal environments, and that the outcomes are likely to be highly complex.

Approaches that focus directly on landslide processes, rather on a proxy such as erosion rate, are much less common. Moore *et al.* (2010) examined the large Ventnor landslide system in southern England using detailed monitoring to identify the relationships between movement rates and weather conditions (Figure 13.8). In common with Brooks and Spencer (2012), they used these relationships and projections of future climate to extrapolate likely future behaviour, finding that, in the words

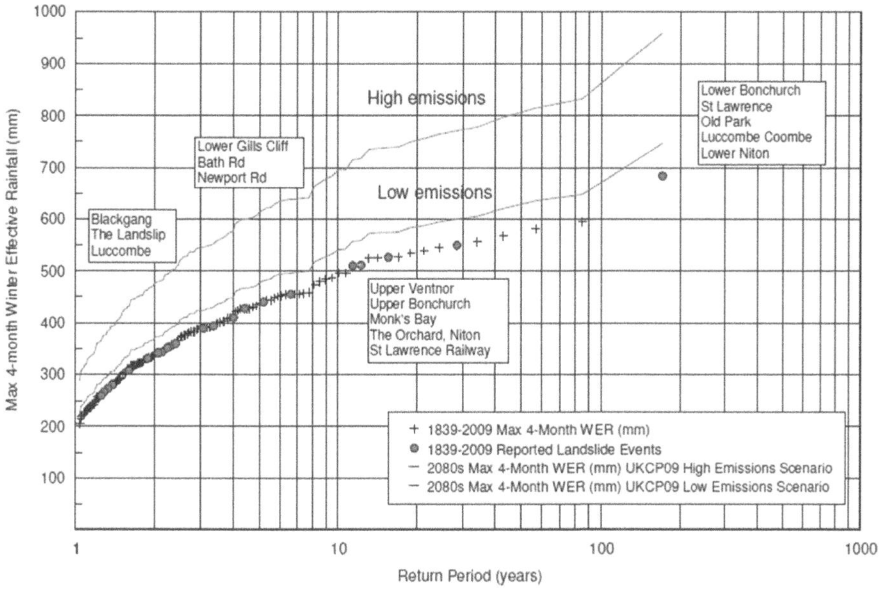

Figure 13.8 An analysis of the potential effects of climate change on the return period of a number of landslides on the Isle of Wight (from Moore *et al.*, 2010).

of the authors, "climate change poses a very real threat and significant challenge to the future management and mitigation of the ground instability risks in the Ventnor Undercliff, which may be echoed for other similarly marginally stable ancient landslides in the UK". However, it should be noted that uncertainties in climate models at this scale in this sort of environment are so large that there is considerable range in the projections generated.

In harder rock areas there is less research and data available. Again, various approaches have been used varying from direct measurement to modelling and even process-based investigations. In the latter camp, Lawrence *et al.* (2013) demonstrated that sea saltwater weakens chalk samples by up to 55% compared with identical inland specimens saturated with non-saline groundwater. They hypothesised that the growth of salt crystals in the pore spaces in the chalk progressively disrupts the structure of the rock, inducing weakening. This process may be a major factor in the occurrence of large-scale collapses in the chalk. Lawrence *et al.* (2013) suggested that the effects of increased storminess may be that the wetting area of the chalk will increase, in turn raising the amount of salt crystals in the pore spaces in the rock, and thus increasing the rate of collapse events. This is in contrast to observations in coastal clay slopes where salt spray has been linked with enhanced shear strength and reduced mobility (Brunsden 1996; Brunsden and Edmonds, 2010).

Process-based understanding of cliff behaviour has been revolutionised by the availability in the last decade of terrestrial laser scanners, which permit detection of changes to slope morphology at high levels of resolution over large areas. Lim *et al.* (2011) used laser scanning and seismometers mounted on the cliff top to measure energy delivery to coastal cliffs, and the resultant occurrence of rockfalls, on hard rock cliffs in eastern England. They found that inundation time is a factor that determines the rate of rockfall activity. This suggests that sea level rise, by increasing inundation time, is likely to result in higher rates of mass movements in such environments. However, the relationship was found to be far more complex than had been anticipated, suggesting that there is a need for increased levels of data collection in a range of environments. Barlow *et al.* (2012) proposed that power law distributions may present a tool through which this can be achieved, whilst Norman *et al.* (2013) presented a more detailed analysis of seismic datasets recording energy delivery to coastal cliffs, providing a key approach to collecting such data.

Climate change and coastal landsliding at The Isle of Wight

The Great Slip, a landslide in the Ferruginous Sands Formation, Compton Bay, Isle of Wight.

It is forecast that climate change will lead to a gradual fall in summer precipitation and a rise in summer temperatures. This is likely to have a consequence for soil moisture; this may reduce up to 40%. However, Moore *et al.* (2010) argue that this is unlikely to affect deeper groundwater flow significantly and that therefore winter rainfall is of much greater concern when analysing the reactivation of mass movements in the Isle of Wight. They plotted the return period of reported landslide events against the 4-month winter effective rainfall and then projected the forecast change of this relationship as climate change progresses. It was observed that, for landslides such as Upper Ventnor, the return period reduces from 10 to 8 years (Low Emissions Scenario) or even to just over 2 years (High Emissions Scenario and the more likely outcome given present developments; Figure 13.8). In coastal environments, additional complexities exist that determine the deformation patterns of landslides. For example, Hutchinson *et al.* (2002) analysed the deformation patterns of the St Catherine's point landslide,

Isle of Wight. They concluded that both coast erosion (influenced by storminess and sea level rise) and effective rainfall interplay to generate movements of typically 35 to 90 mm/year, but with great inter-annual fluctuations, largely driven by the episodic nature of coastal recession.

13.4.2 Climate change and engineered slopes

The UK's transport infrastructure is one of the most heavily used in the world; the UK rail network takes 50% more daily traffic than the French network; the M25 between junctions 15 and 14 carries 165000 vehicles per day; and London Underground is Europe's largest metro subway system (Glendinning *et al.*, 2009). The performance of these networks is critically dependent on the performance of cutting and embankment slopes which make up £20Bn of the estimated £60Bn asset value of major highway infrastructure alone. However, London Underground is the world's oldest underground system and much of the rail network is over 100 years old. Many of these slopes are now suffering from excess deformation and high incidents of instability (which is showing evidence of increasing with time). These problems cause delays and subsequent economic loss to UK industry, and the general public, and pose a safety hazard. Derailment from slope failures is the greatest risk faced by the railway infrastructure. Between August 2003 and December 2009, a total of 429 incidents of failure of Network Rail earthworks were recorded across the rail network, causing 66 days in delays. It is not only safety but economics that is important: emergency repairs cost 10 times that of planned works. Therefore the impact of reducing delays caused by slope failure will be highly significant in both direct economic and indirect social and economic terms. It is thus clear that engineered slope stability varies in a spatial context and is driven by a hierarchy of processes and parameters that respond to climate change (Dijkstra and Dixon, 2010). Current understanding of the impact that age-related degradation and climate change will have on geotechnical transport infrastructure, both now and in the future, can only provide generalised advice on maintenance and cannot be used to justify any repair option other than hard structural solutions. Therefore further research is (and has been) funded by research councils in the UK. Two such projects are FUTURENET (Bouch *et al.*, 2011; Dijkstra *et al.*, 2014) and iSMART, both large collaborative projects stemming from network-based collaborations that were facilitated by the multi-disciplinary CLIFFS (CLimate Impact Forecasting For Slopes) network (Dijkstra and Dixon, 2010). These projects concluded, among others, that in concept quantifying the resilience of infrastructure slopes is possible. However, the present understanding of processes, properties and parameters is still incomplete and carries large uncertainties. There is a need to continue developing the physical-based models that drive the transport infrastructure condition assessments and this requires:

- more detailed and better accessible datasets. Most datasets are currently affected by incomplete and missing data, limiting their usefulness for corridor-wide physical process modelling. It has become apparent that data across several agencies has become fragmented over time and has highlighted a need for a transparent approach to asset data management and data accessibility to enable more detailed analysis of transport infrastructure asset conditions.

- further investigation of climate forecasts (e.g. Kay & Jones, 2012) looking at higher resolution, spatial coherence, and downscaling extremes
- improvements in modelling capability, including modelling platforms capable of managing large volumes of data generated by sophisticated, high-resolution physical-based models (e.g. Clarke *et al.*, 2006; Smethurst *et al.*, 2006, 2012; Davies *et al.*, 2008; Rouainia *et al.*, 2009).
- development of detailed deterministic, physical-based models capable of dealing with dynamic equilibria and threshold exceedence
- further analyses of complex feedbacks and interactions between physical processes.

The impact of earth structures failure has increased in terms of loss of serviceability (insurance claims), human casualties, and reconstruction costs to the point where we need to reconsider our understanding of the whole life-cycle and failure process. Seasonal wetting and drying of engineered slopes results in progressive cycles of heave and crack development leading to degradation, creep, and the acceleration of near-surface moisture ingress and drying (egress). These processes contribute to the deterioration of condition, which is defined by property ranges and their distribution within the slope. Instability in clay-rich slopes typically occurs due to strength reductions in response to moisture content and pore pressure changes which, in turn, are often largely driven by weather event sequences.

Defining instability using surface manifestations, such as subsidence or deformation has reinforced responsive monitoring solutions (or applications) based upon surface observation (Figure 13.9). However, these manifestations are events that occur within a longer term, instability process associated with progressive, subsurface geotechnical property changes (Skempton, 1964; Chambers *et al.*, 2014; Glendinning *et al.*, 2014); from which failure pre-cursors and triggers may be identified.

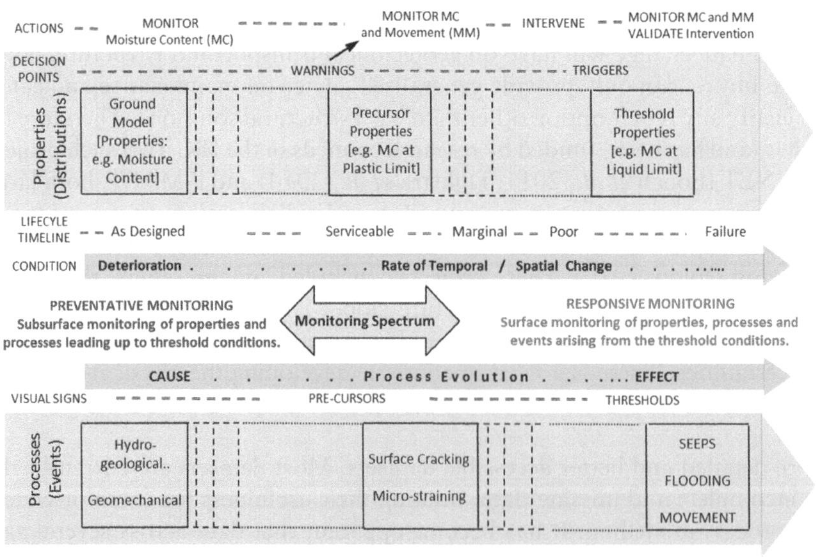

Figure 13.9 An illustration of the monitoring spectrum between preventative and responsive actions.

Choice of monitoring solutions (and technology) includes consideration of the condition of the slope (or asset) relative to its whole life cycle and hence, establishment of the state and rate of deterioration. Maintenance strategy will determine the optimum monitoring solution and its operational point within the preventative to responsive monitoring spectrum shown in Figure 13.9. Advances in computing, remote surveying technologies (including terrestrial and air-based laser scanning, radar interferometry) and geographic information systems have led to responsive solutions dominating monitoring in the UK. Responsive solutions such as movement (effect) monitoring using inclinometers or tiltmeters of geomorphological features are routinely used to design works to mitigate the effects of instability or to renew infrastructure. However, early intervention implicitly requires a preventative solution monitoring causal processes leading to the identification of instability pre-cursor and threshold conditions. This enables a more expansive intervention timeframe, but is more likely to require holistic monitoring of subsurface processes to predict future failure outcomes.

Groundwater movement and stiffness changes can be characterised using subsurface, time lapse, volumetric images provided by electrical or seismic monitoring (Chambers *et al.*, 2014; Donohue *et al.*, 2014; Glendinning *et al.*, 2014), information that can both inform and validate intervention solutions. Whether the solution is preventative or responsive, the key challenge to all monitoring relates to the establishment of a single, integrated ground model that is consistent with the geology, geomorphology and all the process (cause) and the response (effect) data streams. This procedure is becoming collectively known as 'upscaling' and methods are becoming more refined due to the introduction of dynamic, volumetric images providing more insight into subsurface processes. These images also provide vital updates and feedback relating to the property distributions that enables further linkage between integrated monitoring and numerical modelling approaches. A consistent and comprehensive ground model that is developed from site observation is vital as the basis for forward modelling future scenarios using real-time (or near real-time) monitored data to constrain the modelled outcomes, particularly where additional uncertainties such as the potential effects of climate change need to be considered.

Shrink-swell effects on engineered slopes – Network Rail

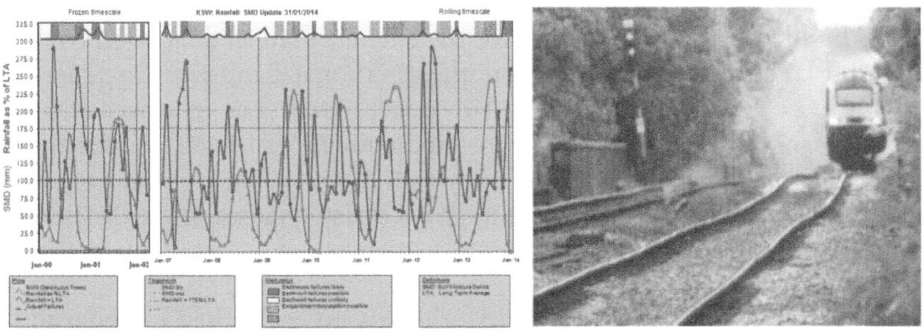

Precipitation and soil moisture deficit as indicators for the triggering of different alerts for railway embankments in Kent, Sussex and Wessex (diagram and photograph courtesy Graham Birch, Network Rail).

An example of the use of precipitation and soil moisture data to inform asset managers about the state of engineered earthworks is provided by 'Earthwork's Watch', a monthly bulletin compiled and circulated by Graham Birch of Network rail. It uses a rainfall expressed as a percentage of the Long Term Average (LTA) for each month and combines this with an assessment of soil moisture deficit (SMD). Both are used to produce a traffic light warning system that reflect 'alert' conditions for very wet sequences where earthwork failures are likely and for drier conditions where swell shrink deteriorates track quality. The Rainfall/SMD charts provide a 'rolling' plot of rainfall and ground moisture trends over the past 7 years for comparison against the exceptionally wet winter of 2000/01, which remains 'frozen'. Separate charts are given for Kent, Sussex, Wessex and Anglia with a fifth chart for the Kent Sussex Wessex (KSW) overview. Elevated SMD can lead to desiccation of clay-cored embankments, with associated deterioration in track quality. The Ground Condition 'Alert status' is shown in the 'traffic light' bar, above the graphs, set by the following criteria; the threshold for SMD (red line) is set at 50 mm for wet weather and 200 mm for dry weather. This reflects the actual earthwork failures (violet line in 'Alert status' bar) in winter, and poor track quality experienced through desiccation of clay embankments in autumn, respectively. Rainfall (blue line) is shown as % of long-term average (LTA), which is normalised to a straight line (broken black line). The rainfall threshold is set at 175% of LTA. As this analysis continues it builds up vital information for evaluation of network response to climate drivers (and possible adjustment of trigger thresholds).

13.5 LANDSLIDE HAZARD AND RISK MANAGEMENT AND ADAPTATION STRATEGIES

13.5.1 Regulatory framework

The legislative and planning framework for landslides is developed around an understanding of policy makers that landslides in the UK tend to be small, localised problems that rarely result in fatality (Gibson *et al.*, 2013). A National Landslide Database established the first national evidence base for the extent of landsliding as a problem with nearly 9000 landslides entered onto the then National Landslide Database. (Foster *et al.*, 2012; Lee and Moore, 1996). Subsequent work has increased this number to over 16 500. Previously, records were predominantly of landslide deposits rather than events (Foster *et al.*, 2008) but now events are being captured regardless of whether or not landslide deposits remain. Work is also ongoing at the British Geological Survey to establish links between event occurrence and climate change (Pennington *et al.*, 2009).

As discussed by Gibson *et al.* (2013), there is no centralised mechanism for the management or mitigation of landslides in the UK. In England, ground movement problems fall under Common Law and the primary responsibility is with the landowner (Department of the Environment, 1996). In terms of future planning, the Town and Country Planning Act (1990), regulates the development of land and its powers are exercised through Local Authorities who also have limited powers related to safety through the Housing Acts (various) and Building Regulations Act. Similarly, the Construction (Design and Management) Regulations (Anon., 2007) are relevant to safe construction. Further considerations of public safety are made in the specific laws related to coal mining, construction regulation; emergency planning and personal liability (Gibson *et al.,* 2013 and references therein).

An effective landslide hazard strategy requires two components; an awareness and understanding of the hazard and a policy framework to implement appropriate planning and management strategies. In the UK, landslides are recognised as a major hazard and the planning context is set out in the government Planning Policy Guidance Note 14 (PPG14) and its Annexes (Anon 1990, 1994; and see also Jones and Lee, 1994). Even though the PPG14 has been replaced by the 'National Planning Policy Framework 2012' the main guidance for planners is still derived from PPG14. The Planning Note clearly identifies that ground movements can lead to loss of life and expensive remedial action or complete loss of functionality. It distinguishes three broad categories:

1. "the effects of underground cavities; these may be of natural origin or due to mining or to civil engineering works;
2. unstable slopes; these may be natural, in both coastal and inland situations; or man-made, whether excavated, as in quarries or cuttings, or constructed, as in tips and embankments;
3. ground compression; this may be of natural origin due to peat, alluvial, estuarine or marine soils; or due to human activities, e.g. made ground, landfill or restored opencast mines; and ground subject to movement due to shrinking and swelling clays." (PPG14; paragraph 13).

Advice is given to consider the following acts:

- Coal Mining (Subsidence) Act 1957
- Coal Industry Act 1975
- Health and Safety at Work Act 1974
- Occupiers Liability Acts 1957 and 1984
- Town and Country Planning Act 1971
- Building Regulations and the Housing Acts

Current planning policy states that policies and decisions should also ensure that sites are suitable for its new use taking account of ground conditions and land instability, including from natural hazards. Local authorities are not responsible for investigation of the ground conditions of any particular development site unless they propose to develop it, but they have the power and the duty to take into account the stability of the ground when deciding a planning application (Town and Country Planning Act, 1971, Building Regulations and the Housing Acts; Dashwood *et al.*, 2014).

In 2012, in England and Wales specific and detailed Planning Policy Guidance Notes (Department of the Environment, 1990, 1996, 1996a; Department for Transport, Local Government and the Regions, 2002) dealing with planning issues in areas affected by instability were replaced with briefer guidance (12 paragraphs) which basically require Local Authorities to insist upon an appropriate ground investigation prior to development in areas of instability. No guidance is provided as to what is appropriate. Although some have cautioned against this approach (for instance Brook, 2007; Brook & Marker, 2008), it is not known as yet what the implications of less guidance (and supposed development of local reliance) will have. It is noted however that climate change is not mentioned in current guidance. Some local districts, notably the Isle of Wight have produced their own guidance documents for landslide management, incorporating consideration of climate change (McInnes, 2000, 2007).

In Scotland, The Town and Country Planning (Scotland) Act 1997 is the main document governing land use and development. Broader national guidance on unstable land Scotland has also been produced (Scottish Government 2014), with greater emphasis on local knowledge of landslide problems and solutions. As in England and Wales, guidance no longer specifically includes reference to landslides and climate change The Scottish Government does however, provide specific advice for peat slides (Scottish Executive, 2006), and debris flows affecting the road network (Winter *et al.*, 2005, 2009), both of which consider landslides and climate change to be an important issue.

Across the UK, there is some compulsion for landowners or industrial operators to consider land instability and climate change in their operation. Under the above framework, primarily through safety legislation, there is a requirement for all infrastructure operators and those in certain industries to consider risks to their operation, and to third parties resulting from landslides now and in the future.

Private losses to landslide events are currently covered for most individuals by their home insurance. In the UK, this is generally combined with other ground movements and termed 'subsidence'. Although figures are difficult to obtain, it is thought that insurance pay-outs in the UK average £300–400 m per year (Culshaw & Harrison, 2010). Research currently underway by the authors indicates that other financial penalties (as measured by variance in household transaction value) in landslide prone areas may actually be negligible in the medium to long term.

13.5.2 Adaptation strategies; The Natural Hazards Partnership (NHP)

Through the consortium of public bodies that forms the Natural Hazards Partnership (NHP; mainly government departments and agencies, trading funds and public sector research establishments;) the partnership aims to build on partners' existing natural hazard science, expertise and services to deliver fully coordinated impact-based natural hazard advice provided for civil contingencies and responder communities and governments across the UK. It provides a platform in which the possible effects of climate change can be addressed. This joined-up research into multi-hazard assessments based on agreed hazard impact matrices will result in consistent, better presented advice to the disaster planning and management community and thereby facilitate the development of improved natural hazard related 'prevention and preparation' strategies. Since its creation in 2011, the NHP has achieved the following:

- Improved cross hazard discipline working resulting in more joined up scientific advice and analysis to UK Government Departments including the provision of scientific advice to support the review of the National Risk Assessment in 2011 and 2012 and coordinated NHP input into the Commission on Climate Change's work on the Adaptation Sub-Committee's report on the UK's infrastructure resilience

- Daily Hazard Assessment (DHA) and underpinning hazard impact matrix providing a coordinated overview of natural hazards likely to affect UK over the next 5-days, available to 12000+ users from the resilience community through 'Hazard Manager'
- Hazard Impact Modelling (HIM) collaborative research and development programme to 'pull through' integrated hazard science and vulnerability assessments into hazard impact services – Wind-induced vehicle over-turning hazard impact model prototype developed with a dedicated focus to develop a near operational prototype surface water flooding hazard impact module by December 2015
- Establishment of the Hazard Impact Production System (HIPS) project comprising an integrated, community of open platforms whose collective outputs are delivered to partners – customer specific visualisation systems
- Establishment of a Science Strategy Group to develop an overarching NHP science strategy that looks across all hazards in a much more coordinated way to better define NHP science needs and priorities
- Establishment of the NHP Advisory Group to maintain user focus
- Development of natural hazard Science Notes for landslides, drought, snow/ice, wildfire, wind and inland flooding (space weather, air quality, and coastal flooding in the pipeline) providing more detailed information about each hazard and likely impacts – available through the NHP micro-site
- Creation of the NHP micro-site at http://www.metoffice.gov.uk/nhp/ providing top level information about the NHP

13.5.3 Changes in vulnerability and risk

The nature of risks, vulnerability and adaptive capacity all change over time in response to both climate and socio-economic drivers. From a historical perspective it is clear that social and economic trends can have a far greater influence than natural climate variability on many risks. Insured losses from weather disasters have generally risen over the last few decades (Munich Re, 2013), but these headline figures include factors such as a greater population, higher GDP and increased development along vulnerable European coastlines, as well natural climate variability. Improvements in services (health, water, energy, etc.), advances in technology and good regulation may reduce risks. Increased population size, uncontrolled development and social changes may mean greater exposure to hazards, increases in vulnerability and greater pressure on the natural environment. Increasing wealth will always inflate damage and disruption costs.

The UK Climate Change Risk Assessment (CCRA) that appeared in 2012 listed socio-economic changes that will influence vulnerability to a changing climate as one of the remaining key uncertainties over the magnitude and timing of climate change impacts in the UK. The CCRA used a standard set of population projections to make an assessment of future risks across a number of sectors. In order to capture the wide range of uncertainty in future population, three projection variants were used (Low Population, Principal and High Population) which were derived from the Office of National Statistics (ONS) 2008-based National Population Projections. These variants combine high and low assumptions in the main demographic components of change. The Principal Population Projection provides the baseline for future growth and indicates a rise in UK population to 69 million in the 2020s, 78 million in the 2050s and 86 million in the 2080s. Under the Low (High) Population projections these numbers are 66 (72), 66 (90) and 63 (113) million, respectively. Recent mid-year estimates show that the UK population (61.4 million in 2008) is increasing with the population getting older (currently 16% of the UK population is over 65 years of age), broadly in line with the ONS Principal Projection. A larger population overall may place greater pressure on the environment, in terms of land use and consumption of natural resources, and greater pressure on health, water, energy, transport and waste services to meet higher demands (CCRA, 2012).

Projections of population and socio-economic trends are ideal for considering risks in the short and medium term (to 2050s) but are less helpful for longer term assessment. This is because the population is just one of many social and economic factors that may influence risks in the long term and other drivers cannot be quantified in the same way as population. As an alternative approach the CCRA sector assessments considered a number of other dimensions, providing some qualitative discussion on how these may affect future risks. These dimensions include population needs and demands (e.g., pressures on the country in terms of housing, education etc.), global stability, distribution of wealth, consumer driven values and wealth (i.e., moving towards sustainable or unsustainable consumption), level of government decision making (i.e., local versus national), and land use change and management (relating to aspects of urbanisation versus rural development). For each detailed risk considered in the CCRA, the relevance of each of these socio-economic dimension and the effects of the extremes of each dimension were briefly discussed in the report, but with the exception of population growth, societal change was otherwise not included in the assessment of future risks (CCRA, 2012).

13.6 CONCLUDING REMARKS

Climate change models predict an increasing rate of change of seasonal and inter-annual variations in precipitation and temperature during the next 100 years. As a consequence, the evapotranspiration-precipitation balance will change. This will affect the hydrological environment governing the majority of mass movements in the UK through, for example, changes in antecedent pore pressures and alteration of trigger event magnitudes. It is recognised that this, in turn, will lead to a change in the frequency, distribution and mode of landsliding in the UK. Landscape sensitivity, in terms of the degree to which it can cope with these rates of change, should therefore be considered as a consequence of combined changes in the triggers (e.g. precipitation events) and preparatory factors (e.g. the antecedent groundwater conditions). It is clear that climate change will significantly modify these and other controlling variables (such as land-use, vegetation cover and soil water chemistry).

To date, studies have assessed the consequences for triggering first-time landslides at individual sites (both natural and constructed). However, the general process-response issues are still not well understood, but this is being addressed by increasing communication between groups involved in research, consultancy and the stakeholders. Current limited understanding and the dearth of data has major implications for both land use and construction. At present, planners use, for example, geological maps produced by the BGS that show areas of past slope instability to help inform decisions on future land use. At the detailed design stage of a project, slope stability assessment is carried out using past records of groundwater levels and/or measurements made during a site investigation (often over a limited time period). These assessments of stability are valid only as long as the groundwater conditions are relevant (i.e. a steady state is assumed). With the projected climate change scenarios this is not a valid assumption because the time period over which designs and land use decisions are being used can readily exceed 50 years. It is thus questionable whether the information is relevant at best, or misleading at worst. Forecast scenarios of slope instability and distribution must therefore be developed. These must be communicated as widely as possible to make stakeholders and the general public aware of the scientifically feasible refinements in confidence level and potential societal impact. Improved understanding of slope dynamics under changing climate conditions will assist the development of appropriate engineering solutions for the economic design and the optimum management of constructed and natural slopes (Dijkstra and Dixon, 2010).

13.6.1 Uncertainties still to be addressed

It can be concluded that the effects of climate change on slopes in the UK are, as yet, uncertain. Climate change may lead to a deterioration of slope stability, for example if we were to follow a scenario where drier summers followed by heavy autumn rain leads to an increased likelihood of deformation in plastic clays. However, it is also plausible that higher evapotranspiration could lead to a reduction in infiltration and more intense rainfall could lead to more runoff, thus resulting in drier, more stable slopes. Until a comprehensive set of narratives is formulated it is only possible, at this stage, to conclude that, as a consequence of these observations, failure mechanisms active in UK slopes are likely to change. Whether this will lead to a higher frequency and magnitude of reactivation of existing unstable slopes or the initiation of slope movement at hitherto stable sites still remains to be determined. Due to the episodic nature of slope stability events and a limited understanding of parameters and processes to which these events respond, current levels of uncertainty are so great that long-term forecasts are still of limited use to those responsible for the management of slopes. There is therefore an urgent need for further research to achieve a better understanding of the following (summarized from Dijkstra and Dixon (2010))

- Spatial and temporal (including strain dependence) characterization of soil parameters. This requires comprehensively characterized sites with long-term datasets.
- The enhanced ability to convert site-specific data into usable regional-scale information.
- Characterization of permeability, in particular near the surface, and the links with infiltration and hence pore pressure variations in the slopes. There is an urgent need for a fundamental understanding of the relationships between temporal and spatial pore pressures in complex geological environments and the link between these and slope movements.
- Collation of historical datasets is needed to aid understanding of processes and to test hypotheses in the field.
- The effect of vegetation, generally poorly addressed in site investigation and project development, on temporal and spatial pore pressures, and their potential mechanical benefits also require further work.
- Key critical climate scenarios and the integration of these into a multi-hazard systems model.

13.6.2 Communicating for adaptation and design in a changing climate

Communicating the message effectively is of paramount importance. Traditional deterministic models appear to have worked well thus far. However, a case is made that the community involved in addressing slope stability needs to put a greater focus on probabilistic approaches, particularly if long-term forecasts are to be achieved that are compatible with the best available science of climate change (e.g. UKCP09). If research produces forecasts that indicate change in slope stability, the underpinning science

must be robust to inform slope managers how they could adapt their assessment and management strategies to ensure long-term resilience. As stated in the introduction, communication is currently poor between the various groups that are actively involved in slope stability assessments. The activities of multi-disciplinary networks (such as CLIFFS) have addressed this to some extent. However, there is still some way to go to achieve progress in the effectiveness of communication of scientific developments to the wider public, including stakeholders such as infra-structure asset managers. Continuing the dialogue is therefore imperative to deliver the insights required to shape an environment that is as resilient as can be reasonably expected from the potential long-term changes in slope stability as these are driven by climate change.

It is clear that the predisposition of slopes to excessive deformation varies in a spatial context and is influenced by a range of variables forming part of a hierarchy of processes and parameters that are largely interdependent and respond at different time scales to the effects of climate change. At many levels, uncertainties and limitations still exist that require further investigation through laboratory testing, field investigations and computer modelling. These uncertainties require quantification (and reduction), and the temporal and spatial variability at each level needs to be clearly defined. This would range from achieving a better understanding of relevant processes, via assessments of seasonal and inter-annual slope surface states (including crack depth, permeability and vegetation cover) to modifications of the hydrogeology driven by changes in the long-term precipitation-evapotranspiration balance. Coupled with additional information on the spatial and temporal changes in susceptibility this should result in improved analyses of long-term forecasts of slope stability in a context of climate change. It must be realized that a successful outcome can be achieved only through multidisciplinary research and a continuing engagement between science-based and engineering-based approaches (Dijkstra and Dixon, 2010).

ACKNOWLEDGEMENTS

This report captures information from a large number of people and their inputs and contributions are gratefully acknowledged, including the landslides response team at the BGS. The authors are also aware that the community involved in dealing with landslides and climate change in the UK is much greater and more diverse than could be captured in this report and apologise for any omissions. We are very grateful for the constructive comments on earlier drafts of this report that we received from Martin Culshaw and Vanessa Banks. Dijkstra, Jenkins, Dashwood and Gunn publish with permission from the Director of the British Geological Survey.

REFERENCES

Alexander, L.V., Zhang, X., Peterson, T.C., Caesar, J., Gleason, B., Klein Tank, A.M.G., *et al.* (2006) Global observed changes in daily climate extremes of temperature and precipitation. *Journal of Geophysical Research*, 111, D05109.

Anonymous. (1990) *Planning Policy Guidance 14: Development on Unstable Land.* London, Department of the Environment, Welsh Office, HMSO.

Anonymous. (1994) *Planning Policy Guidance 14 (Annex 1): Development on Unstable Land: Landslides and Planning*. London, Department of the Environment, Welsh Office, HMSO.

Anonymous. (2007) *The Construction (Design and Management) Regulations*. Health and Safety Executive.

Anonymous. (2011) *Scottish Road Network Climate Change Study: UKCP09 Update*. Edinburgh, Transport Scotland.

Barlow, J., Lim, M., Rosser, N.J., Petley, D.N., Brain, M., Geer, M. & Norman, E. (2012) Modelling cliff erosion using negative power law scaling of rockfalls. *Geomorphology*, 139–140, 416–424.

Barnett, C., Perry, M., Hossell, J., Hughes, G. & Procter, C. (2006) *A Handbook of Climate Trends Across Scotland; Presenting Changes in the Climate Across Scotland Over the Last Century*. Sniffer Project CC03. Edinburgh, Scotland and Northern Ireland Forum for Environmental Research. 62 pp.

Bouch, C., Jaroszweski, D. & Baker, C. (2011) Future resilient transport networks (FUTURENET): An overview of the FUTURENET project with particular reference to railway aspects. In: *Proceedings Ninth World Congress on Railway Research, 22–26 May 2011, Lille, France*. 12 pp.

Bouch, C., Ryley, T., Baker, C., Avery, K., Chapman, L., Dijkstra T., Dingwall, R., Dixon, N., Goulden, M., Gunn, D., Hooper, L., Jaroszweski, D., Lawley, R., Paulley, N., Quinn, A., Reeves, S., Sivell, P., Wade, S., Wilks, J. & Zanni, A. (2012) Future Resilient Transport Networks (FUTURENET): Assessing transport network security in the face of climate change. In: *Transportation Research Board 91st Annual Meeting (No. 12-0279)*. 11 pp.

Briggs, K., Smethurst, J. & Powrie, W. (2014) Modelling the influence of tree removal on embankment slope hydrology. In: *Landslide Science for a Safer Geoenvironment*. Cham, Springer International Publishing. pp. 241–246.

Brohan, P., Kennedy, J.J., Harris, I., Tett, S.F.B. & Jones, P.D. (2006) Uncertainty estimates in regional and global observed temperature changes: A new dataset from 1850. *Journal of Geophysical Research*, 111, D12106.

Brook, D. (2007) The planning response to climate change. In: McInnes, R.G., Jakeways, J., Fairbank, H. & Mathie, E. (eds.) *Landslides and Climate Change*. London, Taylor & Francis. pp. 497–504.

Brook, D. & Marker, B. (2008) Planning for development on land that is potentially prone to subsidence in England. *Quarterly Journal of Engineering Geology and Hydrogeology*, 41, 403–408.

Brooks, S.M. & Spencer, T. (2012) Shoreline retreat and sediment release in response to accelerating sea level rise: Measuring and modelling cliffline dynamics on the Suffolk Coast, UK. *Global and Planetary Change*, 80–81, 165–179.

Brooks, S.M., Spencer, T. & Boreham, S. (2012) Deriving mechanisms and thresholds for cliff retreat in soft-rock cliffs under changing climates: Rapidly retreating cliffs of the Suffolk coast, UK. *Geomorphology*, 153, 48–60.

Brunsden, D. (1996) Landslides of the Dorset Coast: Some unresolved questions. *Proceedings of the Ussher Society*, 9, 1–11.

Brunsden, D. & Edmonds, R. (2010) The Dorset and east Devon Coast: England's geomorphological world heritage site. In: *Geomorphological Landscapes of the World*. Netherlands, Springer. pp. 211–221.

Cannon, S.H., Gartner, J.E., Wilson, R.C., Bowers, J.C. & Laber, J.L. (2008) Storm rainfall conditions for floods and debris flows form recently burned area in southwestern Colorado and southern California. *Geomorphology*, 96, 250–269.

Castedo, R., Murphy, W., Lawrence, J. & Paredes, C. (2012) A new process-response coastal recession model of soft rock cliffs. *Geomorphology*, 177, 128–143.

Chambers, J.E., Gunn, D.A., Wilkinson, P.B., Meldrum, P.I., Haslam, E., Holyoake, S., Kirkham, M., Kuras, O., Merritt, A. & Wragg, J. (2014) 4D electrical resistivity tomography monitoring of soil moisture dynamics in an operational railway embankment. *Near Surface Geophysics*, 12 (1), 61–72.

Clarke, G.R.T., Hughes, D.A.B., Barbour, S.L. & Sivakumar, V. (2006) The implications of predicted climate changes on the stability of highway geotechnical infrastructure: A case study of field monitoring of pore water response. In: *EIC Climate Change Technology, 2006*. IEEE. pp. 1–10.

Climate Change Risk Assessment (CCRA) (2012) *The UK Climate Change Risk Assessment 2012: Evidence Report*. London, UK, Department for Environment, Food and Rural Affairs (DEFRA). 488 pp.

Collison, A., Wade, S., Griffiths, J. & Dehn, M. (2000) Modelling the impact of predicted climate change on landslide frequency and magnitude in SE England. *Engineering Geology*, 55 (3), 205–218.

Creighton, R. (ed.) (2006) *Landslides in Ireland. A Report of the Irish Landslide Working Group*. 125 pp.

Culshaw, M.G. & Harrison, M. (2010) Geo-information systems for use by the UK insurance industry for 'subsidence' risk. In: Williams, A.L., Pinches, G.M., Chin, C.Y., McMorran, T.J. & Massey, C.I. (eds.) *Geologically Active. Proceedings of the 11th Congress of the International Association for Engineering Geology and the Environment, Auckland, New Zealand, September 2010 (on CD-ROM, 1043-1051)*. Leiden, CRC Press/Balkema.

Dashwood, C., Diaz Doce, D. & Lee, K.A. (2014) *GeoSure Version 7 Methodology: Landslides (Slope Instability). British Geological Survey, Information & Environmental Modelling Programme*, Internal Report IR/14/014. 34 pp.

Davies, O., Rouainia, M., Glendinning, S. & Birkinshaw, S.J. (2008) Assessing the influence of climate change on the progressive failure of a railway embankment. In: *Proceedings of the 12th International Association for Computer Methods and Advances in Geomechanics Conference, Goa, CD-Rom*.

Davies, O., Rouainia, M., Glendinning, S., Cash, M. & Trento, V. (2014) Investigation of a pore pressure driven slope failure using a coupled hydro-mechanical model. *Engineering Geology*, 178, 70–81.

Dawson, R.J., Dickson, M.E., Nicholls, R.J., Hall, J.W., Walkden, M.J., Stansby, P.K., *et al.* (2009) Integrated analysis of risks of coastal flooding and cliff erosion under scenarios of long term change. *Climatic Change*, 95 (1–2), 249–288.

Department for Transport, Local Government and the Regions (2002) *Planning Policy Guidance 14 (Annex 2): Development on Unstable Land: Subsidence and Planning*. London, TSO.

Department of the Environment (1990) *Planning Policy Guidance 14: Development on Unstable Land*. London, Department of the Environment, Welsh Office, HMSO.

Department of the Environment (1996a) *Landslide Investigation and Management in Great Britain: A Guide for Planners and Developers*. London, HMSO.

Department of the Environment (1996b) *Planning Policy Guidance 14 (Annex 1): Development on Unstable Land: Landslides and Planning*. London, Department of the Environment, Welsh Office, HMSO.

Dijkstra, T.A. & Dixon, N. (2010) Climate change and slope stability in the UK: Challenges and approaches. *Quarterly Journal of Engineering Geology and Hydrogeology*, 43 (4), 371–385.

Dijkstra, T.A., Dixon, N., Crosby, C., Frost, M., Gunn, D., Fleming, P. & Wilks, JH (2014) Forecasting infrastructure resilience for climate change. *Proceedings of the Institution of Civil Engineering (Transport)*, 167, 269–280.

Dixon, N. & Brook, E. (2007) Impact of predicted climate change on landslide reactivation: Case study of Mam Tor, UK in Landslides. *Journal of the International Consortium on Landslides*, 4 (2), 137–147.

Donohue, S., Gunn, D.A., Bergamo, P., Hughes, E., Dashwood, B., Uhlemann, S., Chambers, J.E. & Ward, D. (2014) Assessing climate effects on railway earthworks using MASW. In: *20th Euro. Meet Environmental and Engineering Geophysics, Athens, Greece, 14–18 September 2014*.

Evans Grant Consulting Structural and Civil Engineers (1994) *Report on Geotechnical Investigation at Fort Warden, Totland, Isle of Wight, 15 April 1994*.

Fletcher, C.N. & Siddle, H.J. (1998) Development of glacial Lake Teifi, West Wales: Evidence for lake-level fluctuations at the margins of the Irish Sea ice sheet. *Journal of the Geological Society*, 155, 389–399.

Foster, C., Gibson, A.D. & Wildman, G. (2008) The new national landslide database and landslide hazards assessment of Great Britain. In: Sassa, K., Fukuoka, H. & Nagai, H., *et al.* (eds.) *Proceedings of the First World Landslide Forum, Tokyo. International Programme on Landslides (IPL), Tokyo*, Parallel Session Volume. pp. 203–206.

Foster, C., Pennington, C.V.L., Culshaw, M.G. & Lawrie, K. (2012) The national landslide database of Great Britain: Development, evolution and applications. *Environmental Earth Sciences*, 66 (3), 941–953.

Galbraith, R.M., Price, D.J. & Shackman, L. (eds.) (2005) *Scottish Road Network Climate Change Study*. Edinburgh, Scottish Executive. 100 pp.

Gibson, A.D., Humpage, A.J., Culshaw, M.G., Forster, A. & Waters, R.A. (2002) The geology and landslides of Nefyn Bay, Gwynedd. In: Nichol, D., Bassett, M.G. & Deisler, V.K. (eds.) *Landslides and landslides Management in North Wales*. Wales, National Museums and Galleries of Wales.

Gibson, A.D., Culshaw, M.G., Dashwood, C. & Pennington, C.V.L. (2013) Landslide management in the UK – The problem of managing hazards in a 'low risk' environment. *Landslides*, 10, 599–610.

Glendinning, S., Hall, J. & Manning, L. (2009a) Asset-management strategies for infrastructure embankments. *Proceedings of the ICE-Engineering Sustainability*, 162 (2), 111–120.

Glendinning, S., Loveridge, F., Starr-Keddle, R.E., Bransby, M.F. & Hughes, P.N. (2009b) Role of vegetation in sustainability of infrastructure slopes. *Proceedings of the ICE-Engineering Sustainability*, 162 (2), 101–110.

Glendinning, S., Hughes, P., Helm, P., Chambers, J., Mendes, J., Gunn, D., Wilkinson, P. & Uhlemann, S. (2014) Construction, management and maintenance of embankments used for road and rail infrastructure: Implications of weather induced pore water pressures. *Acta Geotechnica*, 9, 799–816.

Gunn, D.A., Chambers, J.E., Hobbs, P.R.N., Ford, J.R., Wilkinson, P.B., Jenkins, G.O. & Merritt, A. (2013) Rapid observations to guide the design of systems for long-term monitoring of a complex landslide in the Upper Lias clays of North Yorkshire, UK. *Quarterly Journal of Engineering Geology and Hydrogeology*, 46 (3), 323–336.

Hackney, C., Darby, S.E. & Leyland, J. (2013) Modelling the response of soft cliffs to climate change: A statistical, process-response model using accumulated excess energy. *Geomorphology*, 187, 108–121.

Hodgetts, D. & Rutter, E. (April 2013). A time-lapse LiDAR survey of the Mam Tor landslide, Derbyshire, UK. In: *EGU General Assembly Conference Abstracts*, Vol. 15. p. 5384.

Hughes, D., McKinley, J., Donohue, S., Lynch, K., Bell, A., Harley, R. & Graham, C. (2013) Geotechnical monitoring using a combination of LIDAR, real-time DGPS, and electrical tomography to assess geotechnical risk of a coastal landslide. In: *Proceedings 6th International Conference on Structural Health Monitoring of Intelligent Infrastructure, Hong Kong, 9–11 December 2013, CD*. 10 pp.

Hulme, M., Jenkins, G.J., Lu, X., Turnpenny, J.R., Mitchell, T.D., Jones, R.G., Lowe, J., Murphy, J.M., Hassell, D., Boorman, P., MacDonald, R. & Hill, S. (2002) *Climate Changes*

Scenarios for the United Kingdom: The UKCIP02 Scientific Report. Norwich, Tyndall Centre for Climate Change research, University of East Anglia. 120 pp.

Hutchinson, J.N. & Bromhead, E.N. (2002) Isle of Wight Landslides. In: McInnes, R.G. & Jakeways, J. (eds.) *Instability Planning and Management: Seeking Sustainable Solutions to Ground Movement Problems. Proceedings of the International Conference, Isle of Wight, 2002*. London, Thomas Telford.

Hutchinson, J.N., Bromhead, E.N. & Chandler, M.P. (2002) Landslide movements affecting the lighthouse at Saint Catherine's Point, Isle of Wight. In: *Instability: Planning and Management: Seeking Sustainable Solutions to Ground Movement Problems: Proceedings of the International Conference Organised by the Centre for the Coastal Environment, Isle of Wight Council, and Held in Ventnor. Isle of Wight, UK on 20-23 May 2002*. London, Thomas Telford. pp. 291–298.

Jenkins, G.J., Murphy, J.M., Sexton, D.M.H., Lowe, J.A., Jones, P. & Kilsby, C.G. (2009a) *UK Climate Projections: Briefing Report*. Exeter, UK, Met Office Hadley Centre. Available from: http://ukclimateprojections.defra.gov.uk.

Jenkins, G.J., Perry, M. & Prior, J. (2009b) *The Climate of the United Kingdom and Recent Trends*. Exeter, UK, Hadley Centre, Met Office. Available from: http://ukclimateprojections.defra.gov.uk.

Jenkins, G.O., Jones, L.D. & Gibson, A.D. (2006) *Analysis of the Hollin Hill Landslide, Low Mowthorpe, North Yorkshire: Field Reconnaissance Survey and Proposed Survey Recommendations*. British Geological Survey Internal Report, IR/06/110.

Jenkins, G.O., Gibson, A.D. & Humpage, A.J. (2007) Climate change and evolution of landslide hazard at Nefyn Bay, North Wales. In: McInnes, R. (ed.) *Landslides and Climate Change: Challenges and Solutions. Proceedings of the International Conference on Landslides and Climate Change*. London, Taylor and Francis. pp. 113–119.

Jones, D.K.C & Lee, E.M. (1994) *Landsliding in Great Britain*. London, Her Majesty's Stationery Office. 360 pp.

Kay, A.L. & Jones, R.G. (2012) Comparison of the use of alternative UKCP09 products for modelling the impacts of climate change on flood frequency. *Climatic Change*, 114 (2), 211–230.

Kilsby, C., Glendinning, S., Hughes, P.N., Parkin, G. & Bransby, M.F. (2009) Climate-change impacts on long-term performance of slopes. *Proceedings of the ICE-Engineering Sustainability*, 162 (2), 59–66.

Klein Tank, A.M.G., Wijngaard, J.B., Können, G.P., Böhm, R., Demarée, G., Gocheva, A., *et al.* (2002) Daily dataset of 20th-century surface air temperature and precipitation series for the European Climate Assessment. *International Journal of Climatology*, 22 (12), 1441–1453.

Lambeck, K. (1995) Late Devensian and Holocene shorelines of the British Isles and North Sea from models of glacio-hydro-isostatic rebound. *Journal of the Geological Society-London*, 152 (3), 437–448.

Lawrence, J.A., Mortimore, R.N., Stone, K.J. & Busby, J.P. (2013) Sea saltwater weakening of chalk and the impact on cliff instability. *Geomorphology*, 191, 14–22.

Lee, E.M. (1999) Coastal planning and management; the impact of the 1993 Holbeck Hall landslide, Scarborough. *East Midlands Geographer*, 21, 78–91.

Lee, E.M. (2002) *Investigation and Management of Soft Rock Cliffs*. London, Thomas Telford. 382 pp.

Lee, E.M. & Moore, R. (1996) *Investigation and Management of Landslides in Great Britain*. London, HMSO. 128 pp.

Lee, E.M. & Clark A.R. (2000) The use of archive records in landslide risk assessment: Historical landslide events on the Scarborough coast, UK. In: Bromhead, E.N., Dixon, N. & Ibsen, M.-L. (eds.) *Proceedings of the 8th International Symposium on Landslides, "Landslides in Research, Theory and Practice"*. London, Thomas Telford. pp. 904–910.

Lee, E.M., Clark, A.R. & Guest, S. (1998) An assessment of coastal landslide risk, Scarborough, UK. In: Moore, D. & Hungr, O. (eds.) *Engineering Geology: A Global View from the Pacific Rim. Proceedings of the 8th Congress of the International Association of Engineering Geology and the Environment*, Vol. 3. Rotterdam, Balkema. pp. 1787–1794.

Lim, M., Rosser, N.J., Petley, D.N. & Keen, M. (2011) Quantifying the controls and influence of tide and wave impacts on coastal rock cliff erosion. *Journal of Coastal Research*, 27 (1), 46–56.

Loveridge, F.A., Spink, T.W., O'Brien, A.S., Briggs, K.M. & Butcher, D. (2010) The impact of climate and climate change on infrastructure slopes, with particular reference to southern England. *Quarterly Journal of Engineering Geology and Hydrogeology*, 43 (4), 461–472.

Lowe, J.A., Howard, T.P., Pardaens, A., Tinker, J., Holt, J., Wakelin, S., Milne, G., Leake, J., Wolf, J., Horsburgh, K., Reeder, T., Jenkins, G., Ridley, J., Dye, S. & Bradley, S. (2009) *UK Climate Projections Science Report: Marine and Coastal Projections*. Exeter, Met Office Hadley Centre. Available from: http://ukclimateprojections.defra.gov.uk.

Maddison, J.D. (2000) St Dogmaels landslide: Deep drainage by wells. In: Siddle, H.J., Bromhead, E.N. & Bassett, M.G. (eds.) *Landslide and Landslide Management in South Wales*. Cardiff, National Museum of Wales. pp. 106–108.

McInnes, R.G. (2000) *Managing Ground Instability in Urban Areas: A Guide to Best Practice*. Newport, Isle of Wight, Isle of Wight Centre for the Coastal Environment.

McInnes, R.G. (2007) *The Undercliff of the Isle of Wight – A Guide to Managing Ground Stability*. Ventnor, Isle of Wight Centre for the Coastal Environment.

McMillan, P., Brown, D.J., Forster, A. & Winter, M.G. (2005) Debris flow information sources. In: Winter, M.G., Macgregor, F. & Shackman, L. (eds.) *Scottish Road Network Landslides Study*. Edinburgh, The Scottish Executive. pp. 25–44.

Met Office (2011) *Climate: Observations, Projections and Impacts, United Kingdom*. Exeter, Met Office. 148 pp. Available from: http://www.metoffice.gov.uk/climate-guide/science/uk/obs-projections-impacts.

Milne, F.D., Werritty, A., Davies, M.C.R. & Browne, M.J. (2009) A recent debris flow event and implications for hazard management. *Quarterly Journal of Engineering Geology and Hydrogeology*, 42, 51–60.

Moore, R., Carey, J.M. & McInnes, R.G. (2010) Landslide behaviour and climate change: Predictable consequences for the Ventnor Undercliff, Isle of Wight. *Quarterly Journal of Engineering Geology and Hydrogeology*, 43 (4), 447–460.

Mott MacDonald (2013) *Totland to Colwell Bay Landslide Assessment*. Report to Isle of Wight Council. December 2013.

Munich Re (2013) *Topics Geo 2012: Natural Catastrophes 2012, Analyses, Assessments, Positions*. Munich, Germany, Münchener Rückversicherungs-Gesellschaft. Available from: http://www.munichre.com/.

Murphy, J.M., Sexton, D.M.H., Jenkins, G.J., Boorman, P.M., Booth, B.B.B., Brown, C.C., Clark, R.T., Collins, M., Harris, G.R., Kendon, E.J., Betts, R.A., Brown, S.J., Howard, T.P., Humphrey, K.A., McCarthy, M.P., McDonald, R.E., Stephens, A., Wallace, C., Warren, R., Wilby, R. & Wood, R.A. (2009) *UK Climate Projections Science Report: Climate Change Projections*. Exeter, Met Office Hadley Centre. Available from: http://ukclimateprojections.defra.gov.uk.

Norman, E.C, Rosser, N.J., Brain, M.J., Petley, D.N. & Lim, M. (2013) Coastal cliff-top ground motions as proxies for environmental processes. *Journal of Geophysical Research: Oceans*, 118 (12), 6807–6823.

Pennington, C.V.L. & Harrison, A.M. (2013) 2012 Landslide year? Geoscientist. *Geological Society of London*, 23 (5):10–15.

Pennington, C.V.L., Foster, C., Chambers, J. & Jenkins, G.O. (2009) Landslide research at the British Geological Survey: Capture, storage and interpretation on a national and site-specific scale. *Acta Geologica Sinica*, 83 (6), 801–840.

Pennington, C.V.L., Dijkstra, T., Lark, M., Dashwood, C., Harrison, A. & Freeborough, K. (2014) Antecedent precipitation as a potential proxy for landslide incidence in south west United Kingdom. In: *Landslide Science for a Safer Geoenvironment*. Cham, Springer International Publishing. pp. 253–259.

Quinn, J.D., Philip, L.K. & Murphy, W. (2009) Understanding the recession of the Holderness Coast, east Yorkshire, UK: A new presentation of temporal and spatial patterns. *Quarterly Journal of Engineering Geology and Hydrogeology*, 42 (2), 165–178.

RAIB (2013) *Derailment of a Freight Train at Barrow Upon Soar, Leicestershire, 27 December 2012*. 45 pp.

Rouainia, M., Davies, O., O'Brien, T. & Glendinning, S. (2009) Numerical modelling of climate effects on slope stability. *Proceedings of the ICE-Engineering Sustainability*, 162 (2), 81–89.

Rutter, E.H., Arkwright, J.C., Holloway, R.F. & Waghorn, D. (2003) Strains and displacements in the Mam Tor landslip, Derbyshire, England. *Journal of the Geological Society of London*, 160 (5), 735–744.

Scottish Executive (2006) *Peat Landslide Hazard and Risk Assessments, Best Practice Guide for Proposed Electricity Generation Developments*. Edinburgh, Scottish Executive.

Scottish Government (2014) *Scottish Planning Policy*.

Skempton, A.W. (1964) The long term stability of clay slopes. *Geotechnique*, 14 (2), 77–101.

Skempton, A.W., Leadbeater, A.D. & Chandler, R.J. (1989) The Mam Tor landslide, north Derbyshire. *Philosophical Transactions of the Royal Society of London*, 329 (1607), 503–547.

Smethurst, J.A., Clarke, D. & Powrie, W. (2006) Seasonal changes in pore water pressure in a grass-covered cut slope in London Clay. *Geotechnique*, 56 (8), 523–537.

Smethurst, J.A., Clarke, D. & Powrie, W. (2012) Factors controlling the seasonal variation in soil water content and pore water pressures within a lightly vegetated clay slope. *Géotechnique*, 62 (5), 429–446.

Statham, I. (2002) Failure on the coastal slope above Lon Gâm, Nefyn, Llyn Peninsula: Mechanism and remedial measures. In: Nichol, D., Bassett, M.G. & Deisler, V.K. (eds.) *Landslides and Landslides Management in North Wales*. Wales, National Museums and Galleries of Wales.

Toll, D.G. (2001) Rainfall-induced landslides in Singapore. *Proceedings, Institution of Civil Engineers (Geotechnical Engineering)*, 149 (4), 211–216.

Toll, D.G., Mendes, J., Gallipoli, D., Glendinning, S. & Hughes, P.N. (2012) Investigating the impacts of climate change on slopes: Field measurements. *Geological Society, London, Engineering Geology Special Publications*, 26 (1), 151–161.

Walking the-Talk (2011) Paths and climate change – An investigation into the potential impacts of climate change on the planning, design, construction and management of paths in Scotland. *Scottish Natural Heritage Commissioned Report No 436*. Inverness, Scottish Natural Heritage.

Walstra, J., Dixon, N. & Chandler, J.H. (2007) Historical aerial photographs for landslide assessment: Two case histories. *Quarterly Journal of Engineering Geology and Hydrogeology*, 40 (4), 315–332.

Waltham, T. & Dixon, N. (2000) Movement of the Mam Tor landslide, Derbyshire, UK. *Quarterly Journal of Engineering Geology & Hydrogeology*, 33 (2), 105–123.

West, L.J. (1994) The Scarborough landslide. *Quarterly Journal of Engineering Geology*, 27, 3–6.

Winter, M.G. (2014) A strategic approach to landslide risk reduction. *International Journal of Landslide and Environment*, 2 (1).

Winter, M.G. & Bromhead, E.N. (2012) Landslide risk – Some issues that determine societal acceptance. *Natural Hazards*, 62, 169–187.

Winter, M.G. & Shearer, B. (2013) *Climate Change and Landslide Hazard and Risk – A Scottish Perspective*. Published Project Report PPR 650. Wokingham, Transport Research Laboratory.

Winter, M.G., Macgregor, F. & Shackman, L. (eds.) (2005) *Scottish Road Network Landslides Study*. Edinburgh, Transport Scotland.

Winter, M.G., Heald, A.P., Parsons, J.A., Macgregor, F. & Shackman, L. (2006) Scottish debris flow events of August 2004. *Quarterly Journal of Engineering Geology and Hydrogeology*, 39 (1), 73–78.

Winter, M.G., Macgregor, F. & Shackman, L. (eds.) (2009) *Scottish Road Network Landslides Study: Implementation*. Edinburgh, Transport Scotland.

Winter, M.G., Dixon, N., Wasowski, J. & Dijkstra, T.A. (2010a) Introduction to land-use and climate change impacts on landslides. *Quarterly Journal of Engineering Geology & Hydrogeology*, 43 (4), 3–6.

Winter, M.G., Dent, J., Macgregor, F., Dempsey, P., Motion, A. & Shackman, L. (2010b) Debris flow, rainfall and climate change in Scotland. *Quarterly Journal of Engineering Geology & Hydrogeology*, 43 (4), 429–446.

Winter, M.G., Harrison, M., Macgregor, F. & Shackman, L. (2013a) Landslide hazard assessment and ranking on the Scottish road network. *Proceedings, Institution of Civil Engineers (Geotechnical Engineering)*, 166 (GE6), 522–539.

Winter, M.G., Kinnear, N., Shearer, S., Lloyd, L. & Helman, S. (2013b) A technical and perceptual evaluation of wig-wag signs at the A83 Rest and be Thankful. *Published Project Report PPR 664*. Wokingham, Transport Research Laboratory.

Chapter 14

Landslide hazards and climate change: A perspective from the United States

Jeffrey A. Coe

U.S. Geological Survey, Denver Federal Center, MS 966, Denver, CO 80225

14.1 INTRODUCTION

> *"Human activities are changing Earth's climate. ... The community of scientists has responsibilities to improve overall understanding of climate change and its impacts. Improvements will come from pursuing the research needed to understand climate change, working with stakeholders to identify relevant information, and conveying understanding clearly and accurately, both to decision makers and to the general public."*

Parts of a statement adopted by the American Geophysical Union in December 2003; Revised and Reaffirmed December 2007, February 2012, and August 2013.

> *"The Geological Society of America supports the scientific conclusions that Earth's climate is changing; the climate changes are due in part to human activities; and the probable consequences of the climate changes will be significant and blind to geopolitical boundaries. Furthermore, the potential implications of global climate change and the time scale over which such changes will likely occur require active, effective, long-term planning."*

Part of a Position Statement on Global Climate Change adopted by the Geological Society of America in 2006.

> *"A decade ago, the National Research Council, an arm of the [United States] National Academies that provides advice on science and technology, recommended a major national landslide assessment program that would have included money for state geological surveys to evaluate or re-evaluate landslide hazards using modern technologies. ... However, the program never materialized, and annual financing for the [United States] Geological Survey to research and evaluate hazards has remained essentially flat over the last decade when adjusted for inflation. ... With extreme weather events expected to become increasingly common, we need to commit to a program to systematically map landslide hazards across the nation and use that information to reduce landslide risks."*

Parts of an Op-ed article by David R. Montgomery (University of Washington) entitled "How to make landslides less deadly" published in the New York Times on

March 20, 2015 following the deaths of 43 people in a landslide near Oso, Washington on March 22, 2014.

These statements provide the perspective and motivation needed for scientific and stakeholder communities in the United States (U.S.) to initiate research studies and use results from those studies to prepare for the effects of climate change. However, as illustrated by the last statement, progress on this goal related to landslide hazards has been slow and reflects the fact that recognition, research, and proactive preparedness for the effects of climate change on landslide hazards is in its infancy in the United States. This infancy is illustrated by the fact that key U.S. National landslide hazard planning, strategy, and review documents from the early 2000s (i.e., the U.S. Landslide Hazards Mitigation Strategy, Spiker and Gori, 2003; a review of Mitigation Strategy by the National Academy of Sciences, NRC, 2004; and Landslide Hazards and Planning by the American Planning Association, Schwab *et al.*, 2005) do not include the phrases "climate change" or "global warming." On the other hand, the most recent Hazards Science Strategy by the U.S. Geological Survey (U.S.G.S., Holmes *et al.*, 2013) contains multiple instances of the phrase "climate change" and states that "understanding the direct effects of climate change in relation to natural hazards is critical in order to identify and predict hazards, and estimate impacts for the development of adaptive management strategies" (Holmes *et al.*, 2013, p. 50).

At least three issues hamper rapid advances in understanding the impact of climate change on landslide activity both in the United States and elsewhere, and cause many researchers to question whether they can make tangible progress towards defining the effects that projected changes in climate will have on slope stability. First, landslide activity is difficult to forecast in even a stationary climate (e.g., van Westen, 2006; Keefer and Larsen, 2007; Baum and Godt, 2010), so forecasting landslide activity under a changing climate seems exceptionally difficult. Second, in mid-latitude areas (i.e., much of the United States, Europe, and Asia), future changes in human activity and land use, either related or not related to climate change, are expected to have an equal or greater impact on landslide activity than actual changes in temperature or precipitation (e.g., Sidle and Burt, 2009; Petley, 2010; Crozier, 2010; Wasowski *et al.*, 2010). Lastly, projections of changes in precipitation, the most common cause of landslides in the contiguous United States, are inconsistent (e.g., Mass *et al.*, 2011) and tend to focus on changes to long-term mean conditions rather than on the frequency and magnitude of extreme events that generate heavy rainfall. In general, projections of changes in mean precipitation correspond poorly with the temporal and spatial scale of landslide activity. This is especially true for shallow (<3 m thick) landslides and debris flows which are sensitive to initial soil moisture conditions and variation in storm-scale precipitation intensity and duration (e.g. Caine, 1980; Reid, 1994; Iverson, 2000; Brunsden, 2001; Godt *et al.*, 2006). Although extreme precipitation events are generally expected to increase in a warming climate (e.g., Walsh *et al.*, 2014), specific precipitation amounts and locations are generally lacking in projections. Compared to shallow landslides, the effects of climate change may be easier to forecast for deep-seated (>3 m thick) landslides because they are sensitive to longer-term patterns of temperature and precipitation (e.g., Sidle and Burt, 2009; Coe, 2012). Additionally, the effects of climate change on the magnitude and frequency of debris flows following wildfires, landslides induced by changes in sea level, and landslides induced by changes in permafrost may be easier to quantify because they depend, at least in part, on long-term changes in air temperature, which can be more readily predicted (e.g., Johnson and Sharma, 2009).

In spite of the substantial challenges, some work on the effects of climate change on landslide hazards in the United States has been performed. In this paper, I provide: (1) an overview of the general problem of landslides in the United States, (2) a brief summary of observed and expected environmental changes (e.g., temperature, ice cover, sea level, precipitation) due to climate change in the United States, (3) an overview of existing United States and worldwide landslide research related to environmental changes, and (4) examples of progressive planning for landslide problems related to climate change by federal, state, and local government agencies in the United States. These four themes form the major sections of this paper. Although there is very little landslide research related to climate change in the United States, I highlight the available United States research by presenting selected figures from the work. I include a review of worldwide landslide research related to climate change to provide context for work in the United States. I limit my review of worldwide research to English language publications.

14.2 LANDSLIDES IN THE UNITED STATES

Landslides occur throughout the United States (U.S.) and are responsible for annual economic losses estimated at 1–2 billion dollars (Schuster, 1996). The areas that are most susceptible to landslides (Fig. 14.1) are the coastal and mountainous areas of Washington, Oregon, California, Alaska, and Hawai'i; mountainous areas of the Intermountain West, and mountainous areas in the eastern United States. Within the contiguous 48 States shown in Figure 14.1, as well as Alaska and Hawai'i, landslides are typically triggered by intense or prolonged rainfall, snowmelt, rain-on-snow events, or earthquakes. For this paper, I focus on climatically induced, subaerial landslides and exclude earthquake-triggered and submarine landslides. Alaska is a unique state in the United States because of its position at high northerly latitudes (~N50° to N72°). In a warming climate, this northerly position causes a variety of slopes in the state to be susceptible to landslides induced by changes in ocean ice-cover and permafrost. Ocean ice cover is decreasing, making shorelines more susceptible to landslides, and permafrost is widespread throughout the state (Fig. 14.2). Of all of the states, Alaska has seen, and is projected to see, the most dramatic increases in both temperature and precipitation (e.g., Wuebbles *et al.*, 2014). In fact, President Obama highlighted the impacts of recent climatic changes in Alaska on a late summer 2015 trip to the state (see https://www.whitehouse.gov/2015-alaska-trip).

Many states in the United States, particularly along and near the coasts and southern border, are impacted by precipitation from large, regional-scale climate phenomena, such as the North American Monsoon, tropical cyclones (hurricanes, e.g., Blake and Gibney, 2011), El Niño/La Niña Southern Oscillation (ENSO, e.g., Higgins *et al.*, 2000), and atmospheric rivers such as the Pineapple Express (e.g., Higgins *et al.*, 2000) and the Maya Express (Dirmeyer and Kinter, 2009). An understanding of how climate change will impact these regional systems is critical to understanding future landslide hazards in the United States. Historically, these systems have caused widespread landslides throughout the United States (e.g., Lajoie and

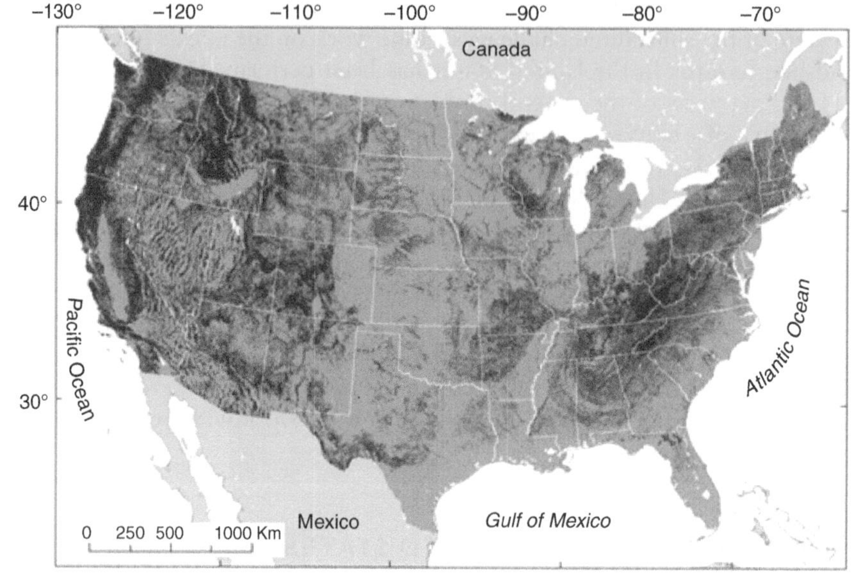

Figure 14.1 Prototype map showing areas of the conterminous United States with negligible (gray) and some (black) susceptibility to landslides (modified from Godt *et al.*, 2012).

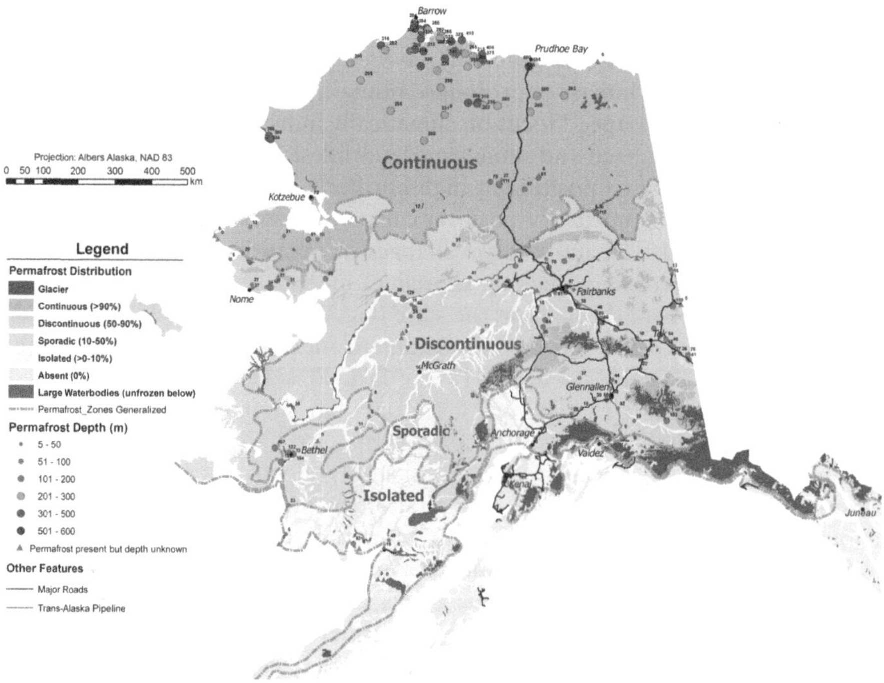

Figure 14.2 Map showing Alaska permafrost and glaciers. Modified from Jorgenson *et al.* (2008).

Mathieson, 1998; Godt *et al.*, 1999; Ashland, 2011; Wilson and Pike, 2005; Godt and Coe, 2007; Wieczorek and Morgan, 2008). When these systems are sustained for long periods of time, they can trigger landslides well within the interior of the United States (e.g., the November 3–5, 1985 storm in West Virginia and Virginia, Jacobson 1993; the Nashville, Tennessee storm of May 1–2, 2010, Ashland, 2011; and the week-long Colorado Front Range storm of September, 2013 (Coe *et al.*, 2014; Gochis *et al.*, 2015). Localized landslide events within the interior of the contiguous United States are often triggered by convective storms, which can sometimes be enhanced with moisture from more regional systems (e.g., Godt and Coe, 2007). Landslides on the islands of Hawai'i can be triggered by regional-scale events or localized storms (e.g., Ellen *et al.*, 1991; Deb and El-Kadi, 2009). Compared to the lower 48 states and Hawai'i, less is known about climatic triggers of landslides in Alaska.

Within the last few years, the United States has had a large number of fatalities due to landslides, with at least 61 people killed by landslides between September 2013 and August 2015 (Table 14.1). These fatalities included 43 people in the Oso, Washington debris avalanche/debris flow (Fig. 14.3, Keaton *et al.*, 2014; Iverson *et al.*, 2015), 3 people in debris flows in the Front Range of Colorado (Coe *et al.*, 2014), 3 people in the West Salt Creek Rock avalanche in western Colorado (Fig. 14.4, Coe *et al.*, 2016), 3 people in a debris flow in Sitka, Alaska (Fig. 14.5, NBC News, 2015), and 2 people in a large rock fall in Utah (Fig. 14.6, Lund *et al.*, 2014). These fatalities have spurred a revival in lidar acquisition and paleolandslide inventory mapping in the United States, and to a lesser extent, interest in the effects that climate change may have on landslide frequency and magnitude (e.g., see opening quote by Dave Montgomery). However, none of the meteorological conditions (rainfall, snowmelt, freezing/thawing of water) that triggered these fatal landslides have been directly associated with climate change. Also, interestingly, only one of these landslide events was caused by a large, regional-scale weather system (the Colorado Front Range event).

Landslide hazards in the United States are managed by a complicated patchwork of state, local, and federal government agencies (e.g., Olshansky and Rogers, 1987). State highway departments are responsible for maintaining most major roads in the United States, and therefore deal with landslides on a regular basis, often through routine maintenance programs and emergency funds reserved for rapid repairs. Cities and counties issue building permits and generally have the authority to specify design and/or mitigation requirements prior to building. However, the rights of individuals are strong in the United States, and, in general, when an individual owns property, it is difficult for local governments to prohibit them from building on that property. Also, many structures built prior to design or mitigation requirements are located in high hazard areas. A few communities in California have formed Geologic Hazards Abatement Districts (GHADs, Schwab *et al.*, 2005) that are designed to prevent and mitigate landslides and other hazards, but, in general, such programs are rare in the United States. Landslide insurance is not a product that is readily available to individual homeowners, although there has been some recent effort to change this situation (Keaton and Roth, 2010; Roth and Keaton, 2010; Godt *et al.*, 2012). When there are damages to private property from landslides, victim compensation is often dependent on the outcome of litigation in the court system, although in some cases, the Federal Emergency Management Agency (FEMA) has compensated homeowners for landslide damages if they had flood insurance (Olshansky and Rogers, 1987). In cases

Table 14.1 Sixty-one fatalities in the United States caused by landslides during the 2-year period between September 2013 and August 2015.

Date	*Location*	*Number of fatalities from landslides*	*Brief Description*	*Source of Information*
September 10–12, 2013	Front Range, Colorado	3	All fatalities from debris flows, 1 in Jamestown, CO and 2 in Pinebrook Hills immediately west of the City of Boulder	Coe *et al.*, 2014; Godt *et al.*, 2014
September 30, 2013	Sawatch Range, Colorado	5	Family of 5 killed by a rockslide in Chalk Creek Canyon near Buena Vista, CO	http://www.denverpost.com/breakingnews/ci_24207249/rock-slide-reported-near-popular-hiking-trail-chaffee
December 12, 2013	Rockville, Utah	2	Couple killed by a large rock mass that fell from a conglomerate cliff and landed on their house	Lund *et al.*, 2014
March 22, 2014	Oso, Washington	43	43 killed by a rapidly moving debris avalanche/debris flow moving through a housing development	Keaton *et al.*, 2014; Iverson *et al.*, 2015
May 25, 2014	Collbran, Colorado	3	3 killed by a rapidly moving rock avalanche on the northern flank of Grand Mesa	Coe *et al.*, 2016; White *et al.*, 2015
July 5, 2014	Lovell, Wyoming	1	4-year old child killed by falling rock while hiking with family	http://billingsgazette.com/news/state-and-regional/wyoming/falling-rock-kills-girl-in-northern-wyoming/article_ce8270af-27cc-50c1-8861-7c4ae3e73183.html
August 12, 2015	North Cascades National Park, Washington	1	1 killed by falling boulder while hiking in North Cascades National Park	http://www.nationalparkstraveler.com/2015/08/rockfall-kills-hiker-north-cascades-national-park
August 18, 2015	Sitka, Alaska	3	3 killed by a debris flow triggered by ~65 mm of rain in 6 hours	http://www.nbcnews.com/news/us-news/final-body-found-after-deadly-landslide-sitka-alaska-n416046

of widespread landslide damages, FEMA has paid for repair and stabilization of public facilities that have been damaged by landslides, if the landslides occurred during an event that resulted in a Presidentially declared emergency or major disaster (FEMA, 2006). Olshansky and Rogers (1987) reason that a coordinated strategy consisting

Figure 14.3 Snohomish County geotechnical engineer surveys damage at the Oso debris avalanche/debris flow in Washington State that killed 43 people. Photo taken April 10, 2014 by J. Coe.

Figure 14.4 May 25, 2014, West Salt Creek Rock avalanche near Collbran, Colorado. The avalanche traveled 4.6 km and killed three people working in the valley. Photo courtesy of Mesa County (Colorado) Sheriff's Office.

Figure 14.5 August, 18, 2015 debris flow in Sitka, Alaska, that killed three people. Triggering rainfall was a 1 in 50 year event (Rick Thoman, U.S. National Weather Service, written communication). Photo by James Poulson, the Daily Sitka Sentinel.

Figure 14.6 House in Rockville, Utah destroyed by a December 12, 2013 rock fall. From Lund *et al.* (2014).

of insurance and regulations supported by increased research and mapping efforts would reduce landslide damages and provide compensation to victims. However, such a coordinated program does not currently (September, 2015) exist in the United States.

14.3 OBSERVED AND PREDICTED ENVIRONMENTAL CHANGES IN THE UNITED STATES

14.3.1 Air temperature

Historical observations show that mean land-surface air temperatures have increased between 1 and 2°C since 1895 in the United States, with most of the increase occurring since 1970 (e.g., Smith and Reynolds, 2005; Trenberth *et al.*, 2007). Walsh *et al.* (2014) shows that nearly all of the United States has experienced this increase in temperature, with the exception of parts of some states in the Southeast (Fig. 14.7). There have been decadal cooling trends (e.g., Easterling and Wehner, 2009; Rangwala and Miller, 2010), within the longer-term trend of substantial warming (Fig. 14.7). Temperature changes have been greatest in high-latitude land areas (i.e., Alaska), with increasing trends ranging from 0.06 to 0.34°C for various decades between 1850 and 2005 (Trenberth *et al.*, 2007). Observations in mountainous areas reveal that these areas have generally warmed at a greater rate than the overall average (e.g., Beniston *et al.*, 1997; Rangwala and Miller, 2010).

Projected changes using median values from a mid-range emissions scenario (RCP4.5, see Moss *et al.*, 2010; van Oldenborgh *et al.*, 2013) indicate that winter (December-February) mean temperatures in the contiguous United States will be between 1 and 4°C higher in 2081–2099 compared to 1986–2005 (Fig. 14.8, van Oldenborgh *et al.*, 2013). Projected changes for summer (June-August) range from 1.5 to 3°C higher (Fig. 14.8) for the same comparison time periods. In mountainous regions, Rangwala and Miller (2010) indicate warming of about 4°C by 2100. In Hawai'i, both winter and summer mean temperatures will be between 1 and 2°C higher in 2081–2099 compared to 1986–2005. Extreme northern regions of Alaska are projected to have the largest increases (5 to 7°C in winter, and 2 to 3°C in summer, Fig. 14.8, van Oldenborgh *et al.*, 2013). Warming in mountainous and high-latitude regions is expected to be amplified because of snow-albedo feedback where snow cover is lost (e.g., Salathé *et al.*, 2008). Temperature extremes (heat waves) are expected to be more intense, frequent, and longer lasting in a warmer climate in the United States (Meehl *et al.*, 2007).

14.3.2 Snow, ice cover, and permafrost

Snow, ice cover, and permafrost have contracted in most regions of the world since the 1970s and projections indicate that they will continue to contract because of increasing air temperatures (e.g., Rignot and Kanagaratnam, 2006; Isaksen *et al.*, 2007; Solomon *et al.*, 2007; Zdanowicz *et al.*, in press). Widespread increases in thaw depth have been observed in most permafrost regions (e.g., Lemke *et al.*, 2007) and glaciers have experienced widespread mass losses that have contributed to sea level rise (e.g., Meier *et al.*, 2007). Changes in the surface albedo

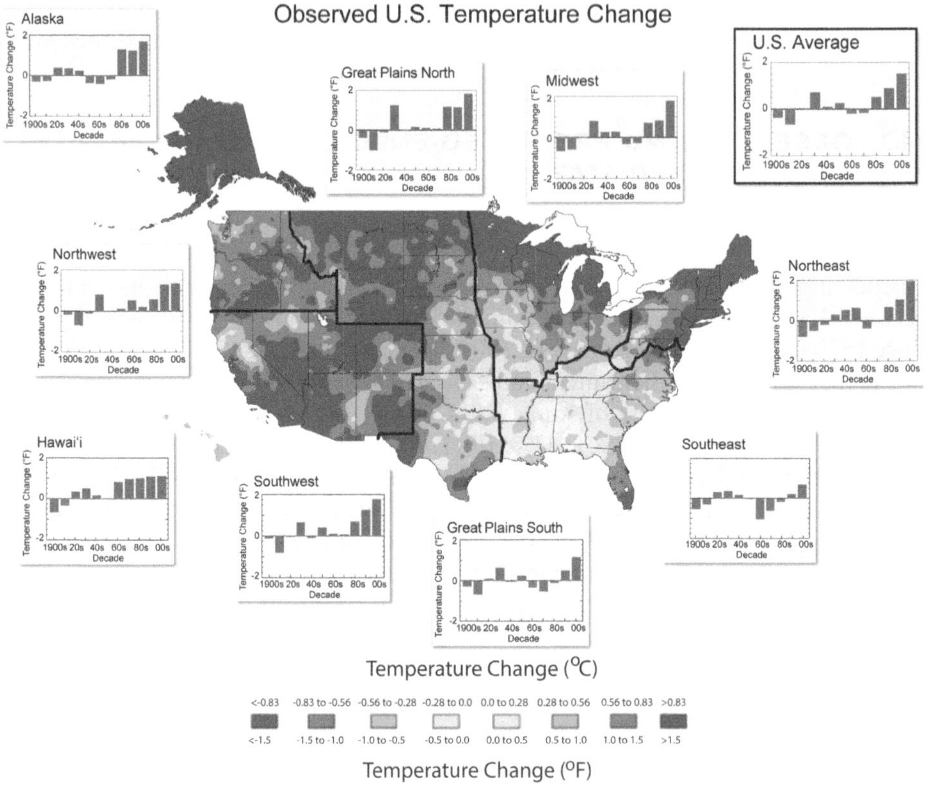

Figure 14.7 Map showing observed change in air temperature between the 1991 to 2012 period average and the 1901 to 1960 period average (for contiguous states) and the 1951 to 1980 period average (for Alaska and Hawai'i). Graphs show changes in average temparature by decade for 1901 to 2012, relative to the 1901 to 1960 period average for each region. Modified from Walsh et *al.* (2014).

associated with melting ice and snow enhance warming, particularly in northern latitudes (Serreze and Francis, 2006). Mellander *et al.* (2007) indicate that in northern latitudes, warming will shorten the period of persistent snowpack by 73–93 days, increase soil temperature by 0.9–1.5°C at 10 cm depth, and lengthen the growing season by 30–43 days. Increasing soil temperatures could increase rates of heterotrophic respiration and decay and enhance rates of CO_2 and CH_4 release (Eliasson *et al.*, 2005; Schuur and Abbott, 2011). Screen and Simmonds (2010) argue that diminishing sea ice cover in the Arctic Ocean has had a primary role in the temperature amplification observed in the Arctic. Their work reinforces indications of a strong positive ice-temperature feedback system in the Arctic. Additionally, decreasing sea-ice cover along the north coast of Alaska has led to a longer open-water season which has increased wind fetch, wave heights, and coastal erosion (e.g., Overeem *et al.*, 2011; Wobus *et al.*, 2011).

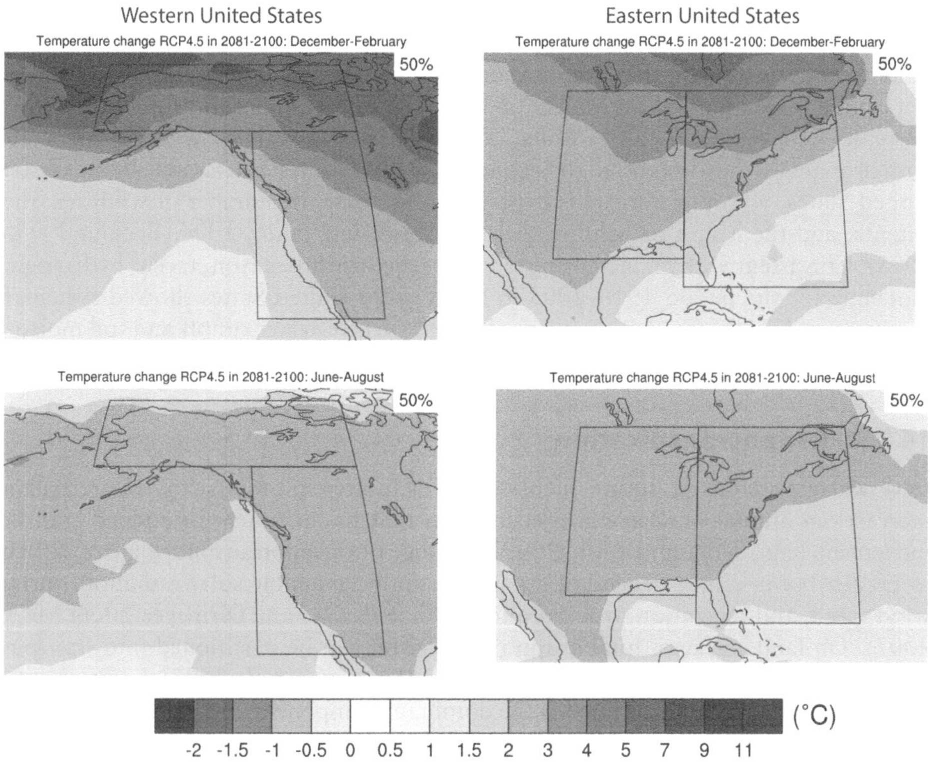

Figure 14.8 Projected changes in median values of mean air temperature for the period 2081–2100 compared to 1986–2005 using the RCP4.5 emissions scenario. Seasonal periods shown are December-February (top row) and June-August (bottom row). The 50% in each panel indicates that these are median values. Simulation results from CMIP5 models (modified from van Oldenborgh *et al.*, 2013).

14.3.3 Sea level

Sea level has risen since the 1960s due to melting of land ice and from the thermal expansion of water due to rising temperatures (Bindoff *et al.*, 2007). Tide gage data indicate that the global average rate of mean sea level rise between 1961 and 2003 was 1.8 ± 0.5 mm/yr (Solomon *et al.*, 2007). Meehl *et al.* (2007) projected a range in total sea level rise from 0.18 to 0.59 m by the 2090–2099 period (relative to 1980–1999), but could not confidently account for the impact of dynamic changes in ice flow from Greenland and Antarctica. Pfeffer *et al.* (2008) accounted for ice dynamics and concluded that a sea-level rise of 0.8 m by 2100 was most likely, with a maximum rise of 2 m physically possible.

14.3.4 Drought and wildfire

Drought and wildfire are important to landslide hazards because they affect the potential for post-wildfire debris flows, which are a major hazard in the western United States (e.g., Westerling *et al.*, 2003; Cannon and Degraff, 2009). By the end of the 21st century, long-term droughts are expected to be three times more common in mid-latitude areas of North America (Sheffield and Wood, 2008). Other recent work indicates that there is some agreement that evaporation and drought will increase in the southwestern United States, but less agreement on locations where evaporation and drought will decrease (Golubev *et al.*, 2001; Wang, 2005; Meehl *et al.*, 2007). Wildfires associated with droughts are also expected to become more prevalent. For example, in the western United States, warming climate is expected to increase the number of wildfires, their extents, and the length of wildfire seasons (Westerling *et al.*, 2006; Backlund *et al.*, 2008). Consistent with an increasing length of the wildfire season, recent hydrological modeling for the period 1916–2003 in the western United States showed systematic increases in the amount of early spring (March) surface-water runoff and soil moisture (Hamlet *et al.*, 2007).

14.3.5 Evapotranspiration

In a warming climate, future increases in evapotranspiration can counterbalance increases in annual or seasonal precipitation that might otherwise induce landslide movement (e.g., Sidle and Burt, 2009). Changes in evapotranspiration are difficult to predict because they depend on moisture supply, available solar radiation, surface wind speed, and vegetation structure and health (e.g., Gao and Dirmeyer, 2006; Meehl, 2007). On land surfaces, total evapotranspiration can be partitioned into transpiration, soil evaporation, and canopy evaporation (Lawrence *et al.*, 2007). Lawrence *et al.* (2007) indicate that transpiration is the dominant component of total evapotranspiration, except in areas where the vegetation canopy is sparse and evaporation from the soil surface dominates. In the United States, future atmospheric potential for evapotranspiration is expected to increase, but much more research is needed to confidently predict future evapotranspiration for specific areas (Georgakakos *et al.*, 2014).

14.3.6 Precipitation

Historical observations show that overall mean annual precipitation in the United States has increased by about 5% since 1900 (Fig. 14.9, Walsh *et al.*, 2014), although much of the Southwest, Southeast, and northern Great Plains states had areas of both increases and decreases. Observations of the top 1% of all daily precipitation events (Fig. 14.10, Walsh *et al.*, 2014) and 2-day precipitation events (Kunkel *et al.*, 2013) between 1901 and 2012 indicate that extreme precipitation has increased in the eastern and central United States, but has not significantly changed in the western United States (Fig. 14.10).

Projected changes in precipitation using a mid-range emissions scenario (RCP4.5, see Moss *et al.*, 2010; van Oldenborgh *et al.*, 2013) indicate that the median values of percent change in mean precipitation in 2081–2099 compared to 1986–2005 for the contiguous United States will range from −10% to +20% in fall/winter (October–March), and −10% to +10% in spring/summer (April-September) (Fig. 14.11). In general, changes in precipitation become greater (more positive) from south to north. In Hawai'i, median values of percent change in mean precipitation for the same comparison periods will range from 0 to −20% for both fall/winter and spring/summer

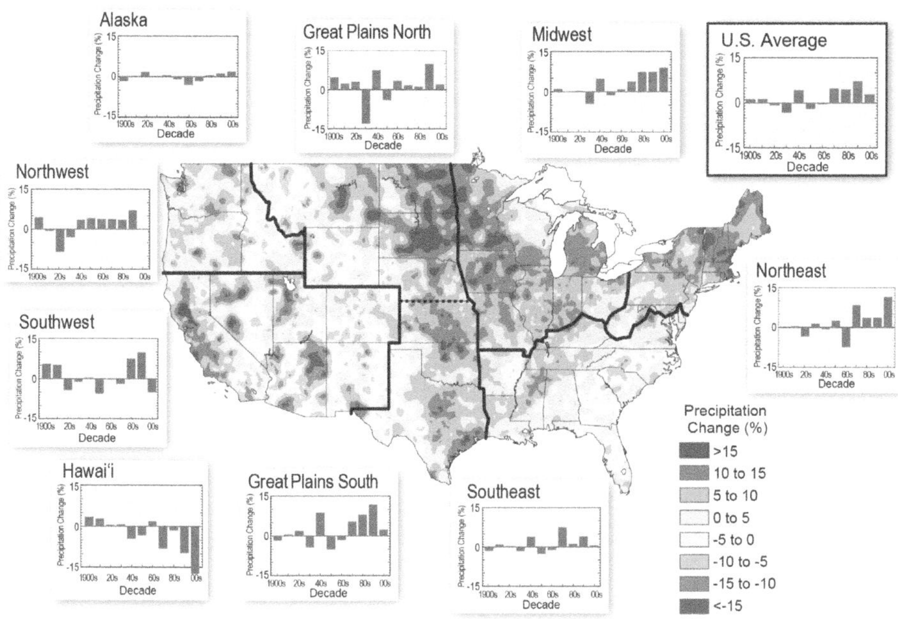

Figure 14.9 Map showing observed percent change in total annual precipitation between the 1991 and 2012 period average and the 1901 to 1960 period average. Graphs show changes in average precipitation by decade for 1901 to 2012, relative to the 1901 to 1960 period average for each region. From Walsh *et al.* (2014).

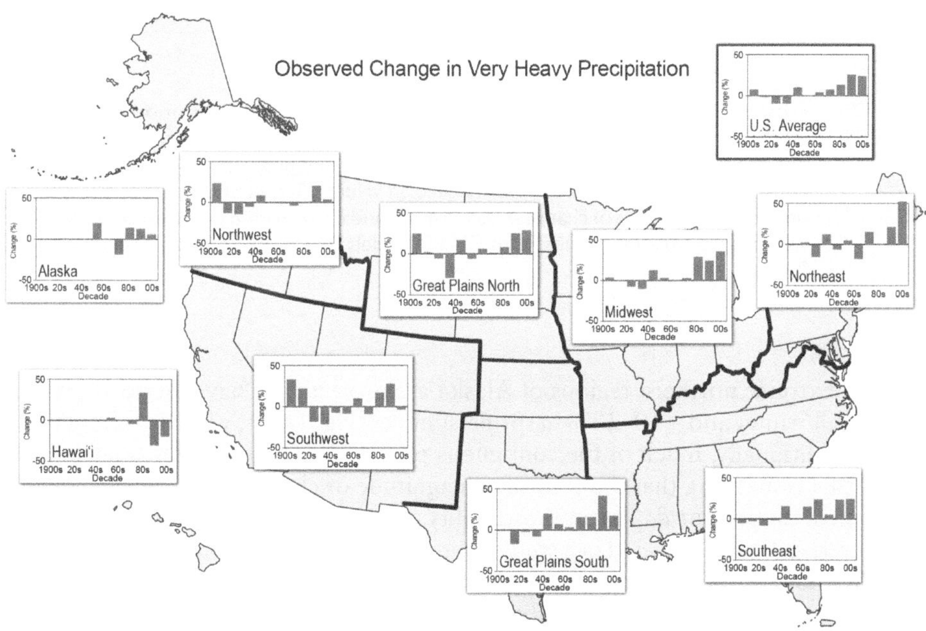

Figure 14.10 Diagram showing the percent changes (averaged by decade) in the annual amount of precipitation falling in extreme events (the top 1% of all daily events) from 1901 to 2012. Changes were derived from a comparison with 1901–1960 period averages for all regions except Alaska and Hawai'i. Changes in Alaska and Hawai'i were derived from a comparison with 1951 to 1980 period averages. From Walsh *et al.* (2014).

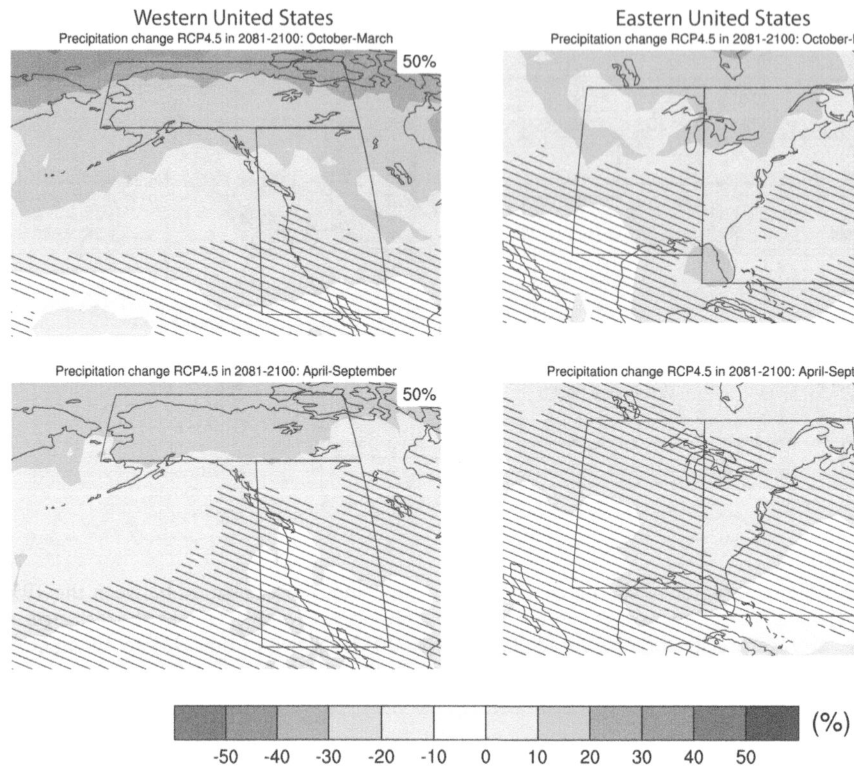

Figure 14.11 Projected changes in median values of percent change in precipitation for the period 2081–2100 compared to 1986–2005 using the RCP4.5 emissions scenario. Seasonal periods shown are October–March (top row) and April-September (bottom row). The 50% in each panel indicates that these are median values. Hachures indicate areas where the projected magnitude of change is less than 1 standard deviation of the present-day natural variability. Simulation results from CMIP5 models (modified from van Oldenborgh *et al.*, 2013).

seasons. Extreme northern regions of Alaska are projected to have increases of +20–30% in fall/winter and +10–20% in spring/summer (Fig. 14.11, van Oldenborgh *et al.*, 2013). Additionally, much of the contiguous part of the United States is hachured in Figure 14.11, meaning that the projected magnitude of change is less than 1 standard deviation of the present-day natural variability (van Oldenborgh *et al.*, 2013). This low level of certainty was reflected in a study of precipitation trends along the West Coast of the U.S. (Mass *et al.*, 2011). Mass *et al.* (2011) examined extreme 2-day precipitation from 1950 to 2009. They found slight increases in the frequency of events in southern California, decreasing frequency in northern California and southern Oregon, and significant increases in northern Oregon and Washington. They compared their results to those from three other studies that used General Circulation Models (GCMs) and Regional Climate Models (RCMs) and found that the distributions of precipitation varied substantially among climate models and that projections from only one of the studies compared favorably with historical trends.

Projections of changes in climate phenomena including storm tracks, tropical cyclone frequency and intensity, oscillatory atmospheric circulation patterns such as ENSO, and convective precipitation are less certain than projected changes in mean precipitation (e.g., Randall *et al.*, 2007), but are particularly relevant to landslide activity. In the northern hemisphere, extra-tropical storm tracks are projected to move poleward in a warming climate (e.g., Yin, 2005; Bengtsson *et al.*, 2006). These findings have particular relevance to changes in landslide activity along the West Coast of the U.S. (Salathé, 2006) where widespread landslide events and flooding are often caused by prolonged intense rainfall associated with atmospheric rivers (Neiman *et al.*, 2008; Dettinger, 2011). Predictions of changes in storms, storm tracks, and circulation phenomena have been attempted (e.g., Tebaldi *et al.*, 2006; Sun *et al.*, 2007), but modeling studies do not typically examine changes in the most extreme events because of the difficulty in dealing with small sample sizes represented by these events (e.g. Tebaldi *et al.*, 2006). This hampers the examination of trends of widespread and destructive events that tend to be associated with the most extreme weather events such as the Nashville, Tennessee area flooding and landslides (Ashland, 2011; Higgins *et al.*, 2011) and the Colorado Front Range flooding and debris flows of September, 2013 (Coe *et al.*, 2014; Gochis *et al.*, 2015).

Several recent papers nicely summarize the status of projecting changes in precipitation from extreme storms and circulation patterns. In a review of the impact of climate change on tropical cyclones, Knutson *et al.* (2012) report substantial recent modeling progress, but make it clear that modeled changes in tropical cyclone characteristics for individual ocean basins show large variations between individual studies. They state that rainfall rates in tropical cyclones are likely to increase, but that the frequency of tropical cyclones will either decrease or remain unchanged. Writing for the U.S. Climate Change Science Program, Karl *et al.* (2008, p. 113) state that "we assess that it is unknown how late 21st century tropical cyclone frequency in the Atlantic and North Pacific basins will change compared to the historical period (approximately 1950–2006)" and "projected effects of global warming on El Niño remain uncertain" (p. 106). Karl *et al.* (2008) go on to state that on average, precipitation is likely to be less frequent but more intense.

14.3.7 Land use

The response of land use and land cover to a warming climate may take decades to centuries to occur and is extremely difficult to predict. Changes in land use and land cover (1) have contributed to climate change through radiative forcing by changes in surface albedo and anthropogenic aerosols and greenhouse gases (e.g., Pielke *et al.*, 2002), and (2) can result from climate change (e.g., Dale, 1997). Here, I focus on land use and cover changes that are likely to result from climate change. Pastor and Post (1988) argue that the land response from climate change is a function of the spatial pattern of soils and the nutrient cycling properties of vegetation. Dale *et al.* (2001) use forests to illustrate the complexity of the problem and make the point that climate change can affect forests by altering the frequency, intensity, duration and the timing of fire, drought, introduced species, insect and pathogen outbreaks, hurricanes, windstorms, ice storms, or landslides. Additionally, these disturbances can interact with air pollution, resource extraction, agricultural needs, residential expansion, and recreation needs. Dale *et al.* (2001) indicate that their analysis led to the realization of a lack of integrated knowledge in many critical areas needed to predict changes to forests in a changing climate. In conclusion, they argue that improved monitoring programs and analytical tools are needed to gain a better understanding of future changes. In general, this example is reflective of the problems with predicting all types of land use and cover changes caused by climate change. The interaction between land use and climate change is currently poorly understood and will require new models that link the physics of climate with the socioeconomic drivers of land use (Mahoney *et al.*, 2003). Remote sensing tools provide quick and relatively inexpensive tools for monitoring land use and cover changes (Mahoney *et al.*, 2003) and will increase our understanding of the topic in the future.

14.4 LANDSLIDE RESEARCH RELATED TO ENVIRONMENTAL CHANGES

Nearly all studies on the impact of climate change on landslide activity have been located in the northern hemisphere (exceptions are New Zealand and Australia) and about two thirds of all studies have been concentrated in Western Europe. I was unable to identify any studies conducted on the African continent. In general, studies are one of three varieties (Coe and Godt, 2012): (1) long-term monitoring of changes in climate variables such as rainfall rates and temperature, and the accompanying geomorphic response at study sites (monitoring studies); (2) approaches to establish links between past climate and landslide occurrence (retrospective studies); and (3) approaches to establish changes in landslide frequency or magnitude in the future using established patterns of historical activity and climate projections (prospective studies). In this paper, I concentrate on results from monitoring and prospective approaches because prehistoric climate reconstructions from retrospective approaches typically yield incomplete information on landslide frequency and do not contain specific information regarding the magnitude of the triggering event (e.g., rainfall intensity, duration, antecedent soil moisture, etc.). Therefore, these retrospective approaches have limited predictive capabilities.

The remainder of this section is organized according to the dominant cause of changes in landslide activity in a changing climate, that is, either temperature or precipitation. The following sections are ordered from high to low confidence that landslide activity will change due to climate change. For example, based on the studies reviewed in the first section on glacial retreat related to changes in temperature, I have a high level of confidence that landslide activity will change and that areas of change are at least somewhat predictable. Whereas for the last section on precipitation, I think that some locations will have changes in landslide activity because of changes in precipitation, but I have low confidence that the locations of these changes can currently be predicted.

One relatively new method for analyzing landslides does not fit neatly into any particular environmental category. That method is the use of multi-temporal landslide inventory data to detect changes in landslide numbers and geometries that may be caused by changing climatic factors. Thus far, three studies have used this approach, with the studies reaching different conclusions about the usefulness of the technique. Schlögel *et al.* (2011) found that the technique was useful for detecting landslide changes in central Asia, whereas Korup *et al.* (2012) used Monte Carlo simulations of statistical models of landslide size-frequency distributions and found that the technique was not useful because of low signal (changes in landslide sizes and occurrence) to noise (numerical errors in model parameters, scaling relationships) ratios. Finally, Wood *et al.* (2015) developed a landslide inventory in the European Alps to provide baseline datasets so that the relationship between landslides and climate change could be tested in the future.

14.4.1 Landslide response related to changes in temperature

14.4.1.1 Glacial retreat

Glacial retreat during the last several decades is well documented in alpine areas throughout the world (e.g., Haeberli and Beniston, 1998; Dyurgerov and Meier, 2000; Luckman and Kavanagh, 2000; Francou *et al.*, 2003; Kaser *et al.*, 2004). The effects of glacial retreat have multiple implications for landslide activity. There is strong evidence that landslide activity will increase in areas affected by glacial retreat because of the removal of toe support (i.e., ice) for oversteepened mountain slopes, moraines, and channel banks (e.g., Blair, 1994; Geertsema *et al.*, 2006; Hugenholtz *et al.*, 2008; Oppikofer *et al.*, 2008; Fischer *et al.*, 2012; Hürlimann *et al.*, 2012; Kochel and Trop, 2012; Fischer *et al.*, 2013; Haeberli, 2013); increased availability of loose debris for debris flows in or near drainages (e.g., Rebetez *et al.* 1997; Chiarle *et al.*, 2007; Legg *et al.*, 2014); more efficient delivery of rainfall to loose debris because of an absence of snow and ice (Lancaster *et al.*, 2012); and failures of moraine-, landslide-, and glacially dammed lakes (e.g., Bovis and Jakob, 2000; Shrestha *et al.*, 2007; Mool *et al.*, 2011; Geertsema *et al.*, 2013).

In the United States, landslides related to glacial retreat will be likely in the Pacific Northwest and Alaska. For example, recent work on debris flows at Mount Rainier, Washington by Legg *et al.* (2014) found that as glaciers recede upslope, they provide steeper and higher elevation areas of loose debris that are more susceptible to debris-flow initiation than debris areas with lower elevations and lower slopes (Fig. 14.12). Much of the early work documenting landslide effects due to glacial retreat was performed in relation to retreat of glaciers in North America following the Little Ice Age (LIA, ~AD 1550 to 1850). O'Connor and Costa (1993) evaluated the effects of warming at the end of the LIA in alpine areas in the northwestern United States, western Canada, and Alaska. They indicated that air temperatures in these areas rose from 0.5 to 1.2°C between the coldest part of the LIA and 1993. Glacial retreat caused by this temperature increase resulted in large quantities of loose, unvegetated sediment on steep slopes. They cite historical examples of this sediment being rapidly mobilized as debris flows caused by outburst floods, precipitation-induced floods, landslides, geothermal head flow, and melting of stagnant ice. Evans and Clague (1994) observed that warming in the mountains of western Canada during the last 100–150 years resulted in glacier loss that has accelerated landslides in oversteeped-deglaciated areas because of debuttressing

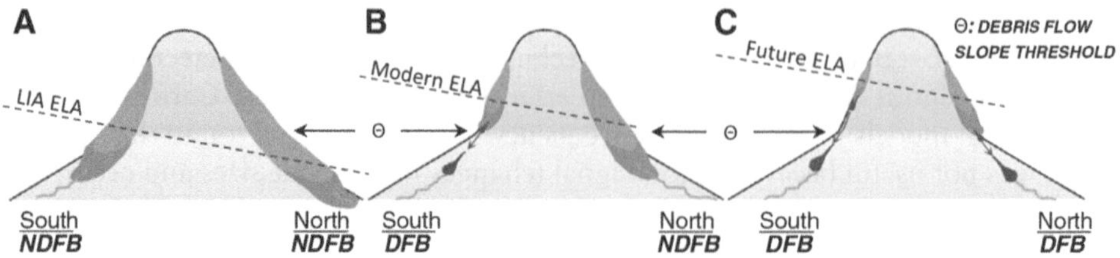

Figure 14.12 Diagram showing spatial and temporal variation in glacier size on Mount Rainier, Washington, and its control on debris-flow occurrence. DFB is debris-flow producing basin. NDFB is non debris-flow producing basin. Equilibrium line altitude (ELA) and corresponding glacial extent are shown for the Little Ice Age (LIA, A), present day (B), and for a future scenario when ELA is higher (C). Results portray how future debris-flow activity may increase as glaciers decrease on slopes steeper than the debris flow slope threshold. From Legg *et al.* (2014).

(removal of toe support) and floods from moraine and glacially dammed lakes. They suggest that effects from recent glacier thinning and retreat will continue hundreds of years into the future, but that these effects will be an order of magnitude smaller than those associated with Pleistocene deglaciation.

In the European Alps, recent work indicates that as glaciers retreat additional sediment is available for debris-flow mobilization in periglacial areas (Rebetez *et al.* 1997; Chiarle *et al.*, 2007; Lugon and Stoffel, 2010) and that source areas are migrating to higher elevations (Jomelli *et al.* 2007; Harris *et al.*, 2009; Fischer *et al.*, 2012). Chiarle *et al.* (2007) indicated that debris flows are increasing in frequency near glaciers because of increased amounts of loose sediment and moraine-dammed lakes and that extraordinarily high sediment yield rates will require a change in presently established hazard assessment procedures.

14.4.1.2 Degradation of permafrost in rock

The degradation of permafrost, snow and ice fields, and glaciers by warming air temperatures could have a multitude of impacts on landslide activity. Permafrost occurs in rock and soil in both alpine areas and high latitude areas and occupies about 24% of the land surface in the northern hemisphere (Brown *et al.*, 1997; Nelson *et al.*, 2003). Temperature changes in these areas by the end of the 21st century are expected to range from +2 to 8°C (see Fig. 14.8). Failures of rock slopes in these areas are typically rock falls and rock avalanches (rapid rock slides, rock flows, and complex rock movements of Varnes, 1978). The general consensus of studies in these areas is that rock falls and rock avalanches are increasing in frequency due to permafrost degradation (e.g., Geertsema *et al.*, 2006). But, as noted by Allen *et al.* (2011) and Huggel *et al.* (2012), attributing specific rock falls and avalanches to permafrost degradation is difficult, especially without monitoring data at the precise location of rock-slope failures, because of difficulties in isolating permafrost degradation from other effects such as precipitation, erosion, seismicity, and uplift (Allen *et al.*, 2011). For example, Huggel *et al.* (2012, p. 87) state that despite abundant observational evidence over the past several decades and many recent alpine rock avalanches without historical precedence (e.g., Deline *et al.*, 2011; Ravanel and Deline, 2011), "There is still no unambiguous evidence that the frequency or the magnitude of landslides has changed over this period."

The processes by which permafrost degrades and leads to rock falls and rock avalanches are a topic of current research. Recent work suggests that movement, and phase change of water resulting from the degradation of permafrost, has an effect on the stability of rock slopes. The change of water from a frozen solid to liquid in a rock mass potentially has the combined effect of reducing rock mass strength and increasing effective stress. Gruber and Haeberli (2007) indicated that permafrost degradation in steep rock slopes occurs by heat conduction and by advection of heat by movement of water in fractures. Advection can create pathways where deep (tens of meters) thawing can occur along fractures leading to rapid destabilization of large volumes of rock.

In the United States, landslides related to degradation of permafrost in rock will be possible in the Pacific Northwest and Alaska. Two very large, recent rock avalanches in Alaska (Mt. Steller in September, 2005, Huggel, 2009, Fig. 14.13A; and Mt. Lituya in June, 2012, Geertsema, 2012 and Repanshek, 2012) provide intriguing research opportunities on degradation of permafrost. But, to my knowledge, research on rock permafrost degradation in relation to landslides in Alaska is limited to a few publications (Huggel, 2009; Huggel *et al.*, 2010). Huggel *et al.* (2010) analyzed short-term air temperature records preceding recent rock and ice avalanches in Alaska (Mt Steller, Fig. 14.13B), New Zealand, and the European Alps, and found that a predictive temperature trigger was not discernible, but that common thermal patterns preceded the events. First, for all of the slope failures, there were unusually warm temperatures (well above freezing and up to 8.5°C above long term average temperatures) in the weeks or days before the avalanches, and second, prior to some of the slope failures, there were sudden drops in temperature (usually below freezing) after the warm periods in the hours to days before the avalanches.

There have been many studies in the European Alps, Scandinavia, and New Zealand that have focused on recent rock fall and rock avalanche activity and make compelling arguments that their occurrence is related to permafrost degradation. For example, a comprehensive review paper by Harris *et al.* (2009) followed earlier conceptual framework papers (Harris *et al.*, 2001; Harris, 2005) and focused on permafrost degradation in Norway, Sweden, and in the European Alps. Harris *et al.* (2009) indicated that changes in the active layer thickness of permafrost are a reliable indicator of the response of permafrost to a warming climate. The active layer is defined as the zone of rock or soil that seasonally freezes and thaws. In terms of the active layer in rock slopes, they also indicated that in 2003, the warmest summer in the historical record in the European Alps (Schär *et al.*, 2004), the depth of thaw penetration (active layer thickness) in a bedrock borehole in the Swiss Alps was about twice the average penetration during the 7-year borehole record. On steep alpine slopes containing discontinuities in rock bound by ice, the scale of active layer thickening observed in the Alps in 2003 significantly increased the probability of rock-fall occurrence. Work by Ravanel and Deline (2011) supports this argument. They studied historical rock falls from rock slopes in the French Alps and documented 42 rock falls between 1947 and 2009, and showed that 70% occurred between 1989 and 2009, with a spike in the rate of occurrence during the 2003 heat wave (11 of 42 rock falls occurred in 2003, 26% of total). They argue that permafrost degradation was responsible for many of the recent

13A

13B

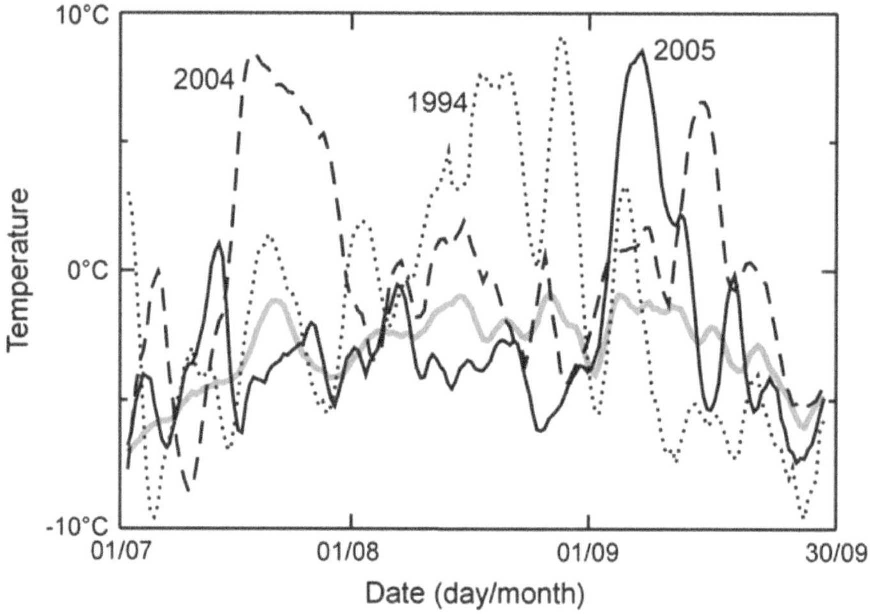

Figure 14.13 Diagram showing records from the Mt. Steller rock avalanche (from Huggel, 2009). A) The source area for the September, 2005 Mt. Steller rock avalanche. Photo by R. Homberger. B) July to September (JAS) temperature records of the Yakutat radiosonde from 3 years with warm summers, 1994, 2004, and 2005. The thick gray line is the 1994 to 2007 JAS average.

falls based on the fact that the average elevation of rock falls (3130 m above mean sea level) was close to the lower permafrost limit (based on an energy balance model), and that most initiated from convex topography (i.e., spurs), where enhanced lateral heat flux from solar radiation was suspected of enhancing thawing of permafrost.

14.4.1.3 Degradation of permafrost in soil

Landslides in soil slopes in permafrost areas include shallow and deep-seated landslides and debris flows. There is evidence of soil permafrost degradation in the northern hemisphere (Brown and Romanovsky, 2008; Harris *et al.*, 2009) and broad concern that the landslide activity in these areas will increase (Slaymaker, 1990; Schlyter *et al.*, 1993; Harris *et al.*, 2009; Francani *et al.*, 2011), but, thus far, there is little evidence that the frequency or magnitude of landslides has changed during the last several decades.

In the United States, the state most likely to be impacted by landslides from degradation of permafrost in soil is Alaska. However, I am aware of only one study on the topic in Alaska. In that study, Daanen *et al.* (2012) found that lobes of frozen debris on permafrost slopes in the Brooks Range have become increasingly active over the last couple decades compared to activity levels from the 1950s to the 1980s (Fig. 14.14).

Elsewhere, in one of the first studies to evaluate the impact of climate change on landslides, Slaymaker (1990) used results from early GCMs that showed at least 1.5 to 4.5°C warming between 1990 and 2050 at high latitudes and focused on the qualitative impacts in five mountain ranges in western Canada. He suggested that glaciers would reflect climate change most rapidly, followed more slowly by timberline, and then mass wasting processes. In terms of landslides, he expected the greatest impact to be in the subarctic, dry, interior mountains where permafrost would be subject to thermokarst (depressions from irregular melting of permafrost) development. Schlyter *et al.* (1993) described a long-term study site in northern Sweden and the expected effects at the site caused by increased air temperatures. These effects included changes to weathering rates, permafrost and snow distribution, frost shattering, surface wash, and soil and rock stability. Schlyter *et al.* (1993) expected the greatest changes to occur at or near treeline due to changes in vegetation cover. In their review paper on permafrost and climate, Harris *et al.* (2009) evaluated temperature changes in boreholes in Norway and Sweden over a 3 to 7 year period and found warming rates in permafrost between 0.02 and 0.06°C/yr at depths of 20 m. They suggested that (1) long-term (decades to centuries) permafrost warming may lead to a higher probability of large, deep-seated landslides and rock avalanches, and (2) that seasonal thawing of ice-rich fine-grained soils will lead to slow (mm to cm/yr), downslope movement of soil (solifluction) and an increase in shallow landslides and debris flows. In the Italian Alps, hydrologic and mechanical modeling by Francani *et al.* (2011) suggested that debris flows will be the most likely type of landslide triggered by thawing soil permafrost and that the flows will often be associated with liquefaction and piping.

14.4.1.4 Increased sea level

If predicted increases in sea level occur, coastal slopes throughout the world will be adversely impacted by landslides. There is general consensus that this will be the case, but surprisingly little work has focused on, or demonstrated this effect.

14A

14B

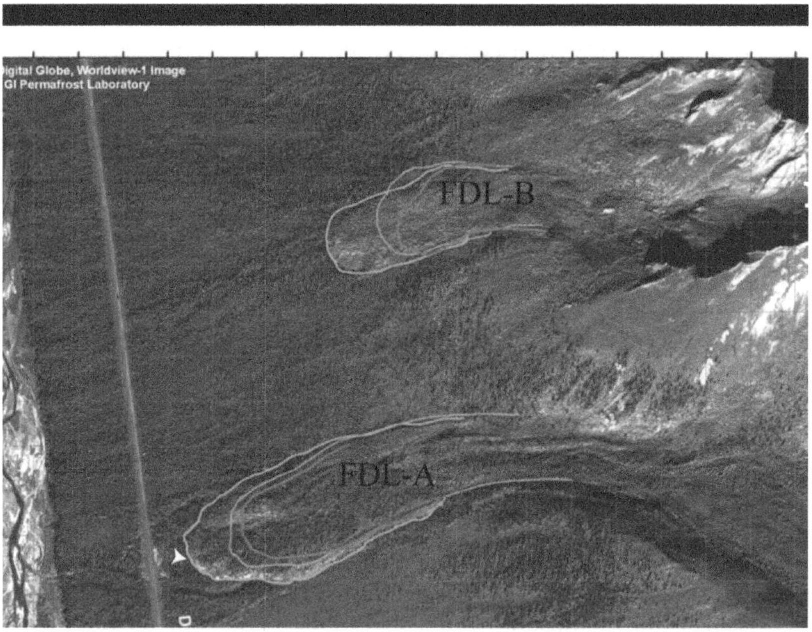

Figure 14.14 Diagram showing Frozen Debris Lobes (FDL) in the Brooks Range of Alaska. From Daanen *et al.* (2012). Available for use through the Creative Commons Attribution 3.0 License. A) Location of lobes (small rectangle near CF, Cold Foot). DH is Deadhorse, FB is Fairbanks. B) Terminal positions of debris lobes from 1955 (red), 1979 (blue), and 2008 (green). Measurements of movement from 2008 to 2010 revealed average movement rates of about 1 cm/day. Ongoing movement may pose a hazard to the Dalton Highway, seen in the left side of the image.

Table 14.2 Prediction of coastal cliff failures in Pacifica, California based on forecast sea level rise from various emissions scenarios. From Collins *et al.* (2007). Collins *et al.* (2007) clearly indicated that failure rate and cliff retreat will increase given any trend of sea level rise.

Sea level trend (m/100 yr)	*Number of points exceeding threshold*	*Number of days of predicted failure*	*Percent increase of failure*	*Global Emissions scenario*
0	434	52	0	Sea level decline
0.2	506	61	17	Present rage
0.4	594	71	37	A2 average
0.6	675	81	56	A2 maximum
0.72	719	86	65	A1fi maximum

In the United States, shorelines in the states of California, Oregon, Washington, Alaska, and Hawaii would be most affected by subaerial landslides due to sea level rise because they are generally steeper, and composed of more susceptible materials, compared to eastern and southern shorelines. However, I am aware of only one study that has evaluated the effect of sea level rise on landslides in these states (Collins *et al.*, 2007). Collins *et al.* (2007) developed short- and long-term predictive thresholds for northern California coastal landslides in a particularly susceptible geologic setting based on correlations of 5 years of historical cliff failures with precipitation and wave action, respectively. They used empirically based equations in combination with predictions of future sea-level rise, and little change in annual precipitation to estimate increasing cliff failure rates and cliff retreat for the next 100 years. They state that the percent increase in failures ranges from 17 percent for the present rate of sea level rise, to 65 percent based on projections from the A1 emissions scenario (see Table 14.2). They also state that the rate of cliff retreat will increase with any trend of sea level rise.

Shorelines in Alaska could be particularly vulnerable to landslides (rotational and toppling failures) and subsequent shoreline recession because the number of the months of each year that they are protected from waves, winds, and currents by sea ice is gradually becoming shorter. Coastal bluffs form much of the northern coast of Alaska, but the land surface inland from these bluffs is mostly flat. An increase in landslides at the coastal bluffs has the potential to affect native Alaskan villages and energy infrastructure (highways, wells, pipelines) near the coast. Recent work by Gibbs and Richmond (2015) documented historical shoreline change along the north coast of Alaska and provided a baseline to compare with future changes. They found rates that averaged −1.4 m/yr (negative indicates shoreline retreat, positive indicates shoreline accretion), with a range of −18.6 to +10.9 m/yr.

Along the north coast of Canada, Ford and Smit (2004) pointed out that instrumental records and visual observations in the Arctic have demonstrated significant warming and increased precipitation in the 20th century and that the future change in polar regions is expected to be among the greatest on Earth. They also described increased coastal erosion and coastal landslides due to storm surges and less extensive sea ice.

Elsewhere, work in the United Kingdom (UK) has yielded inconclusive results. A study by Lee (2011) found that previously predicted accelerations in sea-level rise have not yet occurred. Lee (2011) points out that cliff recession rates at Holderness cliffs have increased with relative sea-level rise (from rates of −1.2 m/yr in the early 1950s to −1.5 m/yr in 2000), but that there has not been a rapid acceleration in either sea-level rise or cliff recession rates since 1990 as suggested in the first Intergovernmental Panel on Climate Change (IPCC) assessment report in 1990. Elsewhere in the UK, Brooks *et al.* (2012) studied long-term (1883-2010) retreat rates of coastal cliffs along the Suffolk coast in the UK. They documented a long-term rate of −3.5 m/yr, but a rate of −4.7 m/yr between 1993 and 2010. The processes responsible for retreat included wave action that caused undercutting and cantilever cliff failures, but also rainfall infiltration and soil suction reduction in cliffs. Soft rock cliffs were highly sensitive to climatic variability on decadal scales and they found the highest retreat rates during strongly positive phases of the North Atlantic Oscillation, when strong low-pressure systems commonly impact the UK. They suggest that predicted increases in sea level and extreme rainfall events should result in more rapid rates of cliff retreat in the UK.

14.4.1.5 Changes in drought and wildfire

Predicting changes in the magnitude and frequency of post-wildfire debris flows is complicated by a lack of confidence in defining areas of increasing drought, but also by unknowns regarding changes in vegetation types and densities and the frequency and intensity of future rainstorms capable of triggering debris flows. Most work related to the climatic influence on wildfires and debris flows comes from retrospective approaches based on detailed studies of debris-fan deposits in Australia and the western U.S. (e.g., Nott *et al.*, 2001; Pierce and Meyer, 2008; Frechette and Meyer, 2009) and tree rings (Bigio *et al.*, 2010). These studies span a diverse range of chronology, from 27 ka to present day, but are fairly consistent in indicating that wildfires and post-wildfire debris flows are most frequent during warm, dry climatic conditions. There is broad consensus that the frequency of wildfires is increasing in the western United States (e.g., see Williams *et al.*, 2014 for a review). Cannon and Degraff (2009) state that climatic variations over inter-annual to decadal scales are a principal control on wildfires in the western U.S., but they also suggest that recent increases in wildfire frequency are likely related to population growth in the urban-wildland interface because many fires are ignited by humans.

14.4.2 Landslide response related to changes in precipitation

In populated areas of the world, precipitation is the most common cause of landslides. Landslides caused by precipitation are triggered by increases in groundwater pore pressures. Evapotranspiration acts counter to precipitation because it removes water from the ground and thus reduces groundwater pore pressures. In a warming climate, the importance of increased evapotranspiration and its effects on soil moisture, vegetation, groundwater pore pressures, and landslide activity should not be ignored. This is particularly true for deep-seated landslides (landslides >3 m thick), because they are more strongly influenced by longer-term precipitation and groundwater conditions compared to shallow landslides and debris flows, which are more dependent on precipitation from extreme storms (e.g., van Asch *et al.*, 1999). Thus, many landslide studies related to climate change incorporate effective precipitation, which is precipitation minus evapotranspiration. Unfortunately, as noted earlier in this review, changes in precipitation and evapotranspiration due to climate change are poorly constrained. Crozier (2010) summed up the problem nicely by stating that existing hydrologic and slope-stability models are capable of predicting landslide response from climate change, but there are large margins of error in global- and downscaled-climate projections that make landslide predictions uncertain.

In this section, I concentrate on research related to landslides triggered by changes in precipitation. I divide the section into two subsections. The first subsection covers studies that account for evapotranspiration, and the second subsection covers studies that do not account for evapotranspiration. Many studies in the first subsection focus on deep-seated landslides, and most studies in the second subsection focus on shallow landslides and debris flows. In general, for deep-seated landslides, existing research indicates decreasing movement or no change in movement in areas with increasing net evapotranspiration (i.e., evapotranspiration is greater than any corresponding increase in precipitation). For shallow landslides and debris flows, there is no general consensus regarding an increase or decrease in future activity.

14.4.2.1 Accounting for changing evapotranspiration

In the United States, the areas impacted by a net increase in evapotranspiration are not well constrained and I know of only one study that focused on the impacts that evapotranspiration and precipitation will have on landslide activity (Coe, 2012). Coe (2012) used a correlation between 12 years of annual movement of the deep-seated Slumgullion landslide and a regional moisture balance index, in combination with downscaled projections of temperature and precipitation, to forecast decreasing landslide movement through the 21st century (Fig. 14.15). Coe forecasted that the upper part of the landslide will stop moving by 2060. He suggested that moisture balance indices that account for evapotranspiration have considerable potential as a tool for forecasting the magnitude of ongoing deep-seated landslide movement, or for assessing the onset of regional landslide activity.

In Europe, the French Alps have been the focus of research on deep-seated landslides, particularly the 6–8 m thick Boisivre landslide near Barcelonnette. Dehn and Buma (1999) were the first to evaluate future movement of the landslide when they used statistically downscaled results from three GCMs in combination with a hydrologic "tank" model and a slope-stability model. For their prediction, they defined the landslide as active when the maximum monthly groundwater level exceeded a critical point. However, they determined that no consistent prediction of future activity could be obtained using the three different GCMs because the GCMs and downscaling models needed to be improved. Buma (2000) used projected precipitation and evapotranspiration to evaluate three slope stability models to assess impacts of climate change on the Boisivre landslide. All models indicated a decrease in landslide activity through 2099. Buma and Dehn (2000) evaluated climate projections from three GCMs and three statistical downscaling methods to simulate the impact on the Boisivre landslide using net precipitation (precipitation minus evapotranspiration) to control landslide movement. They found that the net precipitation model was unable to explain all of the variation in landslide activity, and that variation in climate scenarios

15A

15B

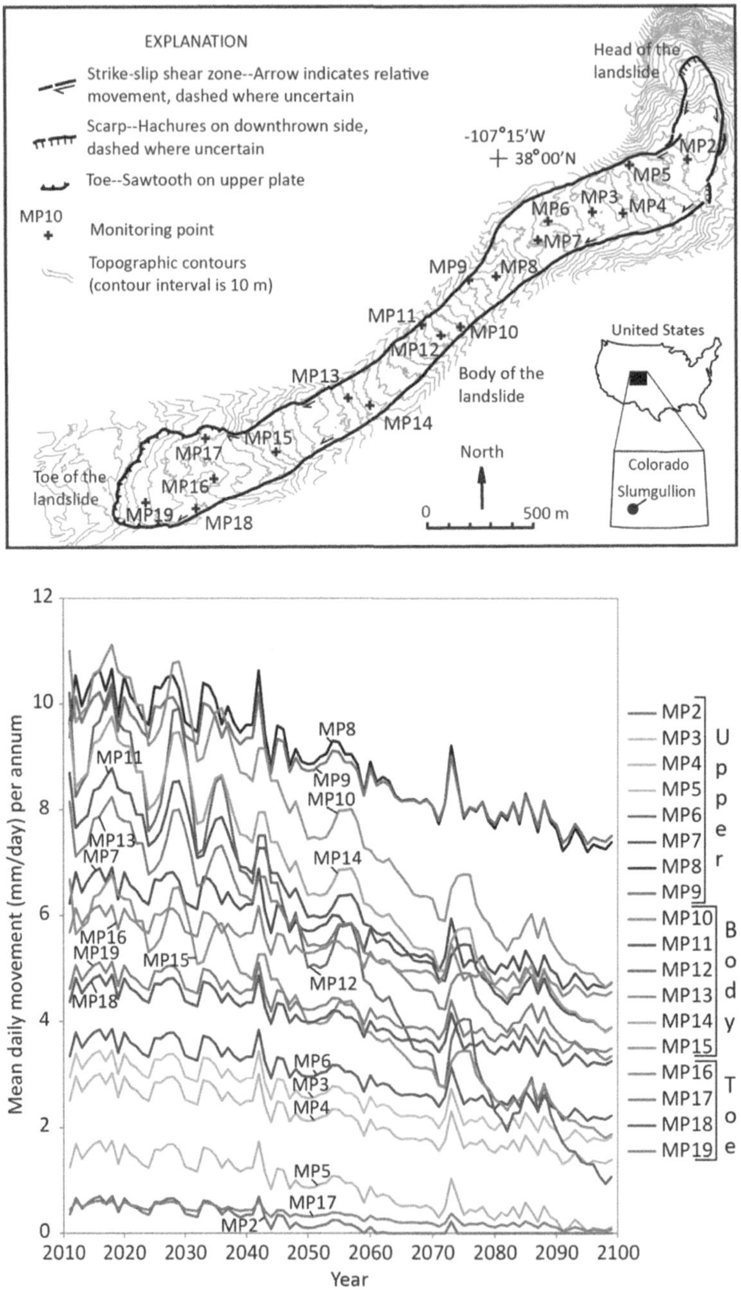

Figure 14.15 Projected movement of monitoring points (MP) on the Slumgullion landslide through 2100 based on projections of precipitation and temperature and a moisture balance index (from Coe, 2012). A) Map showing point locations. B) Projected movement of points. Y-axis shows horizontal movement.

was about the same order of magnitude as that contained in the net precipitation model. Most combinations of the models suggested that the landslide activity would decrease through about 2059, and then slightly increase through 2099. Malet *et al.* (2007) also examined the Boisivre landslide as part of their overall assessment of the influence of climate change on landslide activity in the Ubaye Valley. Malet *et al.* (2007) used downscaled climate projections from the A2 emissions scenario in combination with hydrology and slope stability models to show a decrease in groundwater levels and a decrease in soil moisture in unsaturated zones of both the Boisivre and Super-Sauze landslides for the period 2069–2099 compared to the period 1969–1999. They suggested that this result is related to higher potential evapotranspiration in 2069–2099 and forecast that the activity of the Boisivre landslide will markedly decrease.

In other parts of Europe, studies that incorporated evapotranspiration consistently indicated decreasing or unchanged future landslide activity. In Italy, Dehn *et al.* (2000) used downscaled GCM temperature and precipitation projections and a hydrological tank model to simulate future movement of the deep-seated Alverà landslide in the Dolomite Mountains. Modeling results showed decreased future landslide movement. Comegna *et al.* (2013) used RCM simulations and a geotechnical model (FEM VADOSE/W) to predict that movement of a clay-rich earthflow in the Basento River valley would slowly decrease through 2060. In the Liechtenstein Alps, Bonnard *et al.* (2008) forecasted future movement of the Triesenberg landslide for the year 2050 based on projections of precipitation and temperature, and modeling of hydraulic head and slope stability. Bonnard indicated that landslide movement in 2050 will be similar to what movement was in the year 2000.

In the UK, where total annual precipitation is expected to be slightly greater in the future, four recent studies have yielded mixed conclusions on future landslide activity. Collison *et al.* (2000) studied the Lower Greensand escarpment in southeast England using a GIS-based hydrology and slope stability model and downscaled GCM data. They estimated that increased precipitation in southeast England would be matched by increased evapotranspiration and suggest that the occurrence of large landslides would not change in the future. However, they also conclude that the occurrence of small shallow landslides may increase over the next 80 years because of increased precipitation and shallow soil moisture. Clarke and Smethurst (2010) used a water balance model to simulate changes in the future soil moisture in four engineered clay slopes in England using climate projections from the UK Climate Impacts Programme. They forecasted deficits in annual maximum and average soil moisture and decreased runoff compared to the historical past. Loveridge *et al.* (2010) reviewed existing weather, hydrology, and materials data for modified slopes in southern England and found that clay slopes will be at greater risk of failure because of the expected increased range of soil-moisture levels and accompanying increased rates of strain softening. They indicate that drainage design and maintenance of modified slopes will become increasingly important. Moore *et al.* (2010) evaluated future activity of a coastal landslide complex on the Isle of Wight in the United Kingdom. They established a relation between rainfall and landslide movement and used this relation, in combination with predicted sea level rise to predict future landslide movement. They suggest that the decrease in summer rainfall will have little influence on groundwater levels, but the increase in winter rainfall, and increased storm intensities will lead to an increase in effective antecedent rainfall. They estimate that landslide events will increase in frequency, movement rates will accelerate, and that cliff erosion rates may double by 2080.

One study from the North Island of New Zealand, where annual precipitation is expected to decrease, indicates that shallow landslide activity will also decrease. Schmidt and Glade (2003) used statistically downscaled GCM projections in combination with three empirical landslide models (a daily rainfall model, an antecedent daily rainfall model, and an antecedent soil water status model) to predict future shallow landslide activity. Consistent results from all models suggested that there would be fewer heavy precipitation events, and a decrease in landslide activity for the period 2070–2099 compared to the period 1950–1979.

14.4.2.2 Not accounting for changing evapotranspiration

In the UK, Dixon and Brook (2006) used projections of precipitation to statistically model future return periods of precipitation that exceed established thresholds for initiating movement at the Mam Tor landslide in the United Kingdom. They found that the average return period for landslide-initiating precipitation decreased from about 4 years between 1961 and 1990 to about 3.5 years between 2080 and 2090. They qualify their study by stating that they have not accounted for evapotranspiration and that this fact could reverse their prediction. Winter *et al.* (2010) developed an intensity-duration rainfall threshold for debris flows in Scotland following debris flows that occurred in August of 2004. They observe that the debris flows were controlled by long-term (150 day) antecedent rainfall followed by rainfall from short, intense storms. They suggest that current climate projections indicate that Scotland's climate will not become wetter or drier overall, but that summers will become drier, winters will become wetter, and the intensity of storms may increase. They suggest that to better understand the future debris flows during drier summers (i.e., antecedent conditions that are different from today), an improved understanding of unsaturated soil mechanics is needed.

Two studies from British Columbia, Canada, where annual precipitation is projected to increase, suggested that shallow landslide and debris flow activity will increase. Jakob and Lambert (2009) used an ensemble of 19 GCMs with the B1, A1B, and A2 emission scenarios to estimate changes in future 4-week antecedent rainfall and 24-hour rainfall, as well as shallow landslide and debris flow occurrence in the North Shore Mountains. Their results indicate a 10% increase in 4-week antecedent rainfall and a 6% increase in 24-hour rainfall by the end of the 21st century, which suggest that landslides and debris flows may occur more frequently. However, they acknowledged that their approach assumes that other factors such as land use and forest cover stay constant and that their predicted changes in landslide frequency and magnitude relations were highly speculative. Guthrie *et al.* (2010) analyzed a November 2006 storm and the resulting distribution and characteristics of shallow landslides and debris flows on Vancouver Island. They observed that the landslide distribution did not match the pattern of antecedent or storm rainfall and suggest that snowmelt from rain on snow, with the rate of melt increased by wind, was responsible for the distribution of landslides. They suggested that conventional rainfall thresholds cannot accurately predict impacts from rain-on-snow events. They conclude that a warming climate may prolong spring and fall seasons and increase the likelihood of rain-on-snow events.

One study from Taiwan, where annual precipitation is expected to increase, indicates that shallow landslides will increase in the future. Chiang and Chang (2011) used downscaled annual maximum 24-hr rainfall from one GCM as input to a slope-stability model to forecast shallow landslides in the Biachi catchment in northern Taiwan. They suggested that the average annual maximum rainfall increase by about 50 mm in 2010–2099 (compared to 1960–2008) due to intensified typhoon activity and that the unstable area in the catchment will increase by about 12% for the same period.

14.5 ADAPTATION STRATEGIES IN THE UNITED STATES

Despite the lack of research on the impact of climate change on landslides in the United States, some states and federal agencies have begun to develop plans and adaptive strategies that include landslide hazards. Most adaptation policies have been established by state governments and highway departments in states along the West Coast (Alaska, Washington, Oregon, California, and Hawai'i), and by federal agencies with large areas of infrastructure at risk (the Federal Highway Administration) or large holdings of land in the western United States (the National Park Service and U.S. Forest Service).

Four West Coast states, Alaska, Washington, Oregon, and California, have done the most work on adaptation strategies related to landslide hazards. In a report by the State of Alaska (State of Alaska, 2010), the authors indicate that the state is already being impacted by warming temperatures and that this warming has resulted in increased coastal erosion, landslides, high elevation rock and ice avalanches, and sinking of the ground surface. They indicate that these impacts have disrupted and damaged forests, buildings, infrastructure, and recommend an extensive strategy for adaptation. One of the most important recommendations addresses impacts to infrastructure, which is directly affected by permafrost degradation, landslides, and coastal erosion. Their approach to reduce the impacts to infrastructure include: (1) creating a coordinated system of data collection, analysis and monitoring to identify where problem areas are located; (2) promoting improvements to existing infrastructure to protect and extend its life; and (3) building resiliency into new infrastructure so that it will last in a changing climate. Another recommendation is to improve availability of data from mapping, surveying, and imagery to better assess the magnitude and progression of the impacts of climate change.

The impacts and potential costs of climate change on Alaska's infrastructure and highways are also portrayed in a presentation by the Alaska Department of Transportation & Public Facilities maintenance and operations group (ADOT&PF, see Gregovich, 2010). Gregovich estimates that the northern region of ADOT&PF currently spends U.S. $10 million annually due to permafrost melting and that this cost represents a fraction of the need. He makes the point that ADOT&PF is already implementing mitigation measures to lessen the impacts of climate change including protecting shorelines, relocating villages, building evacuation routes and shelters, improving drainage infrastructure, and protecting against permafrost degradation.

In Washington State, a recent pilot study (one of 19 funded by the Federal Highway Administration (see Federal Highway Administration, 2012)) had a substantial landslide component (Washington State Department of Transportation, see Maurer *et al.*, 2011). Maurer *et al.* (2011) present a summary of expected environmental changes in Washington and highlight the highway segments that they expect will be vulnerable to these changes. In general, they are expecting more wildfires and landslides along roadways through mountainous regions in the western and central parts of the state due to more extreme weather events and a decrease in snowpack. They expect an increase in landslides that close roads for more than 60 days.

Also in Washington State, the Department of Ecology has prepared a climate-change response strategy (Adelsman *et al.*, 2012). They are expecting increased damage costs and disruptions to communities, transportation systems, and other infrastructure along the Washington coastline. They provide specific strategy items that partly relate to expected landslide problems. One of these strategy items is avoiding development in highly vulnerable areas along coastlines. To support this strategy, they recommend specific actions including: (1) providing guidance and tools to local jurisdictions to help them assess risk and vulnerability; (2) identifying incentives and regulatory tools to discourage development in high risk areas; (3) providing incentives to local governments to consider climate change in planning documents; (4) developing policies to guide insurance companies that deal with properties in vulnerable areas; (5) assessing damage costs and discouraging rebuilding of structures in vulnerable areas; and (6) adopting construction techniques and building codes to increase resilience in vulnerable areas.

In Oregon, where damages from landslides exceed U.S. $100 million in some years (McConnell, 2009), the Department of Transportation (ODOT) has developed an adaptation strategy that includes landslides (Oregon Department of Transportation, 2012). ODOT estimates that there will be an approximate 4% annual increase in the number of landslide and rock fall problems, but, because of insufficient funding, it is only able to repair 2% of high-priority landslides each year. Specifically, ODOT proposes to: (1) inventory and rate all landslides that affect infrastructure and highway corridors and consider climate change as a factor in rating and prioritizing landslide areas, and (2) develop new engineering standards for slopes at new and repaired infrastructure to better mitigate against future landslides.

In California, adaptation strategies have been developed by the state (State of California, 2009) and its Department of Transportation (California Department of Transportation, 2013). In the state report, they recognize that the combination of a potentially drier climate, increased sea level, and an increased likelihood of extreme rainfall events will increase the chance of drought and wildfires and cause more landslides that will disrupt major roadways and rail lines. They also acknowledge that landslide problems are already well known in California, but if they become more frequent, they will result in greater costs to the state. Their adaptation strategy related to landslides includes: (1) trying to avoid significant new development in areas that cannot be protected; (2) considering the potential impacts of locating significant new state projects in areas susceptible to hazards; (3) encouraging and assisting local communities to amend their general plans and coastal plans to identify areas most vulnerable to impacts from climate change and develop risk reduction strategies; and (4) requiring state firefighting agencies to include climate change impacts in fire program planning efforts. The California Department of Transportation (2013) adaptation report recognizes the same landslide hazards as the State report (State of California, 2009). In wildfire areas, they plan to incorporate landslide mitigation measures in projects and increase monitoring of slope stability. To adapt to an increase in landslides caused by extreme precipitation events, they plan to identify highway segments affected by past intense events, identify alternative routes, monitor drainage systems during extreme events, incorporate landslide mitigation measures in vulnerable areas, and perform risk assessments for any new roads.

Other states have also adopted climate change action plans and adaptation strategies (e.g., State of Florida, 2008; State of New Hampshire, 2009; State of Vermont, 2010), but because they have experienced fewer historical landslide problems, they have less of a focus on future landslide issues compared to West Coast States. Similarly, many local governments have implemented action plans that do not include landslides because of a lack of historical landslide problems (e.g., City of Houston, 2008; City of Boston, 2011).

In the western United States in general, two federal government agencies, the National Park Service and U.S. Forest Service, have taken active roles in preparing for landslides as a result of climate change. U.S. National Parks and Forests in Washington State are proposing to change their maintenance prioritization process for roads and trails to account for predicted increases in fall and winter rainfall and soil saturation (Strauch *et al.*, 2015). Proposed changes for roads include new construction to avoid areas of elevated landslide hazard, altering road surface types, drainage improvements, and planning for higher maintenance costs. Proposed changes for trails include improving drainage, stabilizing slopes, and restoring vegetation cover. The U.S. Forest Service is very aware of the increasing frequency of wildfires and has published a technical report outlining the problem and presenting possible actions to reduce risks associated with wildfires (Luce *et al.*, 2012). One of these choices is post-fire stabilization, which would be done to suppress post-fire erosion (including debris flows). They acknowledge however, that stabilization efforts are only effective against small rainfall and erosion events, implying that the efforts would not prevent many debris flows.

14.6 SUMMARY

My review of research on landslides and climate change indicates that, within the United States, the impacts of climate change on landslide activity will be greatest in Alaska due to changes in the cryosphere, probably followed by Pacific Northwest States of Washington and Oregon (because of glacial retreat in the Cascade Mountains). The projected increase in sea level will impact all coastal states, but States along the West Coast (Alaska, Washington, Oregon, California, Hawai'i) appear to be most susceptible to increased landslide activity. In the western United States, research indicates that the number of wildfires has increased and may continue to do so in the future. This trend suggests that there will be increased possibilities for post-wildfire debris flows, although their frequency and magnitude is dependent on future precipitation, which is highly uncertain. For the rest of the United States, the impact of climate change on landslide hazards is unclear because of the high uncertainty in projections of future precipitation and evapotranspiration (e.g., Fig. 14.11). In particular, there is large uncertainty in changes in the magnitude and frequency of extreme storms and regional climate phenomena such as the El Niño Southern Oscillation, tropical cyclones, and the North American Monsoon. These phenomena can have a profound influence on the frequency and magnitude of shallow landslides and debris flows. Given this summary, it seems appropriate that the West Coast States have been the most aggressive in developing adaptation strategies that include landslide hazards.

Several efforts are necessary to prompt and facilitate landslide research and adaptation strategies related to climate change in the United States. First, continued development of climate models and downscaling techniques is needed to improve projections of extreme precipitation events and regional climate phenomena. The incorporation of groundwater feedbacks into Regional Climate Models (e.g., Scibek and Allen, 2006; Miguez-Macho *et al.*, 2007) may also improve our ability to forecast future landslide activity. Second, physics-based models are needed to forecast landslide hazards related to glacial retreat and degradation of permafrost. Third, continued efforts should be made to better understand changes in the types and distribution of vegetation to improve estimates of evapotranspiration and predictions of wildfire and the potential for subsequent debris flows. Fourth, continued efforts should be made to better differentiate the impact of population growth and land use from climate-change effects. Last, continued improvement is needed in making downscaled climate projection data readily available to end users and policy makers.

In a more general sense, the traditional problems encountered in analyzing and forecasting landslide activity will continue to be a problem in the future, regardless of the effects of climate change. First and foremost on this list is characterizing the complexity of soil and rock properties and subsurface hydrologic conditions, both of which can vary considerably over short distances. Such characterizations are needed for both site-specific and regional-scale landslide evaluations and are currently very difficult to achieve.

ACKNOWLEDGEMENTS

I thank Brian Collins, Francis Rengers, and Janet Slate for their constructive reviews of this paper. Matt Larsen and two anonymous reviewers provided helpful comments on an earlier manuscript.

REFERENCES

Adelsman, H., Ekrem, J., Binder, L.W., Cook, K., Cooper, K., Helbrecht, L.M., Jamison, R., Phillips, J., Pickett, P., Roalkvam, C.L., Salisbury, S., Siemann, D. & Speaks, P. (2012) *Preparing for a Changing Climate, Washington State's Integrated Climate Response Strategy.* Olympia, WA, Washington Department of Ecology Publication, No. 12-01-004. 203 pp. Available from: https://fortress.wa.gov/ecy/publications/documents/1201004.pdf.

Allen, S.K., Cox, S.C. & Owens, I.F. (2011) Rock avalanches and other landslides in the central Southern Alps of New Zealand: A regional study considering possible climate change impacts. *Landslides*, 8, 33–48.

Ashland, F.X. (2011) *Landslides Triggered by the May 1–2, 2010, Storm in the Nashville Metropolitan Area, Davidson County, Tennesse: Magnitude and Distribution*, Vol. 43. Reston, VA, Geological Society of America Abstracts with Programs. p. 53.

Backlund, P., Janetos, A. & Schimel, D. (2008) The effects of climate change on agriculture, land resources, water resources, and biodiversity in the United States. *Final Report, Synthesis and*

Assessment Product 4.3, A Report by the U.S. Climate Change Science Program and the Subcommittee on Global Change Research. Washington, DC, U.S. Department of Agriculture. 342 pp.

Baum, R.L. & Godt, J.W. (2010) Early warning of rainfall-induced shallow landslides and debris flows in the USA. *Landslides*, 7, 259–272.

Bengtsson, L., Hodges, K.I. & Roeckner, E. (2006) Storm tracks and climate change. *Journal of Climate*, 19, 3518–3543.

Beniston, M., Diaz, H. & Bradley, R. (1997) Climatic change at high elevation sites: An overview. *Climatic Change*, 36, 233–251.

Bigio, E., Swetnam, T.W. & Baisan, C.H. (2010) A comparison and integration of tree-ring and alluvial records of fire history at the Missionary Ridge Fire, Durango, Colorado, USA. *The Holocene*, 20, 1047–1061.

Bindoff, N.L., Willebrand, J., Artale, V., Cazenave, A., Gregory, J., Gulev, S., Hanawa, K., Le Quéré, C., Levitus, S., Norjiri, Y., Shum, C.K., Talley, L.D. & Unnikrishnan, A. (2007) Observations: Oceanic climate change and sea level. In: Solomon, S., Qin, D., Manning, M., Chen, Z., Marquis, M., Averyt, K.B., Tignor, M. & Miller, H.L. (eds.) *Climate Change 2007: The Physical Science Basis, Contribution of the Working Group I to the Fourth Assessment Report of the Intergovernmental Panel on Climate Change*. Cambridge and New York, NY, Cambridge University Press. pp. 385–432.

Blair Jr., R.W. (1994) Moraine and valley wall collapse due to rapid deglaciation in Mount Cook National Park, New Zealand. *Mountain Research and Development*, 14, 347–358.

Blake, E.S. & Gibney, E.J. (2011) *The Deadliest, Costliest, and Most Intense United States Tropical Cyclones From 1851 to 2010 (and Other Frequently Requested Hurricane Facts)*. NOAA Technical Memorandum NWS NHC-6. 47 pp. Available from: http://www.srh.noaa.gov/www/images/tbw/1921/DeadliestHurricanes.pdf.

Bonnard, C., Tacher, L. & Beniston, M. (2008) Prediction of landslide movements caused by climate change: Modelling the behavior of a mean elevation large slide in the Alps and assessing its uncertainties. In: Chen, Z., Zhang, Z., Li, Z., Wu, F. & Ho, K. (eds.) *Landslides and Engineered Slopes, From the Past to the Future, Proceedings of the 10th International Symposium on Landslides, June 30-July 4, 2008*, Vol. 1. Xi'an, China. London, Taylor & Francis Group. pp. 217–227.

Bovis, M.J. & Jakob, M. (2000) The July, 1998 debris flow and landslide dam at Capricorn Creek, southwestern B.C. *Canadian Journal of Earth Science*, 37, 1321–1334.

Brooks, S.M., Spencer, T. & Boreham, S. (2012) Deriving mechanisms and thresholds for cliff retreat in soft-rock cliffs under changing climates: Rapidly retreating cliffs of the Suffolk coast, U.K. *Geomorphology*, 153–154, 48–60.

Brown, J. & Romanovsky, V.E. (2008) Report from the International Permafrost Association: State of permafrost in the first decade of the 21st century. *Permafrost and Periglacial Processes*, 19, 255–260.

Brown, J., Ferrians Jr., O.J., Heginbottom, J.A. & Melnikov, E.S. (1997) *Circum-Arctic Map of Permafrost and Ground-Ice Conditions. Circum-Pacific Map CP-45, 1:10,000,000-Scale*. Washington, DC, U.S. Geological Survey in Cooperation with the Circum-Pacific Council for Energy and Mineral Resources. 1 sheet.

Brunsden, D. (2001) A critical assessment of the sensitivity concept in geomorphology. *Catena*, 42, 99–123.

Buma, J. (2000) Finding the most suitable slope stability model for the assessment of the impact of climate change on a landslide in southeast France. *Earth Surface Processes and Landforms*, 25, 565–582.

Buma, J. & Dehn, M. (2000) Impact of climate change on a landslide in south east France, simulated using different GCM scenarios and downscaling methods for local precipitation. *Climate Research*, 15, 69–81.

Caine, N. (1980) The landslide intensity-duration control of shallow landslides and debris flows. *Geographiska Annaler*, 62A, 23–27.

California Department of Transportation (2013) *Caltrans Activities to Address Climate Change, Reducing Greenhouse Gas Emissions and Adapting to Impacts.* 102 pp. Available from: http://www.dot.ca.gov/hq/tpp/offices/orip/climate_change/documents/Caltrans_ClimateChangeRprt-Final_April_2013.pdf.

Cannon, S.H. & Degraff, J. (2009) The increasing wildfire and post-fire debris-flow threat in western USA, and implications for consequences of climate change (Chapter 9). In: Sassa, K. & Canuti, P. (eds.) *Landslides – Disaster Risk Reduction.* Springer-Verlag Berlin Heidelberg. pp. 177–190.

Chiang, S. & Chang, K. (2011) The potential impact of climate change on typhoon-triggered landslides in Taiwan, 2010-2099. *Geomorphology*, 133, 143–151.

Chiarle, M., Iannotti, S., Mortara, G. & Deline, P. (2007) Recent debris flow occurrences associated with glaciers in the Alps. *Global and Planetary Change*, 56, 123–136.

City of Boston (2011) *A Climate of Progress, City of Boston Climate Action Plan Update.* 43 pp. Available from: http://www.cityofboston.gov/images_documents/A%20Climate%20of%20Progress%20-%20CAP%20Update%202011_tcm3-25020.pdf.

City of Houston (2008) *Houston-Galveston Area Council, Foresight Panel on Environmental Effects.* 43 pp. Available from: http://www.h-gac.com/community/environmental-stewardship/fpee/documents/foresight_panel_on_environmental_effects_report.pdf.

Clarke, D. & Smethurst, J.A. (2010) Effects of climate change on cycles of wetting and drying in engineered clay slopes in England. *Quarterly Journal of Engineering Geology and Hydrogeology*, 43, 473–486.

Coe, J.A. (2012) Regional moisture balance control of landslide motion: Implications for landslide forecasting in a changing climate. *Geology*, 40, 323–326.

Coe, J.A. & Godt, J.W. (2012) Review of approaches for assessing the impact of climate change on landslide hazards. In: Eberhardt, E., Froese, C., Turner, A.K. & Leroueil, S. (eds.) *Landslides and Engineered Slopes, Protecting Society Through Improved Understanding: Proceedings of the 11th International and 2nd North American Symposium on Landslides and Engineered Slopes, Banff, Canada, 3-8 June*, Vol. 1. London, Taylor & Francis Group. pp. 371–377. Available from: http://landslides.usgs.gov/docs/coe/CoeGodt2012.pdf.

Coe, J.A., Kean, J.W., Godt, J.W., Baum, R.L., Jones, E.S., Gochis, D.J. & Anderson, G.S. (2014) New insights into debris-flow hazards from an extraordinary event in the Colorado Front Range. *GSA Today*, 24, 4–10.

Coe, J.A., Baum, R.L., Allstadt, K.E., Kochevar, B.F., Schmitt, R.G., Morgan, M.L., White, J.L., Stratton, B. Hayashi, T.A., and Kean, J.W., 2016, Rock avalanche dynamics revealed by large-scale field mapping and seismic signals at a highly mobile avalanche in the West Salt Creek Valley, western Colorado: *Geosphere*, v. 12, no. 2, p. 607–631, doi:10.1130/GES01265.1

Collins, B.D., Kayen, R. & Sitar, N. (2007) Process-based empirical prediction of landslides in weakly lithified coastal cliffs, San Francisco, California, USA. In: McInnes, R., Jakeways, J., Fairbank, H. & Mathie, E. (eds.) *Landslides and Climate Change – Challenges and Solutions.* London, Taylor & Francis. pp. 175–184.

Collison, A., Wade, S., Griffths, J. & Dehn, M. (2000) Modelling the impact of predicted climate change on landslide frequency and magnitude in SE England. *Engineering Geology*, 55, 205–218.

Comegna, L., Picarelli, L., Bucchignani, E. & Mercogliano, P. (2013) Potential effects of incoming climate changes on the behavior of slow active landslides in clay. *Landslides*, 10, 373–391.

Crozier, M.J. (2010) Deciphering the effect of climate change on landslide activity: A review. *Geomorphology*, 124, 260–267.

Daanen, R.P., Grosse, G., Darrow, M.M., Hamilton, T.D. & Jones, B.M. (2012) Rapid movement of frozen debris-lobes: Implications for permafrost degradation and slope instability in the south-central Brooks Range, Alaska. *Natural Hazards and Earth System Sciences*, 12, 1521–1537.

Dale, V.H. (1997) The relationship between land-use change and climate change. *Ecological Applications*, 7, 753–769.

Dale, V.H., Joyce, L.A., McNulty, S., Neilson, R.P., Ayres, M.P., Flannigan, M.D., Hanson, P.J., Irland, L.C., Lugo, A.E., Peterson, C.J., Simberloff, D., Swanson, F.J., Stocks, B.J. & Wotton, B.M. (2001) Climate change and forest disturbances. *Bioscience*, 51, 723–734.

Deb, S.K. & El-Kadi, A.I. (2009) Susceptibility assessment of shallow landslides on Oahu, Hawaii, under extreme-rainfall events. Geomorphology, 108, 219–233.

Dehn, M. & Buma, J. (1999) Modelling future landslide activity based on general circulation models. *Geomorphology*, 30, 175–187.

Dehn, M., Bürger, G., Buma, J. & Gasparetto, P. (2000) Impact of climate change on slope stability using expanded downscaling. *Engineering Geology*, 55, 193–204.

Deline, P., Alberto, W., Hungr, O., Noetzli, J., Ravanel, L. & Tamburini, A. (2011) The December 2008, Crammont rock avalanche, Mount Blanc massif area, Italy. *Natural Hazards and Earth System Sciences*, 11, 3307–3318.

Dettinger, M. (2011) Climate change, atmospheric rivers, and floods in California – A multimodel analysis of storm frequency and magnitude changes. *Journal of the American Water Resources Association*, 47, 514–523.

Dirmeyer, P.A. & Kinter III, J.L. (2009) The Maya Express – Late spring floods in the U.S. Midwest. Eos, Transactions of the American Geophysical Union, 90, 101–102.

Dixon, N. & Brook, E. (2007) Impact of predicted climate change on landslide reactivation: Case study of Mam Tor, UK. *Landslides*, 4, 137–147.

Dyurgerov, M.B. & Meier, M.F. (2000) Twentieth century climate change: Evidence from small glaciers. *Proceedings of the National Academy of Sciences of the United States of America*, 97, 1406–1411.

Easterling, D.R. & Wehner, M.F. (2009) Is the climate warming or cooling? *Geophysical Research Letters*, 36, L08706.

Eliasson, P.E., McMurtrie, R.E., Pepper, D.A., Strömgren, M., Linder, S. & Ågren, G.I. (2005) The response of heterotrophic CO_2 flux to soil warming. *Global Change Biology*, 11, 167–181.

Ellen, S.D., Iverson, R.M. & Pierson, T.C. (1991) Map showing the distribution of debris flows during the New Year's Eve storm of 1987-1988 in southeastern Oahu, Hawaii. U.S. Geological Survey Open-File Report, 91–129, scale 1:20,000.

Evans, S.G. & Clague, J.J. (1994) Recent climate changes and catastrophic geomorphic processes in mountain environments. *Geomorphology*, 10, 107–128.

Federal Highway Administration (2012) Climate Change & Extreme Weather Vulnerability Assessment Framework. U.S. Department of Transportation, Federal Highway Administration FHWA Publication No. FHWA-HEP-13-005. 51 pp. Available from: http://www.fhwa.dot.gov/environment/climate_change/adaptation/publications_and_tools/vulnerability_assessment_framework/fhwahep13005.pdf.

FEMA (2006) Policy 9524.2 Landslides & Slope Failures. Available from: http://www.fema.gov/9500-series-policy-publications/95242-landslides-slope-failures.

Fischer, L., Purves, R.S., Huggel, C., Noetzli, J. & Haeberli, W. (2012) On the influence of topographic, geological and cryospheric factors on rock avalanches and rockfalls in high-mountain areas. Natural Hazards and Earth System Sciences, 12, 241–254.

Fischer, L., Huggel, C., Kääb, A. & Haeberli, W. (2013) Slope failures and erosion rates on a glacierized high-mountain face under climatic changes. Earth Surface Processes and Landforms, 38, 836–846.

Ford, J.D. & Smit, B. (2004) A framework for assessing the vulnerability of communities in the Canadian Arctic to risks associated with climate change. *Arctic*, 57, 389–400.

Francani, V., Gattinoni, P. & Rampazzo, R. (2011) Slope instability triggered by climate change in periglacial areas. *Italian Journal of Engineering Geology and Environment*, 2, 39–62.

Francou, B., Vuille, M., Wagnon, P., Mendoza, J. & Sicart, J. (2003) Tropical climate change recorded by a glacier in the central Andes during the last decades of the twentieth century: Chacaltaya, Bolivia, 16°S. *Journal of Geophysical Research*, 108, 4154.

Frechette, J.D. & Meyer, G.A. (2009) Holocene fire-related alluvial-fan deposition and climate in ponderosa pine and mixed-conifer forests, Sacramento Mountains, New Mexico. *The Holocene*, 19, 639–651.

Gao, X. & Dirmeyer, P.A. (2006) A multimodel analysis, validation, and transferability study of global soil wetness products. *Journal of Hydrometeorology*, 7, 1218–1236.

Geertsema, M. (2012) Initial Observations of the 11 June 2012 Rock/Ice Avalanche, Lituya Mountain, Alaska. The First Meeting of Cold Region Landslides Network. China, Harbin. 5 pp.

Geertsema, M., Clague, J.J., Schwab, J.W. & Evans, S.G. (2006) An overview of recent large catastrophic landslides in northern British Columbia, Canada. *Engineering Geology*, 83, 120–143.

Geertsema, M., Clague, J.J. & Hasler, A. (2013) Influence of climate change on geohazards in Alaskan Parks. Alaska Park Science, 12 (2), 81–85. Available from: http://www.nps.gov/akso/nature/science/ak_park_science/PDF/Vol12-2/APS_Vol12-Issue2_complete.pdf.

Georgakakos, A., Fleming, P., Dettinger, M., Peters-Lidard, C., Richmond, T.C., Reckhow, K., White, K. & Yates, D. (2014) Chapter 3: Water resources. In Melillo, J.M., Richmond, T.C. & Yohe, G.W. (eds.) Climate Change Impacts in the United States: The Third National Climate Assessment. U.S. Global Change Research Program. pp. 69–112.

Gibbs, A.E. & Richmond, B.M. (2015) National Assessment of Shoreline Change – Historical Shoreline Change Along the North Coast of Alaska, U.S. Canadian Border to Icy Cape. Santa Cruz, CA, U.S. Geological Survey Open-File Report 2015–1048. 96 pp. Available from: http://dx.doi.org/10.3133/ofr20151048, http://pubs.usgs.gov/of/2015/1048/.

Gochis, D., Schumacher, R., Friedrich, K., Doesken, N., Kelsch, M., Sun, J., Ikeda, K., Lindsey, D., Wood, A., Dolan, B., Matrosov, S., Newman, A., Mahoney, K., Rutledge, S., Johnson, R., Kucera, P., Kennedy, P., Sempere-Torres, D., Steiner, M., Roberts, R., Wilson, J., Yu, W., Chandrasekar, V., Rasmussen, R., Anderson, A. & Brown, B. (2015) The great Colorado flood of September 2013. *Bulletin of the American Meteorological Society*, 96, 1461–1487.

Godt, J.W. & Coe, J.A. (2007) Alpine debris flows triggered by a 28 July 1999 thunderstorm in the central Front Range, Colorado. *Geomorphology*, 84, 80–97. Available from: http://dx.doi.org/10.1016/j.geomorph.2006.07.009.

Godt, J.W., Arnal, C.H., Baum, R.L., Brien, D., Coe, J.A., De Mouthe, J., Ellis, W.L., Graymer, R.W., Harp, E.L., Hillhouse, J.W., Houdre, N., Howell, D.G., Jayko, A.S., Lajoie, K.R., Morrissey, M.M., Ramsey, D.W., Savage, W.Z., Schuster, R.L., Wieczorek, G.F. & Wilson, R.C. (1999) Maps Showing Locations of Damaging Landslides Caused by El Niño Rainstorms, Winter Season 1997-98. San Francisco Bay region, California. Oakland, CA, U.S. Geological Survey Miscellaneous Field Studies Map MF-2325-A-J.

Godt, J.W., Baum, R.L. & Chleborad, A.F. (2006) Rainfall characteristics for shallow landslides in Seattle, Washington, USA. *Earth Surface Processes and Landforms*, 31, 97–110.

Godt, J.W., Coe, J.A., Baum, R.L., Highland, L.M., Keaton, J.R. & Roth Jr., R.J., (2012) Prototype landslide hazard map of the conterminous United States. In: Eberhardt, E., Froese, C., Turner, A.K. & Leroueil, S. (eds.) Landslides and Engineered Slopes, Protecting Society

Through Improved Understanding: Proceedings of the 11th International and 2nd North American Symposium on Landslides and Engineered Slopes, Banff, Canada, 3-8 June, Vol. 1. London, Taylor & Francis Group. pp. 245–250.

Godt, J.W., Coe, J.A., Kean, J.W., Baum, R.L., Jones, E.S., Harp, E.L., Staley, D.M. & Barnhart, W.D. (2014) *Landslides in the Northern Colorado Front Range Caused by Rainfall, September 11–13, 2013*. U.S. Geological Survey Fact Sheet 2013-3114. 4 pp.

Golubev, V.S., Lawrimore, J.H., Groisman, P.Y., Speranskaya, N.A., Zhuravin, S.A., Menne, M.J., Peterson, T.C. & Malone, R.W. (2001) Evaporation changes over the contiguous United States and the former USSR: A reassessment. *Geophysical Research Letters*, 28, 2665–2668.

Gregovich, L. (2010) *Alaska DOT&PF Adaptation to Climate Change. American Association of State Highway and Transportation Officials*. Green Streets Conference Power Point Presentation. Available from: http://climatechange.transportation.org/pdf/gregovichadot.pdf.

Gruber, S. & Haeberli, W. (2007) Permafrost in steep bedrock slopes and its temperature-related destabilization following climate change. *Journal of Geophysical Research*, 112, F02S18.

Guthrie, R.H., Mitchell, S.J., Lanquaye-Opoku, N. & Evans, S.G. (2010) Extreme weather and landslide initiation in coastal British Columbia. *Quarterly Journal of Engineering Geology and Hydrogeology*, 43, 417–428.

Haberli, W. & Beniston, M. (1998) Climate change and its impacts on glaciers and permafrost in the Alps. *Ambio*, 27, 258–265.

Haeberli, W. (2013) Mountain permafrost – Research frontiers and a special long-term challenge. *Cold Regions Science and Technology*, 96, 71–76.

Hamlet, A.F., Mote, P.W., Clark, M.P. & Lettenmaier, D.P. (2007) Twentieth-century trends in runoff, evapotranspiration, and soil moisture in the western United States. *Journal of Climate*, 20, 1468–1486.

Harris, C. (2005) Climate change, mountain permafrost degradation and geotechnical hazard. In: Huber, U.M., Bugmann, H.K.M. & Reasiner M.A. (eds.) *Global Change and Mountain Regions*. Netherlands, Springer. pp. 215–224.

Harris, C., Davies, M.C.R. & Etzelmüller, B. (2001) The assessment of potential geotechnical hazards associated with mountain permafrost in a warming global climate. *Permafrost and Periglacial Processes*, 12, 145–156.

Harris, C., Arenson, L.U., Christiansen, H.H., Etzelmüller, B., Frauenfelder, R., Gruber, S., Haeberli, W., Hauck, C., Hölzle, M., Humlum, O., Isaksen, K., Kääb, A., Kern-Lütschg, M.A., Lehning, M., Matsuoka, N., Murton, J.B., Nötzli, J., Phillips, M., Ross, N., Seppälä, M., Springman, S.M. & Mühll, D.V. (2009) Permafrost and climate in Europe: Monitoring and modeling thermal, geomorphological and geotechnical responses. *Earth-Science Reviews*, 92, 117–171.

Higgins, R.W., Schemm, J.-K.E., Shi, W. & Leetma, A. (2000) Extreme precipitation events in the western United States related to tropical forcing. *Journal of Climate*, 13, 793–820.

Higgins, R.W., Kousky, V.E. & Xie, P. (2011) Extreme precipitation events in the south central United States during spring 2010: Historical perspective, role of ENSO and trends. *Journal of Hydrometeorolgy*, 12, 1056–1070.

Holmes Jr., R.R., Jones, L.M., Eidenshink, J.C., Godt, J.W., Kirby, S.H., Love, J.J., Neal, C.A., Plant, N.G., Plunkett, M.L., Weaver, C.S., Wein, A. & Perry, S.C. (2013) *U.S. Geological Survey Natural Hazards Science Strategy – Promoting the Safety, Security, and Economic Well-Being of the Nation*. Reston, VA, U.S. Geological Survey Circular 1383–F. 79 pp.

Hugenholtz, C.H., Moorman, B.J., Barlow, J. & Wainstein, P.A. (2008) Large-scale moraine deformation at the Athabasca glacier, Jasper National Park, Alberta, Canada. *Landslides*, 5, 251–260.

Huggel, C. (2009) Recent extreme slope failures in glacial environments: Effects of thermal perturbation. *Quaternary Science Reviews*, 28, 1119–1130.

Huggel, C., Slazmann, N., Allen, S., Caplan-Auerbach, J., Fischer, L., Haeberli, W., Larsen, C., Schneider, D. & Wessels, R. (2010) Recent and future warm extreme events and high-mountain slope stability. *Philosophical Transactions of the Royal Society A*, 368, 2435–2459.

Huggel, C., Clague, J.J. & Korup, O. (2012) Is climate change responsible for changing landslide activity in high mountains? *Earth Surface Processes and Landforms*, 37, 77–91.

Hürlimann, M., Abancó, C. & Moya, J. (2012) Rockfalls detached from a lateral moraine during spring season. 2010 and 2011 events observed at the Rebaixader debris-flow monitoring site (Central Pyrenees, Spain). *Landslides*, 9, 385–393.

Isaksen, K., Sollid, J.L., Holmlund, P. & Harris, C. (2007) Recent warming of mountain permafrost in Svalbard and Scandinavia. *Journal of Geophysical Research*, 112, F02S04.

Iverson, R.M. (2000) Landslide triggering by rain infiltration. *Water Resources Research*, 36, 1897–1910.

Iverson, R.M., George, D.L., Allstadt, K., Reid, M.R., Collins, B.D., Vallance, J.W., Schilling, S.P., Godt, J.W., Cannon, C.M., Magirl, C.S., Baum, R.L., Coe, J.A., Schulz, W.H. & Bower, J.B. (2015) Landslide mobility and hazards: Broad implications of the Oso disaster. *Earth and Planetary Science Letters*, 412, 197–208.

Jacobson, R.B., McGeehin, J.P., Cron, E.D., Carr, C.E., Harper, J.M. & Howard, A.D. (1993) Landslides triggered by the storm of November 3–5, 1985, Wills Mountain Anticline, West Virginia and Virginia (Chapter C). In: Jacobson, R.B. (ed.) *Geomorphic Studies of the Storm and Flood of November 3–5, 1985, in the Upper Potomac and Cheat River Basins in West Virginia and Virginia*. U.S. Geological Survey Bulletin 1981. pp. C1–C33.

Jakob, M. & Lambert, S. (2009) Climate change effects on landslides along the southwest coast of British Columbia. *Geomorphology*, 107, 275–284.

Johnson, F. & Sharma, A. (2009) Measurement of GCM skill in predicting variables for hydroclimatological assessments. *Journal of Climate*, 22, 4373–4382.

Jomelli, V., Brunstein, D., Grancher, D. & Pech, P. (2007) Is the response of hill slope debris flows to recent climate change univocal? A case study in the Massif des Ecrins (French Alps). *Climatic Change*, 85, 119–137.

Jorgenson, T., Yoshikawa, K., Kanevskiy, M., Shur, Y., Romanovsky, V., Marchenka, S., Grosse, G., Brown, J. & Jones, B. (2008) *Permafrost characteristics of Alaska*. Fairbanks, AK, Institute of Northern Engineering, University of Alaska Fairbanks. Available from: http://permafrost.gi.alaska.edu/sites/default/files/AlaskaPermafrostMap_Front_Dec2008_Jorgenson_etal_2008.pdf.

Karl, T.R., Meehl, G.A., Miller, C.D., Hassol, S.J., Waple, A.M. & Murray, W.L. (eds.) 2008. Weather and climate extremes and in changing climate, Regions of focus: North America, Hawaii, Caribbean, and U.S. Pacific Islands. *Synthesis and Assessment Product 3.3 Report by the U.S. Climate Change Science Program and the Subcommittee on Global Change Research*. Washington, DC, U.S. Climate Change Science Program. 162 pp.

Kaser, G., Hardy, D.R., Mölg, T., Bradley, R.S. & Hyera, T.M. (2004) Modern glacier retreat on Kilimanjaro as evidence of climate change: Observations and facts. *International Journal of Climatology*, 24, 329–339.

Keaton, J.R. & Roth Jr., R.J., (2010) A simple engineering geological approach useful for insuring landslides. In: Williams, A.L., Pinches, G.M., Chin, C.Y., McMorran, T.J. & Massey, C.I. (eds.) *Geologically Active: Proceedings of the 11th IAEG Congress. September 5–10, 2010, CD Paper*. Auckland, CRC Press. 7 pp.

Keaton, J.R., Wartman, J., Anderson, S., Benoît, J., deLaChapelle, J., Gilbert, R. & Montgomery, D.R. (2014) *The 22 March 2014 Oso Landslide*. Snohomish County, Washington, Geotechnical Extreme Events Reconnaissance (GEER) report. 175 pp. Available from: http://www.geerassociation.org/GEER_Post%20EQ%20Reports/Oso_WA_2014/GEER_Oso_Landslide_Report.pdf.

Keefer, D.K. & Larsen, M.C. (2007) Assessing landslide hazards. *Science*, 316 (5828), 1136–1138.

Knutson, T.R., McBride, J.L., Chan, J., Emanuel, K., Holland, G., Landsea, C., Held, I., Kossin, J.P., Srivastava, A.K. & Sugi, M. (2012) Tropical cyclones and climate change. *Nature Geoscience*, 3, 157–163.

Kochel, R.C. & Trop, J.M. (2012) Active processes, morphology, and dynamics of icy debris fans: Landform evolution along rapidly degrading escarpments in alpine regions undergoing recent deglaciation. *Geomorphology*, 151–152, 59–76.

Korup, O., Görüm, T. & Hayakawa, Y. (2012) Without power? Landslide inventories in the face of climate change. *Earth Surface Processes and Landforms*, 37, 92–99.

Kunkel, K.E., Karl, T.R., Brooks, H., Kossin, J., Lawrimore, J., Arndt, D., Bosart, L., Changnon, D., Cutter, S.L., Doesken, N., Emanuel, K., Groisman, P.Ya., Katz, R.W., Knutson, T., O'Brien, J., Paciorek, C.J., Peterson, T.C., Redmond, K., Robinson, D., Trapp, J., Vose, R., Weaver, S., Wehner, M., Wolter, K. & Wuebbles, D. (2013) Monitoring and understanding trends in extreme storms: State of knowledge. *Bulletin of the American Meteorological Society*, 94, 499–514.

Lajoie, K.R. & Mathieson, S.A. (1998) 1982 *-83 El Niño Coastal Erosion*. San Mateo County, CA, U.S. Geological Survey Open File Report. 98–41, 61 pp.

Lancaster, S.T., Nolin, A.E., Copeland, E.A. & Grant, G.E. (2012) Periglacial debris-flow initiation and susceptibility and glacier recession from imagery, airborne LiDAR, and ground-based mapping. *Geosphere*, 8, 417–430.

Lawrence, D.M., Thornton, P.E., Oleson, K.W. & Bonan, G.B. (2007) The partitioning of evapotranspiration into transpiration, soil evaporation, and canopy evaporation in a GCM: Impacts on land-atmosphere interaction. *Journal of Hydrometeorology*, 8, 862–880.

Lee, E.M. (2011) Reflections on the decadal-scale response of coastal cliffs to sea-level rise. *Quarterly Journal of Engineering Geology*, 44, 481–489.

Legg, N.T., Meigs, A.J., Grant, G.E. & Kennard, P. (2014) Debris flow initiation in proglacial gullies on Mount Rainier, Washington. *Geomorphology*, 226, 249–260.

Lemke, P., Ren, J., Alley, R.B., Allison, I., Carrasco, J., Flato, G., Fujii, Y., Kaser, G., Mote, P., Thomas R.H. & Zhang, T. (2007) Observations: Changes in snow, ice and frozen ground. In: Solomon, S., Qin, D., Manning, M., Chen, Z., Marquis, M., Averyt, K.B., Tignor, M. & Miller, H.L. (eds.) *Climate Change 2007: The Physical Science Basis, Contribution of the Working Group I to the Fourth Assessment Report of the Intergovernmental Panel on Climate Change*. Cambridge, and New York, NY, Cambridge University Press. pp. 337–383.

Loveridge, F.A., Spink, T.W., O'Brien, A.S., Briggs, K.M. & Butcher, D. (2010) The impact of climate and climate change on infrastructure slopes, with particular reference to southern England. *Quarterly Journal of Engineering Geology and Hydrogeology*, 43, 461–472.

Luce, C., Morgan, P., Dwire, K., Isaak, D., Holden, Z. & Rieman, B. (2012) *Climate Change, Forests, Fire, Water, and Fish: Building Resilient Landscapes, Streams, and Managers. General Technical Report RMRS-GTR-290*. Fort Collins, CO, U.S. Department of Agriculture, Forest Service, Rocky Mountain Research Station. 207 pp.

Luckman, B. & Kavanagh, T. (2000) Impact of climate fluctuations on mountain environments in the Canadian Rockies. *Ambio*, 29, 371–380.

Lugon, R. & Stoffel, M. (2010) Rock-glacier dynamics and magnitude-frequency relations of debris flows in a high-elevation watershed: Ritigraben, Swiss Alps. *Global and Planetary Change*, 73, 202–210.

Lund, W.R., Knudsen, T.R. & Bowman, S.D. (2014) Investigation of the December 12, 2013, fatal rock fall at 368 West Main Street, Rockville, Utah. *Utah Geological Survey Report of Investigation*, 273. 20 pp. Available from: http://files.geology.utah.gov/online/ri/ri-273.pdf.

Mahoney, J.R., Asrar, G., Leinen, M.S., Andrews, J., Glackin, M., Groat, C., Hohenstein, W., Lawson, L., Moore, M., Neale, P., Patrinos, A., Schafer, J., Slimak, M. & Watson, H.

(2003) *Strategic Plan for the U.S. Climate Change Science Program*. A Report by the Climate Change Science Program and Subcommittee on Global Change Research. Washington, DC, U.S. Climate Change Science Program. 202 pp.

Malet, J.-P., Remaître, A., Maqaire, O., Durand, Y., Etchevers, P., Guyomarch, G., Déqué, M. & van Beek, L.P.H. (2007) Assessing the influence of climate change on the activity of landslides in the Ubaye Valley. In: McInnes, R., Jakeways, J., Fairbank, H. & Mathie, E. (eds.) *Landslides and Climate Change – Challenges and Solutions*. London, Taylor & Francis. pp. 195–205.

Mass, C., Skalenakis, A. & Warner, M. (2011) Extreme precipitation over the west coast of North America: Is there a trend? *Journal of Hydrometeorology*, 12, 310–318.

Maurer, M., Roalkvam, C.L. & Salsibury, S.L. (2011) *Climate Impacts Vulnerability Assessment*. Olympia, Washington State Department of Transportation Report. 70 pp. Available from: http://www.wsdot.wa.gov/NR/rdonlyres/B290651B-24FD-40EC-BEC3-EE5097ED0618/0/WSDOTClimateImpactsVulnerabilityAssessmentforFHWAFinal.pdf.

McConnell, V.S. (2009) *New Maps Reveal Landslides Using Laser Based Imaging: New Technology Is Helping State Geologists Create Some of the Most Accurate Landslide Inventory Maps in the World*. Oregon Department of Geology and Mineral Industries (DOGAMI) News Release, April 20, 2009. Available from: http://www.oregongeology.org/sub/news&events/archives/press-release-2009-04-20.pdf.

Meehl, G.A., Stocker, T.F., Collins, W.D., Friedlingstein, P., Gaye, A.T., Gregory, J.M., Kitoh, A., Knutti, R., Murphy, J.M., Noda, A., Raper, S.C.B., Watterson, I.G., Weaver, A.J. & Zhao, Z.C. (2007) Global climate projections. In: Solomon, S., Qin, D., Manning, M., Chen, Z., Marquis, M., Averyt, K.B., Tignor, M. & Miller, H.L. (eds.) *Climate Change 2007: The Physical Science Basis, Contribution of the Working Group I to the Fourth Assessment Report of the Intergovernmental Panel on Climate Change*. Cambridge and New York, NY, Cambridge University Press. pp. 747–845.

Meier, M.F., Dyurgerov, M.B., Rick, U.K., O'Neel, S., Pfeffer, W.T., Anderson, R.S., Anderson, S.P. & Glazovsky, A.F. (2007) Glaciers dominate eustatic sea-level rise in the 21st century. *Science*, 317, 1064–1067.

Mellander, P., Löfvenius, M.O. & Laudon, H. (2007) Climate change impact on snow and soil temperature in boreal Scots pine stands. *Climatic Change*, 85, 179–193.

Miguez-Macho, G., Fan, Y., Weaver, C.P., Walko, R. & Robock, A. (2007) Incorporating water table dynamics in climate modeling: 2. Formulation, validation and soil moisture simulation. *Journal of Geophysical Research*, 112, d13108.

Mool, P.K., Maskey, P.R., Koirala, A., Joshi, S.P., Lizong, W., Shrestha, A.B., Eriksson, M., Gurung, B., Pokharel, B., Khanal, N.R., Panthi, S., Adhikari, T., Kayastha, R.B., Ghimire, P., Thapa, R., Shrestha, B., Shrestha, S. & Shrestha, R.B. (2011) *Glacial Lakes and Glacial Lake Outburst Floods in Nepal*. Kathmandu, International Center for Integrated Mountain Development. 99 pp. Available from: http://lib.icimod.org/record/27755/files/icimod-glacial_lakes_and_glacial_lake_outburst_floods_in_nepal.pdf.

Moore, R., Carey, J.M. & McInnes, R.G. (2010) Landslide behavior and climate change: Predictable consequences for the Ventnor Undercliff, Isle of Wight. *Quarterly Journal of Engineering Geology and Hydrogeology*, 43, 447–460.

Moss, R.H., Edmonds, J.A., Hibbard, K.A., Manning, M.R., Rose, S.K., van Vuuren, D.P., Carter, T.R., Emori, S., Kainuma, M., Kram, T., Meehl, G.A., Mitchell, J.F.B., Nakicenovic, N., Riahi, K., Smith, S.J., Stouffer, R.J., Thomson, A.M., Weyant, J.P. & Wilbanks, T.J. (2010) The next generation of scenarios for climate change research and assessment. *Nature*, 463, 747–756.

NBC News (2015) Final body found after deadly landslide in Sitka, Alaska. *NBC News Online Article*, August 25, 2015. Available from: http://www.nbcnews.com/news/us-news/final-body-found-after-deadly-landslide-sitka-alaska-n416046.

Neiman, P.J., Ralph, F.M., Wick, G.A., Lundquist, J.D. & Dettinger, M.D. (2008) Meteorological characteristics and overland precipitation impacts of atmospheric rivers affecting the west coast of North America based on eight years of SSM/I satellite observations. *Journal of Hydrometeorology*, 9, 22–47.

Nelson, F.E., Brigham, L.W., Hinkel, K.M., Smith, O., Parker, W., Tucker, W., Romanovsky, V.E., Vinson, T., Shiklomanov, N.I., Cole, D.M., Lee, W., Klene, A.E. & Saxton, A. (2003) *Climate Change, Permafrost, and Impacts on Civil Infrastructure*. Arlington, Virginia, U.S. Arctic Research Commission Permafrost Task Force Special Report 01-03. 62 pp.

Nott, J.F., Thomas, M.F. & Price, D.M. (2001) Alluvial fans, landslides and late quaternary climatic change in the wet tropics of northeast Queensland. *Australian Journal of Earth Sciences*, 48, 875–882.

NRC (National Research Council) (2004) *Partnerships for Reducing Landslide Risk: Assessment of the National Landslide Hazards Mitigation Strategy*. Washington, DC, National Research Council of the National Academies, The National Academies Press. 144 pp.

O'Connor, J.E. & Costa, J.E. (1993) Geologic and hydrologic hazards in glacierized basins in North America resulting from 19th and 20th century global warming. *Natural Hazards*, 8, 121–140.

Olshansky, R.B. & Rogers, J.D. (1987) Unstable ground: Landslide policy in the United States. *Ecology Law Quarterly Article 3*, 13, 939–1006. Available from: http://scholarship.law.berkeley.edu/cgi/viewcontent.cgi?article=1298&context=elq

Oppikofer, T., Jaboyedoff, M. & Keusen, H.R. (2008) Collapse at the eastern Eiger flank in the Swiss Alps. *Nature Geoscience*, 1, 531–535.

Oregon Department of Transportation (2012) *ODOT's Climate Change Adaptation Strategy Report*. Available from: http://www.oregon.gov/ODOT/TD/CLIMATECHANGE/docs/odot_adaptation_strategy_final.pdf.

Overeem, I., Anderson, R.S., Wobus, C.W., Clow, G.D., Urban, F.E. & Matell, N. (2011) Sea ice loss enhances wave action at the Arctic coast. *Geophysical Research Letters*, 38, 1–6.

Pastor, J. & Post, W.M. (1988) Response of northern forests to CO_2 induced climate change. *Nature*, 334, 55–58.

Petley, D.N. (2010) On the impact of climate change and population growth on the occurrence of fatal landslides in South, East and SE Asia. *Quarterly Journal of Engineering Geology and Hydrogeology*, 43, 487–496.

Pfeffer, W.T., Harper, J.T. & O'Neel, S. (2008) Kinematic constraints on glacier contributions to 21st-century sea-level rise. *Science*, 321, 1340–1343.

Pielke, R.A., Marlande Sr., G., Betts, R.A., Chase, T.N., Eastman, J.L., Niles, J.O., Niyogi, D.D.S. & Running, S.W. (2002) The influence of land-use change and landscape dynamics on the climate system: Relevance to climate-change policy beyond the radiative effect of greenhouse gases. *Philosophical Transactions of the Royal Society A*, 360, 1705–1719.

Pierce, J. & Meyer, G. (2008) Long-term fire history from alluvial fan sediments: The role of drought and climate variability, and implications for management of Rocky Mountain forests. *International Journal of Wildland Fire*, 17, 84–95.

Randall, D.A., Wood, R.A., Bony, S., Colman, R., Fichefet, T., Fyfe, J., Kattsov, V., Pitman, A., Shukla, J., Srinivasan, J., Stouffer, R.J., Sumi, A. & Taylor, K.E. (2007) Climate models and their evaluation. In: Solomon, S., Qin, D., Manning, M., Chen, Z., Marquis, M., Averyt, K.B., Tignor, M. & Miller, H.L. (eds.) *Climate Change 2007, The Physical Science Basis, Contribution of Working Group I to the Fourth Assessment Report of the Intergovernmental Panel on Climate Change*. Cambridge, Cambridge University Press. pp. 589–662.

Rangwala, I. & Miller, J.R. (2010) Twentieth century temperature trends in Colorado's San Juan Mountains. *Arctic, Antarctic, and Alpine Research*, 42, 89–97.

Ravanel, L. & Deline, P. (2011) Climate influence on rockfalls in high-Alpine steep rockwalls: The north side of the Aiguilles de Chamonix (Mont Blanc massif) since the end of the 'Little Ice Age'. *The Holocene*, 21, 357–365.

Rebetez, M., Lugon, R. & Baeriswyl, P.A. (1997) Climatic change and debris flows in high mountain regions: The case study of the Ritigraben torrent (Swiss Alps). *Climatic Change*, 36, 371–389.

Reid, M.E. (1994) A pore-pressure diffusion model for estimating landslide-inducing rainfall. *The Journal of Geology*, 102, 709–717.

Repanshek, K. (2012) Massive landslide coats Glacier Bay National Park's Johns Hopkins glacier like chocolate frosting. *National Parks Traveler*, July 11, 2012 Online Article. Available from: http://www.nationalparkstraveler.com/2012/07/massive-landslide-coats-glacier-bay-national-parks-johns-hopkins-glacier-chocolate-frosting10200#comment-41027.

Rignot, E. & Kanagaratnam, P. (2006) Changes in the velocity structure of the Greenland Ice Sheet. *Science*, 311, 986–990.

Roth Jr., R.J. & Keaton, J.R. (2010) Fundamental inputs necessary for insuring landslides. In: Williams, A.L., Pinches, G.M., Chin, C.Y., McMorran, T.J. & Massey, C.I. (eds.) *Geologically Active: Proceedings of the 11th IAEG Congress. September 5-10, CD Paper, 2010.* Auckland, CRC Press. 5 pp.

Salathé Jr., E.P., Steed, R., Mass, C.F. & Zahn, P.H. (2008) A high-resolution climate model for the U.S. Pacific Northwest: Mesoscale feedbacks and local responses to climate change. *Journal of Climate*, 21, 5708–5726.

Schär, C., Vidale, P.L., Luthi, D., Frei, C., Haberli, C., Liniger, M.A. & Appenzeller, C. (2004) The role of increasing temperature variability in European summer heatwaves. *Nature*, 427, 332–336.

Schlögel, R., Torgoev, I., De Marneffe, C. & Havenith, H. (2011) Evidence of a changing size-frequency distribution of landslides in the Kygyz Tien Shan, central Asia. *Earth Surface Processes and Landforms*, 36, 1658–1669.

Schlyter, P., Jönsson, P., Nyberg, R., Persson, P., Rapp, A., Jonasson, C. & Rehn, J. (1993) Geomorphic process studies related to climate change in Kärkevagge, northern Sweden. Status of current research. *Geografiska Annaler*, 75A, 55–60.

Schmidt, M. & Glade, T. (2003) Linking global circulation model outputs to regional geomorphic models: A case study of landslide activity in New Zealand. *Climate Research*, 25, 135–150.

Schuster, R.L. (1996) *Socioeconomic Significance of Landslides, in Landslides: Investigation and Mitigation.* Transportation Research Board, National Research Council, Special Report 247. 675 pp.

Schuur, E.A.G. & Abbott, B. (2011) High risk of permafrost thaw. *Nature*, 480, 32–33.

Schwab, J.C., Gori, P.L. & Jeer, S. (2005) *Landslide Hazards and Planning. American Planning Association.* Planning Advisory Service Report Number 533/534. 208 pp.

Scibek, J. & Allen, D.M. (2006) Modeled impacts of predicted climate change on recharge and groundwater levels. *Water Resources Research*, 42, W11405.

Screen, J.A. & Simonds, I. (2010) The central role of diminishing sea ice in recent Arctic temperature amplification. *Nature*, 464, 1334–1337.

Serreze, M.C. & Francis, J.A. (2006) The Arctic amplification debate. *Climate Change*, 76, 241–264.

Sheffield, J. & Wood, E.F. (2008) Projected changes in drought occurrence under future global warming from multi-model, multi-scenario, IPCC AR4 simulations. *Climate Dynamics*, 31, 79–105.

Shrestha, B., Mool, P.K. & Bajracharya, S.R. (2007) *Impact of Climate Change on Himalayan Glaciers and Glacial Lakes: Case Studies on GLOF and Associated Hazards in Nepal and*

Bhutan. Kathmandu, International Centre for Integrated Mountain Development. 119 pp. Available from: http://lib.icimod.org/record/22442/files/attachment_169.pdf.

Sidle, R.C. & Burt, T.P. (2009) Temperate forests and rangelands. In: Slaymaker, O., Spencer, T. & Embleton-Hamann, C. (eds.) *Geomorphology and Global Environmental Change*. Cambridge, Cambridge University Press. pp. 321–343.

Slaymaker, O. (1990) Climate change and erosion processes in mountain regions of western Canada. *Mountain Research and Development*, 10, 171–182.

Smith, T.M. & Reynolds, R.W. (2005) A global merged land and sea surface temperature reconstruction based on historical observations (1880–1997). *Journal of Climate*, 18, 2021–2036.

Solomon, S., Qin, D., Manning, M., Alley, R.B., Berntsen, T., Bindoff, N.L., Chen, Z., Chidthaisong, A., Gregory, J.M., Hegerl, G.C., Heimann, M., Hewitson, B., Hoskins, B.J., Joos, F., Jouzel, J., Kattsov, V., Lohmann, U., Matsuno, T., Molina, M., Nicholls, N., Overpeck, J., Raga, G., Ramaswamy, V., Ren, J., Rusticucci, M., Somerville, R., Stocker, T.F., Whetton, P., Wood, R.A. & Wratt, D. (2007) Technical summary. In: Solomon, S., Qin, D., Manning, M., Chen, Z., Marquis, M., Averyt, K.B., Tignor, M. & Miller, H.L. (eds.) *Climate Change 2007: The Physical Science Basis, Contribution of the Working Group I to the Fourth Assessment Report of the Intergovernmental Panel on Climate Change*. Cambridge and New York, NY, Cambridge University Press. pp. 19–91.

Spiker, E.C. & Gori, P. (2003) National landslide hazards mitigation strategy – A framework for loss reduction. *U.S. Geological Survey Circular*, 1244, 56 pp.

State of Alaska (2010) *Alaska's Climate Change Strategy: Addressing Impacts in Alaska*. Final Report submitted by the Adaptation Advisory group to the Alaska climate change subcabinet, January 2010. Available from: http://climatechange.alaska.gov/aag/docs/aag_all_rpt_27jan10.pdf.

State of California (2009) *2009 California Climate Adaptation Strategy*. A report to the Governor of the State of California in response to executive order S-13-2008. 197 pp. Available from: http://resources.ca.gov/docs/climate/Statewide_Adaptation_Strategy.pdf.

State of Florida (2008) *Florida's Energy & Climate Change Action Plan*. Available from: http://www.broward.org/NaturalResources/ClimateChange/Documents/phase2report08.pdf.

State of New Hampshire (2009) The New Hampshire climate action plan: A plan for New Hampshire's energy, environmental and economic development future. *New Hampshire Climate Change Task Force*. 74 pp. Available from: http://des.nh.gov/organization/divisions/air/tsb/tps/climate/action_plan/nh_climate_action_plan.htm.

State of Vermont (2010) *Climate Change Adaptation Web Site*. Available from: http://www.anr.state.vt.us/anr/climatechange/Adaptation.html.

Strauch, R.L., Raymond, C.L., Rochefort, R.M., Hamlet, A.F. & Lauver, C. (2015) Adapting transportation to climate change on federal lands in Washington State, U.S.A. *Climatic Change*, 130, 185–199.

Sun, Y., Solomon, S., Dai, A. & Portmann, R.W. (2007) How often will it rain? *Journal of Climate*, 20, 4801–4818.

Tebaldi, C., Hayhoe, K., Arblaster, J.M. & Meehl, G.A. (2006) Going to the extremes: An intercomparison of model-simulated historical and future changes in extreme events. *Climatic Change*, 79, 185–211.

Trenberth, K.E., Jones, P.D., Ambenje, P., Bojariu, R., Easterling, D., Klein Tank, A., Parker, D., Rahimzadeh, F., Renwick, J.A., Rusticucci, M., Soden, B. & Zhai, P. (2007) Observations: Surface and atmospheric climate change. In: Solomon, S., Qin, D., Manning, M., Chen, Z., Marquis, M., Averyt, K.B., Tignor, M. & Miller, H.L. (eds.) *Climate Change 2007, The Physical Science Basis, Contribution of Working Group I to the Fourth Assessment Report of the Intergovernmental Panel on Climate Change*. Cambridge and New York, NY, Cambridge University Press. pp. 235–336.

van Asch, T.W.J., Buma, J. & van Beek, L.P.H. (1999) A view on some hydrological triggering systems in landslides. *Geomorphology*, 30, 25–32.

van Oldenborgh, G.J., Collins, M., Arblaster, J., Christensen, J.H., Marotzke, J., Power, S.B., Rummukainen, M. & Zhou, T. (2013) Annex I: Atlas of global and regional climate projections. In: Stocker, T.F., Qin, D., Plattner, G.-K., Tignor, M., Allen, S.K., Boschung, J., Nauels, A., Xia, Y., Bex, V. & Midgley, P.M. (eds.) *Climate Change 2013: The Physical Science Basis. Contribution of Working Group I to the Fifth Assessment Report of the Intergovernmental Panel on Climate Change*. Cambridge and New York, NY, Cambridge University Press. pp. 1311–1393.

van Westen, C.J., van Asch, T.W.J. & Soeters, R. (2006) Landslide hazard and risk zonation – Why is it still so difficult? *Bulletin of Engineering Geology and the Environment*, 65, 167–184.

Varnes, D.J. (1978) Slope movement types and processes. In: Schuster, R.L. & Krizek, R.J. (eds.) *Landslides: Analysis and Control*. Washington, DC, Transportation Research Board Special Report 176, National Research Council. pp. 11–33.

Walsh, J., Wuebbles, D., Hayhoe, K., Kossin, K., Kunkel, K., Stephens, G., Thorne, P., Vose, R., Wehner, M., Willis, J., Anderson, D., Doney, S., Feely, R., Hennon, P., Kharin, V., Knutson, T., Landerer, F., Lenton, T., Kennedy, J. & Somerville, R. (2014) Chapter 2: Our changing climate. Climate change impacts in the United States: The third national climate assessment. In: Melillo, J.M., Richmond, T.C. & Yohe, G.W. (eds.), *U.S. Global Change Research Program*. pp. 19–67.

Wang, G. (2005) Agricultural drought in a future climate: Results from 15 global climate models participating in the IPCC 4th assessment. *Climate Dynamics*, 25, 739–753.

Wasowski, J., Lamanna, C. & Casarano, D. (2010) Influence of land-use change and precipitation patterns on landslide activity in the Daunia Apennines, Italy. *Quarterly Journal of Engineering Geology and Hydrogeology*, 43, 387–401.

Westerling, A.L., Gershunov, A., Brown, T.J., Cayan, D.R. & Dettinger, M.D. (2003) Climate and wildfire in the western United States. *Bulletin of the American Meteorological Society*, 84, 595–604.

Westerling, A.L., Hidalgo, H.G., Cayan, D.R. & Swetnam, T.W. (2006) Warming and earlier spring increase in western US forest wildfire activity. *Science*, 313, 940–943.

White, J.L., Morgan, M.L. & Berry, K.A. (2015) The West Salt Creek Landslide: A catastrophic rockslide and rock/debris avalanche in Mesa County, Colorado. *Colorado Geological Survey Bulletin*, 55, 45 pp., 2 plates.

Wieczorek, G.F. & Morgan, B.A. (2008) Debris-flow hazards within the Appalachian Mountains of the eastern United States. *U.S. Geological Survey Fact Sheet*, 2008–3070. 4 pp. Available from: http://pubs.usgs.gov/fs/2008/3070/fs2008-3070.pdf.

Williams, C.J., Pierson, F.B., Robichaud, P.R. & Boll, J. (2014) Hydrologic and erosion responses to wildfire along the rangeland-xeric forest continuum in the western US: A review and model of hydrologic vulnerability. *International Journal of Wildland Fire*, 23, 155–172.

Wilson, R. & Pike, R. (2005) Watching for landslide weather in California. *Geotimes*, 50 (11), 28–33.

Winter, M.G., Dent, J., Macgregor, F., Dempsey, P., Motion, A. & Shackman, L. (2010) Debris flow, rainfall and climate change in Scotland. *Quarterly Journal of Engineering Geology and Hydrogeology*, 43, 429–446.

Wobus, C., Anderson, R., Overeem, I., Matell, N., Clow, G. & Urban, F. (2011) Thermal erosion of a permafrost coastline: Improving process-based models using time-lapse photography. *Arctic, Antarctic, and Alpine Research*, 43, 474–484.

Wood, J.L., Harrison, S. & Reinhardt, L. (2015) Landslide inventories for climate impacts research in the European Alps. *Geomorphology*, 228, 398–408.

Wuebbles, D.J., Kunkel, K., Wehner, M. & Zobel, Z. (2014) Severe weather in United States under a changing climate. *EOS, Transactions, American Geophysical Union*, 95 (18), 149–150.

Yin, J.H. (2005) A consistent poleward shift of the storm tracks in simulations of 21st century climate. *Geophysical Research Letters*, 32, L187801.

Zdanowicz, C., Smentny-Sowa, A., Fisher, D., Schaffer, N., Copland, L., Eley, J. & Dupont, F. (2012) Summer melt rates on Penny Ice Cap, Baffin Island: Past and recent trends, and implications for regional climate. *Journal of Geophysical Research – Earth Surface*, 117, F02006.

Questionnaires

Australia

Please provide responses for your country/region to questions below.

1. Particulars of respondent

1.1 Name:
This has been prepared by Dr Kurt Douglas[1] in consultation with Dr Phil Flentje[2]. Andrew Leventhal[3] has also provided advice and is acknowledged

1.2 Organisation:
[1]*UNSW Australia;* [2]*University of Wollongong;* [3]*GHD*

1.3 e-mail:
k.*douglas@unsw.edu.au*

2. Introduction

2.1 Country/region:
General answers have been provided for Australia. Answers have also been provided by Dr Phil Flentje for the case of the Wollongong Coastal Region south of Sydney Australia, an area affected by rainfall induced landslides. The suggestions in Section 6 have been provided by Dr Flentje with respect to his experience in the Wollongong region.

2.2 Do you consider climate change, based presently available information and empirical evidence, an important consideration in respect of landslide risk management?

Please tick 'Yes' ☑ or 'No' ☐. Please elaborate, if any, in the area below.

Climate change is very important and should be taken into account. Unfortunately, it is not currently possible to reliably consider it quantitatively. The Wollongong region is a coastal community, so climate change is attendant with a forecast sea level rise and this will have a huge impact on management related issues.

2.3 Extreme weather events may be due to natural variability and not to climate change. Is it important to distinguish between the impact of climate change and the impact of extreme weather events?

Please tick 'Yes' ☑ or 'No' ☐. Please elaborate, if any, in the area below.

Ideally it would be beneficial to separate the two however, how this could be done accurately is problematic? At some point in the future, we will be able to look back at continuous records and look at historical rainfall/weather data and may be able to do this. But from the current perspective, it is difficult to see when we would be able to achieve such distinctions.

3. Climate change and extreme weather projections

3.1 Please summarise key findings of IPCC studies or other studies (e.g. downscaling studies if applicable/available) on changes in rainfall pattern and other relevant meteorological related events in your country/region that are of relevance to landslide risk management (e.g. in terms of mean annual rainfall, changes in return periods and intensity of shorter duration intense rainfalls for the coming decades).

Australian climate change projections

The Commonwealth Scientific and Industrial Research Organisation (CSIRO) and Australian Bureau of Meteorology (BoM) provide general downscaled climate projection studies for Australia (e.g. CSIRO & Bureau of Meteorology (2015), Climate Change in Australia Information for Australia's Natural Resource Management Regions: Technical Report, CSIRO and Bureau of Meteorology, Australia). Key Website locations for up-to-date research information include:

http://www.bom.gov.au/climate/change/
http://www.climatechangeinaustralia.gov.au/en/

Recent research on short duration intense rainfall

There are a number of researchers in Australia working in the area of rainfall intensity and climate change. The following two key papers of interest, focussing on short-duration, high intensity rainfall, have recently been published.

Westra et al (2014) provide a comprehensive and extensive review of the evidence supporting the current understanding that, with higher temperature, "the greatest increases are likely to occur in short-duration storms lasting less than a day, potentially leading to an increase in the magnitude and frequency of flash floods. The research covered more than just Australia and involved authors from Australia, UK, Sweden and the Netherlands

Wasko & Sharma (2015), focussing on Australia, found at higher temperature, "more intense peak precipitation and weaker precipitation during less intense times, regardless of the climatic region and season". They suggest a "warming climate could lead to increases in the magnitude and frequency of short-duration floods".

REFERENCES

Wasko, C. and Sharma, A. (2015) Steeper temporal distribution of rain intensity at higher temperatures within Australian storms. Nature Geoscience, 8, 527–529 (2015).

Westra, S., Fowler, H.J., Evans, J.P., Alexander, L.V., Berg, P., Johnson, F., Kendon, E.J., Lenderink, G. and Roberts, N.M. (2014) Future changes to the intensity and frequency of short-duration extreme rainfall. Reviews of Geophysics, 52(3), 522–555.

Wollongong

The Department of Environment, Climate Change and Water (DECCW 2009a) included sea level rise of 0.4m by 2015 and 0.9m by 2100. The CSIRO predicts that the Illawarra region will experience substantial increase in summer rainfall.

4. Recent climate impact on slope safety (past 10 to 20 years)

4.1 Please briefly report events of extreme rainfall (or other meteorological) events that gave rise to major landslide damage/casualties in your country/region.

Wollongong

There is a well-developed landslide inventory for the Wollongong region with well documented landslide triggering rainfall thresholds. With comprehensive landslide monitoring stations there are site specific thresholds as well for some sites. Basically the landslides locally are natural rainfall/pore water pressure triggered events, with some anthropogenic factors relevant as well on some sites

4.2 Please report whether any assessment done for landslide hazards and any assessment of risk or changes in the risk profile, due to projected climate impact, was undertaken, and summarise the key findings as appropriate.

Wollongong

The main objective of the Wollongong assessment was to establish a framework or criteria within which current or potential future, geotechnical hazards are deemed to be influenced by coastal processes for planning purposes. This resulted in the definition of a Coastal-Influenced Geotechnical Hazard Zone, defined as follows:-

> *"The Coastal-Influenced Geotechnical Hazard Zone includes areas where coastal processes (including climate change) will directly influence geotechnical hazards for the defined study time period to 2100. Geotechnical assessments of proposed or future development in accordance with Wollongong City Council's Development Control Plan requirements should include specific assessment of coastal processes if located within this Zone." – GHD (2010)*

Rainfall was utilised in the geotechnical and slope stability assessments as it influences the potential for land slip. When considering such events, longer term, 90-days rainfall has been found to correlate to the occurrence of this phenomenon. Furthermore, the potential changes in rainfall intensity due to climate change may affect this likelihood.

A frequency analysis was therefore undertaken on three Bureau of Meteorology (BoM) rainfall gauge sites to define the 90-days rainfall intensities, they being:

- *Woonona (Popes Rd);*
- *Wombarra (Reef Ave); and*
- *Port Kembla (BHP Central Lab).*

A 90-days running sum of rainfall was created from the historical data, ranked and then all non-independent data removed; that is, any data within 90 days of a maximum value was excluded from the next stage of the analysis. The maximum rainfall was then obtained for each calendar year since records were taken and then analysed using a Log-Pearson III Frequency Analysis.

Climate change scenarios were then considered by factoring the rainfall intensities by

+10% (2050) and +20% (2100), in line with rainfall projections.

5. Implications of climate impact on landslide risk management

5.1 Please report relevant strategy or policy on landslide risk management in your country/region to cater specifically for climate impact, including any changes in technical strategies for slope stabilisation works or landslide mitigation works.

The Australian Geomechanics Society (AGS) provide a National Landslide Risk Management Framework for Australia. These documents can be found at:

http://australiangeomechanics.org/public-resources/downloads/#dlACBC2015

In the AGS (2007) Guideline for Landslide Susceptibility, Hazard And Risk Zoning For Land Use Planning, the following paragraph covers climate change:

> *"9.3 POTENTIAL EFFECTS OF CLIMATE CHANGE*
> *There is a developing knowledge of climate change and the effects of this on rainfall and snowfall. It could be anticipated that for example a decreased frequency of high intensity rainfall might reduce the frequency of shallow landslides on steep hill slopes. However the science of prediction of the effects of climate change and the prediction of the frequency of landslides from rainfall is not sufficiently advanced at this time to warrant consideration of climate change when carrying out zoning studies.*
>
> *Those involved in landslide zoning studies should keep informed of developments which might alter this conclusion."*

5.2 Please report specific arrangements or enhancement made to landslide emergency system, public warning systems, and public education in light of climate impact.

Whilst not specific to climate change, property owners within the identified coast zone of Wollongong were notified of landslide risks.

5.3 Please report relevant research and development work on slope safety management in connection with climate impact.

In Wollongong, the coastal zone was identified so that anyone proposing development within this zone, must have their consultants consider the potential impacts of climate change to the proposed developments.

6. Suggestions

6.1 Please provide suggestions on adaptation strategies in landslide risk management in coping with possible climate impact (due to climate change or extreme weather events).

Zones already influenced by slope instability will have to deal with potentially higher frequencies of issues. As the IFD curves move lower or to the left in the figure, increased frequencies of movement can be expected.

6.2 Please provide suggestions on other slope safety related areas that warrant higher priority in research and development because of climate impact.

The forecast 'increased intensity' type of phraseology needs to become more focused and rigorous. The increased intensity should be focused onto Intensity Frequency Duration plots, or rainfall recurrence type curves whereby the changes can be graphed, so that increased frequencies can be quantitatively assessed.

6.3 Do you have any other relevant comments or suggestions?

It is a real scenario which should be taken more seriously and funded accordingly. The sea level rise is interesting, but the evidence is all around us locally. The sea has been at these increased levels recently and it should be easier to manage, but it will be expensive. It will also require immense engineering of coastal protection works.

China

Please provide responses for your country/region to questions below.

1. Particulars of respondent

1.1 Name:
Li Bin

1.2 Organisation:
Institute of Geomechanics, China Geological Survey

1.3 e-mail:
52572706@qq.com

2. Introduction

2.1 Country/region:
China/Beijing

2.2 Do you consider climate change, based presently available information and empirical evidence, an important consideration in respect of landslide risk management?
Please tick 'Yes' ☑ or 'No' ☐. Please elaborate, if any, in the area below.

2.3 Extreme weather events may be due to natural variability and not to climate change. Is it important to distinguish between the impact of climate change and the impact of extreme weather events?
Please tick 'Yes' ☑ or 'No' ☐. Please elaborate, if any, in the area below.

3. Climate change and extreme weather projections

3.1 Please summarise key findings of IPCC studies or other studies (e.g. downscaling studies if applicable/available) on changes in rainfall pattern and other relevant meteorological related events in your country/region that are of relevance to landslide risk management (e.g. in terms of mean annual rainfall, changes in return periods and intensity of shorter duration intense rainfalls for the coming decades).

1. *According to the 5th report of IPCC last year, it attracts more attention in China to strengthen urban and rural risk management. People in the*

rural areas are facing the impact of meteorological disasters such as floods, debris flows and landslides. It is needed to strengthen the impact of early warning, meteorological information dissemination, and to incorporate climate change risk into the new rural construction and rural urbanization planning.

2. *In China, research on rainfall-induced post-earthquake geological hazards in these areas has big significance to the reconstruction of towns and villages and the tourism safety after the earthquake. Due to the ground shaking, the critical amount of accumulated precipitation and the hourly rainfall intensity necessary to initiate debris flow were reduced compared with values before the earthquake.*

4. Recent climate impact on slope safety (past 10 to 20 years)

4.1 Please briefly report events of extreme rainfall (or other meteorological) events that gave rise to major landslide damage/casualties in your country/region.

1. *During July 7 to 12 of 2013, a heavy rainfall occurred at the Wenchuan Ms8.0 Earthquake area. The five-day accumulative precipitation in the Dujiangyan area of Sichuan Province was up to 1,129 mm, which is rare in previous years and exceeds 1109.8mm, the mean annual precipitation of twenty-five years from 1987 to 2012. The heavy rainfall triggered a rapid and long run-out rockslide in Sanxicun which accumulated a volume of 1.9 million cubic meters, caused a sliding distance up to 1,200 meters, and 166 people were killed.*
2. *Many debris flow hazards were triggered recently in the epicentre area of the 2008 Wenchuan Earthquake. The severest debris flow is the debris flow in Wenjia Gully on Aug. 13, 2010.Wenjia Gully was a simply and ordinary gully and had no debris flow at all before the Wenchuan Earthquake. The formation conditions of debris flows were totally changed by a giant landslide caused by the earthquake. The giant landslide formed a landslide-debris deposition in the gully. The debris deposition had a volume of more than 30 million cubic meters. The sediments were easily eroded by flash flood and formed debris flows. In the three raining seasons after the earthquake, a total of five large scale and giant scale debris flow were triggered. The largest and most harmful debris flow occurred on Aug. 13, 2010.The total rainfall during the debris flow was 227 mm, and the duration of debris flow was 2.5 hours. The volume of debris flow was 3.1 million cubic meters. The debris flow resulted in 7 fatalities, 5 persons listed as missing, 39 persons injured, 479 houses buried, and a direct economic loss of RMB 430 million. About16% sediments of the landslide-debris deposition flew down out of the gully during the five large scale debris flows. New debris flows will be triggered by rainfall in the gully in the future. It will be a long-term effort for prevention and mitigation of the debris flow hazards in the gull.*

3. *The reason of giant debris flow triggering was obtained by field investigation and analysis on the debris flow formation conditions: heavy rainfall happened in the upstream of Sanyanyu Gully and Luojiayu Gully. The recorded rainfall was 77.3 mm in one hour from 23 to 24 o'clock, August 7, 2010. The heavy rain caused powerful flash flood in the catchments, which destroyed several natural dams and prevention dams in the two gullies. Then the giant debris flows were formed. The debris flows were high density viscous debris flows. The total volume of debris flows and the volume of sediment in debris flows were estimated at* $144.2 \times 10^4\ m^3$ *and* $97.7 \times 10^4\ m^3$*, respectively. The large boulder involving in debris flows moved with powerful impact force and destroyed more than 5500 houses in depositional fans. The debris flows produced a dam with 550 m in length, 70 m in width, and 10 m in height in Bailongjiang River. The dam blocked the river and formed a lake with 3 km in length and half of Zhouqu city was inundated. As of August 18, 2010, the debris flows claimed 1287 death, with further 457 listed as missing.*
4. *On June 28, 2010, due to sudden heavy rainfall, a high speed and long run-out rockslide-debris flow occurred at Dazhai in Guanling of Guizhou Province. Its run- out distance was about 1.5 km. Its rock debris volume was about 1.749 million* m^3*. It's destroyed two villages and resulted 99 fatalities. The landslide was located at the coal-related sedimentary rock strata area. This rock system commonly presents in southwestern China. The heavy rainfall on June 27 and 28 were the main factor triggering the major disaster. The rainfall was up to 310mm and exceeded the records over the last 60 years. Analysis indicates that the runoff in valley streams were two times more than those during usual rainfall at the region.*
5. *During August 31 to September 2 of 2014, an extreme rainfall occurred at the Three Gorges area. It was considered that this extreme rainfall was a process of heavy rainfall and local rainstorm, which mainly concentrated in northeastern Chongqing city as well as Badong and Zigui cities in Hubei province. The maximum rainfall intensity was daily rainfall of 403.4 mm, significantly more than the daily rainfall critical value of 250 mm, which triggered lots of landslides and debris flows in the region, forming a wide range of chain damages. The landslides were mainly due to the instability of layered slope consisting of Jurassic and Triassic mid-upper strata. The failure mode mainly includes six types, such as nearly horizontal bedding sliding, sliding-tensioning, sliding-buckling, rotational sliding, apparent dip sliding and accumulative landslides.*

4.2 Please report whether any assessment done for landslide hazards and any assessment of risk or changes in the risk profile, due to projected climate impact, was undertaken, and summarise the key findings as appropriate.

A lot of work has been done to assess for landslide hazards and other risks due to projected climate impact, especially in Wenchuan Ms8.0 Earthquake area and Three Gorges Reservoir area. GIS and RS Technologies, fuzzy synthetic discrimination theories are usually used due to annual precipitation data of 30 years.

5. Implications of climate impact on landslide risk management

5.1 Please report relevant strategy or policy on landslide risk management in your country/region to cater specifically for climate impact, including any changes in technical strategies for slope stabilisation works or landslide mitigation works.

1. *Landslides prevention and mitigation system was established including organization, expert teams and relevant legal system. At present, a three-level system of landslides emergency management agencies has been established in 20 provinces (autonomous regions and municipalities) agents, 161 prefecture-level cities (states) agents, and 990 county agents. Meanwhile, 26 provinces, 161 prefecture-level cities and 990 counties have established agencies for landslides emergency technical guidance.*
2. *Landslides monitoring networks are completely covered all the counties by identifying different level potential landslides, carrying out grassroots landslides prevention capacities, training mitigation measures and strengthening landslides prevention and mitigation education. At present, 1:200,000 landslides surveys were already finished at the county level, 645 counties have been carried out detailed landslides surveys in 1:50,000, 10,134 potential landslides had already constructed geotechnical investigations and Surveys.*
3. *Meteorological early warning and forecasting systems for landslides have covered all the country by strict enforcement of monitoring and surveying regulations, completing and improving different levels of landslides emergency plans. Now there are 31 provinces (autonomous regions and municipalities) agents, 323 prefecture-level cities (states) agents, and 1880 county agents have already established the meteorological early warning and forecasting for landslides systems.*
4. *Landslides prevention and mitigation information systems have covered all the country based on the informationization infrastructure, studied on the methods of monitoring and early warning and promoted emergency response system. We have organized thousands times and millions of participants for landslides emergency filed exercises every year all around the country. At the same time, there are nearly 300 thousand people are on real-time monitoring for the potential landslides.*
5. *According to the investigation by the professional group, which constituted by the experts' guidance and combined with the masses' recognition and verification, we have already achieved landslides prevention by the efforts of both specialists and the generalists. There are 3000 landslides emergency professionals in the whole nations.*
6. *We have already completely eliminated a number of large scale potential landslides. Financial funds at all levels involved in the landslides prevention are nearly 20 billion annually, which was mainly used for homeland relocations and engineering project treatments. Over the past 10 years, more than 15 thousand landslides, 530 thousand threatened people were acted to avoid successfully in advance. The effects of landslides prevention and mitigation were literally remarkable in China.*

5.2 Please report specific arrangements or enhancement made to landslide emergency system, public warning systems, and public education in light of climate impact.

5.3 Please report relevant research and development work on slope safety management in connection with climate impact.

1. *National science and technology support program of 'Study on the technical methods of geological hazard monitoring and early warning and risk assessment' during 2010–2015.*

6. Suggestions

6.1 Please provide suggestions on adaptation strategies in landslide risk management in coping with possible climate impact (due to climate change or extreme weather events).

In order to strengthen the landslides risk management and response to global climate changes, we must make our efforts to handle with prevention and mitigate losses and the impacts of major unexpected landslides, the actions need all the international community to work together. Here, we look forward to further enhanced exchanges with international counterparts and organizations.

1. *The first is the technical cooperation involving in some focal and difficult points. We could further jointly study on the prevention theory and techniques methods of high speed and long run-out landslides in high and steep slope area and landslides triggered by reservoir level fluctuations, cooperating on landslides monitoring and early warning theoretical research and devices development under extreme and unusual weather conditions, in order to enhance the prevention level of the major landslides occurred frequently and ubiquitous*
2. *The second is strengthening the support to each other. The landslides susceptibility zones often happened in underdeveloped areas with mountains, canyons and extreme climate affected areas, however, where are the mainland of majority developing regions in the world, landslide prevention technical are lagging behind there. And between different countries have their own unique experience of governance and control technology of landslides, these need to further strengthen mutual support and cooperation.*

6.2 Please provide suggestions on other slope safety related areas that warrant higher priority in research and development because of climate impact.

Strengthening information exchange of landslides prevention is very important. We could take full advantage of government departments, research institutions and non-governmental organizations, promote mutual understanding of landslides prevention works from all the nations access by regular meetings, academic forums, networks and other channels, we also could create good conditions for the personnel training programme, major prevention project implement and significant cross-border hazards emergency disposal.

6.3 Do you have any other relevant comments or suggestions?

No.

Colombia

Please provide responses for your country/region to questions below.

1. Particulars of respondent

1.1 Name:
Juan Montero-Olarte
1.2 Organisation:
Colombian Geotechnical Society- JTC1
1.3 e-mail: juanmontero170@gmail.com

2. Introduction

2.1 Country/region:
Colombia

2.2 Do you consider climate change, based presently available information and empirical evidence, an important consideration in respect of landslide risk management?
Please tick 'Yes' ☑ or 'No' ☐. Please elaborate, if any, in the area below.

One of the possible effects of climate change is the increased frequency and intensity of precipitation with a significant increase in the occurrence of landslides and therefore the subsequent increase in landslide hazard and risk

2.3 Extreme weather events may be due to natural variability and not to climate change. Is it important to distinguish between the impact of climate change and the impact of extreme weather events?
Please tick 'Yes' ☑ or 'No' ☐. Please elaborate, if any, in the area below.

The relevant definitions by IPCC 2013 are:

(1) <u>Extreme weather event or extreme climatic event:</u> "An extreme weather event is an event rare at a particular place and time of year". Definitions of rare vary, but an extreme weather event would normally be as rare as or rarer than the 10th or 90th percentile of a probability density function estimated from observations. By definition, the characteristics of what is called extreme weather may vary from place to place in an absolute sense).

"when a pattern of extreme weather persists for some time, such as a season, it may be classed as an extreme climate event, especially if it yields an average or total that is itself extreme (e.g., drought or heavy rainfall over a season)". (Definitions by Intergovernmental Panel on Climate Change – Climate Change 2013, The Physical Science Basis, eds. Stocker, T.F. et al.).

(2) Climate change: Climate change refers to "a change in the state of the climate that can be identified (e.g., by using statistical tests) by changes in the mean and/or the variability of its properties, and that persists for an extended period, typically decades or longer". Climate change may be due to natural internal processes or external forcings such as modulations of the solar cycles, volcanic eruptions and persistent anthropogenic changes in the composition of the atmosphere or in land use. Note that the Framework Convention on Climate Change (UNFCCC), in its Article 1, defines climate change as: 'a change of climate which is attributed directly or indirectly to human activity that alters the composition of the global atmosphere and which is in addition to natural climate variability observed over comparable time periods'. The UNFCCC thus makes a distinction between climate change attributable to human activities altering the atmospheric composition, and climate variability attributable to natural causes. (IPCC 2013).

(3) Climate variability: Climate variability refers to "variations in the mean state and other statistics (such as standard deviations, the occurrence of extremes, etc.) of the climate on all spatial and temporal scales beyond that of individual weather events". Variability may be due to natural internal processes within the climate system (internal variability), or to variations in natural or anthropogenic external forcing (external variability). (IPCC 2013).

Based on these definitions I conclude that extreme events may be due to natural variability regardless of climate change and that it is important to distinguish between the impact of an extreme event at most in one season and the impact of climate change that persists for long periods (i.e. heavy rains of La Niña in the equatorial Pacific every 6–7 years, with one or two month duration) the effects and damages are different and also the causes are different. In both cases, the risks can be due to natural or anthropogenic causes, but they are different for each case.

3. Climate change and extreme weather projections

3.1 Please summarise key findings of IPCC studies or other studies (e.g. downscaling studies if applicable/available) on changes in rainfall pattern and other relevant meteorological related events in your country/region that are of relevance to landslide risk management (e.g. in terms of mean annual rainfall, changes in return periods and intensity of shorter duration intense rainfalls for the coming decades).

I will refer to results of studies carried out by the Institute of Hydrology, Meteorology and Environmental Studies of Colombia IDEAM on rainfall and other meteorological variables (Cambio Climático en Temperatura, Precipitación y Humedad Relativa para Colombia usando Modelos Meteorológicos de Alta Resolución-Panorama 2011-2100- Mayo 2010) /Climate Change in Temperatura, Precipitation and Relative Humidity for Colombia using High Resolution, Weather Models 2010- May 2010). In this study the evidence of climate change in Colombia is presented and likely climate scenarios are developed for the coming decades from GCMs global climate models to describe and analyze atmospheric processes. at the appropriate scale

Based on appropriate statistical techniques trends of meteorological variables with reduced dynamic range and with known high or low resolution global models considering several proposed by the IPCC, XXI century scenarios were calculated. In the attached document, unfortunately in Spanish, you can see the way the study was done and how its results are presented. Despite language difficulties, the observation of graphs and maps are helpful to visualize the results and some findings related with landslide risk management are presented below:

From high-resolution global and regional climate models of climate change, several scenarios were simulated for Colombia in the coming decades as follows:

1- *The average temperature will continue to rise: 1.4°C for the period 2011–2040; 2.4°C for the period 2041–2070 and 3.2°C for the period 2071–2100. According to this prediction, the glaciers that now dominate our Andes could disappear in 30 or 40 years, with a notable increase in instability in the decreasing periglacial area.*
2- *The annual rainfall will increase in some regions and decrease in others, but on average for the 2011–2070 period the rainfall will decrease between 15% and 36% of the value of 1971–2000, with the most notable changes and possible transition from semi-humid to semi-arid climate in Cundinamarca, Boyacá, Tolima and Huila, which correspond to one of the most populated regions of Colombia including its capital Bogota. This behavior could cause a relative decrease of landslides considering the current scenario of population distribution.*

4. Recent climate impact on slope safety (past 10 to 20 years)

4.1 Please briefly report events of extreme rainfall (or other meteorological) events that gave rise to major landslide damage/casualties in your country/region.

Colombia is an Andean country where erosion and landslides are very frequent due to rainfall, earthquakes and volcanic eruptions.

EVENTS ASSOCIATED WITH CLIMATE CHANGE/El Nino Southern Oscillation (ENSO)

1. *In the periods July and November- December 2010 and March-April-May 2011 heavy rains, unprecedented in the past 40 years in Colombia triggered numerous landslides and flood flows throughout the Andean region*

which concentrates most of the population. Affected population: 2'350.044 victims (7% of the national population), 1,374 dead and 1,016 missing. The estimated cost of the damage was estimated at 6,052 million dollars where the most affected were the habitat (44%) and road infrastructure (29%). (BID_CEPAL 2011).

2. *Two exceptional rainfall events in 1991: on August 7th, 1991 of 84.7 mm of precipitation with duration of two hours and another one on September 9th of 126 mm of rain in three hours, caused accumulated damage from both events that affected a stretch of 20 kilometers of road causing the destruction of five bridges, and work crossing affecting several works and retaining walls with numerous flows, landslides and rockfalls on the Bogota-Villavicencio road, which links the capital Bogotá with the Llanos Orientales (Eastern Plains), source of food and livestock. The road was closed for several weeks and its reconstruction cost 10 million dollars. It was also estimated that the operational and social costs were equivalent to the cost of a dual 100 km road with viaducts and tunnels for the same road, even with some money left.*
3. *In June 1974 in this same road, heavy rains reactivated an old slide at the Quebradablanca site, burying more than 100 people and forcing the closure of this road. A 3 kilometer detour had to be build and losses were very heavy.*
4. *Several small small towns have been destroyed by heavy rains: including San Cayetano in 1999 and Gramalote in 2010.*

4.2 Please report whether any assessment done for landslide hazards and any assessment of risk or changes in the risk profile, due to projected climate impact, was undertaken, and summarise the key findings as appropriate.

There have been many studies about landslide hazard and risk assessment but at the moment I do not have information that could be summarized for this report about assessment of risk changes due to climate impact

5. Implications of climate impact on landslide risk management

5.1 Please report relevant strategy or policy on landslide risk management in your country/region to cater specifically for climate impact, including any changes in technical strategies for slope stabilisation works or landslide mitigation works.

The Colombian territory is large (1,141,748 km^2) and its population also large (47 million people). The developed area is concentrated in the mountainous region of the Andes, which is exposed to high seismic hazard in the Circumpacífic Belt and under tropical climate with heavy rains

To manage and prevent damage and casualties from extreme weather events and climate change, laws and regulations are issued that normally are not very effective in the practice to reduce the risk due to landslides, and other actions can be classified a follows:

1 *<u>Positive:</u> development of landslide hazard inventories, zoning studies are done unfortunately with not enough detail, zoning methodology is being*

developed (SGC.) Despite the shortage of resources construction techniques are improving and in some situations, like in the case of destroyed towns (i.e San Cayetano, Gramalote and others) relocation of the whole town has been done.

2 *Negative: laws and regulations are issued but these provisions are not effective due to the lack of decision power by technicians and experts in these risks and also due to the shortage of resources.*

5.2 Please report specific arrangements or enhancement made to landslide emergency system, public warning systems, and public education in light of climate impact.

- *Some important active landslides are monitored but unfortunately records are not continuous.*
- *IDEAM has an early warning system that operates via Internet to prevent calamities from potential extreme events of flooding and landsliding*

5.3 Please report relevant research and development work on slope safety management in connection with climate impact.
I have not enough information at hand right now.

6. Suggestions

6.1 Please provide suggestions on adaptation strategies in landslide risk management in coping with possible climate impact (due to climate change or extreme weather events).
To Increase research on the subject and to disseminate the results at all levels for both: technical and non-technical people.

6.2 Please provide suggestions on other slope safety related areas that warrant higher priority in research and development because of climate impact.
To improve the national rainfall measurement network system to have data in real time, which it is not an easy task in Colombia with three mountain ranges.

6.3 Do you have any other relevant comments or suggestions?
I have not additional suggestions.

India

Please provide responses for your country/region to questions below.

1. Particulars of respondent

1.1 Name:
Dr R.K. Bhandari

1.2 Organisation:
Chairman, Forum on Disaster Mitigation, Indian National Academy of Engineering, India

1.3 e-mail:
rajmee@yahoo.com

2. Introduction

2.1 Country/region:
India, Asian Region

2.2 Do you consider climate change, based presently available information and empirical evidence, an important consideration in respect of landslide risk management?

Please tick 'Yes' ☑ or 'No' ☐. Please elaborate, if any, in the area below.

Climate Change (CC) is a highly complex global issue, probably feeding to Extreme Weather Events (EWE) which are visibly linked with landslides as a complex local issue. The spotlight on both CC and EWE is thus critical to landslide risk management

2.3 Extreme weather events may be due to natural variability and not to climate change. Is it important to distinguish between the impact of climate change and the impact of extreme weather events?

Please tick 'Yes' ☑ or 'No' ☐. Please elaborate, if any, in the area below.

Climate change is a much bigger and much more complex issue linked with global warming which, Interalia, leads to melting of glaciers, formation of GLOF and landsliding. Extreme Weather Events are the least understood, random events which defy any meaningful predictions at this time.

3. Climate change and extreme weather projections

3.1 Please summarise key findings of IPCC studies or other studies (e.g. downscaling studies if applicable/available) on changes in rainfall pattern and other relevant meteorological related events in your country/region that are of relevance to landslide risk management (e.g. in terms of mean annual rainfall, changes in return periods and intensity of shorter duration intense rainfalls for the coming decades).

In India there have been numerous catastrophic landslides visibly linked to EWE. Rainfall data for those individual events have been analysed in many cases to demonstrate changing rainfall patterns, new rainfall and rainfall intensity peaks and triggering of landslides. There is no in-depth scientific study so far, to the best of my knowledge, which could provide a definitive, holistic view and conclusive explanation based on hard field evidences discounting for various other co-existing causative factors.

4. Recent climate impact on slope safety (past 10 to 20 years)

4.1 Please briefly report events of extreme rainfall (or other meteorological) events that gave rise to major landslide damage/casualties in your country/region.

Landslides are caused by interplay of a large number of causative factors and their synergistic effects spread over long periods of time and, in my view, the highly visible connection between EWE and landslides should never be analysed and reported in isolation. Such reporting has two great disadvantages: (1) it reflects only half-truth (2) it overshadows and even kill the curiosity of insightful investigations and scientific understanding of the composite effect of multitude of causative factors by giving a deceptively simple tool, of forecasting/ explaining away landslides, in the hands of professionals in hurry. I have written about it and one of my Guest Editorial in the Journal of Current Science is attached to convey my view on this.

4.2 Please report whether any assessment done for landslide hazards and any assessment of risk or changes in the risk profile, due to projected climate impact, was undertaken, and summarise the key findings as appropriate.

There are a number of event-specific reports but it would be unscientific to draw conclusions from them without a serious research study. Non-peer reviewed studies, if published, will do more harm than good.

5. Implications of climate impact on landslide risk management

5.1 Please report relevant strategy or policy on landslide risk management in your country/region to cater specifically for climate impact, including any changes in technical strategies for slope stabilisation works or landslide mitigation works.

Enhanced incidences of landslides due to EWE have been instrumental in denser network of weather forecasting stations and instrumentation, rainfall monitoring and real time dissemination of rainfall data, development of location-specific weather and landslide prediction attempts, strengthening of decision support systems, improved community participation, early warning, preparedness and tightening of techno-legal regime

5.2 Please report specific arrangements or enhancement made to landslide emergency system, public warning systems, and public education in light of climate impact.

Establishments of National Disaster Management Authority by an act of Parliament, State Disaster Management Authorities, District level Disaster

Management Authorities, National Disaster Response Force, encouraging and funding projects and programmes on community-centric early warning and training at all levels are parts of the government initiative. India has also created a National Institute of Disaster Management with public education and training function supported by other knowledge institutions.

5.3 Please report relevant research and development work on slope safety management in connection with climate impact.

The work is developing on the following fronts (a) integration of slope safety management with development planning (b) slope safety management in the multihazard context which includes earthquake-induced and flood-induced landslides (c) New set of slope safety management problems in high altitude areas including snow avalanches and GLOF (d) slope safety management in urban and built environments (e) early warning and evacuation and (f) learning of lessons and strengthening of capacities at all levels

6. Suggestions

6.1 Please provide suggestions on adaptation strategies in landslide risk management in coping with possible climate impact (due to climate change or extreme weather events).

It is difficult to stop the global tide of publications of unverified data, event reports and papers without peer-review most of which, at times, hinder rather than help progress. This is because they tend to somehow fit limited field observations with climate change data, and sway opinions. The need of the hour is to fund and support specialized short and long term studies which can take a holistic view of the multitude of causative factors including EWE and Climate change. The task cannot be completed unless it is entrusted to impendent teams of eminent experts.

6.2 Please provide suggestions on other slope safety related areas that warrant higher priority in research and development because of climate impact.

It is unfortunate that landslide hazard mapping programmes for hazard and risk assessments are developing on the convention lines totally disregarding EWE, anthropogenic factors, mindless urbanization, violence against mountain systems and abuse of the sloping land masses. The assessments of hazard and risk based on neglect of vital manmade factors are therefore highly deceptive and even wasteful calculations. This is just one example of disappointment and one could list many more

6.3 Do you have any other relevant comments or suggestions?

Please devote a small part of the technical session, if possible, to address the question of Ethics in scientific reporting. This is critical because, in many cases, professionals begin their studies with the objective of finding correlations and fitting theories to the observed facts without open mind and insightful holistic perceptions. The role of engineering judgment has virtually disappeared from the decision making process because of the advent of powerful tools of modelling and the blind use of high-end computational techniques. Let us remember: Those in hurry must walk slowly

Subject index